Cell and Matrix Mechanics

Cell and Matrix Mechanics

EDITED BY

Roland Kaunas

Texas A&M University, College Station, USA

Assaf Zemel

Hebrew University of Jerusalem, Israel

CRC Press
Taylor & Francis Group
Boca Raton London New York

CRC Press is an imprint of the
Taylor & Francis Group, an **informa** business

CRC Press
Taylor & Francis Group
6000 Broken Sound Parkway NW, Suite 300
Boca Raton, FL 33487-2742

First issued in paperback 2017

ISBN-13: 978-1-4665-5381-1 (hbk)
ISBN-13: 978-1-138-07333-3 (pbk)

Library of Congress Cataloging-in-Publication Data

Cell and matrix mechanics / editors, Roland Kaunas, Assaf Zemel.
   p. ; cm.
   Includes bibliographical references and index.
   ISBN 978-1-4665-5381-1 (hardcover : alk. paper)
   I. Kaunas, Roland, editor. II. Zemel, Assaf, editor.
   [DNLM: 1. Cell Physiological Processes. 2. Biomechanical Phenomena. 3. Cell-Matrix Junctions--physiology. 4. Extracellular Matrix--physiology. 5. Tissue Engineering. QU 375]

   QH631
   571.6--dc23
                                       2014034063

Visit the Taylor & Francis Web site at
http://www.taylorandfrancis.com

and the CRC Press Web site at
http://www.crcpress.com

Contents

Preface

The field of cell mechanics has seen extensive development over the past few decades. New technologies have been invented to finely manipulate the mechanical and geometrical characteristics of the cellular environment, to apply precise doses of mechanical load, and to image the induced cellular response at both subcellular and molecular levels. Mechanical stimulation has been shown to be as potent and significant as chemical stimuli in affecting virtually all aspects of cell function. This includes effects on individual cell processes such as adhesion, migration, division, and gene expression regulation, as well as on collective behaviors such as cell–cell interactions, matrix remodeling, morphogenesis, and disease. A comprehensive understanding of the complex mechanical interplay between cells and their surrounding extracellular matrix is a multiscale problem in time and space, ranging from the molecular level and millisecond time scale up to the multicell or tissue levels involving a time scale of days to months.

This book brings together experts in the field of cell mechanics to summarize cutting-edge research at the molecular, cellular, and tissue levels with a focus on cell–matrix interactions. At each length scale, key experimental observations and corresponding quantitative theoretical models are presented. The book is organized in three sections that correspond to those three hierarchical levels. In Section I, which is focused on the molecular level, the passive and active mechanical properties of cytoskeletal polymers and associated motor proteins are discussed along with the behavior of polymer networks in Chapters 1 and 3. The mechanical properties of the cell membrane are presented in Chapter 2, with an emphasis on membrane protein activation caused by membrane forces. The hierarchical organization of collagen fibrils is discussed in Chapter 4, revealing that a delicate balance exists between specific and nonspecific interactions to result in a structure with semicrystalline order as well as loose associations. Section II focuses on mechanics at the cellular level. In Chapter 5, the roles of matrix mechanical properties on cell adhesion and function are presented along with different mechanical mechanisms of cell–cell interactions. The effects of mechanical loading on cell cytoskeletal remodeling are presented in Chapter 6, summarizing various modeling approaches that explain possible mechanisms regulating the alignment of actin stress fibers in response to stretching. Mechanical testing of cell-populated collagen matrices is described in Chapter 7, along with theory relating the passive and active mechanical properties of the engineered tissues. Chapters 8 and 9 focus on cell migration behavior in 3-D matrices and in collective cell motility. Section III of the book concerns cellular mechanics at the tissue level. Starting with a summary of the pioneering studies of cell–matrix mechanics by Albert Harris, Chapter 10 is devoted to the role of mechanics in cartilage development. Chapters 11 and 12 describe the roles of both cellular and external forces on tissue morphogenesis. Finally, the roles of mechanical forces on tumor growth and cancer metastasis are presented in Chapters 13 and 14, respectively.

These chapters are not meant to provide a complete summary of current research in the field, but instead they provide a thorough description of the roles of mechanical forces in cell and tissue biology. Despite the range of concepts presented, the material in each chapter relates to that of the others. For example, tumor cell metastasis involves cell–cell interactions, cell migration in 3-D tissues, and collective cell migration. Similar connections can be drawn between tissue morphogenesis and cell behavior in mechanically loaded tissues. All of these phenomena depend on the mechanical properties of active cytoskeletal polymer networks, cell membranes, and collagen fibrils. It is our hope that specific examples described herein, along with descriptions of associated theoretical models, will inspire future development of multiscale models that elucidate the roles of mechanical forces in cell–matrix biology.

Editors

Roland Kaunas is an associate professor of biomedical engineering and the director of the Cell Mechanobiology Laboratory at Texas A&M University. He earned his BS in chemical engineering at the University of Wisconsin, his MS in biomedical engineering at Northwestern University, and his PhD in bioengineering with Shu Chien from the University of California, San Diego. He joined the faculty at Texas A&M in 2005, where his research focuses on experimental and computational modeling of cell reorganization and mechanotransduction in response to matrix stretching, fluid shear stress mechanotransduction in sprouting angiogenesis, and the development of collagen-based scaffolds for adult stem cell delivery for osteoregenerative therapies. His research is funded by the National Institutes of Health (NIH), National Science Foundation, American Heart Association, and the Center for the Advancement of Science in Space.

Assaf Zemel is a senior lecturer of theoretical biophysics and head of the theoretical biophysics laboratory at the Institute of Dental Sciences at the Hebrew University of Jerusalem. He is also affiliated with the Fritz Haber Research Center for Molecular Dynamics at the Center for Bioengineering of the Hebrew University. Assaf earned his PhD in theoretical chemistry from the Hebrew University. He then shifted to the field of cell mechanics for his postdoctoral research at the Weizmann Institute and the University of California at Davis. His current research focuses on understanding the physical mechanisms underlying the morphology, dynamics and internal structure of cells, and the mechanics of cell–cell interactions and morphogenesis.

Contributors

José Alvarado studied biophysics at the Universität Leipzig in Germany, where he earned his MS in 2008. He subsequently did his PhD research at AMOLF studying contractility and confinement-induced ordering of actin networks using in vitro model systems. He earned his PhD in 2013 at the VU University of Amsterdam and now performs his postdoctoral training at MIT (Cambridge, Massachusetts) in the group of Peko Hosoi, studying soft condensed matter in relation to microfluidics and robotics.

Anne Bernheim-Groswasser is an associate professor at the Chemical Engineering Dept. at Ben-Gurion University of the Negev in Israel. She is also a member of the Ilse Katz Institute for Nanoscale Science & Technology. She received her BSc and PhD from the Technion (Israel Institute of Technology), where she studied the properties of bicontinous phases of amphiphilic and microemulsion systems. She completed postdoctoral studies at the Curie Institute in Paris with Cecile Sykes and Jacques Prost before joining BGU in 2002, where she is the head of the nano-biophysics group. Her work focuses on the reconstitution of processes associated with the active remodeling of the cell cytoskeleton, in particular effects associated with the collective work of groups of motor proteins and actin polymerization forces. Her work is funded by the Israel Science Foundation, Deutsche Forschungsgemeinschaft, and Israeli Planning and Budgeting Committee.

Peter J. Butler is an associate professor of biomedical engineering at Penn State University. His laboratory focuses on molecular scale mechanotransduction and its role in endothelial cell biology and vascular health. He earned a BA in English and biology from Fordham University, a BS in mechanical engineering from the City College of New York, and a PhD in mechanical engineering from the City University of New York. His graduate advisors were Daniel Lemons and Sheldon Weinbaum from the Departments of Biology and Mechanical Engineering, respectively. After earning his PhD in 1999, he went for postdoctoral training in bioengineering under the guidance of Shu Chien at the University of California, San Diego. His research at Penn State spans cell and cell membrane mechanics and molecular dynamics, single-molecule spectroscopy and microscopy, and drug delivery. His research is funded by the National Institutes of Health (NIH), the National Science Foundation, the American Heart Association, and the Woodward Foundation.

Elliot L. Elson is the alumni endowed professor of biochemistry and molecular biophysics at the Washington University School of Medicine. He received degrees from Harvard and Stanford, performed postdoctoral research at University of California at San Diego (UCSD), and began his academic career at Cornell University. Professor Elson's laboratory focuses on optical and mechanical techniques for biophysical measurement and has been involved in the development of technologies such as fluorescence correlation spectroscopy, fluorescence photobleaching recovery (FRAP), and a series of techniques for assessing cellular mechanics, electrophysiology, and drug sensitivity in 2-D and 3-D culture. He currently serves on the scientific boards of InvivoSciences, Inc., and CibaPure, LLC. Professor Elson is a fellow of the Biophysical Society and of the American Academy of Arts and Sciences.

Sara Farag earned her BSc (Hons) from Penn State Bioengineering. As part of her honors thesis, she investigated the force transaction mechanisms of endothelial cell membranes using computational and experimental methods. After graduation, she attended the College of Medicine in the Hershey Medical Center where she earned her MD. She continued her training during a residency in the Mt. Sinai hospital system in New York City, where she specializes in obstetrics and gynecology.

Krishna Garikipati is a professor of mechanical engineering and mathematics at the University of Michigan. His research is in the broad area of computational science and draws from applied mathematics, nonlinear mechanics, and thermodynamics. Of specific interest to him are problems in mathematical biology, biophysics, and materials physics. His group works on models of tumor growth and cell mechanics in mathematical biology and biophysics. In materials physics, they are particularly interested in phase transformations in structural and battery materials. In all of these problems, they develop physical models, their mathematical forms, and numerical methods for their solution and also write open-source computer code with these algorithms. He earned his PhD from Stanford University in 1996 and joined the faculty at the University of Michigan in 2000. His research is supported by the National Science Foundation, Department of Energy, and the Advanced Research Projects Agency-Energy.

Guy M. Genin studies interfaces and adhesion in nature, physiology, and engineering. His current research focuses on interfaces between tissues at the attachment of tendon to bone, between cells in cardiac fibrosis, and between subcellular components in plant defenses. He is a professor in the Department of Mechanical Engineering and Materials Science at Washington University (WU) and in the Department of Neurological Surgery at WU School of Medicine. His training includes BSCE and MS from Case Western Reserve University, SM and PhD in applied mechanics and solid mechanics from Harvard, and postdoctoral research at Cambridge and Brown. Professor Genin is the recipient of several awards for engineering design, teaching, and research, including a Research Career Award from the National Institutes of Health (NIH), the Skalak Award from the American Society of Mechanical Engineers (ASME), and Professor of the Year from the WU Student Union.

Barak Gilboa is a postdoctoral fellow in the label-free microscopy group and in the experimental soft matter group in the School of Chemistry at Tel Aviv University. He received his BA in physics at the Technion (Israel Institute of Technology) and his MSc and PhD in physics at Ben-Gurion University of the Negev under the supervision of Prof. Oleg Krichevsky and Prof. Anne Bernheim. He is currently researching the methods of super-resolution microscopy using label-free Raman scattering in the lab of Prof. Ori Cheshnovsky and also developing a method for single-cell tomography in the lab of Dr. Yael Roichman.

Nir S. Gov started his BS in physics at the Imperial College (London) and then graduated in physics at the Tel Aviv University. After serving in the army, he earned his MS and PhD in physics at the Technion (Israel Institute of Technology) (1998). His first postdoctoral work, on quantum systems, was at the University of Illinois at Urbana–Champaign, with Tony Leggett and Gordon Baym. In 2002, he switched to soft-matter and biological physics with postdoctoral training at the Weizmann Institute of Science, in the group of Sam Safran. Since 2004, he has been a professor at the Department of Chemical Physics at the Weizmann Institute.

Albert Harris is the son of the artist Kenneth Harris. His undergraduate program is from Swarthmore College and, his PhD is from Yale University, where he was a student of J. P. Trinkaus. He was a Damon Runyon fellow in cancer research at Cambridge, England, in the laboratory of Michael Abercrombie (the discoverer of contact inhibition). Since 1972, he has been professor of biology at the University of North Carolina, and he was distinguished visiting professor at the University of California at Davis. He has discovered focal adhesions and invented the use of elastic gels and rubber layers to map traction exerted by individual cells. His PhD students included David Stopak, with whom he discovered that cell traction not only causes cell locomotion but also rearranges collagen to form tendons, skeletal muscles, and organ capsules; Calhoun Bond, with whom he codiscovered locomotion by intact sponges and that sponge cells constantly sort out even when not dissociated; and Barbara Danowski, who discovered weakening of cell traction in cancerous cells and induction of stronger traction by microtubule poisons. He also writes finite element computer simulations of cell rearrangements and has recently won an award for excellence in teaching.

Dr. Wonmuk Hwang is an associate professor in the Department of Biomedical Engineering at Texas A&M University. He uses biomolecular simulations and theoretical analysis to study mechanics of motor proteins and biofilament assemblies, which make up the molecular *hardware components* of the body. They are responsible for generating subcellular structures and the extracellular matrix, and biological motion including intracellular transport and cell division. Understanding motors and filaments at the atomistic level provides an opportunity to rationally control their behaviors that may lead to novel therapeutic and biotechnology applications. His current research focus includes kinesin motors, microtubule dynamics, collagen mechanobiology, dynamics of water (the most fundamental biomolecule), and developing computational tools for extracting quantitative information from bioimaging data for constructing in silico models.

Dr. Hwang received a BS in physics at Seoul National University and an MA and a PhD in theoretical condensed matter physics at Boston University and had postdoctoral training in biomedical engineering at Massachusetts Institute of Technology, before joining Texas A&M in 2004.

Yaron Ideses finished his PhD studies at the Chemical Engineering Department at Ben-Gurion University of the Negev (BGU) under the guidance of Prof. Anne Bernheim. He also received his BSc in chemical engineering and chemistry and his MSc in chemical engineering from BGU under the guidance of Prof. Anne Bernheim. During his PhD, Yaron has investigated the flow and contractile behavior of actomyosin networks in vitro. His research was supported in part by the Negev scholarship from BGU. Yaron is currently a faculty member in the Pharmaceutical Engineering Dept. at the Azrieli College for Engineering in Jerusalem and also works as an associate researcher in the biophysics laboratory of Professor Anne Bernheim.

Ralf Kemkemer is the head of a research group at the Max Planck Institute for Intelligent Systems, Stuttgart, and professor for biomaterials at Reutlingen University. His research areas are biophysics, cell mechanics, biomaterials, and materials sciences. His particular interests are in the investigation of biological systems interacting with materials and principles of cell migration and mechanoresponses. He currently works on microfabrication techniques and biologically inspired materials and systems. He graduated in physics and biology, earned his PhD in physics from Ulm University, and worked several years in industry building optical devices before he joined the MPI in 2004.

Gijsje Koenderink (1974) is a scientific group leader at the FOM Institute AMOLF and professor at the VU University. Her research focuses on experimental biophysics of the cell, at the interface of biophysics, and soft condensed matter. She originally studied chemistry at Utrecht University (1998) and earned her PhD at the same university in the area of phase behavior and dynamics of colloidal systems. After postdoctoral work at the VU University (2003–2004) and Harvard University (2004–2006), she started her own group at AMOLF. Current research topics in her group include cytoskeletal self-assembly and mechanics, cell membrane interactions, and cellular mechanosensing. She is the recipient of the Dutch NWO-VIDI personal investigator grant (2008) and an ERC Starting Grant (2013).

Kristen L. Mills is a postdoctoral fellow in the Department of New Materials and Biosystems at the Max Planck Institute for Intelligent Systems in Stuttgart, Germany. Her research interests are in experimental mechanics of materials and biomechanics. In particular, she has worked in the areas of thin film mechanics and biomechanics of solid tumor progression. Kristen earned her PhD in mechanical engineering from the University of Michigan in 2008. She has been the recipient of an NSF Graduate Research Fellowship and a Humboldt Postdoctoral Fellowship.

Hari S. Muddana is currently a scientific software engineer at Dart NeuroScience in San Diego. He earned his BTech in computer science and engineering and his MS in computer science from Texas A&M University. He then joined the lab of Dr. Peter Butler at Penn State to develop research in molecular modeling and single-molecule dynamics measurements. After earning a PhD in 2010 in bioengineering under the supervision of Dr. Butler, he went on for a postdoctoral work under Michael Gilson at the University of California San Diego in the Skaggs School of Pharmacy and Pharmaceutical Sciences.

Celeste M. Nelson is an associate professor of chemical and biological engineering at Princeton University, where she directs the Tissue Morphodynamics Laboratory. She earned her BS in chemical engineering and biology from the Massachusetts Institute of Technology (MIT) and her PhD in biomedical engineering with Christopher Chen from the Johns Hopkins University School of Medicine. She completed postdoctoral studies in the laboratory of Mina Bissell at Lawrence Berkeley National Laboratory before joining the faculty at Princeton University in 2007. Her laboratory specializes in using engineered tissues and computational models to understand how mechanical forces direct developmental patterning events during tissue morphogenesis. Her work is funded by the National Institutes of Health (NIH), the Packard Foundation, the Sloan Foundation, the Dreyfus Foundation, and the Burroughs Wellcome Fund.

Paolo P. Provenzano is an assistant professor in the Department of Biomedical Engineering and the Masonic Cancer Center at the University of Minnesota, where he directs the Laboratory for Engineering in Oncology. He earned his BS in mechanical engineering from the University of Wisconsin, followed by an MS and a PhD in biomedical engineering focused on cell and tissue mechanics and tissue engineering at the University of Wisconsin. Dr. Provenzano completed a cell and molecular cancer biology postdoctoral fellowship as a CDMRP Breast Cancer Research Program postdoctoral fellow in the laboratory of Dr. Patricia Keely. Following his postdoctoral fellowship, Dr. Provenzano became a research associate at the Fred Hutchinson Cancer Research Center in Seattle, Washington, studying pancreatic cancer. In 2012, he joined the faculty at the University of Minnesota. His current research focuses on tumor–stroma interactions in breast and pancreatic cancers. His interests are in utilizing advanced quantitative imaging, mechanics, modeling, and cell and molecular biology techniques to elucidate the physical and molecular mechanisms associated with disease progression and resistance to therapy within 3D microenvironments and to develop novel platforms and strategies for cancer diagnosis and treatment.

Shiva Rudraraju is an assistant research scientist in the Department of Mechanical Engineering at the University of Michigan. His research interests are focused around coupled materials physics, nonlinear mechanics, and numerical methods. His primary interest is in mathematical and computational modeling of physical phenomena in materials and biology involving mechanics and transport. He also works on multiscale modeling of deformation and fracture in materials. He earned his PhD from the University of Michigan in 2011.

Yasha Sharma is a PhD candidate in biomedical engineering at Boston University under Muhammad Zaman. Her research focuses on experimental and computational modeling of collective motility of cells in 3D constructs and its relevance to cancer metastasis. She earned her BS in bioengineering at the University of California, San Diego, where she did undergraduate research under Amy Sung on computational modeling of nanomechanics of red cell membranes.

Michael J. Siedlik is a PhD student in the Tissue Morphodynamics Laboratory and the Department of Chemical and Biological Engineering at Princeton University. He earned his BS in chemical engineering at the University of Washington and his MA in chemical and biological engineering at Princeton. Under the guidance of Dr. Celeste M. Nelson, he is currently investigating the endogenous mechanics and electrochemical transport within tissues undergoing epithelial branching morphogenesis. His research is supported in part by a graduate fellowship from the National Science Foundation Graduate Research Fellowship Program.

Orit Siton-Mendelson is an associate researcher in the biophysics laboratory of Professor Anne Bernheim and the Department of Chemical Engineering at Ben-Gurion University of the Negev (BGU). She received her BSc, MSc, and PhD in chemical engineering from BGU. Orit performed her PhD under the supervision of Prof. Anne Bernheim. Currently, Orit is investigating the mechanisms of actin-based motility. She investigates the dynamics and regulation of actin-based processes using reconstituted model systems, where she combines single moleculse experiments with multicomponent multiscale collective studies.

Larry A. Taber is the Dennis and Barbara Kessler Professor of biomedical engineering and professor of mechanical engineering and materials science at Washington University (WU) in St. Louis. He moved to WU in 1997, after working for 4 years at the General Motors Research Laboratories and 15 years at the University of Rochester. Although his formal training is in aerospace engineering (BAE, Georgia Tech, 1974; PhD, Stanford University, 1979), he has published more than 90 journal articles on a wide range of topics, including cochlear mechanics, nonlinear shell theory, cardiovascular mechanics, and the mechanics of growth and development. His ongoing research efforts integrate theoretical modeling with experiments on embryos to study the mechanics of heart, brain, and eye morphogenesis. Dr. Taber is a fellow of the ASME and the American Institute for Medical and Biological Engineering. Twice he won the ASME Richard Skalak Award for the best paper published in the *Journal of Biomechanical Engineering* (2004 and 2007). Currently, he serves as coeditor-in-chief of the journal *Biomechanics and Modeling in Mechanobiology*.

Xiaojing Teng is a graduate student in Dr. Wonmuk Hwang's lab in the Department of Biomedical Engineering at Texas A&M University. He earned a BE and an ME at Beihang University in China. His research focuses on molecular dynamics simulation of collagens, with the aim of understanding the atomistic mechanism for collagen mechanoregulation.

Muhammad H. Zaman is an associate professor of biomedical engineering at Boston University. Prof. Zaman also holds appointments in the Department of Medicine and the Department of International Health at Boston University School of Medicine. Professor Zaman is also the associate chair of biomedical engineering and associate director of Kilachand Honors College at Boston University. He earned his PhD in physical chemistry from the University of Chicago in 2003 where he was a Burroughs Wellcome graduate fellow in interdisciplinary sciences. After his PhD, he was a Herman and Margaret postdoctoral fellow at MIT from 2003 to 2006. Professor Zaman is actively involved in two areas of research, the first is developing new tools and quantitative understanding of tumor formation and tumor metastasis and the second is in developing robust and affordable diagnostic technologies for the developing world, and he is also working on capacity building and engineering education in the developing countries. His research is supported by the National Institutes of Health (NIH), National Science Foundation (NSF), Australian Research Council, National Research Foundation (NRF) in the United Arab Emirates, United States Agency for International Development (USAID), United Nations Economic Commission for Africa (UNECA), Center for Integration of Medicine and Innovative Technology (CIMIT), Saving Lives at Birth Consortium, and a number of other private foundations.

1

Active Mechanics of the Cytoskeleton

José Alvarado and Gijsje Koenderink

CONTENTS

1.1 Introduction

Cells can resist but also actively exert mechanical forces. This ability allows them to perform many essential tasks. Some animal cells can crawl across surfaces and through small pores, pulling themselves forward while pushing against their environment. Some cells such as sperm cells swim by beating long appendages, which push the surrounding fluid. Most cells proliferate by dividing into two daughter cells, requiring drastic changes in cell shape. Many cells maintain their internal components organized by a combination of internal pushing and pulling forces. Combinations of cell growth, divisions, and shape changes allow embryos to develop into organisms with a well-defined anatomy. Given the importance of mechanical forces in life, understanding the mechanisms that determine how cells exert and withstand forces is crucial.

In order to accomplish mechanical tasks, cells rely on *polymers*. The kind of polymer used depends on cell type. Many plant, yeast, and bacterial cells maintain relatively constant, rodlike shapes. These cells possess an outer cell wall composed of rigid polymers, which provide robust mechanical stability. In contrast, animal cells are usually soft and deformable. This allows them to move and change shape. Rather than possessing a static, rigid cell wall, animal cells rely on the cytoskeleton to provide resistance to external forces. However, at the same time, the cytoskeleton is dynamic and adaptable and actively generates forces. Understanding the physical properties of biological polymers like cytoskeletal filaments is thus crucial in order to resolve the role of forces in cell mechanics.

One popular approach to study how biological polymers regulate cell shape and mechanics is to reconstitute purified biological polymers in a simplified, cell-free environment (Bausch and Kroy 2006; Fletcher and Geissler 2009). The advantage of such biomimetic systems is that their molecular and structural complexity can be precisely controlled. The reduced complexity compared to living cells makes it easier to perform quantitative experiments that can be linked to quantitative physical theories that predict the macroscopic physical properties in terms of the molecular properties and interactions of the components.

In this chapter, we will provide a background on theory used in describing the mechanical properties of cytoskeletal polymers and summarize results of experiments with a focus on biomimetic approaches. First, we introduce some examples of cytoskeletal structures found in cells. We then discuss the mechanical properties of single polymers, networks of polymers, and cross-linked networks. Finally, we will discuss how molecular motors allow these networks to actively generate force, which results in fascinating, self-deforming materials known as active gels (Joanny and Prost 2009).

1.2 Cytoskeleton

The *cytoskeleton* is a network of biological polymers that provides cells with mechanical strength and the ability to generate active forces. Cytoskeletal polymers associate with a variety of accessory proteins to form different supramolecular structures that are tailored for distinct tasks. Despite the large number of possible cytoskeletal structures, the cytoskeleton primarily comprises only three types of polymers. In this section, we will introduce these cytoskeletal polymers and highlight some of the structures they form. In later sections, we will investigate the properties of these polymers and some of their accessory proteins in more detail.

Actin filaments are somewhat flexible polymers that can form either fine random meshworks, branched networks, or stiff bundles. In many cases, actin filaments form contractile networks or bundles, in cooperation with myosin motor proteins. The most well-known example of such a contractile array is found in muscle, where actin and myosin form extremely well-organized arrays known as sarcomeres (Rayment et al. 1993). Nonmuscle cells also possess contractile actin–myosin structures, but these tend to be less well ordered. Right underneath the membrane, actin and myosin form a thin *cortex* composed of a random meshwork of cross-linked actin filaments. Myosin motors actively generate contractile stress within this cortex, which can change cell shape (Salbreux et al. 2012). These shape changes can take place at the individual cell level as well as on the tissue scale. During cytokinesis in yeast and animal cells, cortical actin filaments and molecular motors transiently organize into a *contractile ring*, which constricts to pinch off the mother cell into two daughter cells (Guertin et al. 2002; West-Foyle and Robinson 2012; McMichael and Bednarek 2013). The actin cortex also assists yeast and animal cells during *endocytosis*, the process whereby cells internalize foreign objects or fluids (Engqvist-Goldstein and Drubin 2003). One example is *phagocytosis*, where immune cells engulf and destroy invasive pathogens like bacteria (May and Machesky 2001). During *embryogenesis*, epithelial cell monolayers collectively generate forces, maintaining tissue integrity (Cavey and Lecuit 2009), homeostasis (Guillot and Lecuit 2013), and shape (Rauzi and Lenne 2011). Apart from a thin cortex, some large cells such as oocytes additionally have a 3D, cytoplasmic network of actin filaments (Field and Lénárt 2011), which can be used for transporting chromosomes (Lénárt et al. 2005). Crawling cells like fish keratocytes, amoebas, and metastatic cancer cells can move across surfaces using a combination of different actin-based structures (Abercrombie 1980; Rafelski and Theriot 2004; Ananthakrishnan and Ehrlicher 2007). At the front of crawling cells, a thin, 2D branched array of actin filaments called the *lamellipodium* pushes the cell membrane forward. Membrane protrusions reinforced by tightly bundled actin filaments called *filopodia* often extend from the lamellipodium, which are thought to sense environmental cues and guide the direction of cell motion (Davenport et al. 1993; Mattila and Lappalainen 2008). Similar actin-filled membrane protrusions are also found in specialized cell types, such as inner hair cells in the inner ear, which project *stereocilia* that participate in the transduction of sound waves to neuronal impulses (Manor and Kachar 2008). At the back of crawling cells, a network of actin filaments and molecular motors exerts retraction forces, which allows cell detachment from the substrate and forward motion of the cell body. In strongly adherent cells, actin and myosin form contractile *stress fibers* that span the cell and connect to focal adhesions (Naumanen et al. 2008). There is mounting evidence that cells can switch between different migration mechanisms. In dense tissues, some cells use mechanisms based on myosin-induced membrane *blebbing* for migration instead of polymerization-based motility (Paluch and Raz 2013).

Interestingly, the actin cytoskeleton in plant cells has a completely different organization and function than in animal cells. The most prominent actin structure is provided by cytoplasmic actin cables composed of actin filaments bundled by accessory proteins. These *actin cables* usually radiate from the nucleus toward the cell membrane (Hussey et al. 2013) and assist in properly positioning the nucleus (Starr and Han 2003). Moreover, actin cables are used as tracks for transport by molecular motors, for instance, for pollen tube growth (Kroeger and Geitmann 2012). Intriguingly, these roles are reminiscent of positioning and transport roles of microtubules in animal cells (see Section 1.5.1).

Microtubules are much stiffer polymers than actin filaments. In animal cells, they play a crucial role in organizing the cell interior. Microtubules act as tracks for kinesin and dynein motors, which move along microtubules to transport intracellular cargo. In interphase cells, microtubules usually emanate from a *microtubule-organizing center* positioned near the nucleus and grow radially outward toward the cell membrane, enabling long-range transport through the crowded cytoplasm (Vale 2003; Barlan et al. 2013). A striking example is provided by cells called *melanophores*, which allow many amphibians and fish to change color (Tuma and Gelfand 1999). This is accomplished by motor-driven transport of vesicles containing the pigment melanin across microtubules. When cells divide, the microtubules reorganize to form the *mitotic spindle*, a specialized assembly of microtubules, molecular motors, and many other accessory proteins that reliably separate chromosomes to the two daughter cells (Walczak and Heald 2008). Similarly to animal cells, fission yeast cells also use microtubules and molecular motors to separate chromosomes and transport cargo (Hagan 1998). However, unlike in animal cells, interphase microtubules are organized into a small set of bundles that extend to the two poles of the rod-shaped yeast cell. This allows molecular motors to deliver growth factors specifically to these two ends, maintaining yeast cells' rodlike shape (Chang and Martin 2009). Some cells, such as algae and certain bacteria, swim by beating *flagella* or *cilia*. These are long appendages that comprise an ordered arrangement of microtubules that slide past one another, causing the entire appendage to lash back and forth (Brokaw 1994; Kantsler et al. 2013). In plant cells, microtubules are again very differently organized than in animal cells. They usually form an ordered *cortex* underneath the cell membrane that helps guide the ordered production of the cell wall, which is essential in maintaining plant cells' elongated shape (Gutierrez et al. 2009; Bringmann et al. 2012). Intriguingly, the cortical localization of microtubules mirrors the cortical localization of actin filaments in animal cells. When plant cells divide, the cortical microtubule array transforms into a *preprophase band* with the nucleus anchored at the center, thus determining the division plane (Van Damme 2009).

Intermediate filaments are a third set of cytoskeletal filaments that are present in animal cells, but not in yeast or plant cells. They are so named because their diameter (~10 nm) is intermediate between the diameters of actin filaments (~6 nm) and microtubules (~25 nm). Surprisingly, they are 10-fold more flexible than actin filaments (see Section 1.2.4), indicating that their internal structure affects their mechanical properties. Intermediate filaments (IF's) are encoded by 70 genes in the human genome, which are often categorized into six sequence homology classes (class I–VI) or, alternatively, into three assembly groups (I, II, III) that can coexist as three separate IF systems within the same cell (Szeverenyi et al. 2008). One type of intermediate filaments, the *lamins*, forms a basket that provides the cell nucleus with mechanical strength and also regulates nuclear events such as chromosome replication and cell death (Gruenbaum et al. 2000). The other intermediate filaments are found in the cytoplasm and are expressed in a tissue-dependent and developmentally regulated manner (Helfand et al. 2003; Herrmann et al. 2007). For example, epithelial cells such as skin cells resist deformation by a network of *keratin* filaments (Omary et al. 2009), neuronal axons are reinforced with *neurofilaments* (Lepinoux-Chambaud and Eyer 2013), and astrocytes have a cytoplasmic network of *glial fibrillary acidic protein* (Middeldorp and Hol 2011). Fiber cells of the vertebrate eye lens contain *beaded filaments*, which not only provide the lens with mechanical strength but also maintain its transparency (Song et al. 2009). Intermediate filaments are generally thought to provide mechanical protection against large deformations (Fudge et al. 2009), and also to serve as a platform for signal transduction.

Septin filaments have only recently begun to gain recognition as a fourth component of the cytoskeleton (Mostowy and Cossart 2012). Septins were originally discovered in budding yeast, where they form rings at the bud neck that form a diffusion barrier between the membranes of the mother and daughter cell (Hartwell 1971; Byers 1976). Septins were later found to be conserved across the animal kingdom. All eukaryotes have multiple septins, ranging in number from 2 in nematodes to 13 in humans. Functional studies of cells involving septin deletion or mutation suggest that septin assemblies play three key roles, which are likely interrelated: they maintain cortical integrity, act as diffusion barriers for membrane proteins, and serve as scaffolds for cytoplasmic proteins. During cytokinesis, septins are core components of the contractile ring, where they may act both as a diffusion barrier and as a scaffold (Glotzer 2005). In motile cells such as

T cells, septins form cortical arrays, which contribute to cell rigidity and to regulation of cell motility (Gilden and Krummel 2010). However, given the complex composition of the cortex, much of the evidence that septins fulfill these functions is indirect. It is, for instance, unknown whether septin exerts its functions independent of or in concert with actin. Moreover, quantitative biophysical studies of septins are still lacking.

The amino acid sequences of actin and tubulin proteins are surprisingly well conserved across many eukaryotic species (Sheterline and Sparrow 1994; Mitchison 1995). Intermediate filaments and septin filaments also maintain a large degree of evolutionary conservation, although species-specific variation is greater than with actin and tubulin. Throughout this chapter, we will focus on these eukaryotic cytoskeletal filaments. However, we note that various actin, tubulin, and intermediate filament homologues have now been identified in prokaryotes (Shih and Rothfield 2006), whose structural and mechanical roles are starting to become more clear.

1.2.1 Single Cytoskeletal Filaments

Cells employ biological polymers that are built up from different building blocks, including sugars, nucleic acids, and amino acids. The cell wall of plant cells is made up of *cellulose fibers*, built from linked glucose chains (Somerville 2006). Bacterial colonies secrete extracellular polysaccharide chains, which maintain cohesion and contribute to the formation of *biofilms* such as dental plaque (Costerton et al. 1999). Cells store their genetic information in the form of *deoxyribonucleic acid* (DNA), which are chains built from four different types of interchangeable nucleic acids (Alberts 2008). Cells express DNA to produce *proteins*, which are macromolecules composed of one or more polypeptide chains, which themselves are long chains of up to 20 interchangeable amino acids (Alberts 2008). Due to the covalent bonds that link subunits together, macromolecular polymers typically form stable, long-lasting structures, which maintain their structure and shape. In the case of cell walls or biofilms, these sturdy polymers allow cells or colonies to resist mechanical deformation.

Cytoskeletal polymers are supramolecular polymers built up from many individual protein subunits. These subunits are typically linked via weak interactions such as electrostatic interactions, hydrophobic interactions, and hydrogen bonding. The specificity of these interactions results in highly ordered structures, while their noncovalent nature allows cytoskeletal protein polymers to assemble and disassemble dynamically in response to biochemical or mechanical signals. This adaptability is crucial for enabling cells to form dynamic and adaptable structures for cell movement and shape change. This dynamic character contrasts with the covalent bonds that usually link the subunits of biological polymers such as cellulose. Another interesting consequence of the dynamic nature of cytoskeletal filaments is that it allows them to actively exert forces (in conjunction with the consumption of chemical energy).

1.2.2 Cytoskeletal Polymer Structure

Actin filaments comprise globular actin protein monomers, which themselves comprise two domains separated by a cleft that binds a divalent cation together with either adenosine triphosphate (ATP) or adenosine diphosphate (ADP) (Carlier et al. 1994). Monomers assemble head to tail to form linear filaments. The ligand-binding cleft is directed toward the so-called minus end or pointed end of the filament. The opposite side is directed toward the *plus end* or *barbed end*. Apart from assembling head to tail, actin monomers also associate via side-by-side contacts, forming a double-stranded helical structure with a 37 nm pitch (Selby and Bear 1956; Galkin et al. 2012a). Actin monomers incorporated in filaments exhibit multiple conformational states (Galkin et al. 2010b), potentially allowing them to act as tension sensors (Galkin et al. 2012a).

Microtubules comprise α- and β-tubulin proteins, which form stable heterodimers. β-tubulin proteins bind guanine triphosphate (GTP) or guanine diphosphate (GDP) (Gardner et al. 2013). Although α-tubulin proteins bind GTP, this binding site is buried at the dimer interface. Dimers of α–β-tubulin assemble head to tail to form linear protofilaments, with α-tubulin at the *plus end* and β-tubulin at the *minus end* (Nogales et al. 1998). Typically, 13 protofilaments associate side by side to form a hollow, cylindrical microtubule. This stable, tubular structure makes microtubules stiffer than actin filaments by a factor of approximately 300 (Gittes et al. 1993).

Intermediate filaments have a rather different structure from actin filaments and microtubules, because their subunits are fibrous rather than globular. IF proteins are rod-shaped, double-stranded parallel dimers with a length of 40–50 nm and a diameter of 2 nm. The dimers possess a central, largely α-helical rod domain consisting of four coiled-coil segments, flanked by non-α-helical N- and C-terminal end domains. Cytoplasmic IF proteins assemble into filaments via a multistep pathway, where dimers first associate to form antiparallel, approximately half-staggered tetramers, which laterally aggregate into unit-length filaments, which in turn longitudinally anneal to form filaments (Herrmann et al. 2007). The fibrils subsequently mature by a compaction process. Nuclear lamins follow a different assembly pathway, involving head-to-tail polymerization (Figure 1.1).

The notion of head-to-tail assembly indicates a special property of actin filaments and microtubules. *Structural polarity* refers to the fact that the two ends of the filament can be distinguished from each other. This is in strong contrast to intermediate filaments and septin filaments, which exhibit *structural symmetry* (Herrmann et al. 2007): both ends of the filament are identical and indistinguishable. The structural polarity of actin and microtubules has two important functional consequences: they can polymerize in a directional manner, and they are recognized by molecular motors that take advantage of their

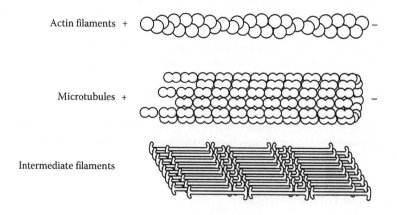

FIGURE 1.1

Three kinds of cytoskeletal polymers. Actin monomers (white circles) assemble to form actin filaments with a double-stranded helical structure. The α–β-tubulin dimers (joined white circles) form long protofilaments, which assemble laterally into a hollow tube. Intermediate filaments comprise elongated tetramer subunits (I-shaped structures), which associate laterally and stagger to form fibrils. Note that actin filaments and microtubules are structurally polar and therefore have two different ends, designated + and –. Intermediate filaments are structurally symmetric and their ends are thus indistinguishable.

polarity to move in a processive manner. Given the structural symmetry of IF filaments and septins, it is believed that there are no motors interacting with them.

1.2.3 Polymerization and Enzymatic Activity

The process by which monomer subunits join a polymer is called *polymerization*. For supramolecular polymers such as actin, this process is characterized by the rate of monomer addition, k_{on}, as well as the rate of monomer dissociation, k_{off}. In equilibrium, these rates are identical. For septins and intermediate filaments, this equilibrium description is sufficient. Actin and microtubules, however, are nonequilibrium polymers because their assembly is coupled with enzymatic activity. Shortly after ATP-bound actin or GTP-bound tubulin joins a growing actin filament or microtubule, hydrolysis of the bound nucleotide occurs. This hydrolysis provides chemical energy that maintains different on/off rates at the two filament ends. The end with the higher on rate is conventionally called the *plus end*, while *minus end* refers to the end with the slower on rate.

Continuous addition of fresh ATP- or GTP-bound monomers results in a so-called *ATP cap* or *GTP cap* at the plus end, while the rest of the filament contains ADP- or GDP-bound monomers. The presence of such a cap allows for interesting nonequilibrium processes such as actin *treadmilling* (Pollard and Borisy 2003) and microtubule *dynamic instability* (Gardner et al. 2011). These dynamic processes allow cytoskeletal filaments to exert polymerization forces, which we will discuss later in this chapter. With the help of specialized tip-binding molecules, actin filaments and microtubules can also generate depolymerization forces. Microtubule plus ends are targeted by so-called end-binding (or tip-tracking) proteins (Maurer et al. 2012; Seetapun et al. 2012; Bowne-Anderson et al. 2013), while actin filament plus ends are targeted by formins (Jegou et al. 2013).

Although intermediate filaments lack structural polarity and enzymatic activity, evidence suggests that intermediate filaments exhibit fast polymerization and depolymerization kinetics, which is regulated by (de)phosphorylation (Helfand et al. 2003). Septin filaments also lack structural polarity but do exhibit enzymatic activity (Weirich et al. 2008). Septin subunits bind and slowly hydrolyze GTP, and septin subunits can form filaments via alternating interfaces of the GTP-binding domains and interfaces containing the N- and C-termini. Fluorescence recovery after photobleaching experiments in cells has revealed that septins are rather dynamic. This may be related to their enzymatic activity, but posttranslational modifications may also play a role.

1.2.4 Wormlike Chain Model

So far, we have seen how molecular structure can determine many of the special properties of cytoskeletal filaments. Yet physical models of cytoskeletal mechanics often ignore the fine structural details of cytoskeletal filaments. The wormlike chain model is the most common coarse-grained model used to describe the mechanical properties of cytoskeletal polymers. This model was originally developed by Kratky and Porod (Kratky and Porod 1949). It approximates polymers by a smooth linear contour that resists bending with a quantity κ called the *bending modulus*. High values of κ indicate a stiff polymer that resists deformation. Polymers with low κ deform more easily and appear soft. In the absence of thermal fluctuations, linear polymers would assume a straight shape. At finite temperatures, however, random forces from thermal fluctuations will cause cytoskeletal polymers to bend. These thermal bending undulations have been observed experimentally for actin filaments, microtubules, and intermediate filaments by fluorescence microscopy

(Gittes et al. 1993; Ott et al. 1993; Isambert et al. 1995; Noding and Köster 2012) and have been used to measure κ. This was achieved by measuring a length scale l_p called the *persistence length*, which is defined as the decay length of angular correlations along the polymer contour. Roughly speaking, the persistence length is the distance over which the polymer contour appears approximately straight. The persistence length is related to the bending modulus by the relation $\kappa = kT\, l_p$, where k is Boltzmann's constant and T is temperature.

Based on the persistence length l_p relative to the contour length L, we can distinguish between three classes of polymers. If the polymer backbone offers little resistance to bending ($l_p \ll L$), thermal fluctuations dominate, bending the polymer so strongly that it crumples to a highly bent conformation well described by a fractal contour (de Gennes 1979). Such polymers are called *flexible polymers* and are suitable for describing many synthetic macromolecular polymers. In the opposite scenario ($l_p \gg L$), *stiff polymers* strongly resist thermal fluctuations and can be modeled as rigid rods (Landau et al. 1986). A third, and intermediate, regime occurs when $l_p \sim L$. In this regime, thermal fluctuations cannot be neglected, though the polymer retains a well-defined, mostly straight shape with long, wavelike undulations. Polymers in this intermediate regime are called *semiflexible polymers*. Cytoskeletal filaments fall in this intermediate regime, having contour lengths of several μm and relatively large persistence lengths that range from 0.5–1 μm for intermediate filaments to 8–15 μm for actin filaments and several mm for microtubules (Kasza et al. 2007). The small persistence length of intermediate filaments is surprising, given that their diameter is intermediate between those of actin filaments and microtubules. This suggests that the bundle-like architecture of intermediate filaments plays a key role in their mechanics. Sliding between subunits may, for instance, act to lower the bending rigidity (Figure 1.2).

1.2.5 Response to Pulling Forces

So far, we have seen that the mechanical properties of semiflexible filaments can be characterized by the persistence length. This quantity describes how filaments respond to thermal forces. But how do semiflexible polymers respond to external pulling forces? Given an infinitely strong force, we should expect the polymer to assume a straight shape: such a taut filament would not bend due to thermal forces. Theoretical models have accounted for the reduction of thermally induced bends due to external pulling forces (MacKintosh et al. 1995). The amplitude of thermally induced bends in the polymer depends on wavelength, typically quantified through the wave vector $q = n\,\pi/L$, where $n = 1, 2, 3\ldots$ If the polymer experiences tension due to an external pulling force f, the amplitude u_q of bending mode q is given by

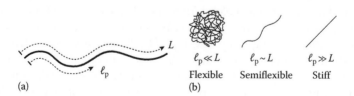

(a)

$\ell_p \ll L$ $\ell_p \sim L$ $\ell_p \gg L$
Flexible   Semiflexible   Stiff
(b)

FIGURE 1.2
The wormlike chain. (a) Polymers can be modeled as linear contours in space (bold black line). They possess contour length L and persistence length l_p. (b) The relationship between L and l_p determines whether the polymer is classified as flexible, semiflexible, or rigid.

$$\left\langle \left| u_q \right|^2 \right\rangle = \frac{2kT}{L\left(\kappa q^4 + fq^2 \right)}$$

Long-wavelength bends (lower q) have the largest bending amplitudes, while short-wavelength bends (higher q) decay quickly, as q^{-4}. This formula also shows that applying a pulling force f to the polymer reduces the transverse bending amplitudes u_q. This reduction in thermal modes results in an effective restoring force:

$$f \sim \frac{l_p \kappa}{L^4} x$$

where x denotes the displacement of the end-to-end-distance vector of the polymer contour from its equilibrium length. The effective spring constant of a semiflexible polymer is thus $l_p \kappa L^{-4}$. Because semiflexible polymers bend in response to thermal forces, their response to a pulling force is entropic in origin. This effect is often called the *entropic spring*. The force–extension relation was experimentally verified with optical tweezers for DNA (Bustamante et al. 1994) as well as for actin and microtubules (van Mameren et al. 2009).

The earlier equations are only valid under the assumption of small forces and *linear responses*. For strong pulling forces, once the thermal undulations are pulled out, this assumption breaks down. In this case, we must account for stretch deformations of the polymer backbone itself. This has been accomplished by introducing in the wormlike chain model an enthalpic quantity µ called the *stretch modulus* (Odijk 1995; Storm et al. 2005). The result is the emergence of two different force-response regimes with distinct spring constants. For small deformations, the effective spring constant is dominated by the bending modulus κ according to the entropic spring. For large deformations, the effective spring constant is dominated by the stretching modulus µ, corresponding to the enthalpic stretch of the polymer contour. Experimental evidence suggests that actin filaments and microtubules break before they are substantially extended, whereas intermediate filaments are highly stretchable (Kreplak and Fudge 2007; Lin et al. 2010).

1.2.6 Response to Pushing Forces

We have investigated how individual semiflexible polymers respond to pulling forces. But how do they respond to pushing forces? Given small forces and linear responses, the stretch modulus µ determines the compressive deformation of elastic rods. Long, semiflexible rods can readily undergo a buckling instability when pushing forces exceed the critical *Euler force* f_c (Landau et al. 1986):

$$f_c \sim \kappa L^{-2}$$

This force sets the maximal protrusive force that a rod can exert. Since microtubules are rather stiff at the length scale of the cell, they can withstand high compressive forces (Dogterom and Yurke 1997). Actin filaments buckle at force levels that are 300-fold lower, owing to their smaller bending stiffness (Gittes et al. 1993; Footer et al. 2007). This buckling response results in an asymmetric mechanical response: single actin filaments can withstand and propagate pulling forces but not pushing forces. However, bundling can overcome this limitation, and actin bundles can exert substantial pushing forces

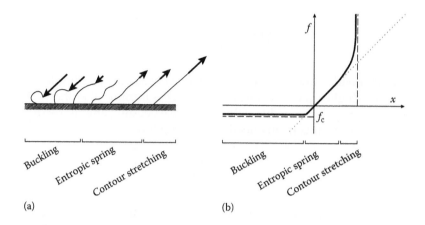

FIGURE 1.3
Response of semiflexible polymers to pushing and pulling. (a) Schematic of a polymer bound at one end and deformed at the other end, undergoing buckling, linear response (entropic spring), and contour stretching. (b) Force–extension curves for the three types of response. Polymers cannot exert compressive forces higher than the critical Euler buckling force f_c.

(see Section 1.5). In cells, coupling to the surrounding cytoskeleton can further reinforce filaments against compressive loads (Brangwynne et al. 2006; Das and MacKintosh 2010) (Figure 1.3).

1.3 Filament Networks

In cells, cytoskeletal polymers are generally present at high density where filaments overlap and entangle. In this section, we will describe the collective properties of materials composed of many filaments. We will begin by summarizing theoretical models for solutions of rigid rods, a limit that has been theoretically well characterized.

1.3.1 Rigid Rod Solutions

Consider a rigid rod of length L and thickness a diffusing freely in solution. As it translates and rotates, it sweeps out a volume $\sim L^3$. Thus, for a suspension of rods with concentration (number density) $c \ll L^{-3}$, neighboring rods are spaced far enough apart that they do not significantly interfere with each other's motions. In this *dilute regime*, the rotational diffusion constant D_{rot} scales with rod length (with a prefactor that depends on temperature and viscosity) (Riseman and Kirkwood 1950):

$$D_{rot,dilute} \sim L^{-3} \ln\left(\frac{L}{a}\right)$$

If the rod concentration c increases beyond $\sim L^{-3}$, rods start to interact via *steric repulsion* (excluded volume): two rods cannot overlap in space and therefore repel each other upon contact. In this *entangled regime*, the diffusion of a rod is constrained by its neighbors.

Early theory by Doi and Edwards modeled the effect of entanglements for concentrated suspensions of rods by proposing the *tube model* (Doi and Edwards 1978). In this model, a rod of interest cannot diffuse freely in a volume $\sim L^3$, but is rather confined to an elongated virtual tube formed by the presence of neighboring rods. This results in a drastically reduced rotational diffusion constant in the entangled concentration regime:

$$D_{\text{rot,entangled}} \sim L^{-6} D_{\text{rot,dilute}}$$

In the dilute and entangled concentration regimes, the rod orientations are isotropically distributed to maximize rotational entropy. If the rod concentration c increases further toward $c^* \sim a^{-1}L^{-2}$ (where aL^2 is the volume of a single rod), however, packing effects cause spontaneous rod alignment. Although this orientational alignment decreases the rotational entropy, this loss is compensated by an increase in translational entropy due to a decrease in the mutual excluded volume. This entropically driven isotropic to nematic phase transition was first predicted by Onsager (1945). In the limit of infinitely long rods, the critical concentration c^* depends only on the inverse of the rod aspect ratio a/L. The orientational anisotropy of rods in the nematic phase results in an anisotropy in optical properties, giving the appearance of a crystal. Hence, materials in the nematic phase form one of many possible *liquid-crystalline regimes*.

1.3.2 Polymer Networks

The equations given earlier were derived assuming rods that are perfectly rigid and of uniform diameter and length. These assumptions have been experimentally validated for monodisperse rod-shaped viruses, specifically tobacco mosaic virus (Graf and Löwen 1999) and bacteriophage fd (Dogic and Fraden 2006). However, cytoskeletal polymers are neither rigid nor uniform in length. Semiflexibility and length polydispersity cause quantitative changes to the phase behavior and dynamics, which can be accounted for theoretically (Khokhlov and Semenov 1982; Odijk and Lekkerkerker 1985; Odijk 1986; Glaser et al. 2010). However, qualitatively, the earlier description for rigid rods is consistent with the dynamics and phase behavior observed for purified actin filaments and microtubules. For entangled actin filament networks with actin concentrations between 0.1 and 2 mg mL^{-1}, the tube model was confirmed experimentally (Käs et al. 1994): labeled filaments were observed to fluctuate within a virtual, confining tube formed by the unlabeled surrounding filaments. Filaments slide back and forth along the tube in a snakelike motion called *reptation*. The width of the virtual tube decreases with increased actin concentration (Käs et al. 1996). Above concentrations of ~ 2 mg mL^{-1}, networks of actin filaments can form nematic phases (Suzuki et al. 1991; Käs et al. 1996). Shortening actin filaments by adding the capping protein gelsolin increases c^*, consistent with Onsager's theory (Suzuki et al. 1991) (Figure 1.4).

1.3.3 Mechanical Properties

Entangled polymer networks are *viscoelastic materials*, which exhibit behavior characteristic of both fluids (viscous) and solids (elastic). The viscoelasticity of soft materials such as polymer networks can be conveniently measured by *rheology* experiments. In these experiments, networks are allowed to form between two large, flat surfaces, which are moved relative to one another to apply a shear stress. The response of a viscoelastic material to a shear stress is given by the complex *shear modulus*, $G = G' + i\,G''$ (Meyers and

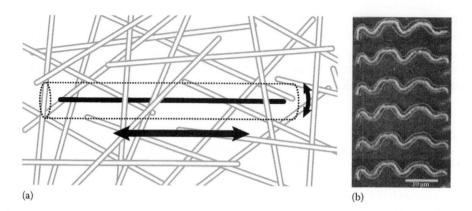

(a)   (b)

FIGURE 1.4

Entangled rods and polymers. (a) Schematic of an entangled rod solution. The black rod cannot rotate freely because it cannot penetrate the surrounding gray bars. Rather, it moves in a virtual *tube* (black dotted line), whose width decreases with increasing rod density. The rod cannot rotate freely, but can still diffuse back and forth (bold arrows). (b) Time-lapse image of a fluorescently labeled actin filament entangled by neighboring unlabeled filaments. The filament is confined to a virtual tube (thin white lines) and slides back and forth inside this tube in a snakelike motion called reptation (note the location of the filament ends with respect to the tube). (From Käs, J. et al., *Biophys. J.*, 70(2) (February), 609, 1996.)

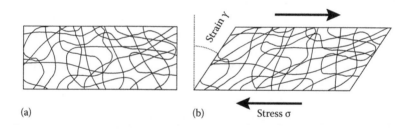

(a)   (b)   Stress σ

FIGURE 1.5

Shearing an entangled network, (a) the undeformed network and (b) applying a shear stress σ results in a shear strain γ.

Chawla 2009). The complex modulus has two components: the *storage modulus G′*, which measures elastic or solid-like behavior, and the *loss modulus G″*, which measures viscous or fluid-like behavior. The shear moduli are usually determined by applying a small oscillatory shear stress of controlled frequency and measuring the resultant oscillatory shear strain response. The magnitude of the shear modulus follows from the stress/strain ratio, while the extent of viscous dissipation is reflected in a phase shift between the stress and strain signals (Figure 1.5).

For entangled actin networks, the primary determinant of the shear modulus is the filament density. The storage modulus was measured to scale with concentration according to $G' \sim \varphi^{7/5}$, where φ is the polymer volume fraction (Hinner et al. 1998; Gisler and Weitz 1999). Theoretical scaling arguments, which consider two length scales, can account for this experimental result (MacKintosh 2011). The first length scale is the *mesh size* ξ, which is defined as the typical spacing between filaments. For random networks of rigid filaments, the mesh size scales as $\xi \sim \varphi^{-1/2} a$, where a is the thickness of a single filament (Schmidt et al. 1989). The second relevant length scale is the *entanglement length* l_e, which describes the typical length over which filament entanglements restrict thermal fluctuations. It scales as $l_e \sim (a^4\, l_p)^{1/5}\, \varphi^{-2/5}$ (Isambert and Maggs 1996; Hinner et al. 1998). Together, these two length scales determine the storage modulus according to $G' \sim kT/(\xi^2\, l_e)$. Substituting this expression with the expressions for ξ and l_e yields the experimentally validated scaling relation $G' \sim \varphi^{7/5}$. Interestingly, the two length scales, and thus the storage modulus G', do not depend strongly on the bending stiffness of the filaments, given by l_p. At small frequencies, below a timescale set by the reptation time of the filaments, entangled solutions of actin filaments behave as viscous liquids. This timescale is on the order of minutes to hours, depending on the filament length.

1.3.4 Physical Forces

In cells, the organization of actin filaments is affected not only by mutual repulsion but also by steric interactions with other cytoplasmic components and with the cell membrane. Soluble cytoplasmic components such as globular proteins can in principle act to introduce an effective *depletion attraction* between filaments. From a modeling point of view, globular proteins can be thought of as diffusing, impenetrable spheres that exclude some volume. Above a critical protein concentration c^*, filaments spontaneously bundle in order to maximize the free volume available to the globular polymers, thereby maximizing translational entropy (Lekkerkerker and Tuinier 2011). Experiments have shown that actin filament networks indeed become bundled when sufficient amounts of inert polyethylene glycol (PEG) polymers are added (Hosek and Tang 2004). Depletion forces can effectively cross-link actin filament networks and increase the shear modulus (Tharmann et al. 2006).

External boundaries such as the plasma membrane can affect filament organization by *spatial confinement*. Models of rigid rods predict that rods in an isotropic suspension in contact with an impenetrable planar surface will align along the surface, forming a so-called orientational wetting layer (van Roij et al. 2000). This entropic effect only occurs for rods close to the surface, with an effective layer depth on the order of one rod length. However, walls also give rise to a depletion layer, again due to entropic (volume exclusion) effects. For actin filaments, such a depletion zone with reduced actin density was experimentally shown (Fisher and Kuo 2009). Yet, when filaments are confined in three dimensions to emulsion droplets or liposomes, they form a cortex-like layer when droplets are smaller than the persistence length of actin filaments (Limozin et al. 2003; Claessens et al. 2006). Similarly, microtubules grown in confining microchambers were found to coil and wrap around the chamber edges (Cosentino Lagomarsino et al. 2007). This effect can be explained by enthalpic effects: cortical localization minimizes the energy penalty associated with filament bending. Microrheology measurements showed that confinement also affects the mechanical properties of entangled actin solutions, inducing stiffening (Claessens et al. 2006). It is still poorly understood how entropic and enthalpic effects together determine the organization of entangled (and liquid crystalline) solutions of cytoskeletal filaments in confinement.

Note that the depletion attraction and confinement effects induce only effective interactions between filaments, mediated by the maximization of entropy and/or the minimization of bending enthalpy. These indirect, physical forces likely contribute to the organization of cytoskeletal structures inside cells. The environment inside most cells is crowded with soluble proteins, which comprise 20%–30% of the cytoplasmic volume (Ellis 2001). For this reason, the depletion interaction (often referred to as *crowding*) has been suggested to contribute to actin filament bundling, amyloid fibril formation, and DNA looping (Marenduzzo et al. 2006). Similarly, confinement effects play a significant role in cytoskeletal organization (Chen et al. 1997). But the physical mechanisms underlying confinement effects remain poorly understood. Even simplified experiments of dense reconstituted actin filament networks remain challenging to predict theoretically (Soares e Silva et al. 2011a). Cytoskeletal filaments have contours and persistence lengths that are often comparable to cellular dimensions, especially in thin compartments such as lamellipodia and filopodia. Understanding how these length scales interact should lead to a physical framework, which quantifies confinement effects in cells.

1.4 Cross-Links

The extent to which physical forces determine the organization and mechanics of cytoskeletal networks remains poorly understood. One reason for this is that cytoskeletal organization is regulated by a myriad of *accessory proteins*, including cytoskeletal cross-links, cytoskeleton–membrane linkers, and nucleating and length-controlling factors. By tightly regulating the density and activity of these accessory proteins, cells can create different cytoskeletal structures without significantly affecting the physical properties of the filaments themselves. In this section, we will focus on actin-binding proteins that bind to actin filaments and connect them into cross-linked networks or bundles. We will describe the molecular properties of these *cross-links* and the consequences of these molecular properties for the organization and mechanical properties at the network level.

Cross-link proteins create connections between actin filaments by binding to two separate filaments with two different actin-binding domains. Different types of actin-binding domains have been identified. The most common type is the *calponin homology domain*, which is found across a broad class of cross-link proteins, including *spectrin, filamin, fimbrin*, and *α-actinin* (Korenbaum 2002). These cross-links are all homodimeric. *Fascin* proteins are unusual: they are monomers with two actin-binding domains and bind actin filaments through β-trefoil domains (Jansen et al. 2011). Fascin and fimbrin are compact, globular proteins that prefer to bind to tightly apposed filaments under a small angle. As a result, they generate tight, unipolar bundles. In contrast, larger cross-link proteins such as α-actinin and filamin can bind actin filaments over a wide range of angles, forming isotropic networks at low cross-link density and mixed network/bundle phases at high cross-link density (Stossel et al. 2001; Courson and Rock 2010).

1.4.1 Cross-Link Binding

Except for the acrosomal protein *scruin* (Gardel et al. 2004), actin cross-links bind transiently, with typical dissociation constants of 0.1–3 µM (Meyer and Aebi 1990; Goldmann and Isenberg 1993; Wachsstock et al. 1993; Yamakita et al. 1996; Ono et al. 1997; Chen et al. 1999; Skau et al. 2011). This corresponds to binding free energies of 32–42 kJ mol^{-1}, or 13–17 kT at room temperature. (For comparison, the energy released by hydrolyzing ATP to ADP is approximately 50 kJ mol^{-1}, or 20 kT. One kT roughly corresponds to the average energy from thermal agitation.) In equilibrium, cross-links unbind with typical timescales of several seconds (Courson and Rock 2010). Stresses acting on cross-links usually accelerate cross-link unbinding (Evans and Ritchie 1997). Such cross-links are known as *slip bonds*. However, α-actinin 4 exhibits different stable conformations (Galkin et al. 2010a), which can expose cryptic actin-binding domains (Volkmer Ward et al. 2008) when subject to stress (Yao et al. 2011). Remarkably, these cross-links therefore bind more tightly under tension, which is known as *catch bond* behavior (Thomas et al. 2008).

1.4.2 Cross-Linked Meshworks and Bundles

Adding cross-links to entangled actin networks can result in a variety of structures, including fine cross-linked meshworks, pure bundle networks, bundle cluster networks, and composite meshwork–bundle networks (Lieleg et al. 2010). However, predicting network structure given the cross-link molecular structure and its binding kinetics remains elusive. This challenge is further compounded by the observation that slow kinetics can give rise to nonequilibrium network structures (Lieleg et al. 2011; Falzone et al. 2012).

1.4.3 Mechanical Properties

Introducing cross-links in an actin filament network introduces a new length scale that determines the network mechanics, called the *cross-link distance* l_c. In particular, l_c determines whether a macroscopically imposed network deformation locally results in filament stretching or filament bending. When a macroscopic shear strain results predominantly in filament stretching, the network experiences *affine deformations* (or *uniform deformations*). In this case, provided that thermal fluctuations are appreciable, the storage modulus of an isotropic, randomly cross-linked network depends on the concentration (molar ratio of cross-links to actin monomers) c_x of cross-links according to Gardel et al. (2004):

$$G'_{\text{affine}} \sim c_x \kappa l_p l_c^{-3}$$

When the filaments are very stiff (for instance, when the network consists of bundled actin (Lieleg et al. 2007)), or when the network connectivity is low, the filaments (or bundles) can locally bend when a macroscopic shear stress is applied, resulting in *nonaffine deformations*. In this case, the storage modulus of the network is insensitive to the concentration of cross-links and instead depends strongly on the concentration of actin filaments c (Kroy and Frey 1996) (Figure 1.6):

$$G'_{\text{nonaffine}} \sim \kappa \xi^4 \sim c^2$$

1.4.4 Nonlinear Response

A unique property of actin networks as well as other protein polymer networks such as extracellular collagen is their strongly nonlinear response to an applied stress. This nonlinearity is evident when pulling on human skin (for instance, the earlobe). For small deformations, skin appears soft and deforms easily. Yet after pulling past a certain amount, skin resists deformation and therefore appears stiffer. Cross-linked actin networks also exhibit this *stiffening* effect. Past a critical stress σ_{crit}, the network becomes stiffer as the stress increases. The strain-stiffening response of actin networks contrasts with the rather linear

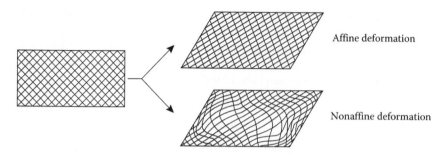

FIGURE 1.6
Shearing a network can result in two kinds of deformations. When a network deforms affinely, filaments stretch and compress in such a way that the strain field at every scale matches the macroscopically applied strain (top right). Filament bending or rotations result in deviations from affinity (bottom right).

response of conventional synthetic polymers, which are flexible and act as linear springs under both compression and tension (Storm et al. 2005). However, recent work demonstrated that supramolecular synthetic polymers can be designed to mimic the remarkable nonlinear rheology of cytoskeletal networks (Kouwer et al. 2013).

The highly nonlinear viscoelastic response of actin networks poses technical challenges for measuring quantitative rheological properties. When a sinusoidal stress is applied, the oscillatory strain response tends to deviate strongly from a sinusoidal shape, and the shear modulus extracted from the stress–strain ratio represents only the first harmonic component of the response. Most studies on actin networks therefore report instead the *differential storage modulus*, $K' = [\delta\sigma/\delta\gamma] \mid \sigma_0$, which is the local tangent of the stress–strain curve. K'-values can be directly compared with theoretical predictions based on semiflexible polymer models (Gardel et al. 2004). K' is usually obtained by a differential prestress protocol, where small amplitude oscillations are superimposed on a steady-state shear flow. An alternative protocol that can be more suitable for materials exhibiting creep is a strain rate ramp protocol, where K' is extracted by differentiating the stress/strain curve measured at different strain rates (Semmrich et al. 2007; Broedersz et al. 2010). For cross-linked actin networks, differential moduli obtained from strain ramp and prestress protocols have been shown to agree, whereas for entangled solutions of actin filaments, deviations have been reported.

The origin of the strong nonlinearity of protein polymer networks is the large thermal persistence length of the polymers. Strain stiffening can result from the entropic force–extension behavior of the filaments, which are easily buckled but strongly resist stretching (MacKintosh et al. 1995; Storm et al. 2005). Alternatively, strain stiffening can emerge as a consequence of a shear-induced transition from bending- to stretching-dominated elasticity (Onck et al. 2005; Broedersz et al. 2011). For actin networks, the critical stress where stiffening first occurs and the stress dependence of the modulus in the nonlinear regime depend on the type of cross-link. With the permanent and rigid cross-link scruin, the elastic modulus increases with stress as according to $K' \sim \sigma^{3/2}$ as a direct consequence of entropic elasticity (Gardel et al. 2004). In this case, networks likely fail due to the rupture of actin filaments. The larger and more flexible cross-link protein filamin A causes stiffening with a markedly different stress dependence ($K' \sim \sigma^1$) (Gardel et al. 2006), which has been ascribed to the entropic compliance of this cross-link molecule (Kasza et al. 2009). Bundle networks obtained by polymerizing actin filaments in the presence of fascin strain stiffen by a different mechanism based on nonaffinity (Lieleg et al. 2007) (Figure 1.7).

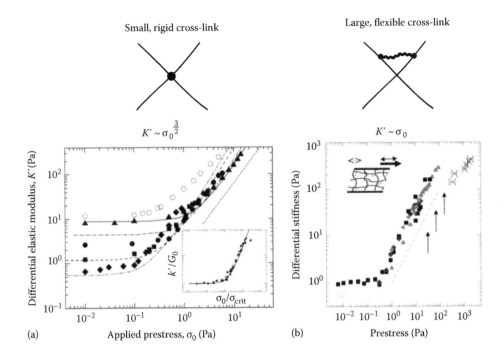

FIGURE 1.7
The type of cross-link determines the nonlinear response of entangled actin filament networks. (a) Networks cross-linked by the small, rigid cross-link scruin result in power law stiffening as $K' \sim \sigma_0^{3/2}$. (From Gardel, M.L. et al., *Proc. Natl. Acad. Sci. USA*, 103(6): 1762, 2006.) (b) Networks cross-linked by the large, flexible cross-link filamin stiffen with a different power law according to $K' \sim \sigma_0$. (From Gardel, M.L. et al., *Science*, 304(5675) (May 28), 1301, 2004.)

1.5 Force Generation

Passive physical forces already lead to a rich mechanical response of cytoskeletal polymer networks. Networks of actin filaments and microtubules gain additional complexity because these polymers are inherently out of equilibrium due to the continuous consumption of chemical energy in the form of the nucleotides ATP and GTP. Directional (de) polymerization allows actin filaments and microtubules to exert forces as they grow or shrink. Furthermore, certain motor proteins can slide filaments past one another, leading to generation of pushing and pulling forces. In this section, we review the mechanisms whereby cytoskeletal polymers can actively exert forces.

1.5.1 Polymerization and Depolymerization Forces

Actin filaments and microtubules polymerize asymmetrically due to differences in the free energy of monomer binding at the plus end and the minus end. This free energy difference can be harnessed as filaments grow against a barrier to exert *pushing forces* (Hill 1981; Theriot 2000). Single microtubules are stiff and can exert forces of up to 3–4 pN as they polymerize (Dogterom and Yurke 1997), though theoretical arguments suggest that forces of up to ~50 pN should be possible (Dogterom et al. 2005). These forces are essential for maintaining the internal organization of the cell (Tolić-Nørrelykke 2008; Laan et al. 2012), including the proper positioning of the kinetochore and chromosomes in the mitotic spindle in animal cells (Inoué and Salmon 1995). Actin, too, exerts pushing forces. Despite the fact that single actin filaments alone are more flexible than microtubules and buckle readily under compressive loads, actin filaments can exert polymerization forces of up to 1 pN provided that they are sufficiently short (Footer et al. 2007). The branched architecture of actin networks in the *lamellipodium* of migrating cells such as crawling fish keratocytes indeed ensures the presence of a dense array of short actin filaments right underneath the leading edge (Mogilner and Oster 1996; Mogilner and Oster 2003b). Crawling cells can thus exert forces of about 100 pN (du Roure et al. 2005). A similar polymerization-based propulsion mechanism is used by the bacterium *Listeria monocytogenes* (Tilney and Portnoy 1989; Cameron et al. 2001). This pathogen uses the actin machinery of infected cells to propel itself with forces of 10–100 pN (McGrath et al. 2003; Wiesner 2003). Bundling of actin by cross-link proteins allows for even larger pushing forces. *Filopodia* in the growth cones of migrating neurons, which contain fascin-mediated bundles that consist only of about 10 actin filaments, can exert pushing forces of up to 3 pN (Cojoc et al. 2007). These forces may enable filopodia to sense mechanical cues and guide preferential extension of neuronal growth cones along soft substrates (Betz et al. 2011). More dramatically, during the *acrosomal process* of the horseshoe crab *Limulus polyphemus*, a stiff bundle of 15–79 actin filaments cross-linked by the protein scruin extends from sperm cells to break open the egg cell wall with a force of 2 nN (Tilney 1975; Shin et al. 2003, 2007). The actin homologue *ParM* in *Escherichia coli* bacteria can similarly polymerize and generate forces to push the chromosomes apart to the cell poles before division (Bork et al. 1992; Garner et al. 2007; Wickstead and Gull 2011; Galkin et al. 2012b).

Surprisingly, there are also mechanisms by which structurally symmetric cytoskeletal polymers can exert pushing forces. Sperm cells of the nematode *Ascaris suum* and also sperm cells of *Caenorhabditis elegans* migrate using *major sperm protein* (MSP) polymers, which elongate and pack in a similar fashion as actin filaments in the lamellipodium (Roberts and Stewart 2000; Miao et al. 2008; Batchelder et al. 2011). A *push–pull* model has been proposed that relies on a pH gradient that regulates gelation and solation of MSP filaments, but the molecular details remain poorly understood (Miao et al. 2003; Mogilner and Oster 2003a).

Apart from exerting pushing forces during polymerization, actin filaments and microtubules can also exert *pulling forces* during depolymerization. In the case of microtubules, these forces are transmitted through *tip-tracking proteins*, which selectively bind the plus end of microtubules (Schuyler and Pellman 2001). As microtubules shrink, tip-tracking proteins can remain bound to the retreating plus end (Lombillo et al. 1995). The resulting pulling forces are believed to contribute to proper positioning of chromosomes during cell division (Hill 1981; Dickinson et al. 2004; Joglekar et al. 2010; McIntosh et al. 2010). In the case of actin filaments, depolymerization was shown to be essential for *actomyosin ring constriction* in dividing budding yeast *Saccharomyces cerevisiae* (Mendes Pinto et al. 2012). Recent experimental evidence suggests that formins anchored at actin plus ends allow for the generation of pulling forces (Jegou et al. 2013).

Molecular motors. In addition to the intrinsic ability of actin filaments and microtubules to exert forces via (de)polymerization, cells also possess specialized proteins called molecular motors. These proteins can exert forces by again coupling the free energy of ATP hydrolysis to mechanical work. This mechanical work can be harnessed for a wide variety of tasks, including DNA replication and expression, protein translocation, cell migration, chromosome separation, and cytokinesis (Bustamante et al. 2004).

Here, we focus on the cytoskeletal motor proteins, which can exert forces while moving along cytoskeletal filaments. There are three classes of cytoskeletal motor proteins (Howard 1997). *Myosin* motors bind actin filaments and most of the *ca.* 20 different types of myosins move toward the plus end (Korn and Hammer 1988). *Kinesin* and *dynein* motors bind microtubules and move toward the plus and minus end, respectively. Although there can be considerable variation among molecular motor types (Goodson et al. 1994; Thompson and Langford 2002), cytoskeletal motor proteins share a few common design principles (Howard 1997; Schliwa and Woehlke 2003). They possess one or two head domains that bind filaments as well as ATP or ADP. Upon ATP hydrolysis, motor proteins undergo conformational changes, manifested in a *power stroke* that results in stepwise motion of the motor along the filament. Step sizes typically vary between 8 and 30 nm, generating forces of up to ~10 pN (Finer et al. 1994; Ishijima et al. 1998; De La Cruz et al. 1999; Mehta et al. 1999; Visscher et al. 2000; Burgess et al. 2003). Motor proteins also possess tail domains, which can bind to the tail domains of other motors to form oligomeric motor complexes (Bresnick 1999), or to the cell cortex (Dujardin and Vallee 2002), or to intracellular cargo (Hirokawa 1998).

Many cells and organisms rely on molecular motors to exert forces that are stronger than by polymerization or depolymerization alone. Unicellular organisms such as the alga *Chlamydomonas reinhardtii* beat two long *flagella* composed of microtubules and dynein and kinesin motors (Bernstein and Rosenbaum 1994), allowing the cell to propel itself with a force of 30 pN (McCord et al. 2005). Fish keratocytes glide on surfaces powered by myosin contraction, exerting traction forces of 45 nN (Harris et al. 1980; Oliver et al. 1995). Similar traction forces between kidney epithelial cells maintain tissue integrity and reach 100 nN (Maruthamuthu et al. 2011). Even higher forces can be achieved by muscle cells, which organize actin filaments and myosin motors in a *sarcomeric structure* dedicated to integrating the power strokes of many myosin motors (Huxley and Hanson 1954; Huxley and Niedergerke 1954; Gautel 2011). Individual cardiac muscle cells have been measured to exert forces of 10 μN (Tarr et al. 1983; Lin et al. 2000; Yin et al. 2005).

1.6 Active Gels

The ability of molecular motors to exert forces on cytoskeletal polymers allows for the existence of a fascinating class of materials called *active gels*. These gels can deform themselves by coupling internal enzymatic activity to mechanical work. Such internal driving allows cells to move and change shape without relying on external forces. In this section, we will explore some of the properties of active gels.

1.6.1 Motor Activity and Spatial Organization

Apart from exerting forces on their surroundings, cells use molecular motors to organize transient internal structures such as the mitotic spindle (Tolić-Nørrelykke 2008; Dumont and Mitchison 2009). Understanding how forces produced by single motors translate into cell-scale forces and cell-scale spatial organization remains an enormous challenge. Forces reorganize the cytoskeleton, but the spatial organization of the cytoskeleton in turn influences force generation. Addressing this feedback in living cells is hindered by their inherent complexity. Recent experiments with reconstituted cytoskeletal networks driven by molecular motors have started to address the feedback between spatial organization and force generation.

Microtubules driven by kinesin or dynein motors exhibit fascinating structural patterns in solution, including vortices and asters (Nédélec et al. 1997) and active liquid crystals (Sanchez et al. 2012). Similar asters have also been reported in the case of actin bundles driven by myosin motors (Backouche et al. 2006). In confined geometries, microtubule asters can be reliably centered by a delicate combination of pushing forces from microtubule polymerization and pulling forces from dynein motors (Holy:1997uq; Laan et al. 2012). In all these cases, self-organization arises from a feedback between force generation and the motion of stiff filaments. Compared to microtubules, single actin filaments are relatively flexible and readily buckle under compressive forces. This property is likely the reason why actin filament meshworks driven by myosin motor complexes have not been reported to exhibit the same pattern formation as microtubules (Soares e Silva et al. 2011b). Buckling of actin filaments under compressive loads leads to an asymmetry in the response of actin networks to local internal forces generated by motors (Lenz et al. 2012; Murrell and Gardel 2012; Vogel et al. 2013). Such an asymmetry biases motor forces in favor of pulling rather than extensile forces, thus causing network contraction (Liverpool et al. 2009). Other effects, such as rearrangements of myosin motors within the actin network, may also contribute to bias activity toward contraction (Dasanayake et al. 2011).

1.6.2 Material Properties of Active Gels

Apart from exerting forces and affecting spatial organization, molecular motors can also strongly affect the material properties of the polymer systems with which they interact. Rheology experiments have shown that the stresses induced by myosin motors stiffen cross-linked actin networks by a factor of 100 or more (Mizuno et al. 2007). In the case of actin networks cross-linked by filamin, it was shown that internally generated stresses by myosin motors stiffen the network in a similar way as an externally applied stress from a rheometer. In both cases, the stiffness increased with stress according to the scaling relationship $K' \sim \sigma^1$ (Koenderink et al. 2009). Remarkably, the same scaling relationship was also observed in whole fibroblasts, which were stretched axially (Fernández et al. 2006) (Figure 1.8).

Myosin activity furthermore causes enhanced fluctuations in cross-linked actin networks that violate the fluctuation–dissipation theorem (Mizuno et al. 2007) and cause strong non-Gaussian displacements of embedded probe particles (Stuhrmann et al. 2012). These enhanced fluctuations are typically observable at frequencies below *ca.* 10 Hz, which reflects the typical on time of the transiently binding motors (MacKintosh and Levine 2008). Microrheology experiments on whole cells have revealed very similar violations of the fluctuation–dissipation theorem (Lau et al. 2003; Balland et al. 2004; Wilhelm 2008). In suspensions of clusters of actin bundles, myosin motors can maintain dynamic steady states where clusters continuously grow and shrink (Köhler et al. 2011a) and actin bundles move superdiffusively (Köhler et al. 2011b). Similar superdiffusive behavior was recently observed in active microtubule–kinesin solutions (Sanchez et al. 2012).

1.6.3 Active Gel Theories

The ability of force-generating elements to bring active gels out of equilibrium has led to a lot of recent theoretical effort in predicting the phase behavior of actively driven matter using generalized statistical–mechanical frameworks (Joanny and Prost 2009).

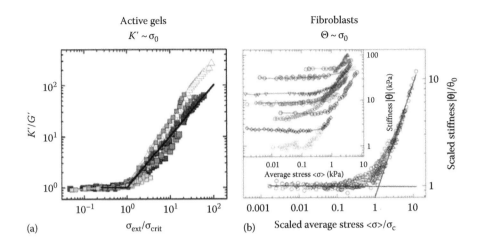

FIGURE 1.8
The nonlinear response of reconstituted active actomyosin gels and living fibroblasts exhibits comparable scaling relations. (a) Actin networks cross-linked by filamin and actively stressed by muscle myosin II motors stiffen with stress according to $K' \sim \sigma_0$. (From Koenderink, G.H. et al., *Proc. Natl. Acad. Sci. USA*, 106(36), 15192, 2009.) (b) A similar stiffening behavior was observed for fibroblasts that were individually stretched between two glass plates. A linear relationship between the stretching modulus Θ and the applied stress σ_0 was found. (From Fernández, P. et al., *Biophys. J.*, 90(10) (May), 3796, 2006.)

These frameworks derive from general principles (force balance, conserved quantities, and constitutive relations) to describe behavior over long length and timescales. This feature often carries a drawback in that microscopic details such as filament semiflexibility are neglected, complicating a direct comparison with experiments on in vitro model systems as well as living cells. Also, these models generally assume small, linear perturbations from equilibrium. Whether and when this assumption holds is unclear. Nevertheless, these models have been successfully applied to several different biological contexts. For instance, a 2D model of lamellipodia has predicted gel thicknesses, flow profiles, and cell velocity consistent with experiment (Kruse et al. 2006), and models of active fluids have been applied to explain the origin of cortical actomyosin flows that establish polarity in *C. elegans* zygotes (Mayer et al. 2010). A similar combination of active gel models and experiment has also revealed that actomyosin contraction and ensuing membrane blebbing at the cell poles stabilize cleavage furrow positioning during cytokinesis (Sedzinski et al. 2011).

1.7 Toward a Mechanical Understanding of Cellular Forces

Cytoskeletal polymers allow cells to actively move, change shape, and exert forces. In this chapter, we have primarily focused on recent insights into the origin of the passive and active mechanical properties of cells gained from in vitro model systems. The advantage of such systems is that they allow one to directly, and quantitatively, compare experimental findings to theoretical models. Increasingly, biophysicists turn to experiments on living cells and model organisms and develop coarse-grained theoretical models. In future years, there will be an important challenge to bridge from an understanding of these simplified systems to an understanding of cells in their full complexity. One interesting route to create such a bridge is to study systems of intermediate complexity, such as cell extracts (Pinot et al. 2012). Another interesting and important avenue is to reconstitute composite cytoskeletal networks. Studying such composites can uncover how the highly disparate bending rigidities of the three cytoskeletal filament systems can affect the passive and active mechanics of the cytoskeleton (Lin et al. 2011).

References

Abercrombie, M. 1980. The Croonian Lecture, 1978: The crawling movement of metazoan cells. *Proceedings of the Royal Society B: Biological Sciences* 207 (1167) (February 29): 129–147. doi:10.1098/rspb.1980.0017.

Alberts, B. 2008. *Molecular Biology of the Cell*. Garland Science, New York.

Ananthakrishnan, R and A Ehrlicher. 2007. The forces behind cell movement. *International Journal of Biological Sciences* 3 (January 1): 303–317.

Backouche, F, L Haviv, D Groswasser, and A Bernheim-Groswasser. 2006. Active gels: Dynamics of patterning and self-organization. *Physical Biology* 3 (264): 264–273. doi:10.1088/1478-3975/3/4/004.

Balland, M, A Richert, and F Gallet. 2004. The dissipative contribution of myosin II in the cytoskeleton dynamics of myoblasts. *European Biophysics Journal* 34 (3) (December 18): 255–261. doi:10.1007/s00249-004-0447-7.

Barlan, K, M J Rossow, and V I Gelfand. 2013. The journey of the organelle: Teamwork and regulation in intracellular transport. *Current Opinion in Cell Biology* 25 (4) (August): 483–488. doi:10.1016/j.ceb.2013.02.018.

Batchelder, E L, G Hollopeter, C Campillo, X Mezanges, E M Jorgensen, P Nassoy, P Sens, and J Plastino. 2011. Membrane tension regulates motility by controlling lamellipodium organization. *Proceedings of the National Academy of Sciences of the United States of America* 108 (28) (January 12): 11429–11434.

Bausch, A R and K Kroy. 2006. A bottom-up approach to cell mechanics. *Nature Physics* 2 (4): 231–238.

Bernstein, M and J L Rosenbaum. 1994. Kinesin-like proteins in the flagella of *Chlamydomonas*. *Trends in Cell Biology* 4 (7) (July): 236–240. doi:10.1016/0962-8924(94)90115-5.

Betz, T, D Koch, Y-B Lu, K Franze, and J A Käs. 2011. Growth cones as soft and weak force generators. *Proceedings of the National Academy of Sciences of the United States of America* 108 (33) (January 16): 13420–13425.

Bork, P, C Sander, and A Valencia. 1992. An ATPase domain common to prokaryotic cell cycle proteins, sugar kinases, actin, and Hsp70 heat shock proteins. *Proceedings of the National Academy of Sciences of the United States of America* 89 (16) (August 15): 7290–7294.

Bowne-Anderson, H, M Zanic, M Kauer, and J Howard. 2013. Microtubule dynamic instability: A new model with coupled GTP hydrolysis and multistep catastrophe. *BioEssays* 35 (5) (March 27): 452–461. doi:10.1002/bies.201200131.

Brangwynne, C P, F C MacKintosh, S Kumar, N A Geisse, J Talbot, L Mahadevan, K K Parker, D E Ingber, and D A Weitz. 2006. Microtubules can bear enhanced compressive loads in living cells because of lateral reinforcement. *The Journal of Cell Biology* 173 (5) (June): 733–741. doi:10.1083/jcb.200601060.

Bresnick, A R. 1999. Molecular mechanisms of nonmuscle myosin-II regulation. *Current Opinion in Cell Biology* 11 (1) (February): 26–33. doi:10.1016/S0955-0674(99)80004-0.

Bringmann, M, B Landrein, C Schudoma, O Hamant, M-T Hauser, and S Persson. 2012. Cracking the elusive alignment hypothesis: The microtubule-cellulose synthase nexus unraveled. *Trends in Plant Science* 17 (11) (November): 666–674. doi:10.1016/j.tplants.2012.06.003.

Broedersz, C P, K E Kasza, L M Jawerth, S Münster, D A Weitz, and F C MacKintosh. 2010. Measurement of nonlinear rheology of cross-linked biopolymer gels. *Soft Matter* 6 (17) (January 1): 4120. doi:10.1039/c0sm00285b.

Broedersz, C P, X Mao, T C Lubensky, and F C MacKintosh. 2011. Criticality and isostaticity in fibre networks. *Nature Physics* 7 (12) (October 30): 983–988. doi:10.1038/nphys2127.

Brokaw, C J. 1994. Control of flagellar bending: A new agenda based on dynein diversity. *Cell Motility and the Cytoskeleton* 28 (3): 199–204. doi:10.1002/cm.970280303.

Burgess, S A, M L Walker, H Sakakibara, P J Knight, and K Oiwa. 2003. Dynein structure and power stroke. *Nature* 421 (6924) (February 13): 715–718. doi:10.1038/nature01377.

Bustamante, C, Y R Chemla, N R Forde, and D Izhaky. 2004. Mechanical processes in biochemistry. *Annual Review of Biochemistry* 73 (1) (June): 705–748. doi:10.1146/annurev. biochem.72.121801.161542.

Bustamante, C, J Marko, E Siggia, and S Smith. 1994. Entropic elasticity of lambda-phage DNA. *Science* 265 (5178) (September 9): 1599–1600. doi:10.1126/science.8079175.

Byers, B. 1976. A highly ordered ring of membrane-associated filaments in budding yeast. *The Journal of Cell Biology* 69 (3) (June 1): 717–721. doi:10.1083/jcb.69.3.717.

Cameron, L A, T M Svitkina, D Vignjevic, J A Theriot, and G G Borisy. 2001. Dendritic organization of actin comet tails. *Current Biology* 11 (2) (January 1): 130–135.

Carlier, M F, C Valentin-Ranc, C Combeau, S Fievez, and D Pantoloni. 1994. Actin polymerization: Regulation by divalent metal ion and nucleotide binding, ATP hydrolysis and binding of myosin. *Advances in Experimental Medicine and Biology* 358: 71–81.

Cavey, M and T Lecuit. 2009. Molecular bases of cell-cell junctions stability and dynamics. *Cold Spring Harbor Perspectives in Biology* 1 (5) (November 2): a002998. doi:10.1101/cshperspect.a002998.

Chang, F and S G Martin. 2009. Shaping fission yeast with microtubules. *Cold Spring Harbor Perspectives in Biology* 1 (1) (July): a001347. doi:10.1101/cshperspect.a001347.

Chen, B, A Li, D Wang, M Wang, L Zheng, and J R Bartles. 1999. Espin contains an additional actin-binding site in its N terminus and is a major actin-bundling protein of the sertoli cell-spermatid ectoplasmic specialization junctional plaque. *Molecular Biology of the Cell* 10 (12) (December): 4327–4339.

Chen, C S, M Mrksich, S Huang, G M Whitesides, and D E Ingber. 1997. Geometric control of cell life and death. *Science* 276 (5317): 1425–1428.

Claessens, M M A E, R Tharmann, K Kroy, and A R Bausch. 2006. Microstructure and viscoelasticity of confined semiflexible polymer networks. *Nature Physics* 2 (3) (February 26): 186–189. doi:10.1038/nphys241.

Cojoc, D, F Difato, E Ferrari, R B Shahapure, J Laishram, M Righi, E M Di Fabrizio, and V Torre. 2007. Properties of the force exerted by filopodia and lamellipodia and the involvement of cytoskeletal components. *Plos One* 2 (10): e1072. doi:10.1371/journal.pone.0001072.

Cosentino Lagomarsino, M, C Tanase, J W Vos, A M Emons, B M Mulder, and M Dogterom. 2007. Microtubule organization in three-dimensional confined geometries: Evaluating the role of elasticity through a combined in vitro and modeling approach. *Biophysical Journal* 92 (3) (January 1): 1046–1057. doi:10.1529/biophysj.105.076893. http://linkinghub.elsevier.com/retrieve/pii/biophysj.105.076893.

Costerton, J W, P S Stewart, and E P Greenberg. 1999. Bacterial biofilms: A common cause of persistent infections. *Science* 284 (5418) (January 21): 1318–1322. doi:10.1126/science.284.5418.1318.

Courson, D S and R S Rock. 2010. Actin cross-link assembly and disassembly mechanics for alpha-actinin and fascin. *Journal of Biological Chemistry* 285 (34): 26350–26357. doi:10.1074/jbc. M110.123117.

Das, M and F C MacKintosh. 2010. Poisson's ratio in composite elastic media with rigid rods. *Physical Review Letters* 105 (13) (September): 138102.

Dasanayake, N L, P J Michalski, and A E Carlsson. 2011. General mechanism of actomyosin contractility. *Physical Review Letters* 107 (11) (September): 118101.

Davenport, R W, P Dou, V Rehder, and S B Kater. 1993. A sensory role for neuronal growth cone filopodia. *Nature* 361 (6414) (January 1): 721–724. doi:10.1038/361721a0.

de Gennes, P-G. 1979. *Scaling Concepts in Polymer Physics*. Cornell University Press, Ithaca, NY.

De La Cruz, E M, A L Wells, S S Rosenfeld, E Michael Ostap, and H Lee Sweeney. 1999. The kinetic mechanism of myosin V. *Proceedings of the National Academy of Sciences of the United States of America* 96 (24) (January 23): 13726–13731. doi:10.1073/pnas.96.24.13726.

Dickinson, R B, L Caro, and D L Purich. 2004. Force generation by cytoskeletal filament end-tracking proteins. *Biophysical Journal* 87 (4) (October): 2838–2854. doi:10.1529/biophysj.104.045211.

Dogic, Z and S Fraden. 2006. Ordered phases of filamentous viruses. *Current Opinion in Colloid & Interface Science* 11 (1): 47–55.

Dogterom, M, J W J Kerssemakers, G Romet-Lemonne, and M E Janson. 2005. Force generation by dynamic microtubules. *Current Opinion in Cell Biology* 17 (1) (February): 67–74. doi:10.1016/j.ceb.2004.12.011.

Dogterom, M and B Yurke. 1997. Measurement of the force-velocity relation for growing microtubules. *Science* 278 (5339) (January 31): 856–860.

Doi, M and S F Edwards. 1978. Dynamics of concentrated polymer systems. Part 1. Brownian motion in the equilibrium state. *Journal of the Chemical Society, Faraday Transactions 2* 74 (0): 1789–1801. doi:10.1039/F29787401789.

du Roure, O, A Saez, A Buguin, R H Austin, P Chavrier, P Siberzan, and B Ladoux. 2005. Force mapping in epithelial cell migration. *Proceedings of the National Academy of Sciences of the United States of America* 102 (7) (January 15): 2390–2395. doi:10.1073/pnas.0408482102.

Dujardin, D L and R B Vallee. 2002. Dynein at the cortex. *Current Opinion in Cell Biology* 14 (1) (February): 44–49. doi:10.1016/S0955-0674(01)00292-7.

Dumont, S and T J Mitchison. 2009. Force and length in the mitotic spindle. *Current Biology* 19 (17) (September): R749–R761. doi:10.1016/j.cub.2009.07.028.

Ellis, R J. 2001. Macromolecular crowding: Obvious but underappreciated. *Trends in Biochemical Sciences* 26 (10) (October): 597–604.

Engqvist-Goldstein, Å E and D G Drubin. 2003. Actin assembly and endocytosis: From yeast to mammals. *Annual Review of Cell and Developmental Biology* 19 (1) (November): 287–332. doi:10.1146/annurev.cellbio.19.111401.093127.

Evans, E and K Ritchie. 1997. Dynamic strength of molecular adhesion bonds. *Biophysical Journal* 72 (4) (April 1): 1541.

Falzone, T T, M Lenz, D R Kovar, and M L Gardel. 2012. Assembly kinetics determine the architecture of α-actinin crosslinked F-actin networks. *Nature Communications* 3 (May 29): 861. doi:10.1038/ncomms1862.

Fernández, P, P A Pullarkat, and A Ott. 2006. A master relation defines the nonlinear viscoelasticity of single fibroblasts. *Biophysical Journal* 90 (10) (May): 3796–3805. doi:10.1529/biophysj.105.072215.

Field, C M and P Lénárt. 2011. Bulk cytoplasmic actin and its functions in meiosis and mitosis. *Current Biology* 21 (19) (October): R825–R830. doi:10.1016/j.cub.2011.07.043.

Finer, J T, R M Simmons, and J A Spudich. 1994. Single myosin molecule mechanics: Piconewton forces and nanometre steps. *Nature* 368 (6467) (March 10): 113–119. doi:10.1038/368113a0.

Fisher, C I and S C Kuo. 2009. Filament rigidity causes F-actin depletion from nonbinding surfaces. *Proceedings of the National Academy of Sciences of the United States of America* 106 (1) (January 6): 133–138. doi:10.1073/pnas.0804991106.

Fletcher, D A and P L Geissler. 2009. Active biological materials. *Annual Review of Physical Chemistry* 60 (January 1): 469–486. doi:10.1146/annurev.physchem.040808.090304.

Footer, M J, J W J Kerssemakers, J A Theriot, and M Dogterom. 2007. Direct measurement of force generation by actin filament polymerization using an optical trap. *Proceedings of the National Academy of Sciences of the United States of America* 104 (7) (February 13): 2181–2186. doi:10.1073/pnas.0607052104.

Fudge, D S, T Winegard, R H Ewoldt, D Beriault, L Szewciw, and G H McKinley. 2009. From ultrasoft slime to hard {alpha}-keratins: The many lives of intermediate filaments. *Integrative and Comparative Biology* 49 (1) (July): 32–39. doi:10.1093/icb/icp007.

Galkin, V E, A Orlova, and E H Egelman. 2012a. Actin filaments as tension sensors. *Current Biology* 22 (3) (February): R96–R101. doi:10.1016/j.cub.2011.12.010. http://www.ncbi.nlm.nih.gov/pubmed?cmd=search&term=22321312.

Galkin, V E, A Orlova, and E H Egelman. 2012b. Are ParM filaments polar or bipolar? *Journal of Molecular Biology* 423 (4) (November): 482–485. doi:10.1016/j.jmb.2012.08.006.

Galkin, V E, A Orlova, A Salmazo, K Djinovic-Carugo, and E H Egelman. 2010a. Opening of tandem calponin homology domains regulates their affinity for F-actin. *Nature Structural & Molecular Biology* 17 (5) (April 11): 614–616. doi:10.1038/nsmb.1789.

Galkin, V E, A Orlova, G F Schröder, and E H Egelman. 2010b. Structural polymorphism in F-actin. *Nature Structural & Molecular Biology* 17 (11) (October 10): 1318–1323. doi:10.1038/nsmb.1930.

Gardel, M L, F Nakamura, J H Hartwig, J C Crocker, T P Stossel, and D A Weitz. 2006. Prestressed F-actin networks cross-linked by hinged filamins replicate mechanical properties of cells. *Proceedings of the National Academy of Sciences of the United States of America* 103 (6): 1762–1767.

Gardel, M L, J H Shin, F C MacKintosh, L Mahadevan, P Matsudaira, and D A Weitz. 2004. Elastic behavior of cross-linked and bundled actin networks. *Science* 304 (5675) (May 28): 1301–1305. doi:10.1126/science.1095087.

Gardner, M K, B D Charlebois, I M Jánosi, J Howard, A J Hunt, and D J Odde. 2011. Rapid microtubule self-assembly kinetics. *Cell* 146 (4) (August): 582–592. doi:10.1016/j.cell.2011.06.053.

Gardner, M K, M Zanic, and J Howard. 2013. Microtubule catastrophe and rescue. *Current Opinion in Cell Biology* 25 (1) (February): 14–22. doi:10.1016/j.ceb.2012.09.006.

Garner, E C, C S Campbell, D B Weibel, and R D Mullins. 2007. Reconstitution of DNA segregation driven by assembly of a prokaryotic actin homolog. *Science* 315 (5816) (March 2): 1270–1274. doi:10.1126/science.1138527.

Gautel, M. 2011. The sarcomeric cytoskeleton: Who picks up the strain? *Current Opinion in Cell Biology* 23 (1) (February): 39–46. doi:10.1016/j.ceb.2010.12.001.

Gilden, J K and M F Krummel. 2010. Control of cortical rigidity by the cytoskeleton: Emerging roles for septins. *Cytoskeleton* 67 (8): 477–486.

Gisler, T and D A Weitz. 1999. Scaling of the microrheology of semidilute F-actin solutions. *Physical Review Letters* 82 (7): 1606–1609.

Gittes, F, B Mickey, J Nettleton, and J Howard. 1993. Flexural rigidity of microtubules and actin filaments measured from thermal fluctuations in shape. *The Journal of Cell Biology* 120 (4) (February 1): 923–934.

Glaser, J, D Chakraborty, K Kroy, I Lauter, M Degawa, N Kirchgessner, B Hoffmann, R Merkel, and M Giesen. 2010. Tube width fluctuations in F-actin solutions. *Physical Review Letters* 105 (3) (July): 037801.

Glotzer, M. 2005. The molecular requirements for cytokinesis. *Science* 307 (5716) (March 18): 1735–1739. doi:10.1126/science.1096896.

Goldmann, W H and G Isenberg. 1993. Analysis of filamin and A-actinin binding to actin by the stopped flow method. *FEBS Letters* 336 (3) (December): 408–410. doi:10.1016/0014-5793(93)80847-N.

Goodson, H V, S J Kang, and S A Endow. 1994. Molecular phylogeny of the kinesin family of microtubule motor proteins. *Journal of Cell Science* 107 (7) (January 1): 1875–1884.

Graf, H and H Löwen. 1999. Phase diagram of tobacco mosaic virus solutions. *Physical Review E* 59 (2) (February): 1932–1942. doi:10.1103/PhysRevE.59.1932.

Gruenbaum, Y, K L Wilson, A Harel, M Goldberg, and M Cohen. 2000. Review: Nuclear lamins—Structural proteins with fundamental functions. *Journal of Structural Biology* 129 (2–3) (April): 313–323. doi:10.1006/jsbi.2000.4216.

Guertin, D A, S Trautmann, and D McCollum. 2002. Cytokinesis in eukaryotes. *Microbiology and Molecular Biology Reviews* 66 (2) (June): 155–178.

Guillot, C and T Lecuit. 2013. Mechanics of epithelial tissue homeostasis and morphogenesis. *Science* 340 (6137) (June): 1185–1189. doi:10.1126/science.1235249.

Gutierrez, R, J J Lindeboom, A R Paredez, A M C Emons, and D W Ehrhardt. 2009. Arabidopsis cortical microtubules position cellulose synthase delivery to the plasma membrane and interact with cellulose synthase trafficking compartments. *Nature Cell Biology* 11 (7) (July): 797–806. doi:10.1038/ncb1886.

Hagan, I M. 1998. The fission yeast microtubule cytoskeleton. *Journal of Cell Science* 111 (12) (January 15): 1603–1612.

Harris, A, P Wild, and D Stopak. 1980. Silicone rubber substrata: A new wrinkle in the study of cell locomotion. *Science* 208 (4440) (April 11): 177–179. doi:10.1126/science.6987736.

Hartwell, L H. 1971. Genetic control of the cell division cycle in yeast. IV. Genes controlling bud emergence and cytokinesis. *Experimental Cell Research* 69 (2) (December): 265–276.

Helfand, B T, L Chang, and R D Goldman. 2003. The dynamic and motile properties of intermediate filaments. *Annual Review of Cell and Developmental Biology* 19: 445–467. doi:10.1146/annurev. cellbio.19.111401.092306.

Herrmann, H, H Bär, L Kreplak, S V Strelkov, and U Aebi. 2007. Intermediate filaments: From cell architecture to nanomechanics. *Nature Reviews Molecular Cell Biology* 8 (7) (July): 562–573. doi:10.1038/nrm2197. http://www.nature.com/doifinder/10.1038/nrm2197.

Hill, T L. 1981. Microfilament or microtubule assembly or disassembly against a force. *Proceedings of the National Academy of Sciences of the United States of America* 78 (9) (January 1): 5613–5617.

Hinner, B, M Tempel, E Sackmann, K Kroy, and E Frey. 1998. Entanglement, elasticity, and viscous relaxation of actin solutions. *Physical Review Letters* 81 (12): 2614–2617.

Hirokawa, N. 1998. Kinesin and dynein superfamily proteins and the mechanism of organelle transport. *Science* 279 (5350) (January 23): 519–526. doi:10.1126/science.279.5350.519.

Hosek, M and J X Tang. 2004. Polymer-induced bundling of F actin and the depletion force. *Physical Review E* 69 (5 Pt 1) (May 1): 051907.

Howard, J. 1997. Molecular motors: Structural adaptations to cellular functions. *Nature* 389 (6651) (October 9): 561–567. doi:10.1038/39247.

Hussey, P J, T Ketelaar, and M J Deeks. 2013. Control of the actin cytoskeleton in plant cell growth. *Annual Review of Plant Biology* 57 (1) (January 28): 109–125. doi: 10.1146/annurev. arplant.57.032905.105206.

Huxley, A F and R Niedergerke. 1954. Structural changes in muscle during contraction: Interference microscopy of living muscle fibres. *Nature* 173 (4412) (May 22): 971–973. doi:10.1038/173971a0.

Huxley, H and J Hanson. 1954. Changes in the cross-striations of muscle during contraction and stretch and their structural interpretation. *Nature* 173 (4412) (May 22): 973–976. doi:10.1038/173973a0.

Inoué, S and E D Salmon. 1995. Force generation by microtubule assembly/disassembly in mitosis and related movements. *Molecular Biology of the Cell* 6 (12) (December 1): 1619. http://www.molbiolcell.org/content/23/10/1798.full.

Isambert, H and A C Maggs. 1996. Dynamics and rheology of actin solutions. *Macromolecules* 29 (3) (January): 1036–1040.

Isambert, H, P Venier, A C Maggs, A Fattoum, R Kassab, D Pantaloni, and M F Carlier. 1995. Flexibility of actin filaments derived from thermal fluctuations. Effect of bound nucleotide, phalloidin, and muscle regulatory proteins. *Journal of Biological Chemistry* 270 (19) (May 12): 11437–11444.

Ishijima, A, H Kojima, T Funatsu, M Tokunaga, H Higuchi, H Tanaka, and T Yanagida. 1998. Simultaneous observation of individual ATPase and mechanical events by a single myosin molecule during interaction with actin. *Cell* 92 (2) (January): 161–171. doi:10.1016/S0092-8674(00)80911-3.

Jansen, S, A Collins, C Yang, G Rebowski, T Svitkina, and R Dominguez. 2011. Mechanism of actin filament bundling by fascin. *Journal of Biological Chemistry* 286 (34) (August 26): 30087–30096. doi:10.1074/jbc.M111.251439.

Jegou, A, M-F Carlier, and G Romet-Lemonne. 2013. Formin mDia1 senses and generates mechanical forces on actin filaments. *Nature Communications* 4 (May 21): 1883. doi:10.1038/ncomms2888. http://www.nature.com/doifinder/10.1038/ncomms2888.

Joanny, J-F and J Prost. 2009. Active gels as a description of the actin-myosin cytoskeleton. *HFSP Journal* 3 (2) (March 31): 94. doi:10.2976/1.3054712. http://www.ncbi.nlm.nih.gov/pmc/articles/PMC2707794/.

Joglekar, A P, K S Bloom, and E D Salmon. 2010. Mechanisms of force generation by end-on kinetochore-microtubule attachments. *Current Opinion in Cell Biology* 22 (1) (February): 57–67. doi:10.1016/j.ceb.2009.12.010.

Kantsler, V, J Dunkel, M Polin, and R E Goldstein. 2013. Ciliary contact interactions dominate surface scattering of swimming eukaryotes. *Proceedings of the National Academy of Sciences of the United States of America* 110 (4) (January): 1187–1192. doi:10.1073/pnas.1210548110.

Kasza, K E, G H Koenderink, Y C Lin, C P Broedersz, W Messner, F Nakamura, T P Stossel, F C MacKintosh, and D A Weitz. 2009. Nonlinear elasticity of stiff biopolymers connected by flexible linkers. *Physical Review E* 79 (4) (April 1): 1–5. doi:10.1103/PhysRevE.79.041928.

Kasza, K E, A C Rowat, J Liu, T E Angelini, C P Brangwynne, G H Koenderink, and D A Weitz. 2007. The cell as a material. *Current Opinion in Cell Biology* 19 (1): 101–107. doi:10.1016/j.ceb.2006.12.002.

Käs, J, H Strey, and E Sackmann. 1994. Direct imaging of reptation for semiflexible actin filaments. *Nature* 368 (6468) (March 17): 226–229. doi:10.1038/368226a0.

Käs, J, H Strey, J X Tang, D Finger, R Ezzell, E Sackmann, and P A Janmey. 1996. F-actin, a model polymer for semiflexible chains in dilute, semidilute, and liquid crystalline solutions. *Biophysical Journal* 70 (2) (February): 609–625. doi:10.1016/S0006-3495(96)79630-3.

Khokhlov, A R and A N Semenov. 1982. Liquid-crystalline ordering in the solution of partially flexible macromolecules. *Physica A: Statistical Mechanics and Its Applications* 112 (3): 605–614.

Koenderink, G H, Z Dogic, F Nakamura, P M Bendix, F C MacKintosh, J H Hartwig, T P Stossel, and D A Weitz. 2009. An active biopolymer network controlled by molecular motors. *Proceedings of the National Academy of Sciences of the United States of America* 106 (36): 15192–15197. doi:10.1073/pnas.0903974106.

Köhler, S, V Schaller, and A R Bausch. 2011a. Structure formation in active networks. *Nature Materials* 10 (6) (June): 462–468. doi:10.1038/nmat3009.

Köhler, S, V Schaller, and A R Bausch. 2011b. Collective dynamics of active cytoskeletal networks. *Plos One* 6 (8) (August 26): e23798.

Korenbaum, E. 2002. Calponin homology domains at a glance. *Journal of Cell Science* 115 (18) (September 15): 3543–3545. doi:10.1242/jcs.00003.

Korn, E D and J A Hammer III. 1988. Myosins of nonmuscle cells. *Annual Review of Biophysics and Biophysical Chemistry* 17 (1) (June): 23–45. doi:10.1146/annurev.bb.17.060188.000323.

Kouwer, P H J, M Koepf, V A A Le Sage, M Jaspers, A M van Buul, Z H Eksteen-Akeroyd, T Woltinge et al. 2013. Responsive biomimetic networks from polyisocyanopeptide hydrogels. *Nature* 493 (7434) (January): 651–655. doi:10.1038/nature11839.

Kratky, O and G Porod. 1949. Röntgenuntersuchung Gelöster Fadenmoleküle. *Recueil Des Travaux Chimiques Des Pays-Bas* 68 (12) (September 2): 1106–1122. doi:10.1002/recl.19490681203.

Kreplak, L and D Fudge. 2007. Biomechanical properties of intermediate filaments: From tissues to single filaments and back. *BioEssays* 29 (1) (January): 26–35. doi:10.1002/bies.20514.

Kroeger, J and A Geitmann. 2012. The pollen tube paradigm revisited. *Current Opinion in Plant Biology* 15 (6) (December): 618–624. doi:10.1016/j.pbi.2012.09.007.

Kroy, K and E Frey. 1996. Force-extension relation and plateau modulus for wormlike chains. *Physical Review Letters* 77 (2) (July): 306–309. doi:10.1103/PhysRevLett.77.306.

Kruse, K, J-F Joanny, F Jülicher, and J Prost. 2006. Contractility and retrograde flow in lamellipodium motion. *Physical Biology* 3 (2) (June 1): 130–137. doi:10.1088/1478-3975/3/2/005.

Laan, L, N Pavin, J Husson, G Romet-Lemonne, M van Duijn, M P López, R D Vale, F Jülicher, S L Reck-Peterson, and M Dogterom. 2012. Cortical dynein controls microtubule dynamics to generate pulling forces that position microtubule asters. *Cell* 148 (3) (February): 502–514. doi:10.1016/j.cell.2012.01.007.

Landau, L D, E M Lifshitz, A M Kosevich, and L P Pitaevskiĭ. 1986. *Theory of Elasticity*. Butterworth-Heinemann, Oxford, U.K.

Lau, A, B Hoffman, A Davies, J Crocker, and T Lubensky. 2003. Microrheology, stress fluctuations, and active behavior of living cells. *Physical Review Letters* 91 (19) (November): 198101. doi:10.1103/PhysRevLett.91.198101.

Lekkerkerker, H N W and R Tuinier. 2011. *Colloids and the Depletion Interaction*. Springer, New York.

Lénárt, P, C P Bacher, N Daigle, A R Hand, R Eils, M Terasaki, and J Ellenberg. 2005. A contractile nuclear actin network drives chromosome congression in oocytes. *Nature* 436 (7052) (August 11): 812–818. doi:10.1038/nature03810.

Lenz, M, T Thoresen, M Gardel, and A Dinner. 2012. Contractile units in disordered actomyosin bundles arise from F-actin buckling. *Physical Review Letters* 108 (23) (June 1): 238107. doi:10.1103/PhysRevLett.108.238107.

Lepinoux-Chambaud, C and J Eyer. 2013. Review on intermediate filaments of the nervous system and their pathological alterations. *Histochemistry and Cell Biology* 140 (1) (July): 13–22. doi:10.1007/s00418-013-1101-1.

Lieleg, O, M Claessens, C Heussinger, E Frey, and A Bausch. 2007. Mechanics of bundled semiflexible polymer networks. *Physical Review Letters* 99 (8) (August): 088102. doi:10.1103/PhysRevLett.99.088102.

Lieleg, O, M M A E Claessens, and A R Bausch. 2010. Structure and dynamics of cross-linked actin networks. *Soft Matter* 6 (2): 218–225. doi:10.1039/b912163n.

Lieleg, O, J Kayser, G Brambilla, L Cipelletti, and A R Bausch. 2011. Slow dynamics and internal stress relaxation in bundled cytoskeletal networks. *Nature Materials* 10 (3) (March): 236–242. doi:10.1038/nmat2939.

Limozin, L, M Bärmann, and E Sackmann. 2003. On the organization of self-assembled actin networks in giant vesicles. *The European Physical Journal E* 10: 319–330. http://www.springerlink.com/index/4MPMDPMTP2M5TMA4.pdf.

Lin, G, K S J Pister, and K P Roos. 2000. Surface micromachined polysilicon heart cell force transducer. *Journal of Microelectromechanical Systems* 9 (1): 9–17. doi:10.1109/84.825771.

Lin, Y-C, G H Koenderink, F C MacKintosh, and D A Weitz. 2011. Control of non-linear elasticity in F-actin networks with microtubules. *Soft Matter* 7 (3): 902–906. doi:10.1039/c0sm00478b.

Lin, Y-C, N Y Yao, C P Broedersz, H Herrmann, F C Mackintosh, and D A Weitz. 2010. Origins of elasticity in intermediate filament networks. *Physical Review Letters* 104 (5) (February): 058101.

Liverpool, T B, M C Marchetti, J-F Joanny, and J Prost. 2009. Mechanical response of active gels. *Europhysics Letters* 85 (1) (January 13): 18007. doi:10.1209/0295-5075/85/18007.

Lombillo, V A, R J Stewart, and J R McIntosh. 1995. Minus-end-directed motion of kinesin-coated microspheres driven by microtubule depolymerization. *Nature* 373 (6510) (January): 161–164. doi:10.1038/373161a0.

MacKintosh, F C. 2011. Elasticity and dynamics of cytoskeletal filaments and networks of them. In *New Trends in the Physics and Mechanics of Biological Systems: Lecture Notes of the Les Houches Summer School: Volume 92*, Martine, B A, G Alain, M M Martin, and C Leticia (eds.) *July 2009*. Oxford University Press, Oxford, U.K.

MacKintosh, F C, J Käs, and P A Janmey. 1995. Elasticity of semiflexible biopolymer networks. *Physical Review Letters* 75 (24) (December 11): 4425–4428.

MacKintosh, F C and A J Levine. 2008. Nonequilibrium mechanics and dynamics of motor-activated gels. *Physical Review Letters* 100 (1) (January 11): 018104.

Manor, U and B Kachar. 2008. Dynamic length regulation of sensory stereocilia. *Seminars in Cell & Developmental Biology* 19 (6) (December): 502–510. doi:10.1016/j.semcdb.2008.07.006.

Marenduzzo, D, K Finan, and P R Cook. 2006. The depletion attraction: An underappreciated force driving cellular organization. *The Journal of Cell Biology* 175 (5) (December 4): 681–686. doi:10.1083/jcb.200609066.

Maruthamuthu, V, B Sabass, U S Schwarz, and M L Gardel. 2011. Cell-ECM traction force modulates endogenous tension at cell–cell contacts. *Proceedings of the National Academy of Sciences of the United States of America* 108 (12) (January 22): 4708–4713. doi:10.1073/pnas.1011123108.

Mattila, P K and P Lappalainen. 2008. Filopodia: Molecular architecture and cellular functions. *Nature Reviews Molecular Cell Biology* 9 (6) (June): 446–454. doi:10.1038/nrm2406.

Maurer, S P, F J Fourniol, G Bohner, C A Moores, and T Surrey. 2012. EBs recognize a nucleotide-dependent structural cap at growing microtubule ends. *Cell* 149 (2) (April): 371–382. doi:10.1016/j.cell.2012.02.049.

May, R C and L M Machesky. 2001. Phagocytosis and the actin cytoskeleton. *Journal of Cell Science* 114 (Pt 6) (March): 1061–1077.

Mayer, M, M Depken, J S Bois, F Jülicher, and S W Grill. 2010. Anisotropies in cortical tension reveal the physical basis of polarizing cortical flows. *Nature* 467 (7315) (September 21): 617–621. doi:10.1038/nature09376.

McCord, R P, J N Yukich, and K K Bernd. 2005. Analysis of force generation during flagellar assembly through optical trapping of free-swimming *Chlamydomonas reinhardtii*. *Cell Motility and the Cytoskeleton* 61 (3) (July): 137–144. doi:10.1002/cm.20071.

McGrath, J L, N J Eungdamrong, C I Fisher, F Peng, L Mahadevan, T J Mitchison, and S C Kuo. 2003. The force-velocity relationship for the actin-based motility of *Listeria monocytogenes*. *Current Biology* 13 (4) (February): 329–332. doi:10.1016/S0960-9822(03)00051-4.

McIntosh, J R, V Volkov, F I Ataullakhanov, and E L Grishchuk. 2010. Tubulin depolymerization may be an ancient biological motor. *Journal of Cell Science* 123 (Pt 20) (October 15): 3425–3434. doi:10.1242/jcs.067611.

McMichael, C M and S Y Bednarek. 2013. Cytoskeletal and membrane dynamics during higher plant cytokinesis. *The New Phytologist* 197 (4) (March): 1039–1057. doi:10.1111/nph.12122.

Mehta, A D, R E Cheney, R S Rock, M Rief, J A Spudich, and M S Mooseker. 1999. Myosin-V is a processive actin-based motor. *Nature* 400 (6744) (August 5): 590–593. doi:10.1038/23072.

Mendes Pinto, I, B Rubinstein, A Kucharavy, J R Unruh, and R Li. 2012. Actin depolymerization drives actomyosin ring contraction during budding yeast cytokinesis. *Developmental Cell* 22 (6) (June): 1247–1260. doi:10.1016/j.devcel.2012.04.015.

Meyer, R K and U Aebi. 1990. Bundling of actin filaments by alpha-actinin depends on its molecular length. *The Journal of Cell Biology* 110 (6) (January 1): 2013–2024. doi:10.1083/jcb.110.6.2013.

Meyers, M A and K K Chawla. 2009. *Mechanical Behavior of Materials*. Cambridge University Press, Cambridge, MA.

Miao, L, O Vanderlinde, J Liu, R P Grant, A Wouterse, K Shimabukuro, A Philipse, M Stewart, and T M Roberts. 2008. The role of filament-packing dynamics in powering amoeboid cell motility. *Proceedings of the National Academy of Sciences of the United States of America* 105 (14) (April): 5390–5395. doi:10.1073/pnas.0708416105.

Miao, L, O Vanderlinde, M Stewart, and T M Roberts. 2003. Retraction in amoeboid cell motility powered by cytoskeletal dynamics. *Science* 302 (5649) (January 21): 1405–1407. doi:10.1126/science.1089129.

Middeldorp, J and E M Hol. 2011. GFAP in health and disease. *Progress in Neurobiology* 93 (3) (March): 421–443. doi:10.1016/j.pneurobio.2011.01.005.

Mitchison, T J. 1995. Evolution of a dynamic cytoskeleton. *Philosophical Transactions of the Royal Society of London. Series B, Biological Sciences* 349 (1329) (September): 299–304. doi:10.1098/rstb.1995.0117.

Mizuno, D, C Tardin, C F Schmidt, and F C MacKintosh. 2007. Nonequilibrium mechanics of active cytoskeletal networks. *Science* 315 (5810) (January 19): 370–373. doi:10.1126/science.1134404. http://www.sciencemag.org/cgi/content/full/315/5810/370.

Mogilner, A and G Oster. 1996. Cell motility driven by actin polymerization. *Biophysical Journal* 71 (6) (December): 3030–3045. doi:10.1016/S0006-3495(96)79496-1.

Mogilner, A and G Oster. 2003a. Shrinking gels pull cells. *Science* 302 (5649) (January 21): 1340–1341. doi:10.1126/science.1092041.

Mogilner, A and G Oster. 2003b. Force generation by actin polymerization II: The elastic ratchet and tethered filaments. *Biophysical Journal* 84 (3) (March 1): 1591–1605.

Mostowy, S and P Cossart. 2012. Septins: The fourth component of the cytoskeleton. *Nature Reviews Molecular Cell Biology* 13 (3): 183–194.

Murrell, M P and M L Gardel. 2012. F-actin buckling coordinates contractility and severing in a biomimetic actomyosin cortex. *Proceedings of the National Academy of Sciences of the United States of America* 109 (51) (January 18): 20820–20825. doi:10.1073/pnas.1214753109.

Naumanen, P, P Lappalainen, and P Hotulainen. 2008. Mechanisms of actin stress fibre assembly. *Journal of Microscopy* 231 (3) (September): 446–454. doi:10.1111/j.1365-2818.2008.02057.x.

Nédélec, F J, T Surrey, A C Maggs, and S Leibler. 1997. Self-organization of microtubules and motors. *Nature* 389 (6648) (September 18): 305–308. doi:10.1038/38532. http://www.nature.com/nature/journal/v389/n6648/abs/389305a0.html.

Noding, B and S Köster. 2012. Intermediate filaments in small configuration spaces. *Physical Review Letters* 108 (8) (February): 088101.

Nogales, E, S G Wolf, and K H Downing. 1998. Structure of the |[Alpha]| |[Beta]| tubulin dimer by electron crystallography. *Nature* 391 (6663) (January 8): 199–203. doi:10.1038/34465.

Odijk, T. 1986. Theory of lyotropic polymer liquid crystals. *Macromolecules* 19 (9): 2313–2329. doi:10.1021/ma00163a001.

Odijk, T. 1995. Stiff chains and filaments under tension. *Macromolecules* 28 (20) (September): 7016–7018. doi:10.1021/ma00124a044.

Odijk, T and H N W Lekkerkerker. 1985. Theory of the isotropic-liquid crystal phase separation for a solution of bidisperse rodlike macromolecules. *The Journal of Physical Chemistry* 89 (10) (May): 2090–2096. doi:10.1021/j100256a058.

Oliver, T, K Jacobson, and M Dembo. 1995. Traction forces in locomoting cells. *Cell Motility and the Cytoskeleton* 31 (3): 225–240. doi:10.1002/cm.970310306.

Omary, M B, N-O Ku, P Strnad, and S Hanada. 2009. Toward unraveling the complexity of simple epithelial keratins in human disease. *Journal of Clinical Investigation* 119 (7) (July 1): 1794–1805. doi:10.1172/JCI37762.

Onck, P, T Koeman, T van Dillen, and E van der Giessen. 2005. Alternative explanation of stiffening in cross-linked semiflexible networks. *Physical Review Letters* 95 (17) (October): 178102. doi:10.1103/PhysRevLett.95.178102.

Ono, S, Y Yamakita, S Yamashiro, P T Matsudaira, J R Gnarra, T Obinata, and F Matsumura. 1997. Identification of an actin binding region and a protein kinase C phosphorylation site on human fascin. *Journal of Biological Chemistry* 272 (4) (January 24): 2527–2533. doi:10.1074/jbc.272.4.2527.

Onsager, L. 1945. Theories and problems of liquid diffusion. *Annals of the New York Academy of Sciences* 46 (5) (January 1): 241–265. doi:10.1111/j.1749-6632.1945.tb36170.x.

Ott, A, M Magnasco, A Simon, and A Libchaber. 1993. Measurement of the persistence length of polymerized actin using fluorescence microscopy. *Physical Review E* 48 (3) (September 1): R1642–R1645.

Paluch, E K and E Raz. 2013. The role and regulation of blebs in cell migration. *Current Opinion in Cell Biology* (June). doi:10.1016/j.ceb.2013.05.005.

Pinot, M, V Steiner, B Dehapiot, B-K Yoo, F Chesnel, L Blanchoin, C Kervrann, and Z Gueroui. 2012. Confinement induces actin flow in a meiotic cytoplasm. *Proceedings of the National Academy of Sciences of the United States of America* 109 (29) (January 17): 11705–11710.

Pollard, T D and G G Borisy. 2003. Cellular motility driven by assembly and disassembly of actin filaments. *Cell* 112 (4) (February): 453–465. doi:10.1016/S0092-8674(03)00120-X. http://www.sciencedirect.com/science/article/pii/S009286740300120X.

Rafelski, S M and J A Theriot. 2004. Crawling toward a unified model of cell mobility: Spatial and temporal regulation of actin dynamics. *Annual Review of Biochemistry* 73 (January 1): 209–239. doi:10.1146/annurev.biochem.73.011303.073844.

Rauzi, M and P-F Lenne. 2011. Cortical forces in cell shape changes and tissue morphogenesis. *Current Topics in Developmental Biology* 95: 93–144. doi:10.1016/B978-0-12-385065-2.00004-9.

Rayment, I, H Holden, M Whittaker, C Yohn, M Lorenz, K Holmes, and R Milligan. 1993. Structure of the actin-myosin complex and its implications for muscle contraction. *Science* 261 (5117) (July 2): 58–65. doi:10.1126/science.8316858.

Riseman, J and J G Kirkwood. 1950. The intrinsic viscosity, translational and rotatory diffusion constants of rod-like macromolecules in solution. *The Journal of Chemical Physics* 18 (4): 512. doi:10.1063/1.1747672.

Roberts, T M and M Stewart. 2000. Acting like actin. The dynamics of the nematode major sperm protein (msp) cytoskeleton indicate a push-pull mechanism for amoeboid cell motility. *The Journal of Cell Biology* 149 (1) (April 3): 7–12.

Salbreux, G, G Charras, and E Paluch. 2012. Actin cortex mechanics and cellular morphogenesis. *Trends in Cell Biology* 22 (10): 536–545. doi:10.1016/j.tcb.2012.07.001. http://linkinghub.elsevier.com/retrieve/pii/S0962892412001110.

Sanchez, T, D T N Chen, S J DeCamp, M Heymann, and Z Dogic. 2012. Spontaneous motion in hierarchically assembled active matter. *Nature* 491 (7424) (November 7): 431–434. doi:10.1038/nature11591.

Schliwa, M and G Woehlke. 2003. Molecular motors. *Nature* 422 (6933) (April 17): 759–765. doi:10.1038/nature01601.

Schmidt, C F, M Baermann, G Isenberg, and E Sackmann. 1989. Chain dynamics, mesh size, and diffusive transport in networks of polymerized actin: A quasielastic light scattering and microfluorescence study. *Macromolecules* 22 (9) (September): 3638–3649. doi:10.1021/ma00199a023.

Schuyler, S C and D Pellman. 2001. Microtubule 'plus-end-tracking proteins': The end is just the beginning. *Cell* 105 (4) (May 18): 421–424.

Sedzinski, J, M Biro, A Oswald, J-Y Tinevez, G Salbreux, and E Paluch. 2011. Polar actomyosin contractility destabilizes the position of the cytokinetic furrow. *Nature* 476 (7361): 462–466. doi:10.1038/nature10286. http://www.nature.com/doifinder/10.1038/nature10286.

Seetapun, D, B T Castle, A J McIntyre, P T Tran, and D J Odde. 2012. Estimating the microtubule GTP cap size in vivo. *Current Biology* 22 (18) (September): 1681–1687. doi:10.1016/j.cub.2012.06.068.

Selby, C C and R S Bear. 1956. The structure of actin-rich filaments of muscles according to x-ray diffraction. *The Journal of Biophysical and Biochemical Cytology* 2 (1) (January 25): 71–85. doi:10.1083/jcb.2.1.71.

Semmrich, C, T Storz, J Glaser, R Merkel, and A Bausch. 2007. Glass transition and rheological redundancy in F-actin solutions. *Proceedings of the National Academy of Sciences of the United States of America* 104 (January 1): 20199–20203.

Sheterline, P and J C Sparrow. 1994. Actin. *Protein Profile* 1 (1): 1–121.

Shih, Y-L and L Rothfield. 2006. The bacterial cytoskeleton. *Microbiology and Molecular Biology Reviews* 70 (3) (September): 729–754. doi:10.1128/MMBR.00017-06.

Shin, J H, L Mahadevan, G S Waller, K Langsetmo, and P Matsudaira. 2003. Stored elastic energy powers the 60-microm extension of the *Limulus polyphemus* sperm actin bundle. *The Journal of Cell Biology* 162 (7) (September): 1183–1188.

Shin, J H, B K Tam, R R Brau, M J Lang, L Mahadevan, and P Matsudaira. 2007. Force of an actin spring. *Biophysical Journal* 92 (10) (January 1): 3729–3733. doi:10.1529/biophysj.106.099994.

Skau, C T, D S Courson, A J Bestul, J D Winkelman, R S Rock, V Sirotkin, and D R Kovar. 2011. Actin filament bundling by fimbrin is important for endocytosis, cytokinesis, and polarization in fission yeast. *Journal of Biological Chemistry* 286 (30) (January 29): 26964–26977.

Soares e Silva, M, J Alvarado, J Nguyen, N Georgoulia, B M Mulder, and G H Koenderink. 2011a. Self-organized patterns of actin filaments in cell-sized confinement. *Soft Matter*, 7 (22), 10631–10641.

Soares e Silva, M, M Depken, B Stuhrmann, M Korsten, F C Mackintosh, and G H Koenderink. 2011b. Active multistage coarsening of actin networks driven by myosin motors. *Proceedings of the National Academy of Sciences of the United States of America* 108 (23) (June 7): 9408–9413. doi:10.1073/pnas.1016616108.

Somerville, C. 2006. Cellulose synthesis in higher plants. *Annual Review of Cell and Developmental Biology* 22: 53–78. doi:10.1146/annurev.cellbio.22.022206.160206.

Song, S, A Landsbury, R Dahm, Y Liu, Q Zhang, and R A Quinlan. 2009. Functions of the intermediate filament cytoskeleton in the eye lens. *Journal of Clinical Investigation* 119 (7) (January 1): 1837–1848.

Starr, D A and M Han. 2003. ANChors away: An actin based mechanism of nuclear positioning. *Journal of Cell Science* 116 (2) (January 1): 211–216. doi:10.1242/jcs.00248.

Storm, C, J J Pastore, F C MacKintosh, T C Lubensky, and P A Janmey. 2005. Nonlinear elasticity in biological gels. *Nature* 435 (7039) (May 12): 191–194. doi:10.1038/nature03521.

Stossel, T P, J Condeelis, L Cooley, J H Hartwig, A Noegel, M Schleicher, and S S Shapiro. 2001. Filamins as integrators of cell mechanics and signalling. *Nature Reviews Molecular Cell Biology* 2 (2) (February): 138–145. doi:10.1038/35052082.

Stuhrmann, B, M Soares e Silva, M Depken, F C MacKintosh, and G H Koenderink. 2012. Nonequilibrium fluctuations of a remodeling in vitro cytoskeleton. *Physical Review E* 86: 020901(R).

Suzuki, A, T Maeda, and T Ito. 1991. Formation of liquid crystalline phase of actin filament solutions and its dependence on filament length as studied by optical birefringence. *Biophysical Journal* 59 (1) (January 1): 25–30. doi:10.1016/S0006-3495(91)82194-4.

Szeverenyi, I, A J Cassidy, C W Chung, B T K Lee, J E A Common, S C Ogg, H Chen et al. 2008. The human intermediate filament database: Comprehensive information on a gene family involved in many human diseases. *Human Mutation* 29 (3): 351–360. doi:10.1002/humu.20652.

Tarr, M, J W Trank, and K K Goertz. 1983. Effect of external force on relaxation kinetics in single frog atrial cardiac cells. *Circulation Research* 52 (2) (February): 161–169.

Tharmann, R, M M A E Claessens, and A R Bausch. 2006. Micro- and macrorheological properties of actin networks effectively cross-linked by depletion forces. *Biophysical Journal* 90 (7) (April 1): 2622–2627. doi:10.1529/biophysj.105.070458.

Theriot, J A. 2000. The polymerization motor. *Traffic* 1 (1) (January): 19–28. doi:10.1034/j.1600-0854.2000.010104.x.

Thomas, W E, V Vogel, and E Sokurenko. 2008. Biophysics of catch bonds. *Annual Review of Biophysics* 37: 399–416. doi:10.1146/annurev.biophys.37.032807.125804.

Thompson, R F and G M Langford. 2002. Myosin superfamily evolutionary history. *The Anatomical Record* 268 (3) (October 15): 276–289. doi:10.1002/ar.10160.

Tilney, L G. 1975. Actin filaments in the acrosomal reaction of *Limulus* sperm. Motion generated by alterations in the packing of the filaments. *The Journal of Cell Biology* 64 (2) (February): 289–310.

Tilney, L G and D A Portnoy. 1989. Actin filaments and the growth, movement, and spread of the intracellular bacterial parasite, *Listeria monocytogenes*. *The Journal of Cell Biology* 109 (4 Pt 1) (October): 1597–1608.

Tolić-Nørrelykke, I M. 2008. Push-me-pull-you: How microtubules organize the cell interior. *European Biophysics Journal* 37 (7) (April 11): 1271–1278. doi:10.1007/s00249-008-0321-0.

Tuma, M and V I Gelfand. 1999. Molecular mechanisms of pigment transport in melanophores. *Pigment Cell Research* 12 (5): 283–294. doi:10.1111/j.1600–0749.1999.tb00762.x.

Vale, R D. 2003. The molecular motor toolbox for intracellular transport. *Cell* 112 (4) (February): 467–480. doi:10.1016/S0092-8674(03)00111-9.

Van Damme, D. 2009. Division plane determination during plant somatic cytokinesis. *Current Opinion in Plant Biology* 12 (6) (December): 745–751. doi:10.1016/j.pbi.2009.09.014.

van Mameren, J, K C Vermeulen, F Gittes, and C F Schmidt. 2009. Leveraging single protein polymers to measure flexural rigidity. *Journal of Physical Chemistry B* 113 (12) (February 16): 3837–3844. doi:10.1021/jp808328a.

van Roij, R, M Dijkstra, and R Evans. 2000. Orientational wetting and capillary nematization of hard-rod fluids. *Europhysics Letters* 49 (3) (January 2): 350–356. doi:10.1209/epl/i2000-00155-0.

Visscher, K, M J Schnitzer, and S M Block. 2000. Force production by single kinesin motors. *Nature Cell Biology* 2 (10) (September 14): 718–723. doi:10.1038/35036345.

Vogel, S K, Z Petrasek, F Heinemann, and P Schwille. 2013. Myosin motors fragment and compact membrane-bound actin filaments. *eLife* 2 (0) (January 8): e00116. doi:10.7554/eLife.00116.019.

Volkmer Ward, S M, A Weins, M R Pollak, and D A Weitz. 2008. Dynamic viscoelasticity of actin cross-linked with wild-type and disease-causing mutant A-actinin-4. *Biophysical Journal* 95 (10) (November): 4915–4923. doi:10.1529/biophysj.108.131722.

Wachsstock, D H, W H Schwarz, and T D Pollard. 1993. Affinity of alpha-actinin for actin determines the structure and mechanical properties of actin filament gels. *Biophysical Journal* 65 (1) (July): 205–214. doi:10.1016/S0006-3495(93)81059-2.

Walczak, C E and R Heald. 2008. Mechanisms of mitotic spindle assembly and function. *International Review of Cytology* 265: 111–158. doi:10.1016/S0074-7696(07)65003-7.

Weirich, C S, J P Erzberger, and Y Barral. 2008. The septin family of GTPases: Architecture and dynamics. *Nature Reviews Molecular Cell Biology* 9 (6) (May 14): 478–489. doi:10.1038/nrm2407.

West-Foyle, H and D N Robinson. 2012. Cytokinesis mechanics and mechanosensing. *Cytoskeleton (Hoboken, N.J.)* 69 (10) (October): 700–709. doi:10.1002/cm.21045.

Wickstead, B and K Gull. 2011. The evolution of the cytoskeleton. *The Journal of Cell Biology* 194 (4) (January 22): 513–525.

Wiesner, S. 2003. A biomimetic motility assay provides insight into the mechanism of actin-based motility. *The Journal of Cell Biology* 160 (3) (January 27): 387–398. doi:10.1083/jcb.200207148.

Wilhelm, C. 2008. Out-of-equilibrium microrheology inside living cells. *Physical Review Letters* 101 (2) (July): 028101. doi:10.1103/PhysRevLett.101.028101.

Yamakita, Y, S Ono, F Matsumura, and S Yamashiro. 1996. Phosphorylation of human fascin inhibits its actin binding and bundling activities. *Journal of Biological Chemistry* 271 (21) (January 24): 12632–12638.

Yao, N Y, D J Becker, C P Broedersz, M Depken, F C MacKintosh, M R Pollak, and D A Weitz. 2011. Nonlinear viscoelasticity of actin transiently cross-linked with mutant α-actinin-4. *Journal of Molecular Biology* 411 (5) (September 2): 1062–1071. doi:10.1016/j.jmb.2011.06.049.

Yin, S, X Zhang, C Zhan, J Wu, J Xu, and J Cheung. 2005. Measuring single cardiac myocyte contractile force via moving a magnetic bead. *Biophysical Journal* 88 (2) (February): 1489–1495. doi:10.1529/biophysj.104.048157.

Waterman-Storer, C.M., W.H. Schwartz, and E.D. Salmon. 1993. Actin filament dynamics for dynamics and mechanical properties of actin filament gels. Biophysical Journal 65 (1) December: 205–214. doi:10.1016/S0006-3495(93)81059-6.

Watanabe, T. and E. Heald. 2005. Microtubule-medioble spindle assembly and function. International Review of Cytology 265 (1) 1–36. doi:10.1016/S0074-7696(08)60010-X.

Weber, G.S., J.P. Trotter, and S.J. Herral. 2013. The septin 2013 in CE Paso. Archive of Cite and Systems Biology. Molecular Cell Biology 4 (3). Myer 141–173. doi:10.1016/j.yexcr.2007.

Weeds-Davis, H. and D.S. Robinson. 2017. Actin gene structure, regulation and gene expression. Cytoskeleton 74 (1–2) (October): 700–709. doi:10.1002/cm.21367.

Wegmann, B. and S. Grill. 2011. The evolution of the cytoskeleton. The Journal of Cell Biology 194 (4) (January 23): 513–525.

Wegner, S. 2003.1. Microtubule mobility assay provides insight into the mechanism of actin based motility. Biophysical Journal of Biology. 180 D. Harvard Cell. 363. doi:10.1016/j.bpj.2003.07.

Wilhelm, C. 2005. Particle equilibrium microstructure transfer inside cells. Physical Review Letters 101 (1) July (0.8099): doi:10.1103/PhysRevLett.101.028101.

Woolley, V.V.C.V. Measures 2003. Vorsábleu. Cen. These deviation of bal and Brain inflate to actin fragile grant formations in these planos of physical Geometry. 41 (3). 141–151.

Yao, S.-Y., D.T. Brangwynne, D. Dogterom, F.C. MacKintosh, M.T. Stuhrmann, and D.A. Weitz. 2011. Nonlinear viscoelasticity of actin transiently crosslinked with mutant scruins. Journal of Molecular Biology 411 (5) (September 2): 1062–1071. doi:10.1016/j.jmb.2011.06.049.

Yao, N.-Y., Zhou, C. Zhang, J. Wu, L. Wen, and J. Cheung. 2008. Measuring single cardiac myocyte contractile force to track cells during a maturation. Biophysical Journal 95 (2) (February 1): 939–1001. doi:10.1529/biophysj.107.119750.

2

Mechanobiology of the Cell Membrane

Peter J. Butler, Hari S. Muddana, and Sara Farag

CONTENTS

2.1 Introduction

In this chapter, we introduce the main concepts in membrane mechanics through an exploration of the relationship between mechanics and protein activation. Proteins can be activated by bringing them together, or by changing their conformation. So the question arises, "Can forces on membranes cause changes in the transport of membrane components, such as proteins and lipids, and can forces cause changes in integral membrane protein conformation that are equivalent to those caused by ligand binding, a process well known to initiate biochemical signaling?" The answers to these questions have significant implications in the field of mechanobiology because many of the important proteins that transduce external cellular signals to intracellular changes in signaling pathways and genetic expression reside in the plasma membrane.

The chapter begins with an overview of membrane composition and how this composition leads to unique mechanical properties of membranes. These mechanical properties are then described in terms of moduli, and the principal equations describing the relationship between stress, strain, curvature, and membrane material properties are presented. We then present a link between physiological membrane stress and protein activation using a series of theoretical models that quantify the free energy arising from hydrophobic mismatch between proteins and test whether this free energy is sufficient for protein activation. Next, we show how continuum membrane mechanics emerges from interactions of molecules using an example wherein membrane tension is applied to a molecular dynamics (MD) simulation of a lipid bilayer. Such changes in diffusion and lipid packing are measured experimentally using single-molecule fluorescence. With this background in mind, we present experimental evidence that membrane stress can lead to protein activation and changes in transport. Such studies represent a confluence of membrane mechanics, membrane composition, lateral transport, and mechanotransduction in focal adhesions (FAs).

2.2 Membrane Organization Arises from the Amphiphilic Nature of Lipids

Lipids are amphiphilic molecules that self-assemble into ~5 nm thick cellular membranes. Such membranes function as semipermeable barriers between cell cytosol and extracellular material and compartmentalize internal organelles. Biological membranes harbor many proteins (nearly 30% of the genes expressed in animal cells[1]) that play essential roles in cellular function. Such functions include maintaining ion gradients across the membrane (e.g., ion channels), transduction of information from outside the cell to the cytosol through receptor activation (e.g., G-protein signaling), formation of structural connections between cell cytoskeleton and extracellular matrix (e.g., integrins), and communication between cells through junctional proteins (e.g., cadherins and connexins). In the famous fluid-mosaic model,[2] Singer and Nicholson defined the lipid bilayer as a 2D viscous solution with integral proteins. Since that original insight, our understanding of the lipid bilayer structure and its active role in regulating protein activity has developed significantly.[3–5] Lipids are no longer considered as passive structural elements of cell membranes. Rather, they govern membrane protein activity through direct action on proteins, through their influence on structural properties of the bilayer, and through lipid metabolism. Importantly, understanding the interrelationship between structure and function of lipid bilayers and downstream signal transduction underlies the development of novel biotherapeutic agents[6] against diseases that find their origins in lipid–protein interaction.

Membrane lipids in eukaryotic cells can be broadly classified into glycerophospholipids and sphingolipids. The majority of the lipids in the eukaryotic cell membranes belong to the class of glycerophospholipids, which include phosphatidylcholines (PCs), phosphatidylserine (PS), phosphatidylethanolamine (PE), and phosphatidylinositol (PI), of which PC is the major class comprising more than 50% of the total lipid content in cells.[7] These lipid types have a common glycerol backbone and differ in the chemical nature of the headgroup. As a general feature, all lipids are amphiphilic with a polar (hydrophilic) headgroup region and a nonpolar (hydrophobic) hydrocarbon tail region (schematic shown in Figure 2.1). Due to the amphiphilic nature, most lipids spontaneously aggregate to form double layers (i.e., bilayers) under aqueous conditions resulting in a 2D planar structure. Depending on the geometry of the molecule (e.g., diacyl vs. monoacyl lipids), they might also form micellar structures (Figure 2.1).

Another key molecular component of biological membranes is cholesterol. Cholesterol is unique to eukaryotic cell membranes and is the principal sterol synthesized in animal cells. Cholesterol is also an amphiphilic molecule that is largely hydrophobic with a rigid ring structure and a polar hydroxyl headgroup (Figure 2.1). Cholesterol plays several structural and functional roles in biological membranes.[8] Cholesterol regulates membrane protein activity through specific sterol–protein interactions, by altering the bilayer's physical properties and through self-organization of the bilayer into domains.[9]

Specific cholesterol-binding sites have been identified for several membrane proteins including G-protein-coupled receptors,[10–12] ligand-gated ion channels,[13] inwardly rectifying potassium channels,[14,15] and large conductance calcium and potassium channels.[16] Specific sterol–protein interactions are thought to modulate the functioning of these membrane proteins.[8,9] Cholesterol can also indirectly modulate protein activity by maintaining certain physical properties of the lipid bilayer. For example, membrane cholesterol content has been shown to modulate volume-regulated anion current in endothelial cells (ECs), and this effect was reversed by replacing cholesterol with its chiral analogues,[17,18] indicating modulation of protein activity through bilayer fluidity changes. Finally, cholesterol

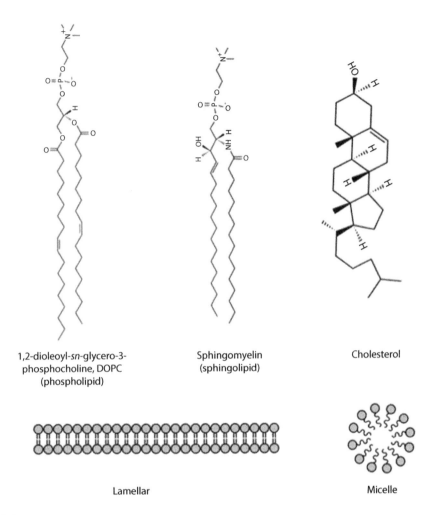

FIGURE 2.1
Chemical structures of phospholipids, sphingolipids, and cholesterol. Lipids adopt lamellar or micellar phases depending on whether their shape is optimal for planar or highly curved membranes as shown in the bottom panel.

is an essential ingredient for lateral organization of lipid membranes into membrane domains (or rafts).[19]

Even though, macroscopically, the lipid bilayer is thought of as a 2D continuous fluid, microscopically, it exhibits high asymmetry from one leaflet to the other and significant heterogeneity in the lateral direction. The inner and outer leaflets have asymmetric distribution in lipid composition with the outer leaflet enriched in PCs and sphingomyelins and the inner leaflet enriched in PSs and PEs[20] with much of this asymmetry maintained by fli- pases, proteins that enable the "flipping" of polar headgroups across the inner membrane hydrophobic barrier to the opposite leaflet. Cell membranes also exhibit highly complex self-organization in the lateral dimension, with regions enriched in sphingomyelin and cholesterol and regions enriched with polyunsaturated PCs and depleted of cholesterol.[21] Such compositional heterogeneity implies that mechanical properties are heterogeneous as well, with areas of high and low compressibility moduli.[22]

The wide variety of lipids gives rise to bilayers of various crystalline phases including gel, ripple, and liquid (liquid disordered [L_d] and liquid ordered [L_o]).[23] In the gel phase, the hydrocarbon chains are arranged in an all-trans state, whereas in the liquid state, they undergo fast *trans* to *gauche* transformations resulting in high fluidity. The transition of the gel phase to liquid phase happens at a specific temperature, referred as the phase transition temperature. The phase transition temperature of a lipid is governed by its chemical structure and is predominantly affected by the length and degree of saturation of the hydrocarbon chains. For example, the phase transition temperature of dipalmitoylphosphatidylcholine (DPPC) (16:0 PC) is 42°C and that of 1,2-Dimyristoyl-*sn*-Glycero-3-Phosphocholine (DMPC) (14:0) is 24°C. Thus, a decrease in chain length by two carbon atoms decreases the melting temperature by 18°C. In a mixture of two or more lipids with different phase transition temperatures, the nonideal mixing behavior of the lipids results in heterogeneous phase domains.[24,25] The miscibility of a mixture is typically represented by phase diagrams. Depending on the composition and temperature, the mixture can result in solid/solid, liquid/liquid, or solid/liquid, coexistence regions. Coexisting liquid–liquid phase regions are typically observed in ternary mixtures of two lipids and cholesterol. These regions are referred to as "liquid disordered" (cholesterol poor) and "liquid ordered" (cholesterol rich) and are thought to resemble lipid rafts in cell membranes. While the existence of lipid domains under physiological conditions is still debated, recent state-of-the-art measurement techniques confirm their presence in cell membranes with sizes on the order of a few tens of nanometers.[26] Moreover, these domains are transient in nature, forming and disappearing on the timescales of a few milliseconds.[27]

Giant unilamellar vesicles prepared from simple lipid mixtures (two or three components) exhibit microscopically observable phase domains, permitting their characterization under regular fluorescence microscopy.[28,29] These simplified model membrane systems closely mimic realistic biological membranes and thus provide a platform on which to build fundamental physical and chemical principles for lipid–lipid interactions. One step closer to the biological membranes is to study giant plasma membrane vesicles derived from cells that include all the membrane proteins.[30] In fact, recent studies report microscopic lateral phase separation in such giant plasma membrane vesicles, and several membrane proteins show preferential partitioning to these domains.[30,31]

2.3 Lipid Amphiphilicity Gives Rise to Mechanical Properties

Membranes are very flaccid materials and the bonds between molecules are dynamic and loose. So how do they resist tension? It turns out that exposure of water in the aqueous environment to the hydrophobic tails of lipids requires a significant amount of energy. The amount of mechanical energy it takes to separate lipids in the face of exposure of hydrophobic tails to water constitutes the main mechanical resistance of membranes to tension. Furthermore, moduli of lipid bilayers (bending, areal expansion, etc.) arise to a large degree from the interaction of the aqueous environment with lipids. The overall mechanics of the cell surface is also affected by the membrane-associated cytoskeleton, but this component will not be considered in this chapter.

Often membranes are considered as shells surrounding a liquid medium. But a significant result of this assumption is that the resistance to expansion with tension depends on bilayer thickness.[32] However, since molecules in the membrane can rapidly rearrange in the bilayer plane in response to tension, with very little shear between molecules, the only resistance to expansion is the steric hindrance between molecules and the surface tension of the water/headgroup interface, which arises from the hydrophobic effect. When lipids are crowded together, the energy penalty to bring them closer together scales as $1/a$ where a is the mean area occupied by the lipid.[33] The energy penalty to move them further apart is proportional to a. If we add these two energies together, we get to total energy for a lipid, E_l, given by

$$E_l = \gamma a + \frac{K}{a} \qquad (2.1)$$

The proportionality constant to bring lipids apart is the surface tension, γ, since surface tension arises from the exposure of hydrocarbon tails to water. For steric repulsion, K is an undefined proportionality constant that governs the energy to bring molecules close together. It arises from steric contribution, a hydration force contribution, and an electrostatic double layer.[34] At equilibrium, we would expect to be able to find an area, a_0, at which E_l is minimized such that

$$\left.\frac{dE_l}{da}\right|_{a_0} = 0 \qquad (2.2)$$

Carrying out this derivative leads to the relation $a_0 = \sqrt{K/\gamma}$

resulting in (after some simplifications)

$$E_l = 2\gamma a_0 + \frac{\gamma}{a}\left(a - a_0\right)^2 \qquad (2.3)$$

If we take the derivative of E_l with respect to a and solve at $a = a_0$, we get the energy fluctuation around equilibrium. From this, we conclude that the elastic energy density fluctuates by $\gamma((a - a_0)^2/a_0^2)$ around a minimum area per lipid a_0 (note that we have divided elastic energy by a_0 to get the elastic energy density). This change in energy comes from the fact that small increments of elastic expansion of the membrane are countered by surface tension arising from the hydrophobic effect. Such tension (T)–strain relationships are governed by the compressibility modulus, such that

$$T = K_a \frac{\left(a - a_0\right)}{a_0} \qquad (2.4)$$

where K_a is the compressibility modulus.[35] The energy density from tension, E_T/a_0, of this relationship is derived from the integral of the tension over area. Therefore,

$$\frac{E_T}{a_0} = \frac{1}{2} K_a \frac{\left(a - a_0\right)^2}{a_0^2} \qquad (2.5)$$

Equating the energy density from tension (Equation 2.5) with the energy density penalty from expansion arising from exposure of hydrophobic lipid tails, $\gamma((a - a_0)^2/a_0^2)$, we see that $K_a = 2\gamma$ for a monolayer and $K_a = 4\gamma$ for a bilayer. This remarkably simple yet elegant result allows us to assess the continuum property of modulus from the molecular property of surface tension.

Thus, membranes resist expansion when under stress via surface tension, and surface tension, a molecular quantity, can be related to the compressibility modulus. Because bending of a bilayer requires compression of one surface and expansion of another surface, we would expect there to be a relationship between the bending modulus and compressibility modulus. We can connect the bending modulus and the compressibility modulus by considering the energy it takes to deform an isotropic volume of the material. For example, the energy density of bending is

$$E_b = \frac{K_b}{2}\left(\frac{1}{R_1}+\frac{1}{R_2}\right)^2 + \frac{K_G}{R_1 R_2} \tag{2.6}$$

where

K_b is the bending modulus for average curvature $(1/R_1 + 1/R_2)$
K_G is the bending modulus for Gaussian curvature $(1/(R_1 R_2))$[32,36]

To bend this material in one direction, R_2 goes to ∞ and therefore, $E_b = K_b/(2R_1^2)$. If we are bending a monolayer of thickness h, and we assume that the midplane of the monolayer is the neutral plane with strain $=0$ at $z=0$, and the strain (u) increases linearly with distance z from the midplane (i.e., $u = z \cdot (u_{hg}/(h/2))$), we can show the strain on the lipid headgroups is $u_{hg} = (h/(2R_1))$. We can also compute the energy of expansion of one monolayer E_m as

$$E_m = \frac{1}{2} K_a \left\langle u_m^2 \right\rangle \tag{2.7}$$

where $\left\langle u_m^2 \right\rangle$ is the average of the square of the monolayer strain. To compute this quantity, we integrate u_m^2 from $z=0$ to $z=h/2$ (where $u_m = u_{hg}$) and divide by h to get $\left\langle u_m^2 \right\rangle = u_{hg}^2/3 = 1/12(h/R_1)^2$. Substituting this result into E_m (Equation 2.7), we get

$$E_m = \frac{1}{24} K_a \left(\frac{h}{R_1}\right)^2 \tag{2.8}$$

Equating E_m (energy from expanding the monolayer) and E_b (energy of bending), we get

$$K_b = \frac{1}{12} K_a \left(h\right)^2 \tag{2.9}$$

This is the relation between bending moduli and compressibility moduli for an elastic sheet of thickness h. If we have two elastic sheets that slide past each other, we can change h to $h_b/2$ and arrive at (for a bilayer)

$$K_b = \frac{1}{48} K_a \left(h_b\right)^2 \tag{2.10}$$

While we have ignored changes in thickness of the bilayer with bending, a feature that would introduce a Poisson's ration, this result shows that bending rigidity is much lower than compressibility and the bending modulus depends on thickness of the bilayer. It is, of course, important to keep in mind that K_a is a true modulus that relates tension to strain. The bending modulus, K_b, has the units of energy and, since it depends on the thickness of the membrane, it is not a true material property. Nevertheless, it can be seen that the membrane is much softer in bending than it is in tension and that the bending modulus depends on the membrane thickness, which depends on acyl chain length. In short, membranes easily fold and bend but do not easily expand. Thus, we might expect that activation of proteins in the bilayer might be easier upon bending of the membrane rather than upon stretching it. If bending causes membrane thickness changes, then proteins embedded in membranes would experience hydrophobic mismatch upon membrane bending. In addition, there are proteins that prefer bent membranes. As a consequence, membrane bending fluctuations could be a significant source of protein activation and could be a mechanism of dynamic protein sorting.[37]

2.4 Can Stretching or Bending Appreciably Alter Hydrophobic Mismatch?

Membrane-mediated protein activity can be classified into two categories: specific interactions of lipids with proteins inducing conformational changes and nonspecific interaction in which lipid bilayer physical properties modulate protein conformational changes by indirectly altering the energetic state of proteins. Various physical properties of the lipid bilayers including viscosity, hydrophobicity, compressibility, curvature, and lateral pressure are thought to play an essential role in modulating integral membrane protein activity.[38] A key property of the lipid bilayer that has a strong influence on the protein conformation is its hydrophobic thickness.[39] Matching the hydrophobic thickness of the lipids to that of the embedded proteins avoids the energetic costs associated with exposing to water the hydrophobic side chains of these integral proteins.[40,41] Mismatching the hydrophobic thickness of the lipid bilayer to the protein can result in a change in the lipid or protein's conformation or both.[42,43] Hydrophobic mismatch has been shown to influence the opening and closing of transmembrane stretch-activated ion channels.[44] Moreover, such hydrophobic mismatch can also drive aggregation or oligomerization of the membrane proteins.[45–46] Interestingly, lipid–lipid mismatch can also drive membrane segregation into domains,[47] which in turn can facilitate segregation of membrane proteins. Another well-studied property of the lipid bilayer is its fluidity (i.e., inverse of viscosity). Fluidity of the lipid bilayer can be modulated through compositional changes or through application of external mechanical forces.[48–50] Though changes in fluidity cannot directly influence the activation barrier for proteins,[38] it could potentially alter the kinetics of ligand–receptor interactions or protein–protein interactions in the membrane.[51,52] Nicolau et al.[53] have shown that the diffusion characteristics of raft and nonraft regions can also determine the residence time of the protein in raft regions and thus the protein concentrations in rafts. Moreover, several anesthetics and drugs are known to alter cellular function through interactions with the plasma membrane and alteration in membrane fluidity.[54]

While bilayer thickness and fluidity have been investigated for a long time, more recent studies show that other properties of the lipid bilayer such as curvature and lateral pressure profiles might also influence protein sorting and activity.[55,56] While theories for how hydrophobic mismatch induces protein conformational change tend to focus on free energy transfer, alteration of depth-dependent lateral pressure can provide a means to induce conformational changes nonuniformly along the transverse direction of the protein.[57] So far, this mechanism has been investigated only theoretically or through the use of computational MD, as lateral pressure profiles cannot be obtained experimentally. Membrane curvature is a well-observed phenomenon that is critical for membrane budding and fusion and plays an important role in intracellular trafficking. Bilayer curvature affects lipid–protein interactions and vice versa. Lipid curvature-induced protein sorting and protein-induced lipid curvature are highly interlinked phenomena.[58]

To begin, the motivation for proposing that membrane thinning occurs due to curvature changes will be shown. Thinning due to curvature changes will be compared to thinning due to tension on the membrane. Thinning of the membrane from h_0 to h will be determined by looking at a simplified model that calculates thickness changes from areal strain (assuming bilayer incompressibility):

$$\epsilon_a = \frac{\sigma}{Y} = \frac{(a - a_0)}{a_0} = \frac{h_0}{h} - 1 \tag{2.11}$$

or

$$h = \frac{h_0}{\dfrac{\sigma}{(K_a / h_0)} + 1} \tag{2.12}$$

Note that we let Young's modulus, $Y = K_a / h_0$.

Let us consider the shearing of a membrane by flow. Such a problem may be relevant to the effects of blood flow on vascular endothelium.[59] For simplicity, it is assumed that a rectangular piece of membrane is anchored on one side with the dimensions of 10 μm by 100 μm from the top and a thickness of 4 nm (note that we have neglected the role of the glycocalyx in this example; a more thorough treatment of this type of problem can be found in Ref. [60]). In addition, it is assumed that a shear stress of 10 dynes/cm² (1 Pa) is applied to the surface of a membrane with Young's modulus of 250×10^6 Pa (Y)[61] or $K_a = 1$ N/m, which is the order of magnitude of the modulus of membranes with cholesterol.[61] For this shear, the integrated force on the top surface stress will be 1 Pa × 1000 μm² yielding 10^{-9} N. This force will be balanced by a membrane cross section of area 4 nm × 10 μm. The resulting stress (σ) in the membrane is then 0.025×10^6 Pa (force/area). Substituting this stress into Equation 2.12 yields a decrease in membrane thickness of <0.001 nm. Thus, lateral tension from fluid flow yields an almost negligible change in membrane thickness. Whereas this example is related to shear stress on the vascular endothelium, other forces on the membrane from cilia, or those experienced during cell stretching, may be much larger.[44]

As we saw before, membranes bend more easily than they stretch. Recently, Nir Gov[62] showed that the thickness and static curvature of membranes are related by

$$h \cong h_0 \left(1 - h_0 H + 2h_0^2 H^2\right) \tag{2.13}$$

in which Gaussian curvature is assumed to be zero. Here, the mean curvature is H, the initial thickness of the membrane is h_0, and the final thickness is h. The derivation for this equation is modified from Ref. [33].

Using Equation 2.13 relating membrane thickness and curvature, we can construct the following graphs.

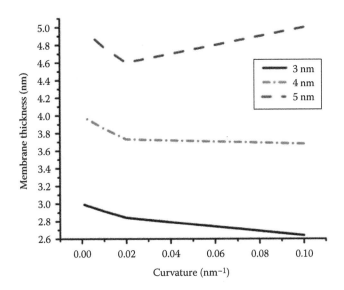

FIGURE 2.2
Membrane thickness changes with mean curvature for three different initial membrane thicknesses. The radii of curvature range from 10 nm to 1 μm.

Figure 2.2 illustrates that the relationship between membrane thickness and mean curvature is quadratic. The radii of curvature used to construct these graphs range from 10 nm (very small vesicles) to 1 μm (cell scale). Hence, when the membrane is thick, curvature causes either a decrease or increase in thickness, depending on the initial curvature. With thin membranes, curvature changes tend to cause a thinning of the membrane. Thus, the starting thickness will affect the magnitudes of the final thickness. In addition, the thicker membranes (e.g., 5 nm) will see a larger percentage change in thickness than thinner membranes (e.g., 3 nm) upon bending. Such insight may be relevant for domains that are separated due to hydrophobic mismatch arising from thickness differences. Conversely, bending can differentially change membrane thickness depending on the variability in the membrane's original thickness resulting from heterogeneous lipid composition. Such an occurrence could lead to further phase separation or mixing.

We can now try to assess if the curvature-induced membrane thickness is sufficient to activate integral membrane proteins. According to Ref. [63], free energy (G_U) from hydrophobic mismatch can be quantified using the relation

$$G_U = \frac{1}{2} K_{eff} U^2 \cdot 2\pi R \qquad (2.14)$$

Figure 2.3 shows a representation of the model proposed by Wiggins and Phillips in which R is the radius from the center of the inclusion to the end of the mismatch region and U is half the hydrophobic mismatch. In Equation 2.14, K_{eff} represents an effective elastic modulus $K_{eff} = \sqrt{2}\left(K_a^3 K_b / a^6\right)^{1/4}$ where a is the monolayer thickness. For $K_a = 2.33$ k_BT Å^{-2} ($K_a = 1$ N/m at 37°C), $K_b = 78$ k_BT (from Equation 2.10), and $2a = 40$ Å, $K_{eff} = 8.86 \times 10^{-2}$ k_BT Å^{-3}.

In order to determine whether or not protein activation is feasible with membrane thickness changes, the elastic energy from hydrophobic mismatch was calculated using the values $K_{eff} = 8.86 \times 10^{-2}$ k_BT·Å^{-3}, $U = ½$ * (hydrophobic mismatch), and R = radius of the protein

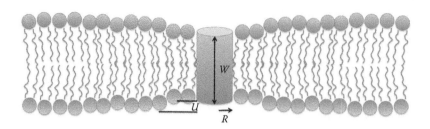

FIGURE 2.3
Protein inclusion in bilayer experiencing hydrophobic mismatch.

inclusion. In addition, the hydrophobic mismatch was calculated as (hydrophobic mismatch) = h − (length of protein hydrophobic region).

In order to judge whether the thinning of the membrane from bending is sufficient to provide the energy of activation for an embedded protein, we can consider the gramicidin system, which has been used as a binary (on–off) indicator of membrane thinning. Gramicidin dimerizes under appropriate condition of membrane thickness.[64,65] It is an antibiotic ion channel, sensitive to univalent cations such as Na⁺, H⁺, and K⁺, derived from *Bacillus brevis*. This tryptophan-rich protein is made up of an all single-stranded right-handed β-helical dimer that has an alternating L-D-amino acid sequence and a mass of ~4 kD. For gramicidin to activate in membranes, the two subunits undergo transmembrane dimerization. Dimerization occurs to minimize unfavorable hydrophobic–hydrophilic interactions in order to match the hydrophobic regions of the protein with the hydrophobic regions of the membrane. Stabilization of the activated molecule occurs through six hydrogen bonds between the end CO and NH groups from either subunit. In order for activation to occur, membrane deformation must take place to thin the membrane to accommodate the specific thickness of the activated channel. The membrane thickness needed for dimerization is not exactly known due to membrane undulations and fluctuations over time.[66] However, estimates of the thickness are around 21.7 Å.[67] Such a dimension can only be achieved through membrane thinning[66] since normal membrane hydrophobic thickness dimensions range from about 25 to 35 Å[68] (for a 40 Å thick bilayer in our example).

We first address the question of whether the thinning of a membrane due to curvature changes might be sufficient to be detected with gramicidin. The length of the hydrophobic region of the gramicidin channel is estimated to be 21.7 Å. In an example where shear forces cause a membrane to go from flat to exhibiting folds with diameters of 50 nm, the curvature would change from 0.001 to 0.02 nm⁻¹ and the thickness (Equation 2.13 and Figure 2.2) would change from 3.73 to 3.98 nm (assuming a totally flat membrane is 4 nm). If we subtract 8 Å for each leaflet headgroup region, the hydrophobic mismatch can be calculated to range from −0.4 to 2.1 Å. Therefore, it is possible that such membrane thinning could lead to gramicidin activation if the initial membrane thickness were within range. Alternatively, we can calculate the energy of hydrophobic mismatch and compare that value to the activation energy of integral membrane proteins. For a protein of hydrophobic thickness of 21.7 Å, let the radius, R, of a channel (e.g., gramicidin) be ~10 Å. If we insert these values into Equation 2.14 to determine changes in G_U, we obtain a resulting free energy change from 4 to 0.01 $k_B T$, for a change of ~4 $k_B T$. The gramicidin activation energy is 10.4 $k_B T$ per channel.[69] Therefore, although the activation energy is on the right order of magnitude, it may not be enough to activate this channel. However, this activation energy may be sufficient to activate other bilayer spanning proteins. For membranes with a low initial thickness (30 Å), their hydrophobic mismatch is small; hence, the available free energy may not be sufficient to activate a gramicidin channel. For all other thicknesses, the free energy from changes in hydrophobic mismatch is either equal to or greater than the activation energy of gramicidin. Despite the many simplifications in this analysis, it can be concluded that forces that alter curvature on the cell surface, such as shear flow,[70] may be sufficient to activate proteins via changes in membrane curvature and the attendant change in bilayer thickness. Further analysis is necessary to determine if the cortical spectrin or actin cytoskeleton that supports membrane in many eukaryotic cells has sufficient compliance in the face of physiological stresses, to allow the alterations in lipid bilayer curvature.

2.5 Stress on the Membrane Continuum Alters the Molecular Dynamics of Lipids

Mechanical forces modulate cell growth, differentiation, signal transduction, transport, and migration, through biochemical signaling pathways,[71] which may be related to membrane molecular organization and dynamics.[49,72,73] For example, lateral membrane tension causes conformational changes in integral membrane proteins[73] and affects membrane permeability,[74,75] lipid lateral diffusion,[48,49] and organization of lipid rafts.[47,76] These effects are believed to be mediated by bilayer thickness changes that result in hydrophobic mismatch between the lipid acyl chains and transmembrane region of proteins, leading to distortion of the lipid bilayer and concomitant protein conformational changes.[38–40] As we showed earlier, tension is most likely to alter membrane thickness through alterations in membrane curvature rather than lateral tension per se.

Despite the importance of lipid dynamics in cell signaling, to date, the only experimental studies quantifying the relationship between lipid dynamics and force have been conducted in sheared ECs[48–50] and in hair cells.[77,78] In these studies, a lipoid dye, such as 1,1′-dioctadecyl-3,3,3′,3′-tetramethylindocarbocyanine perchlorate (DiI), 9-(dicyanovinyl)-julolidine (DCVJ), or di-8-ANEPPS, was used to infer lipid dynamics from fluorescence intensity or fluorescence recovery after photobleaching (FRAP). Because these studies probed lipid dynamics indirectly and because the precise membrane tensions, at the molecular level, were unknown, there is a need to quantify directly the relationship between membrane tension and lipid dynamics.

The most prominent methods to assess lipid dynamics, including FRAP, fluorescence correlation spectroscopy (FCS), fluorescence anisotropy, and fluorescence lifetime (FL) imaging,[79–81] probe membrane lipid dynamics by analyzing the dynamics of lipophilic fluorescent dyes (e.g., DiI, 1,6-diphenyl-1,3,5-hexatriene [DPH], and Laurdan). In particular, DiI is popular because of its structural similarity to phospholipids and its ability to selectively partition into different lipid phases (gel or fluid) depending on the matching between the length of its alkyl chains and the lipid acyl chain length.[82] Spectroscopic investigations employing DiI have been used to study membrane organization and dynamics.[82,83] The FL of DiI depends on the accessibility to water[84] and on the viscosity of the local microenvironment,[82] offering a useful tool to detect lipid rafts in cells and phase separation in model membranes. However, proper interpretation of these fluorescence measurements requires precise knowledge of location, orientation, and interactions of dye with lipids and water, which are difficult to obtain experimentally.[85,86] Examples of the utility of using MD simulation as a tool to answer these questions include predictions of the location of drug-like small molecules in lipid bilayers along with validation by small-angle neutron scattering experiments.[87,88]

The aim of a recent computational modeling study was to determine the effects of membrane tension on mechanotransduction-related structural and dynamical properties of the bilayer.[89] In addition, we wished to understand the fidelity with which DiI, a popular membrane probe, reflects lipid dynamics, so that DiI photophysics could be used as a readout for tension effects on stressed membranes. To accomplish this goal, we performed a series of atomistic MD simulations of fluid-phase DPPC/DiI bilayers under various physiological tensions. The main readouts from this study are as follows: First, we characterized the effects of tension on bilayer thickness, acyl chain packing, and leaflet interdigitation. Second, we determined the relationship between area per lipid and lipid lateral diffusion and compared these results to predictions from free area diffusion theory. Third, we compared the DiI probe dynamics to the dynamics of the native lipids, leading to an analysis of the relationship between lipid packing and FL of DiI in terms of hydration and local viscosity.

Our first observation was that tension induces bilayer thinning and interleaflet interdigitation. Surface tension was estimated from the pressure tensor, as described in Ref. [90]. As expected, the surface tension increased linearly with an increase in area, from -2.6 mN/m at $\alpha = 0.635$ nm^2 to 15.9 mN/m at $\alpha = 0.750$ nm^2, above which rupturing of the bilayer is observed. While this rupture tension is in good agreement with values from micropipette aspiration of lipid vesicles (ranging from 10 to 20 mN/m),[74] MD-simulated rupture and experimental rupture tensions often differ because rupture/pore tension depends strongly on the loading rate, which is effectively larger in MD simulations.[91,92] Zero surface tension corresponded to $\alpha = 0.646$ nm^2, close to the experimental value of 0.64 nm^2 for DPPC.[93] In addition, the area compressibility modulus calculated from the tension–area plot was 105 mN/m, in good agreement with the previous simulation value of 107 mN/m for DPPC bilayer at 50°C.[90] Experimentally, compressibility modulus values of 145 mN/m[8] and 234 mN/m[94] were reported for DPPC at 50°C and DMPC at room temperature, respectively. Using an identical force field to the current simulations, Lindahl and Edholm[95] reported a simulated value of 250–300 mN/m for a larger membrane patch (1024 lipids), suggesting that the lower value in the current study is likely due to the finite size effect. Considering the empirical nature of the force field parameters, these results indicate that the simulation methodology is sufficiently accurate in determining the microscopic and macroscopic properties of the lipid bilayer over an extended range of simulated tensions.

Bilayer thickness, defined as the distance between water and lipid density crossover points on either side of the bilayer, was directly computed from the mass density profiles (Figure 2.4).[86] The bilayer thickness decreased linearly with increases in area per lipid, consistent with volume incompressibility. The density profile of the bilayer is highly reminiscent of a confined film rather than a constant density bulk fluid—and thus changes in bilayer thickness are expected to result in structural reorientations within the bilayer.[96] In support of this interpretation, it was observed that increasing the surface area resulted in a decrease of the lipid density at the headgroup region and a concurrent increase in the local density at the midplane of the bilayer (Figure 2.4). This indicates increased interdigitation of the acyl chains of the opposing leaflets due to extension of the chains beyond the bilayer midplane. Increased interdigitation has physiological implications; for example, acyl interdigitation has been proposed to result in the formation of membrane microdomains.[8]

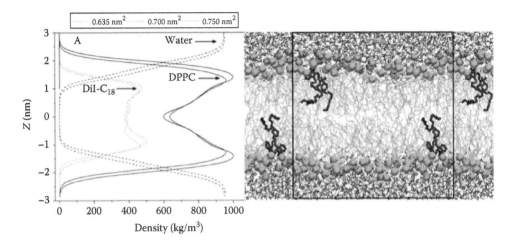

FIGURE 2.4

Mass density profiles of lipid (solid), water (dashed), and DiI-C_{18} (dotted) across the lipid bilayer at selected values of area per lipid (the center of the bilayer was set at $z = 0$; DiI density is at 20× for clarity). A snapshot of the simulation box is also shown. Molecules in the right hand figure are identified in the density profile on the left.

Also, interdigitation of the acyl chains can alter the hydrophobic interactions and lateral pressure profile of the bilayer, which in turn can alter protein conformation.[57,97]

Moderate tension increases lipid lateral diffusion by increasing free area, but free area theory does not hold for large tensions. Lateral diffusion coefficients (D) were computed from the mean-squared displacement (MSD) of the center-of-mass (COM) motion of the molecules. The MSD was ensemble averaged and calculated for multiple time origins, and D was quantified through Einstein's equation:

$$D = \lim_{t \to \infty} \frac{1}{2dt} \left\langle \left[\vec{r_i}(t+t') - \vec{r_i}(t') \right]^2 \right\rangle \tag{2.15}$$

where

   r_i are the x,y positions of the COM of a lipid i at a given time t' and after a time interval t (i.e., at time $t + t'$)

   d is the dimensionality of the motion considered (here, $d = 2$ for the inplane lateral diffusion)

   the brackets denote ensemble average (over molecules and time) and also over multiple time origins t'

The MSDs were corrected for the COM motion of the membrane (i.e., removing any net leaflet translation). MSDs of DPPC at different area-per-lipid values are shown in Figure 2.5. Simulation-measured diffusion coefficient of DPPC at $\alpha = 0.635$ nm² was 8.1×10^{-12} m²/s, which is close to the values obtained using FCS.[83]

Changes in lipid packing are reflected in changes in DiI diffusion and rotation. Experimentally, membrane dynamics are often assessed using measurements of dynamics of fluorescent probe molecules.[79,81–83] Such spectroscopic measurements assume that the probe molecules faithfully reflect lipid dynamics. Interpretation of the obtained data necessitates knowledge of the microenvironment factors such as hydration and viscosity, which are dictated by the location and orientation of the chromophore.

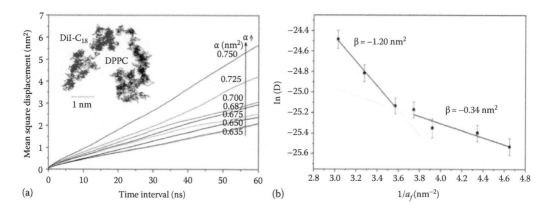

FIGURE 2.5

(a) MSDs of lipid molecules under different tensions. Representative xy-trajectories of DPPC and DiI molecules are shown in the inset ($\alpha = 0.635$ nm^2). (b) The plot of $\ln(D)$ versus $1/a_f$, where two different linear regimes were identified, represented by solid lines, with slopes β. Error bars represent standard errors, $n = 124$. (From Muddana, H.S. et al., *Phys. Chem. Chem. Phys.*, 13, 1368, 2011.)

We then found that fluorescence dynamics of DiI were sensitive to lipid packing and compared DiI dynamics to the native lipid dynamics. The lateral diffusion coefficient of DiI has been shown to be in the same range but slightly lower than that of DPPC.[85] In this study, we could not test the sensitivity of long-time lateral diffusion coefficient of DiI to lipid packing due to lack of sufficient statistics; there exist only two DiI molecules in the simulation box compared to 124 DPPC molecules. Nevertheless, based on the earlier observations, we conclude that the lateral diffusion mechanism of DiI is similar to that of the native lipid and that tension induces increases in DiI diffusion that are quantitatively similar to lipid diffusion.

Key findings from the simulations are as follows: First, physiologically relevant tensions in the range of 0–15 mN/m caused decreases in bilayer thickness in a linear fashion consistent with volume incompressibility. Second, tension induced a significant increase in acyl chain interdigitation and a decrease in lipid order. Third, the observed lateral diffusion coefficient of DPPC cannot be described satisfactorily using the free area theory, across all tensions applied, due to a significant change in molecular shape and friction at high tensions. Finally, DiI has systematically lower lateral and rotational diffusion coefficients compared to DPPC, but the increase in each with tension is quantitatively similar for DiI and DPPC. Similarly, FL of DiI, which depends on lipid order near the headgroups, appears to be a good indicator of tension in membranes.

Experimentally, DiI sensitivity to membrane tension may be revealed in FL measurements. Although the present classical MD simulations cannot simulate fluorescence, which is a quantum mechanical process, they do enable one to assess the local physical factors that govern fluorescence. In general, FL of carbocyanine chromophores is sensitive to water accessibility and to the local microviscosity. Cyanine dyes exhibit weak fluorescence in water and a dramatic increase in quantum yield upon incorporation into lipid membranes.[84] Viscosity-dependent FL of cyanine dyes has been shown to be related to changes in the *trans–cis* photoisomerization dynamics of the central methine bridge.[98,99] Moreover, Packard and Wolf have shown that FL of DiI increases with an increase in the order of the lipid acyl chains.[82]

Thus, we present some preliminary data suggesting that tension can change the diffusion of molecules in the membrane. Giant unilamellar vesicles (>10 μm) were used to assess

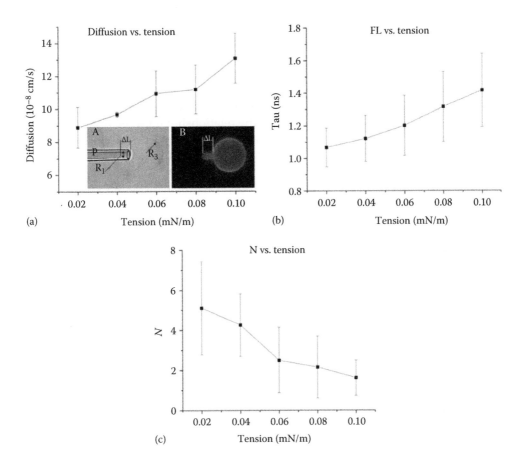

FIGURE 2.6
Membrane tension–induced: (a) increase in diffusion coefficient, (b) increase in FL, and (c) decrease in apparent number of molecules (N) in observation volume (mean ± SD, $n = 4$). Inset in (a) are images of aspirated vesicle.

mechanism of tension-induced reorganization by measuring changes in molecular motion of domain-sensitive fluorescent lipoid dyes using time-correlated single photon counting (TCSPC). Figure 2.6 shows the relationship between tension and measured MD in stressed membrane of uniform composition.

Figure 2.6c, in which the number of molecules in the confocal volume decreases with tension, suggests that the membrane undergoes curvature fluctuations that flatten out with increased tension. The apparent diffusion coefficient should increase if the path of a diffusing molecule is a flat plane rather than over a hilly terrain.[62] According to Equation 2.13, such changes in curvature would thin or thicken the membrane, depending on the initial thickness. If the bilayer were thinned, then the area per lipid would increase according to bilayer incompressibility.[62] If the bilayer were made thicker, then area per lipid would decrease. Such decreases in area per lipid are consistent with an increase in FL (Figure 2.6b) with tension. Decreased area per lipid is expected to provide less opportunity for *trans–cis* isomerization of the DiI chromophore leading to a decrease in the nonradiative decay rate. Such a decrease in nonradiative decay should increase the FL, as is observed. Although more work is needed to understand this trend, it is clear that TCSPC analysis of lipoid dyes in membranes provides an unprecedented window into the dynamics of lipids in membranes under stress.

These results have potential physiological implications. For instance, hydrophobic mismatch between lipids and proteins causes opening and closing of transmembrane stretch-activated ion channels.[44] Altered lipid mobility, due to force-induced changes in lipid packing, can lead to changes in protein molecular mobility and change the kinetics of enzymatic reactions that require protein complex formation (e.g., dimerization).[51,52] Force-induced changes in lipid mobility are also associated with regulation of mitogen-activated protein kinase (MAPK) activity.[6,49] To explain the relationship between lipid mobility and membrane protein-mediated signaling, Nicolau et al.[53] proposed that a local decrease in lipid viscosity, reflected in lipid mobility, temporarily corrals membrane proteins and increases their residence time and interaction kinetics leading to initiation of MAPK signaling pathways once a threshold residence time is reached.[100] Studies on model membranes have demonstrated that membrane tension promotes formation of large domains from microdomains in order to minimize line tension developed at microdomain boundaries,[7,8] and there exists a critical pressure at which lipid phase separation into liquid-ordered and liquid-disordered domains is observed.[101] Taken together, these studies point to changes in bilayer structure and dynamics as a mechanism of force-induced biochemical signaling.

Future research will be needed to develop a new theory for tension–diffusion relationship that takes into account frictional and molecular shape changes. The simulations described here not only provide additional quantitative insights into some of the well-studied bilayer properties (e.g., bilayer thickness, diffusion coefficient) but also lead to novel hypotheses related to membrane-mediated mechanotransduction in cells (e.g., interdigitation) that can be tested experimentally.

In addition, we tested which DiI fluorescence spectroscopic properties have potential as reporters of membrane tension effects on lipids. We observed that although DiI exhibited slower lateral and rotational diffusion compared to DPPC, its lateral and rotational diffusion increased with tension in a manner quantitatively similar to DPPC. This suggests that changes in DiI dynamics are good indicators of membrane tension. We also showed that hydration of the dye does not vary with packing, whereas the local viscosity experienced by the dye changes significantly. These results support the utility of DiI as a reporter of lipid packing and validate the use of DiI to label membrane cellular microdomains based on underlying heterogeneity in lipid order.[80] Thus, these findings offer new insights into the interpretation of fluorescence dynamics of DiI and lipids in lipid bilayer systems.

2.6 Membrane Force Affects Transport of Proteins to Focal Adhesions

To conclude this chapter, we summarize a series of investigations intended to detect the role of the membrane in transduction of shear stress through alterations in the coalescence of lipid rafts and associated protein recruitment to FAs. We first show how these levels of force are indeed sensed by the cells. We then show how FA formation is preceded by lipid raft recruitment, highlighting the role of the membrane in transport of proteins to FAs, and FA activation, as evidence by talin recruitment. Finally, we show, using MD simulations and imaging of phase separation in vesicles, how chemical additives alter line tension in membranes to control phase separation. We also provide preliminary data suggesting these same agents alter FA formation in ECs. Taken together, we propose a line of research in which membrane tension alters curvature leading to clustering of lipid-based proteins (Figure 2.7).

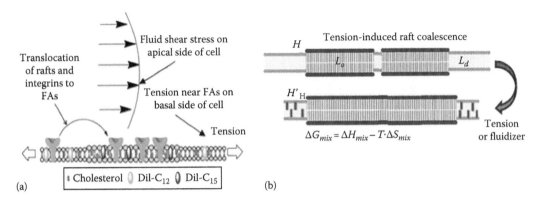

FIGURE 2.7
Hypothesis that membrane tension causes coalescence of membrane microdomains. (a) Fluid shear stress can increase in-plane tension in membranes and influence organization of integrin molecules. (b) Tension may cause coalescence of membrane rafts. Such coalescence can also be induced by molecules that fluidize the liquid-disordered part of the membrane.

We first note that integrin ligation and clustering are major events in vascular tone regulation and shear-induced gene expression. Jalali et al., showed that shear stress caused an increase in new ligand binding of β_1-integrins in and around FAs of ECs plated on fibronectin and an increase in ligand binding of β_3-integrins in ECs plated on vitronectin.[102] In ex vivo arteriolar preparations, activation of the vitronectin receptor, $\alpha_v\beta_3$-integrin, and fibronectin receptor, $\alpha_5\beta_1$-integrin, induced coronary arteriolar dilation by stimulating endothelial production of cyclooxygenase-derived prostaglandins,[103] which dilate blood vessels.[59,104] Thus, integrin–matrix interactions at FAs are required to initiate the signaling pathway leading to shear stress–induced vasodilation and blood pressure regulation.

We also note that integrins are associated with rafts. Lipid rafts are 10–200 nm cholesterol- and sphingomyelin-enriched liquid-ordered (L_o) membrane domains that are involved in signaling and nucleate actin polymerization (reviewed in Ref. [105]) by concentrating phosphatidylinositol 4,5-bisphosphate (PIP$_2$).[19] FAs are cholesterol-rich microdomains, as are caveolae and rafts,[106] and β_1-integrins are required for raft formation[107] and signaling through Rac-1.[108] Wang and colleagues found that Src-activation colocalized with Lyn, a raft marker[109] supporting an emerging picture of rafts as dynamic nanodomains that cluster the necessary critical mass of receptors[110] for downstream signaling of important pathways such as MAPK[111] with timescales of formation of 20 ms and length scales of tens of nanometers.[26] The dynamic formation and dissolution of rafts may be related to the dynamics of membrane bending and protein sorting.[37]

It is believed that rafts can coalesce with force due to enhanced hydrophobic mismatch between liquid-ordered (L_o) and liquid-disordered (L_d) membrane domains. Mismatch of the hydrophobic thickness of various lipids in the membrane bilayer drives aggregation of lipid domains,[28] which, in turn, facilitates segregation or aggregation of membrane proteins.[45] Membrane tension induces raft clustering[26,47] with a time course on the order of seconds.[26] These studies demonstrate that rafts are poised to coalesce at physiological temperatures[31] or with minor alterations in the force landscape (Figure 2.7).

To determine if the forces experienced by sheared ECs rise to sufficient magnitude for protein activation, Ferko et al. developed a 3D mechanical model of an EC, which predicts membrane stress distribution due to fluid flow[112,113] (Figure 2.8). Steady-state shear-induced

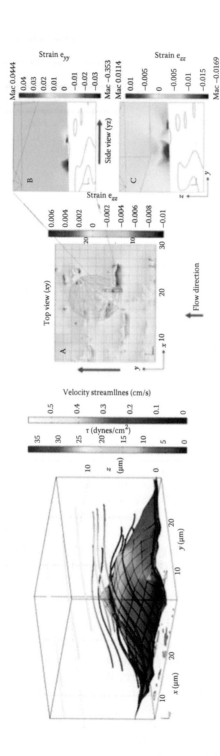

FIGURE 2.8

Shear induces stress concentrations around FAs (with compression upstream and tension downstream) and in areas where there is juxtaposition of stiff organelles and soft cytoplasm (e.g., nucleus). Results suggest two mechanisms of FA growth, downward deformation on the downstream side toward the ECM (negative e_{zz}) leading to new integrin ligation and lateral tension in the downstream side (positive e_{yy}) leading to raft coalescence. (From Ferko, M.C. et al., *Ann. Biomed. Eng.*, 35, 208, 2007.)

stress, strain, and displacement distributions were determined from finite-element stress analysis of a cell-specific, multicomponent elastic continuum model developed from multimodal fluorescence images of confluent EC monolayers and their nuclei. FA locations and areas were determined from quantitative total internal reflection fluorescence (TIRF) microscopy and verified using green fluorescence protein-focal adhesion kinase (GFP-FAK). The model predicted that shear stress induces small heterogeneous ~100 nm deformations of the EC cytoplasm and that strain and stress were amplified 10–100-fold over apical values near FAs with magnitudes sufficient to alter domain line tension[28] and induce domain coalescence.[28,47,114–116]

In order to study the dynamics of lipids in membrane under stress, Gullapalli et al. developed a system for integrated multimodal microscopy, time-resolved fluorescence, and optical-trap rheometry for single-molecule mechanobiology[79] (Figure 2.9). To enable experiments that determine the molecular basis of mechanotransduction over large time and length scales, they constructed a confocal MD microscope. This system integrates TIRF, epifluorescence, differential interference contrast (DIC), and 3D deconvolution with TCSPC instrumentation and an optical trap.

Using this apparatus, Tabouillot et al. measured shear stress–induced modulation of single-molecule diffusion, order (viscosity), and membrane surface topography[117] (Figure 2.10 left and right). From experiments on sheared ECs stained with phase domain-specific DiI-C_{18} and DiI-C_{12}, they found that (1) shear stress induces an early and transient decrease in L_d lifetime and a later and sustained decrease in L_o lifetime (Figure 2.10 left); (2) shear stress induces a rapid increase in number of molecules in DiI-C_{12} domains and a decrease in DiI-C_{18} domains (Figure 2.10 right) likely due to changes in membrane curvature; and (3) shear stress induced an increase in lateral diffusion of DiI-C_{18} but not DiI-C_{12} (not shown). This study demonstrated that L_d and L_o domains are differentially sensitive to fluid shear stress.

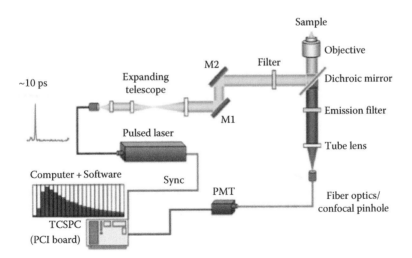

FIGURE 2.9
Confocal MD microscope using TCSPC and pulsed laser excitation: ps pulses of laser light are directed and focused onto cells. Time stamps of laser pulse time and fluorescence photon arrival time are recorded and routed to the TCSPC electronics and analyzed for FL, molecular brightness, and FCS in a computer. An optical trap has been integrated into this setup with a spring constant of 13 pN/μm. (From Gullapalli, R.R. et al., *J. Biomed. Opt.*, 12, 014012, 2007.)

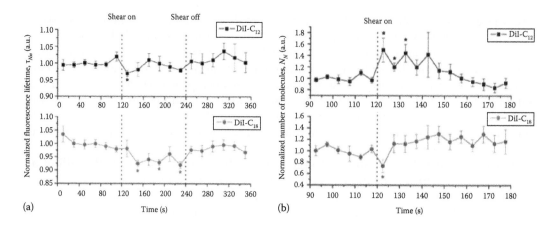

(a)   (b)

FIGURE 2.10

Shear causes a transient change in FL of DiI (a) and change in the number of DiI molecules (b) in confocal volume. (From Tabouillot et al. *Cell Mol Bioeng.* 4, 169, 2011.)

More recently, Fuentes et al. measured membrane-dependent kinetics of GM-1 (rafts), integrin, and talin activation in and around nascent and mature FAs. Recent studies on apical formation of FAs, the effect on basal FAs, and the kinetics of assembly of GM-1, actin, and talin (Figure 2.11) suggest a model of FA assembly under force that is a complex interaction of force transmission, membrane perturbation, protein dynamics, and FA reinforcement.[118–120]

In these studies, mechanical coupling between FAs and GM-1 labeled rafts was discovered as were highly mobile GM-1 suggesting that there are two populations of rafts.[118] Modeling the cell membrane as a 5 nm thick elastic sheet resulted in good agreement between model predictions and experimental measurements of raft displacement thus allowing the determination role of force propagation in raft and protein mobility.

In a recent study, Muddana et al. conducted coarse-grained MD simulations of the additives vitamin E (VE), benzyl alcohol (BA), and Triton X (TX) to uncover the mechanisms by which these additives induce domain formation or disperse them (see Figure 2.12). In that study, it was found that each of the three additives affected one phase or another by changing its thickness. Depending on the original heterogeneity of the membrane, the thickness increased domain separation if there was an increase in mismatch of thicknesses and decrease separation if the thickness of different lipids were brought into close agreement.

In preliminary data, we used these additives to see if they had corresponding effects on FA formation. As shown in Figure 2.13, these additives had similar effect on FA number and size as they did on phase separation in ternary lipid mixture. Such data provide circumstantial evidence that lipid domains are important modulators of FA formation.

The hypothesis that tension in membrane phases plays a role in FA formation is consistent with an emerging view of FAs as cholesterol-rich, liquid-ordered domains,[106] much like caveolae and lipid rafts, which are stabilized by integrin attachment to the cytoskeleton and extracellular matrix. While there exist models of FA assembly in response to force, none take into account lipid control of this process, and there have not been methods to measure membrane tension or methods to delineate kinetics of assembly with and without force. These techniques are necessary to understand how lipid domains control nascent FA formation, a prerequisite for FA-mediated mechanosensing. Despite their importance,

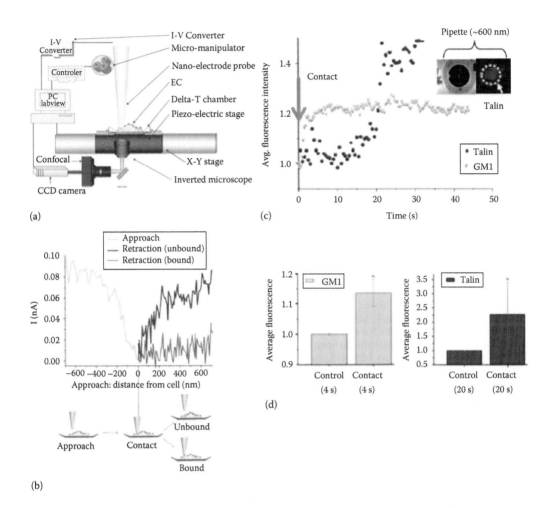

FIGURE 2.11
(a) Pipette is functionalized with fibronectin and brought close to the cell using computer control. (b) Distance from the cell and timing of contact are determined from electronic signature. (c) High-speed confocal microscopy images at the point of pipette contact with the cell. *Inset:* Electron micrograph of 600 nm pipette tip shows conducting fiber, also red fluorescent protein (RFP) talin at tip. (d) Increase in GM-1 (membrane raft marker and talin, indicator of integrin activation). (From Fuentes, D.E. et al., *Cell. Mol. Bioeng.*, 4, 616, 2011; Fuentes, D.E. and Butler, P.J., *Cell Mol. Bioeng.*, 5, 143, 2012.)

determining membrane stresses in vivo (in cells) has been hampered by the membrane's complex interaction between its fluid nature and dynamic constituents (e.g., proteins and lipids) and its solid-like ability to support tension, deform, and recoil. It may now be possible to measure membrane tension using time-resolved spectroscopy of DiI, a lipoid dye that intercalates into cell membrane domains in a chain-length-specific manner. Also, tools based on ion-conductance spectroscopy[121] can now be used to initiate nascent FAs at zero force and to measure local kinetics of lipid raft, talin, and actin accumulation using high-speed confocal microscopy.[118,119] Finally, manipulation of the hydrophobic thickness of liquid-disordered domains by nonlipid amphiphiles can be used to tune phase separation (e.g., lipid raft formation and coalescence) in model membranes.[122] Combined with previous studies on force distribution in sheared and focally adhered ECs, it may

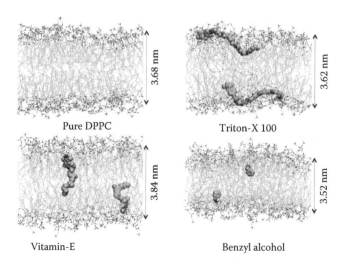

FIGURE 2.12

Left: Compared to pure DPPC bilayer, VE thickens and TX and BA thin the liquid-disordered domain. With respect to Figure 2.13, thickening L_d (red) region with VE abolishes phase separation. Thinning L_d with TX and BA restores phase separation. Mechanism of phase separation is by hydrophobic mismatch between L_o and L_d phases. Thus, MD simulations provide quantitative insight into effects of nonlipid amphiphiles on L_d control of phase separation. (From Muddana, H.S. et al., *Biophys. J.*, 102, 489, 2012.)

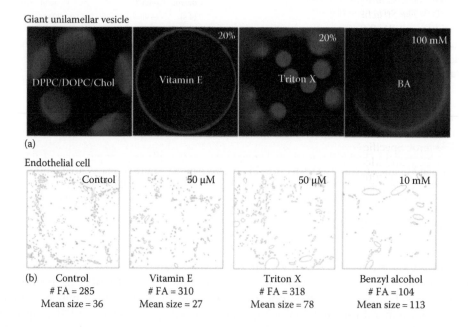

FIGURE 2.13

(a) In vesicles with L_d (red) and L_o (black) coexistence, VE disperses domains by making the L_d domain match the thickness of the L_o domain; TX induces raft formation by decreasing thickness of L_d and increasing hydrophobic mismatch; BA more strongly induces domain formation by further reducing L_d thickness (see Figure 2.12 for thickness changes).[122] (b) Cells were transfected with GFP focal adhesion kinase (FAK) and imaged under TIRF (images of single cells were thresholded and analyzed for FA size and number). Consistent with vesicles, in cells, VE increased FA number and decreased size, TX increased FA number and increased size, and BA decreased FA number and increased size (sizes are in pixels). Scale bar is 10 µm. (From Butler, P.J., unpublished data.)

now be possible to uncover the relationship between force and membrane control of FA assembly and attendant signaling. Such studies may uncover the nature of lipid control of FA function and identify the lipid bilayer as a new target for biophysical regulation of mechanotransduction.

References

1. Alberts, B., Johnson, A., Lewis, J., Raff, M., Roberts, K., and Walter, P. *Molecular Biology of the Cell*. Garland Science, New York, 2002.
2. Singer, S. J. and Nicolson, G. L. The fluid mosaic model of the structure of cell membranes. *Science* **175**, 720–731 (1972).
3. Simons, K. and Toomre, D. Lipid rafts and signal transduction. *Nat. Rev. Mol. Cell Biol.* **1**, 31–39 (2000).
4. Jacobson, K., Mouritsen, O. G., and Anderson, R. G. W. Lipid rafts: At a crossroad between cell biology and physics. *Nat. Cell Biol.* **9**, 7–14 (2007).
5. Edidin, M. Lipids on the frontier: A century of cell-membrane bilayers. *Nat. Rev. Mol. Cell Biol.* **4**, 414–418 (2003).
6. Mollinedo, F. et al. Lipid raft-targeted therapy in multiple myeloma. *Oncogene* **29**, 3748–3757 (2010).
7. Nelson, D. L. *Lehninger Principles of Biochemistry*. W.H. Freeman & Company, New York, 2008.
8. Lazar, T. *The Structure of Biological Membranes*, 2nd edn., P. Yeagle, ed. CRC Press, Boca Raton, FL, 540pp., ISBN 0-8493-1403-8 (2004). *Cell Biochem. Funct.* **23**, 294–295 (2005).
9. Burger, K., Gimpl, G., and Fahrenholz, F. Regulation of receptor function by cholesterol. *Cell. Mol. Life Sci.* **57**, 1577–1592 (2000).
10. Hanson, M. A. et al. A specific cholesterol binding site is established by the 2.8 A structure of the human beta2-adrenergic receptor. *Structure* **16**, 897–905 (2008).
11. Paila, Y. D., Tiwari, S., and Chattopadhyay, A. Are specific nonannular cholesterol binding sites present in G-protein coupled receptors? *Biochim. Biophys. Acta* **1788**, 295–302 (2009).
12. Paila, Y. D. and Chattopadhyay, A. The function of G-protein coupled receptors and membrane cholesterol: Specific or general interaction? *Glycoconj. J.* **26**, 711–720 (2009).
13. Jones, O. T. and McNamee, M. G. Annular and nonannular binding sites for cholesterol associated with the nicotinic acetylcholine receptor. *Biochemistry* **27**, 2364–2374 (1988).
14. Singh, D. K., Rosenhouse-Dantsker, A., Nichols, C. G., Enkvetchakul, D., and Levitan, I. Direct regulation of prokaryotic Kir channel by cholesterol. *J. Biol. Chem.* **284**, 30727–30736 (2009).
15. Singh, D. K., Shentu, T.-P., Enkvetchakul, D., and Levitan, I. Cholesterol regulates prokaryotic Kir channel by direct binding to channel protein. *Biochim. Biophys. Acta* **1808**, 2527–2533 (2011).
16. Dopico, A. M., Bukiya, A. N., and Singh, A. K. Large conductance, calcium- and voltage-gated potassium (BK) channels: Regulation by cholesterol. *Pharmacol. Ther.* **135**, 133–150 (2012).
17. Levitan, I., Christian, A. E., Tulenko, T. N., and Rothblat, G. H. Membrane cholesterol content modulates activation of volume-regulated anion current in bovine endothelial cells. *J. Gen. Physiol.* **115**, 405–416 (2000).
18. Romanenko, V. G., Rothblat, G. H., and Levitan, I. Sensitivity of volume-regulated anion current to cholesterol structural analogues. *J. Gen. Physiol.* **123**, 77–87 (2004).
19. Kwik, J. et al. Membrane cholesterol, lateral mobility, and the phosphatidylinositol 4,5-bisphosphate-dependent organization of cell actin. *Proc. Natl Acad. Sci. USA* **100**, 13964–13969 (2003).
20. Devaux, P. F. and Morris, R. Transmembrane asymmetry and lateral domains in biological membranes. *Traffic* **5**, 241–246 (2004).

21. Van Meer, G., Voelker, D. R., and Feigenson, G. W. Membrane lipids: Where they are and how they behave. *Nat. Rev. Mol. Cell Biol.* **9**, 112–124 (2008).
22. Tokumasu, F., Jin, A. J., Feigenson, G. W., and Dvorak, J. A. Nanoscopic lipid domain dynamics revealed by atomic force microscopy. *Biophys. J.* **84**, 2609–2618 (2003).
23. Lipowsky, R. and Sackmann, E. *Structure and Dynamics of Membranes: From Cells to Vesicles.* Elsevier, Amsterdam, the Netherlands, p. 519, 1995.
24. Sankaram, M. B. and Thompson, T. E. Cholesterol-induced fluid-phase immiscibility in membranes. *Proc. Natl Acad. Sci. USA* **88**, 8686–8690 (1991).
25. Veatch, S. L. and Keller, S. L. Separation of liquid phases in giant vesicles of ternary mixtures of phospholipids and cholesterol. *Biophys. J.* **85**, 3074–3083 (2003).
26. Eggeling, C. et al. Direct observation of the nanoscale dynamics of membrane lipids in a living cell. *Nature* **457**, 1159–1162 (2009).
27. Sahl, S. J., Leutenegger, M., Hilbert, M., Hell, S. W., and Eggeling, C. Fast molecular tracking maps nanoscale dynamics of plasma membrane lipids. *Proc. Natl Acad. Sci. USA* **107**, 6829–6834 (2010).
28. Baumgart, T., Hess, S. T., and Webb, W. W. Imaging coexisting fluid domains in biomembrane models coupling curvature and line tension. *Nature* **425**, 821–824 (2003).
29. Baumgart, T., Hunt, G., Farkas, E. R., Webb, W. W., and Feigenson, G. W. Fluorescence probe partitioning between L_o/L_d phases in lipid membranes. *Biochim. Biophys. Acta* **1768**, 2182–2194 (2007).
30. Baumgart, T. et al. Large-scale fluid/fluid phase separation of proteins and lipids in giant plasma membrane vesicles. *Proc. Natl Acad. Sci. USA* **104**, 3165–3170 (2007).
31. Lingwood, D., Ries, J., Schwille, P., and Simons, K. Plasma membranes are poised for activation of raft phase coalescence at physiological temperature. *Proc. Natl Acad. Sci. USA* **105**, 10005–10010 (2008).
32. Deserno, M. Fluid lipid membranes – A primer, pp. 1–29. At http://www.cmu.edu/biolphys/deserno/pdf/membrane_theory.pdf. Accessed June 26, 2014.
33. Safran, S. A. *Statistical Thermodynamics of Surfaces, Interfaces, and Membranes.* Westview Press, Boulder, CO, 284pp., 1994.
34. Leckband, D. and Israelachvili, J. Intermolecular forces in biology. *Q. Rev. Biophys.* **34**, 105–267 (2001).
35. Helfrich, W. Elastic properties of lipid bilayers: Theory and possible experiments. *Z. Naturforsch* **28**, 693–703 (1973).
36. Landau, L. D., Pitaevskii, L. P., Lifshitz, E. M., and Kosevich, A. M. *Theory of Elasticity*, 3rd edn., Vol. 7 (*Theoretical Physics*). Butterworth-Heinemann, Oxford, U.K., p. 195, 1986.
37. Heinrich, M., Tian, A., Esposito, C., and Baumgart, T. Dynamic sorting of lipids and proteins in membrane tubes with a moving phase boundary. *Proc. Natl Acad. Sci. USA* **107**, 7208–7213 (2010).
38. Lee, A. G. Lipid-protein interactions in biological membranes: A structural perspective. *Biochim. Biophys. Acta-Biomembranes* **1612**, 1–40 (2003).
39. Andersen, O. S. and Koeppe, R. E. Bilayer thickness and membrane protein function: An energetic perspective. *Annu. Rev. Biophys. Biomol. Struct.* **36**, 107–130 (2007).
40. Killian, J. A. Hydrophobic mismatch between proteins and lipids in membranes. *Biochim. Biophys. Acta* **1376**, 401–415 (1998).
41. Fattal, D. R. and Ben-Shaul, A. A molecular model for lipid-protein interaction in membranes: The role of hydrophobic mismatch. *Biophys. J.* **65**, 1795–1809 (1993).
42. Kandasamy, S. K. and Larson, R. G. Molecular dynamics simulations of model transmembrane peptides in lipid bilayers: A systematic investigation of hydrophobic mismatch. *Biophys. J.* **90**, 2326–2343 (2006).
43. Park, S. H. and Opella, S. J. Tilt angle of a trans-membrane helix is determined by hydrophobic mismatch. *J. Mol. Biol.* **350**, 310–318 (2005).
44. Martinac, B. Mechanosensitive ion channels: Molecules of mechanotransduction. *J. Cell Sci.* **117**, 2449–2460 (2004).

45. Botelho, A. V., Huber, T., Sakmar, T. P., and Brown, M. F. Curvature and hydrophobic forces drive oligomerization and modulate activity of rhodopsin in membranes. *Biophys. J.* **91**, 4464–4477 (2006).
46. Periole, X., Huber, T., Marrink, S.-J. J., and Sakmar, T. P. G protein-coupled receptors self-assemble in dynamics simulations of model bilayers. *Biophys. J.* **129**, 10126–10132 (2007).
47. Ayuyan, A. G. and Cohen, F. S. Raft composition at physiological temperature and pH in the absence of detergents. *Biophys. J.* **94**, 2654–2666 (2008).
48. Butler, P. J., Norwich, G., Weinbaum, S., and Chien, S. Shear stress induces a time- and position-dependent increase in endothelial cell membrane fluidity. *Am. J. Physiol. Cell Physiol.* **280**, C962–C969 (2001).
49. Butler, P. J., Tsou, T.-C. C., Li, J. Y.-S., Usami, S., and Chien, S. Rate sensitivity of shear-induced changes in the lateral diffusion of endothelial cell membrane lipids: A role for membrane perturbation in shear-induced MAPK activation. *FASEB J.* **16**, 216–218 (2002).
50. Haidekker, M. A., L'Heureux, N., and Frangos, J. A. Fluid shear stress increases membrane fluidity in endothelial cells: A study with DCVJ fluorescence. *Am. J. Physiol. Heart Circ. Physiol.* **278**, H1401–H1406 (2000).
51. Axelrod, D. Lateral motion of membrane proteins and biological function. *J. Membr. Biol.* **75**, 1–10 (1983).
52. Schreiber, G. Kinetic studies of protein–protein interactions. *Curr. Opin. Struct. Biol.* **12**, 41–47 (2002).
53. Nicolau, D. V., Burrage, K., Parton, R. G., and Hancock, J. F. Identifying optimal lipid raft characteristics required to promote nanoscale protein-protein interactions on the plasma membrane. *Mol. Cell. Biol.* **26**, 313–323 (2006).
54. Goldstein, D. B. The effects of drugs on membrane fluidity. *Annu. Rev. Pharmacol. Toxicol.* **24**, 43–64 (1984).
55. Gullingsrud, J. and Schulten, K. Lipid bilayer pressure profiles and mechanosensitive channel gating. *Biophys. J.* **86**, 3496–3509 (2004).
56. McMahon, H. T. and Gallop, J. L. Membrane curvature and mechanisms of dynamic cell membrane remodelling. *Nature* **438**, 590–596 (2005).
57. Cantor, R. S. Lateral pressures in cell membranes: A mechanism for modulation of protein function. *J. Phys. Chem. B* **101**, 1723–1725 (1997).
58. Tian, A. and Baumgart, T. Sorting of lipids and proteins in membrane curvature gradients. *Biophys. J.* **96**, 2676–2688 (2009).
59. Butler, P. J., Weinbaum, S., Chien, S., and Lemons, D. E. Endothelium-dependent, shear-induced vasodilation is rate-sensitive. *Microcirculation* **7**, 53–65 (2000).
60. Fung, Y. C. and Liu, S. Q. Elementary mechanics of the endothelium of blood vessels. *J. Biomech. Eng.* **115**, 1–12 (1993).
61. Needham, D. and Nunn, R. S. Elastic deformation and failure of lipid bilayer membranes containing cholesterol. *Biophys. J.* **58**, 997–1009 (1990).
62. Gov, N. Diffusion in curved fluid membranes. *Phys. Rev. E* **73**, 041918 (2006).
63. Wiggins, P. and Phillips, R. Analytic models for mechanotransduction: Gating a mechanosensitive channel. *Proc. Natl Acad. Sci. USA* **101**, 4071–4076 (2004).
64. Helfrich, P. and Jakobsson, E. Calculation of deformation energies and conformations in lipid membranes containing gramicidin channels. *Biophys. J.* **57**, 1075–1084 (1990).
65. Lundbaek, J. A., Collingwood, S. A., Ingólfsson, H. I., Kapoor, R., and Andersen, O. S. Lipid bilayer regulation of membrane protein function: Gramicidin channels as molecular force probes. *J. R. Soc. Interface* **7**, 373–395 (2010).
66. Andersen, O. S., Koeppe, R. E., and Roux, B. Gramicidin channels. *IEEE Trans. Nanobiosci.* **4**, 10–20 (2005).
67. Huang, H. W. Deformation free energy of bilayer membrane and its effect on gramicidin channel lifetime. *Biophys. J.* **50**, 1061–1070 (1986).
68. Boal, D. H. *Mechanics of the Cell*. Cambridge University Press, Cambridge, MA, 406pp., 2002.

69. Chernyshev, A. and Cukierman, S. Thermodynamic view of activation energies of proton transfer in various gramicidin A channels. *Biophys. J.* **82**, 182–192 (2002).
70. Schmid-Schönbein, G. W., Kosawada, T., Skalak, R., and Chien, S. Membrane model of endothelial cells and leukocytes. A proposal for the origin of a cortical stress. *J. Biomech. Eng.* **117**, 171–178 (1995).
71. Huang, H., Kamm, R. D., and Lee, R. T. Cell mechanics and mechanotransduction: Pathways, probes, and physiology. *Am. J. Physiol. Cell Physiol.* **287**, C1–C11 (2004).
72. Bao, X., Lu, C., and Frangos, J. A. Mechanism of temporal gradients in shear-induced ERK1/2 activation and proliferation in endothelial cells. *Am. J. Physiol. Heart Circ. Physiol.* **281**, H22–H29 (2001).
73. Chachisvilis, M., Zhang, Y.-L., and Frangos, J. A. G protein-coupled receptors sense fluid shear stress in endothelial cells. *Proc. Natl Acad. Sci. USA* **103**, 15463–15468 (2006).
74. Rawicz, W., Smith, B. A., McIntosh, T. J., Simon, S. A., and Evans, E. Elasticity, strength, and water permeability of bilayers that contain raft microdomain-forming lipids. *Biophys. J.* **94**, 4725–4736 (2008).
75. Olbrich, K., Rawicz, W., Needham, D., and Evans, E. Water permeability and mechanical strength of polyunsaturated lipid bilayers. *Biophys. J.* **79**, 321–327 (2000).
76. Garcia-Saez, A. J., Chiantia, S., and Schwille, P. Effect of line tension on the lateral organization of lipid membranes. *J. Biol. Chem.* **282**, 33537–33544 (2007).
77. De Monvel, J. B., Brownell, W. E., and Ulfendahl, M. Lateral diffusion anisotropy and membrane lipid/skeleton interaction in outer hair cells. *Biophys. J.* **91**, 364–381 (2006).
78. Oghalai, J. S., Zhao, H. B., Kutz, J. W., and Brownell, W. E. Voltage- and tension-dependent lipid mobility in the outer hair cell plasma membrane. *Science* **287**, 658–661 (2000).
79. Gullapalli, R. R., Tabouillot, T., Mathura, R., Dangaria, J. H., and Butler, P. J. Integrated multimodal microscopy, time-resolved fluorescence, and optical-trap rheometry: Toward single molecule mechanobiology. *J. Biomed. Opt.* **12**, 014012 (2007).
80. Ariola, F. S., Li, Z., Cornejo, C., Bittman, R., and Heikal, A. A. Membrane fluidity and lipid order in ternary giant unilamellar vesicles using a new bodipy-cholesterol derivative. *Biophys. J.* **96**, 2696–2708 (2009).
81. De Almeida, R. F. M., Loura, L. M. S., and Prieto, M. Membrane lipid domains and rafts: Current applications of fluorescence lifetime spectroscopy and imaging. *Chem. Phys. Lipids* **157**, 61–77 (2009).
82. Packard, B. S. and Wolf, D. E. Fluorescence lifetimes of carbocyanine lipid analogues in phospholipid bilayers. *Biochemistry* **24**, 5176–5181 (1985).
83. Kahya, N., Scherfeld, D., Bacia, K., and Schwille, P. Lipid domain formation and dynamics in giant unilamellar vesicles explored by fluorescence correlation spectroscopy. *J. Struct. Biol.* **147**, 77–89 (2004).
84. Nakashima, N. and Kunitake, T. Drastic fluorescence enhancement of cyanine dyes bound to synthetic bilayer membranes. Its high sensitivity to the chemical structure and the physical state of the membrane. *J. Am. Chem. Soc.* **104**, 4261–4262 (1982).
85. Gullapalli, R. R., Demirel, M. C., and Butler, P. J. Molecular dynamics simulations of DiI-C18(3) in a DPPC lipid bilayer. *Phys. Chem. Chem. Phys.* **10**, 3548–3560 (2008).
86. Repáková, J., Čapková, P., Holopainen, J. M., and Vattulainen, I. Distribution, orientation, and dynamics of DPH probes in DPPC bilayer. *J. Phys. Chem. B* **108**, 13438–13448 (2004).
87. Boggara, M. B. and Krishnamoorti, R. Partitioning of nonsteroidal antiinflammatory drugs in lipid membranes: A molecular dynamics simulation study. *Biophys. J.* **98**, 586–595 (2010).
88. Boggara, M. B. and Krishnamoorti, R. Small-angle neutron scattering studies of phospholipid-NSAID adducts. *Langmuir* **26**, 5734–5745 (2010).
89. Muddana, H. S., Gullapalli, R. R., Manias, E., and Butler, P. J. Atomistic simulation of lipid and DiI dynamics in membrane bilayers under tension. *Phys. Chem. Chem. Phys.* **13**, 1368–1378 (2011).
90. Feller, S. E. and Pastor, R. W. Constant surface tension simulations of lipid bilayers: The sensitivity of surface areas and compressibilities. *J. Chem. Phys.* **111**, 1281 (1999).

91. Tieleman, D. P., Leontiadou, H., Mark, A. E., and Marrink, S.-J. Simulation of pore formation in lipid bilayers by mechanical stress and electric fields. *J. Am. Chem. Soc.* **125**, 6382–6383 (2003).

92. Leontiadou, H., Mark, A. E., and Marrink, S. J. Molecular dynamics simulations of hydrophilic pores in lipid bilayers. *Biophys. J.* **86**, 2156–2164 (2004).

93. Nagle, J. F. and Tristram-Nagle, S. Structure of lipid bilayers. *Biochim. Biophys. Acta-Rev. Biomembr.* **1469**, 159–195 (2000).

94. Rawicz, W., Olbrich, K. C., McIntosh, T., Needham, D., and Evans, E. Effect of chain length and unsaturation on elasticity of lipid bilayers. *Biophys. J.* **79**, 328–339 (2000).

95. Lindahl, E. and Edholm, O. Mesoscopic undulations and thickness fluctuations in lipid bilayers from molecular dynamics simulations. *Biophys. J.* **79**, 426–433 (2000).

96. Manias, E., Hadziioannou, G., and ten Brinke, G. Inhomogeneities in sheared ultrathin lubricating films. *Langmuir* **12**, 4587–4593 (1996).

97. Patra, M. Lateral pressure profiles in cholesterol-DPPC bilayers. *Eur. Biophys. J.* **35**, 79–88 (2005).

98. Muddana, H. S., Morgan, T. T., Adair, J. H., and Butler, P. J. Photophysics of Cy3-encapsulated calcium phosphate nanoparticles. *Nano Lett.* **9**, 1559–1566 (2009).

99. Widengren, J. and Schwille, P. Characterization of photoinduced isomerization and back-isomerization of the cyanine dye Cy5 by fluorescence correlation spectroscopy. *J. Phys. Chem. A* **104**, 6416–6428 (2000).

100. Tian, T. et al. Plasma membrane nanoswitches generate high-fidelity Ras signal transduction. *Nat. Cell Biol.* **9**, 905–914 (2007).

101. Keller, S. L., Anderson, T. G., and McConnell, H. M. Miscibility critical pressures in monolayers of ternary lipid mixtures. *Biophys. J.* **79**, 2033–2042 (2000).

102. Jalali, S. et al. Integrin-mediated mechanotransduction requires its dynamic interaction with specific extracellular matrix (ECM) ligands. *Proc. Natl Acad. Sci. USA* **98**, 1042–1046 (2001).

103. Hein, T. W. et al. Integrin-binding peptides containing RGD produce coronary arteriolar dilation via cyclooxygenase activation. *Am. J. Physiol. Heart Circ. Physiol.* **281**, H2378–H2384 (2001).

104. Frame, M. D., Rivers, R. J., Altland, O., and Cameron, S. Mechanisms initiating integrin-stimulated flow recruitment in arteriolar networks. *J. Appl. Physiol.* **102**, 2279–2287 (2007).

105. Levitan, I. and Gooch, K. J. Lipid rafts in membrane-cytoskeleton interactions and control of cellular biomechanics: Actions of oxLDL. *Antioxid. Redox Signal.* **9**, 1519–1534 (2007).

106. Del Pozo, M. A. and Schwartz, M. A. Rac, membrane heterogeneity, caveolin and regulation of growth by integrins. *Trends Cell Biol.* **17**, 246–250 (2007).

107. Singh, R. D. et al. Gangliosides and beta1-integrin are required for caveolae and membrane domains. *Traffic* **11**, 348–360 (2010).

108. Del Pozo, M. A. et al. Integrins regulate Rac targeting by internalization of membrane domains. *Science* **303**, 839–842 (2004).

109. Lu, S. et al. The spatiotemporal pattern of Src activation at lipid rafts revealed by diffusion-corrected FRET imaging. *PLoS Comput. Biol.* **4**, e1000127 (2008).

110. Van Zanten, T. S. et al. Hotspots of GPI-anchored proteins and integrin nanoclusters function as nucleation sites for cell adhesion. *Proc. Natl Acad. Sci. USA* **106**, 18557–18562 (2009).

111. Rotblat, B. et al. H-Ras nanocluster stability regulates the magnitude of MAPK signal output. *PLoS One* **5**, 1–6 (2010).

112. Ferko, M. C., Bhatnagar, A., Garcia, M. B., and Butler, P. J. Finite-element stress analysis of a multicomponent model of sheared and focally-adhered endothelial cells. *Ann. Biomed. Eng.* **35**, 208–223 (2007).

113. Ferko, M. C., Patterson, B. W., and Butler, P. J. High-resolution solid modeling of biological samples imaged with 3D fluorescence microscopy. *Microsc. Res. Tech.* **69**, 648–655 (2006).

114. Akimov, S. A., Kuzmin, P. I., Zimmerberg, J., and Cohen, F. S. Lateral tension increases the line tension between two domains in a lipid bilayer membrane. *Phys. Rev. E. Stat. Nonlin. Soft Matter Phys.* **75**, 011919 (2007).

115. Méléard, P., Bagatolli, L. A., and Pott, T. Giant unilamellar vesicle electroformation from lipid mixtures to native membranes under physiological conditions. *Methods Enzymol.* **465**, 161–176 (2009).

116. Heberle, F. A. et al. Bilayer thickness mismatch controls domain size in model membranes. *J. Am. Chem. Soc.* **135**, 6853–6859 (2013).
117. Tabouillot, T., Muddana, H. S., and Butler, P. J. Endothelial cell membrane sensitivity to shear stress is lipid domain dependent. *Cell. Mol. Bioeng.* **4**, 169–181 (2011).
118. Fuentes, D. E., Bae, C., and Butler, P. J. Focal adhesion induction at the tip of a functionalized nanoelectrode. *Cell. Mol. Bioeng.* **4**, 616–626 (2011).
119. Fuentes, D. E. and Butler, P. J. Coordinated mechanosensitivity of membrane rafts and focal adhesions. *Cell. Mol. Bioeng.* **5**, 143–154 (2012).
120. Bae, C. and Butler, P. J. Automated single-cell electroporation. *Biotechniques* **41**, 399–402 (2006).
121. Hansma, P. K., Drake, B., Marti, O., Gould, S. A., and Prater, C. B. The scanning ion-conductance microscope. *Science* **243**, 641–643 (1989).
122. Muddana, H. S., Chiang, H. H., and Butler, P. J. Tuning membrane phase separation using non-lipid amphiphiles. *Biophys. J.* **102**, 489–497 (2012).

3

Cellular Reconstitution of Actively Self-Organizing Systems

Orit Siton-Mendelson, Barak Gilboa, Yaron Ideses, and Anne Bernheim-Groswasser

CONTENTS

3.1 Cell Cytoskeleton

Living cells are extremely sophisticated devices that detect specific environmental signals, process this information, and generate specific mechanical responses, such as growth, shape change, or directed movement. The active part of the biodevice is the cell cytoskeleton, a spatially extended network (gel), self-organized, mechanochemical machine that forms via the nucleation and multiscale self-organization of biomolecules (e.g., biopolymers such as filamentous actin [F-actin], microtubules [MTs], accessory proteins, and molecular motors [1,2]), in both the temporal and spatial domains. The cytoskeleton determines the mechanical properties of a cell and plays important roles in many cellular processes, such as division [3–5], motility [6], adhesion [7], and tissue morphogenesis. The multiscale nature of the cytoskeleton enables response times ranging from fast dynamics for individual molecular-sized building blocks to the persistent motion or shape change of whole cells over minutes and hours, well beyond the time range of man-made analogues.

The *activity* of the cell cytoskeleton is essential for accomplishing the routine tasks that a cell has to cope with. *Active* refers to cytoskeletal processes driven by the hydrolysis of adenosine triphosphate (ATP). This activity drives the cytoskeleton network to be intrinsically out of thermodynamic equilibrium. An additional difference between the cytoskeleton and traditional gels lies in the structural polarity of the cytoskeletal filaments. The activity combined with polarity of these filaments produces a host of dynamic phenomena that do not exist in traditional physical gels. These include filament treadmilling (a process in which filaments elongate at one end and shorten at the other end at the same rate) [8] and the generation of movement and contractile stresses by molecular motors [9].

3.1.1 Actin Cytoskeleton and Its Associated Proteins

This chapter focuses on cytoskeletal processes driven by actin polymerization dynamics and reorganization by motor proteins, which is discussed in great detail. Processes associated with MT dynamics and their reorganization by motor proteins appear in Section 3.1.2.

3.1.1.1 *Actin Nucleation and Polymerization Factors*

3.1.1.1.1 *Mechanism of Actin Polymerization*

Actin is a globular (and polar) protein composed of a single polypeptide chain of 375 amino acids and with a molecular weight of ~42 kDa. Each actin monomer (G-actin) contains nucleotide and monovalent and divalent cation-binding sites. Under appropriate conditions, polymerization of the monomers into a two-stranded right-handed helical structure, with a pitch of 36–39 nm, is induced. The filament thickness is 8 nm [10] and its persistence length ranges between 8 and 17 μm [11,12]. Actin polymerization drives several cellular processes such as locomotion, cytokinesis, and adhesion. Forces developed during polymerization are responsible for different forms of cell motility and, in particular, extension of cellular protrusions [13]. Actin polymerization is energetically favorable above a critical actin monomer concentration C^*. Due to the asymmetric (i.e., polar) structure of actin monomers, an actin filament is polar, with two ends having different affinities, different binding constants for monomeric actin, different critical concentrations, as well as different polymerization or depolymerization rates [14]. In addition, the polymerization process is also accompanied by the hydrolysis of the ATP-bound nucleotide along the filament. Altogether, the result is that the barbed end (or *plus* end) of an ATP-bound actin filament (ATP–F-actin) polymerizes faster and has higher binding constants for ATP–G-actin (ATP-bound actin monomer) than the adenosine diphosphate (ADP)-bound actin filament's (ADP–F-actin) pointed end (or *minus* end). At steady state, the rate of polymerization of ATP–G-actin at the barbed end is equal to the depolymerization of ADP–G-actin units from the pointed end. This dynamic state is referred to as *treadmilling* [15]. The ADP–G-actin monomers leaving the pointed end exchange their ADP with ATP present in excess in the solution, refilling the pool of ATP–G-actin monomers. Therefore, at steady state, the concentrations of ATP–G-actin monomers (referred to as C_{ss}) and actin filaments remain constant. Moreover, the filament's length is also kept constant.

In vivo, actin does not function alone but rather interacts with a large number of actin-binding proteins (ABPs). Each subcellular structure is maintained by a specific set of ABPs, which also control its dynamic properties. Under these conditions, the time and length scale of the system corresponds to the assembly/disassembly dynamics of the overall structure, which in many cases do not correspond to the treadmilling of the individual filaments themselves.

3.1.1.1.2 *Actin Nucleators*

While actin polymerization is favorable energetically, nucleation of actin filaments is not; thus, the rate-limiting step for filament polymerization is nucleation. This fact is used by the cell to prevent unwanted (i.e., uncontrolled) polymerization of actin. The regulation of de novo nucleation of actin filaments is done using specific actin nucleators. These proteins enable the cell to control the temporal and spatial (i.e., location) polymerization of their cytoskeletal networks, required for establishment of different cellular processes. The filament nucleators work by different mechanisms to accomplish their tasks, stabilizing small actin oligomers along either the long-pitch helix or the short-pitch helix of the actin filament. So far, two major classes of actin nucleators are known to nucleate actin filaments in vivo: nucleation-promoting factors (NPFs)–Arp2/3 complex [14,16] and the formin homology proteins [16–19]. Recently, new filament nucleators have been discovered, including Spire [20], Cordon bleu (Cobl), *Vibrio parahaemolyticus* and *Vibrio cholerae* (VopL/VopF) factors, leiomodin (Lmob), and junction-mediating regulatory protein (JMY) (reviewed in Refs. [21–23]). With exception of formins, all these nucleators use a Wiskott–Aldrich syndrome protein (WASP) homology 2 (WH2 or W) domain for interaction with actin monomers [24]. Although not directly associated with any class of actin nucleators, enabled/vasodilator-stimulated phosphoproteins (Ena/VASPs) are also WH2-based proteins and function as elongation factors and bundling proteins [25,26].

Each of the aforementioned proteins acts according to a different biochemical mechanism of actin nucleation, resulting in diverse actin assembly-based structures. Arp2/3 complex is activated by NPFs to nucleate branched actin networks. NPFs include the ActA of the bacterium *Listeria monocytogenes* [27], Wiskott-Aldrich syndrome protein (WASP), its brain homologue (N-WASP), three WASP and verprolin homologues (WAVEs), and the more recently identified WASP homologue associated with actin, membranes and MT (WHAMM), WASP and Scar homologue (WASHs), and JMY protein [28]. Cortactin, another known NPF, functioning in cellular motility and cancer invasiveness (reviewed in Refs. [29–31]), is considered a weak NPF compared to WASP and WASP homologues [32].

WASP, N-WASP, Scar/WAVE, and other NPFs usually contain a VCA domain. The VCA domain brings together the Arp2/3 complex and the first actin subunit in the new filament (the daughter filament). The VCA domain consists of a V motif (verprolin homology or WH2) that binds the barbed end of an actin monomer, the C (connecting), and A (acidic) motifs bind to subunits of the Arp2/3 complex and stabilize the activated conformation [14,16,33]. The actin monomer bound to the V motif, together with Arp2 and Arp3, forms a filament-like seed for nucleation of a new filament branch that emerges at a 70° angle from the side of a mother filament. Arp2/3 complex participates in the formation of the branched actin network in the lamellipodia of motile cells [34]. The multifunctional actin nucleator JMY was also shown to localize to lamellipodia of migrating cells, and its depletion slowed migration rate, while its overexpression enhanced migration velocity [35]. JMY has dual nucleation ability: (1) direct nucleation of unbranched actin filaments and (2) activation of the Arp2/3 complex to polymerize branched actin filaments [35].

Formins nucleate linear actin filaments, but, in contrast to Arp2/3 complex, they do so from the barbed end. Formins are a family of structurally related proteins found throughout eukaryotic evolution that are involved in a wide range of actin-based processes (reviewed in Refs. [16–19,36]), including the formation of actin cables [37,38] and stress fibers in mammalian cells [39], the assembly of contractile rings during cytokinesis [40], cell polarization [41,42], endocytosis and endosome motility [43], morphogenesis [44], and the formation of filopodia [45–47]. Filopodia are long, fingerlike extensions that protrude from the cell surface. They are used for sensing the cell environment and guide cell migration. These protrusions are maintained by thick parallel actin bundles that are nucleated and polymerized by Dia2 formins localized at the filopodial tip. In fact, since formins remain associated with barbed ends during filament elongation [48–50] and because they antagonize capping proteins (CPs), they are ideal for efficient generation of unbranched and long actin filaments in the presence of CPs [51]. Similar to formin, Ena/VASP proteins are responsible for the formation of unbranched actin filaments. Moreover, the Ena/VASP protein family was shown to influence the dynamic of lamellipodia and formation of filopodial protrusions in fibroblast and neuronal growth cones [52]. Ena/VASP proteins (reviewed in Refs [25,26,53,54]) play important roles in many cellular processes including actin-based movement of cells and the bacterial pathogen *L. monocytogenes* [55–57], fibroblast migration [58], axon growth and guidance [52,59,60], and cell shape and morphology changes [61]. Ena/VASP is located at the tips of filopodia and lamellipodia and at focal adhesions of migrating cells. The capacity of VASP to act as an actin nucleator remains controversial [58,62–65].

3.1.1.2 Actin Cross-Linkers

Passive cross-linkers aid the formation of networks or bundles in the cell. Cross-linking proteins create connections between actin filaments by binding to two separate filaments with two different actin-binding domains. Different types of actin-binding domains have been identified. The most common type is the calponin homology domain, which is found in numerous cross-linkers, including spectrin, filamin, fimbrin, and α-actinin [66]. These cross-links are all homodimeric. Fascin is an unusual ABP. It is a monomer with two actin-binding domains, and it binds actin filaments through β-trefoil domains [67]. Fascin, and fimbrin and espin are compact, globular proteins that generate tight, unipolar bundles when mixed with actin monomers. The nucleation of the bundle is initiated by the formation of disklike nuclei from which the filaments subsequently elongate and bundle [68]. Coupling the elongation and bundling processes allows perfect mismatch of the orientations of adjacent filaments, which is essential for bundle formation by fascin [68]. The formation and compactness of actin–fascin bundles can be disrupted in the presence of myosin II motors [69]. In contrast to compact cross-linkers, larger cross-link proteins such as α-actinin and filamin can bind actin filaments over a wide range of angles, forming isotropic networks at low cross-link density, and mixed network/bundle phases at high cross-link density [70,71]. VASP tetramers can also cross-link actin filaments through their filament-binding (FAB) domain [62].

Specific actin cross-linkers provide the necessary physical properties for the proper function of different cytoskeletal structures and usually require the synergetic work of several cross-linkers. In filopodia, filaments are bundled by VASP accumulated at the filopodial tip to create an initial bundle nucleus [62]. Other bundling proteins, like fascin, stabilize the emerging bundles in the filopodial shaft and provide it with the stiffness necessary to promote its extension beyond the cell's leading edge [72]. In stereocilia, espin bundle actin filaments hinder depolymerization [73]. In numerous cases, the assembly dynamics of the cytoskeletal structure also involves myosin II motors that act as active cross-linkers. The action of myosin II provides these structures the ability to contract. This is the case for contractile rings in dividing cells, where α-actinin, fimbrin, and myosin II participate in ring formation [74,75]. The motors apply contractile stresses within the ring structure and are required for its constriction. The formation of stress fibers also involves two passive cross-linkers, α-actinin and fascin, and myosin II motors. Finally, the cell cortex is a dense actin network, which localizes just below the plasma membrane of animal cells and plays a central role in cell shape control [76].

Several actin-bundling and cross-linking proteins localize to the cortex, including α-actinin [74], filamin [77], and fimbrin [78], as well as myosin II motors that provide the cortex with its contractile ability.

A unique property of actin networks is their strongly nonlinear response to an applied stress. Cross-linked actin networks exhibit strain-stiffening effect, which was suggested to result from the nonlinear force–extension behavior of the actin filaments, which easily buckle but strongly resist stretching [79]. Past a critical stress, the network becomes stiffer as the stress increases. The critical stress depends on the type of cross-link, notably on its F-actin-binding affinity, size, and flexibility [80–82]. The addition of myosin II motors (active cross-links) strongly affects the mechanical properties of actin solutions. Apart from exerting forces and affecting spatial reorganization of the actin filaments [83,84], they strongly influence the mechanical properties of actin networks and can also induce stress stiffening. In the case of actin networks cross-linked by filamin, it was shown that internally generated stresses by myosin II motors stiffen the network similarly to an externally applied stress [85].

3.1.1.3 Motor Proteins

3.1.1.3.1 Introduction

Cells utilize biological motors for various tasks. Some motors carry cargo by *walking* along a preferred track [2,86–88], while large groups of other motors modify the cytoskeleton using their collective force [2,88]. Motors require specific tracks in order to perform their tasks. Kinesins and dyneins travel along MT, while myosin motors move along actin tracks. Motor proteins utilize ATP to produce work and motion.

3.1.1.3.2 Motor Characteristics

There are a few common characteristics that help differentiate molecular motors such as structures, step size, motor directionality, generated force, and processivity [2]. A typical motor has two heads (dimer) capable of binding to its underlying track. The heads are followed by a neck region. In response to ATP hydrolysis, the conformation of the heads relative to the neck changes and produces force and motion. The neck is followed by a regulatory light chain (RLC) and then by a tail that can be used for motors' self-assembly (aggregation) or cargo binding depending on the functions of the specific motors. The directionality of individual motors stems from interactions between different parts of the motor and from interactions between the motor and the track filament [89,90]. Step sizes of motors differ widely among species from a few nm to tens of nm. Kinesins and dyneins move with steps of 8 nm, while myosin II step size is estimated to be 5–10 nm [91,92]. Processive myosin motors such as myosins V and VI have considerably higher step sizes of 36 and 30 nm, respectively [87], while myosin X has a step size of 18 nm as it moves between neighboring filaments [93]. Motors can produce a range of forces depending on their configuration. Individual motors produce forces in the order of a few pN. Single-molecule experiments in optical traps have shown that myosin II and myosin V produce 1.4–4 pN of force [87,91,92], while kinesins produce 5–8 pN [94], and dyneins are able to produce only ~1 pN [95]. On the other hand, dyneins working in the axoneme of sperms produce a few nN [96], while deformation of a silicone elastomer by fibroblasts demonstrated that groups of myosin II motors assembled into filaments produce forces on the order of a few μN [97] and sarcomeres in smooth muscle cells produce close to 10 μN [98].

3.1.1.3.3 *Processivity*

Distinct differences in the mechanism of action exist between processive and nonprocessive motors. Processivity is the ability of a motor to complete multiple steps and walk a considerable distance along a track before detaching. While processive motors have states where two heads are bound together to the track [99–101], it appears that nonprocessive motors bind only one head most of the time, perform their action, and then are released from the track as the head detaches [99], therefore requiring more than one motor in order to walk on tracks. However, there are several exceptions. Myosin IX binds only one head while being able to move processively [102], while single-molecule experiments with one- and two-headed myosin II have shown a difference in the work and displacement between the species [92], suggesting the second head has a role in creating the power stroke, either directly or indirectly. More recent studies on skeletal muscle response to strain [103,104] have shown that more motors bind a stretched actin track, suggesting the binding of the second head is strain related.

This difference in activity between processive and nonprocessive motors is readily visible in the duty ratio $r = (\tau_{on}/(\tau_{on} + \tau_{off})) = (\tau_{on}/\tau_{total})$ of the different motors. The duty ratio is the fraction of time a motor stays in contact with the track during its ATPase cycle, where τ_{on} and τ_{off} are the times the motor is attached to/detached from the track, respectively. While myosin II under unloaded conditions has a duty ratio of ~0.04 [105], kinesin's duty ratio is at least 0.5 in order to avoid a situation where both heads are detached [2]. This difference in duty ratios helps explain the difference in velocities obtained for motors in motility assays, where a filament glides over a bed of immobilized motors. While the ATP hydrolysis cycle of skeletal muscle myosin II is 50 ms in duration, velocities as high as 6 µm/s were observed in vitro [106], which means that during one hydrolysis cycle, the filament moves 300 nm, about two orders of magnitude larger than the myosin step size. The solution to this paradox comes when one considers the low duty ratio of myosin II. Since the duty ratio is only 0.04, myosin II is able to produce its power stroke in $50 \cdot 0.04 = 2$ ms and then detach, pushing the filament 5–10 nm forward and then allowing other heads to drive it further [107]. On the other hand, kinesin achieves velocities of 800 nm/s. Since kinesin has a high duty ratio and since the time required performing a power stroke is the product of the ATP cycle time and the duty ratio, kinesin requires a faster ATPase cycle in order to increase its velocity [107].

The duty ratio of motors can increase due to (1) *confinement effects*, in which the confinement can *effectively* increase individual motors and motor cluster processivity by prohibiting their escape and increasing their chance to rebind, after terminating their ATPase cycle and detaching from their track [108], and (2) *application of resistive loads*—which increase motors' duty ratio by reducing their unbinding rate [109,110]. In a recent paper, we were able to show that myosin II motor clusters exhibit more than fourfold higher processivity [84] than unconfined myosin II motors [105]. In our experimental system, both effects could act simultaneously, since the motor clusters are embedded within the actin networks and are therefore expected to be subjected to both resistive loads and to large confinements, overall increasing their processivity [84].

3.1.1.3.4 *Myosin II Motors*

In vivo myosin II motors play a major role in cell adhesion and migration [7], cell division [3,4,6], wound healing, morphogenesis, and many contexts of development [111–115]. Three myosin II isoforms include skeletal muscle, smooth muscle, and nonmuscle myosin II motors. They differ in their mechanochemistry as well as in the geometry and number of molecules in a motor cluster. Myosin II motors comprise three pairs of peptides: two heavy chains (230 kDa), two RLCs (20 kDa), and two essential light chains (ELC 17 kDa). The two light chains regulate and stabilize the heavy chain structure, respectively. Myosin II structure includes a head domain that contains a binding site for ATP and actin. Following the head domain is the neck region, which binds the two light chains (RLC and ELC). The neck domain acts as a lever arm to amplify head rotation when ATP hydrolysis takes place and chemical energy is converted to mechanical movement of the head. A single myosin II motor head does not stay in contact with its actin track throughout its motion; it is inherently nonprocessive. However, when assembled into filaments (clusters) [116], at least one myosin molecule remains in contact with the actin track at a given instant, such that the overall myosin/actin connection does not break resulting in processive motion toward the plus ends of actin filaments.

The property of myosin II motors to self-assemble into large clusters renders them unique. The size of the clusters can be tuned by changing the concentration of monovalent salts (Figure 3.1). In a recent paper, we show that the mean cluster size l decreases with KCl concentration with a power law: $l \sim [KCl]^{-0.8}$ (Figure 3.1F) [84]. l can be used to estimate the mean number of myosin II molecules per cluster, and the typical force f a motor aggregate can generate. When assembled into aggregates, myosin II motors act as active cross-linkers, applying contractile stresses at the molecular level. Besides serving as an active cross-linker, myosin II motors can actively depolymerize actin filaments [69,83,84] (Figure 3.2A), thereby regulating actin turnover by increasing the reservoir of actin monomers available for network polymerization [117]. Myosin II was also shown to function in the turnover of actin during cytokinesis, suggesting that there might be a coupling between the contractile forces applied by the motors during ring constriction to the rate of actin disassembly [118–121]. Recent evidence for such a coupling was also demonstrated in vitro [122]. Despite the increasing evidence that myosin II plays a role in actin turnover, the exact mechanism by which these motors function in filament disassembly is not well understood. It is not clear whether they inhibit filament assembly by accumulating at the plus end of actin filaments, whether they enhance filaments' disassembly by inducing conformational change in the actin filament, or whether they actively disassemble actin subunits during their motion from the minus end toward the filaments' plus end. Future studies are needed to elaborate myosin II function on filaments depolymerization. The ability of motor proteins to induce filaments' disassembly is not new. MT can be disassembled by the kinesin MCAK, which was shown to target the MTs' ends and actively induces their depolymerization [123] (Figure 3.2B and C).

3.1.2 Cellular Reconstitution

Cell motility and cell shape changes involve the participation of both the cell cytoskeleton and the plasma membrane. The membrane functions as a fluid boundary of cells and serves as a substrate for enzymatic reactions. The physical linkage between the membrane and the cytoskeleton is established via membrane proteins (e.g., actin nucleators) or by direct physical interactions between filaments and lipids. Through this coupling, the cytoskeleton applies two kinds of forces: (1) protrusive forces that push the membrane outward, mainly through directed actin polymerization, and (2) contractile forces that pull the membrane inward, mainly through the activity of molecular motors. Reciprocally, the membrane shape and elasticity may have an important impact on cytoskeletal self-organization, for example, through the distribution of membrane proteins, as proposed theoretically [124–126].

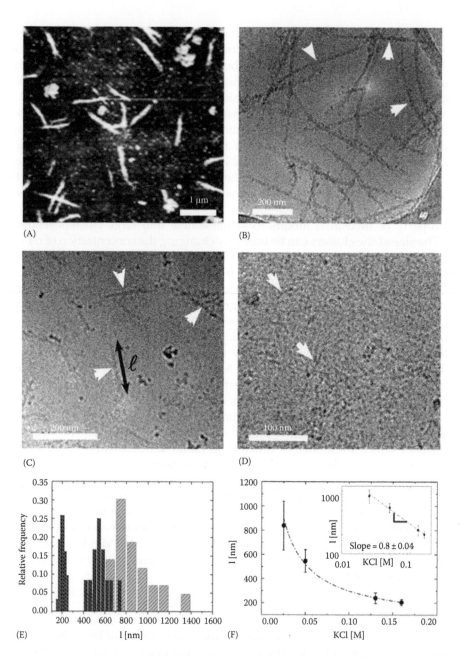

FIGURE 3.1

Mean size of myosin II clusters. (A) AFM and (B–D) high-resolution cryo-transmission electron microscopy (cryo-TEM) imaging of myosin II motor–filaments formed at variable KCl concentrations. (A–C) Bipolar filaments (clusters) formed at low and intermediate KCl concentrations; individual myosin II molecules exist at 0.5 M KCl (D). The white arrowheads mark individual myosin aggregates; the black double-headed arrow marks the length l of such an aggregate. Conditions: (a) 0.025, (b) 0.05, (c) 0.16, and (d) 0.5 M KCl. (e) Distribution of motor aggregate sizes: green, 0.025 M; red, 0.05 M; and blue, 0.16 M KCl. (f) Mean motor cluster size l as a function of [KCl] exhibits a power law dependence of $\sim[\text{KCl}]^{-0.8 \pm 0.04}$ as extracted from the fit of the log–log plot (inset). Values correspond to mean ± SD. (Ideses, Y., Sonn-Segev, A., Roichman, Y., and Bernheim, A., Myosin II does it all: Assembly, remodeling, and disassembly of actin networks are governed by myosin II activity, *Soft Matter*, 9, 7127–7137, 2013. Reproduced by permission of The Royal Society of Chemistry.)

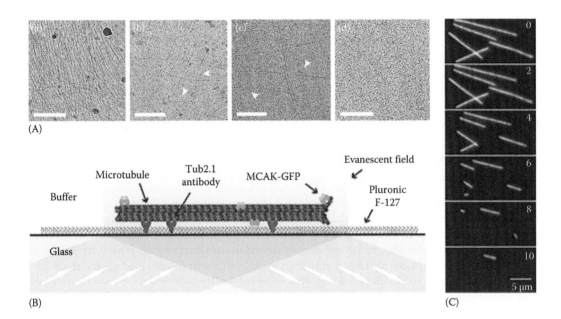

FIGURE 3.2
(A) Depolymerization of actin filaments by myosin II motors. Cryo-TEM micrographs obtained for 2 μM actin and myosin II/actin molar ratio of (a) 0, (b) 0.1, (c) 0.5, and (d) 3. (a) No motors are added, and an entangled network of actin filaments is observed. (b and c) Addition of myosin II motors induces a decrease in the concentration of filaments and an increase in actin monomers (black dots in the background as indicated by arrowheads). (d) Eventually, at a sufficiently high motor concentration, no filaments are observed; the system is composed solely of actin monomers. Bars are 200 nm. (Reprinted from *J. Mol. Biol.*, 375, Haviv, L., Gillo, D., Backouche, F., and Bernheim-Groswasser, A., A cytoskeletal demolition worker: Myosin II acts as an actin depolymerization agent, 325–330. Copyright 2008, with permission from Elsevier.) (B and C) Kinesin MCAK-dependent MT depolymerization. Diagram of the in vitro assay depicting an MT (red) immobilized above the glass surface by anti-tubulin antibodies (dark blue). Excitation by total internal reflection allows the detection of single molecules (viz., MCAK–GFP in green) in the evanescent field (shown in blue). Epifluorescence images of immobilized MTs at different times (shown in minutes). MCAK dimers were added at $t = 2$ min. (Reprinted by permission from Macmillan Publishers Ltd. *Nature*, Helenius, J., Brouhard, G., Kalaidzidis, Y., Diez, S., and Howard, J., The depolymerizing kinesin MCAK uses lattice diffusion to rapidly target microtubule ends, 441, 115–119, copyright 2006.)

Research on cell motility and cell shape changes in the last few decades has focused on uncovering its biochemical basis. Numerous studies are now converging into an overall picture of the identities of the major molecular players and the order of events. With the characterization of the molecular basis of cellular function well underway, a new challenge has come to the fore: understanding the multiscale multicomponent self-organization of the components associated with a biological function. A powerful way to systematic investigation of the hierarchical logic of such systems is to start with a simple, well-characterized model system and to apply a bottom-up synthetic approach. This approach provides a complementary methodology to biochemical and live-cell studies. Such an approach balances the mutually conflicting demands for simplicity, which is required for systematic and quantitative studies, and for a sufficient degree of complexity that allows a faithful representation of biological functions. A significant advantage lies from the fact that these systems are composed of a relatively small number of components, or controlled parameters, which can be systematically varied under controlled conditions. This renders reconstituted systems optimal for physical modeling [127–141], that is, the data extracted from such experiments provide useful input for the models as well as values of the system's parameters.

In this chapter, we focus on two different types of cellular processes: (1) those regulated by the assembly/disassembly dynamics of cytoskeletal filaments and (2) those that also involve motor proteins; we will particularly discuss processes that involve groups of motors, which is relevant to intracellular transport and cytoskeleton remodeling. We will discuss the collective behavior of such systems.

3.1.2.1 Cellular Processes Driven by the Assembly/Disassembly Dynamics of Cytoskeletal Filaments

Cell division and migration is fundamental for proper cellular functioning. Cellular motility is a fundamental process essential for embryonic development, wound healing, immune responses, and development of tissues. Aberrant regulation of cell migration drives progression of many diseases, including cancer invasion and metastasis, which in the latter case utilize their intrinsic migratory ability to invade adjacent tissues and vasculature and ultimately metastasize. Understanding the fundamental mechanism of these processes and investigating the role of the individual proteins that drive abnormal behavior are therefore critical for understanding both basic biology and the pathology of diseases.

The crawling of cells involves a cycle of three steps: protrusion of the leading edge, adhesion to the substratum, and retraction of the rear. Initiation of cellular motility involves the extension of cellular protrusions, for example, *lamellipodia*, at the cell leading edge. The lamellipodium is composed of a branched dense array of short-branched filaments that begins at the leading edge and extends several microns back, and then the lamella takes over, extending from the lamellipodium to the cell body [13]. The driving force for lamellipodial membrane protrusion is the localized nucleation and polymerization of a submembrane branched actin network. At present, these branched networks are known to underlie four basic cellular structures: (1) the leading edge of motile cells, (2) the advancing edges of phagocytic and macropinocytic cups, (3) the actin networks required for the late stages of endocytosis, and (4) the *comet tails* that drive intracellular motility of various pathogens [142].

The key point related to cellular motility is that while pure actin filaments at steady state (in vitro) under physiological ionic conditions treadmill very slowly (~0.04 μm/min), keratocytes (in vivo) move more than two orders of magnitude faster (~10 μm/min). This suggests that additional regulatory proteins are involved in the control of actin treadmilling dynamics in cells [14]. Several regulatory proteins and biochemical processes control actin polymerization dynamics. According to the dendritic nucleation model, the branched nucleation process is initiated by activation of the Arp2/3 complex by NPFs that are activated just beneath the plasma membrane by signaling molecules (Cdc42, PIP2, etc.). The branched nucleation process is followed by filament elongation that pushes the membrane forward. In response to this deformation, the membrane applies resistive elastic forces on the growing filaments. As a whole, the actin networks that are continuously generated at the leading edge display features characteristic of a steady-state process. The actin filaments are oriented with their barbed ends toward the plasma membrane: their growth pushes the plasma membrane forward, while the pointed ends of the filaments are severed and depolymerized at the rear of the lamellipodium in order to refill the pull of actin monomer available for polymerization.

Also at the front are filopodial protrusions. Filopodia are found in many different cell types and typically display active protrusive, retractile, and sweeping motility, which may be necessary for their proposed functions as cellular sensors. Filopodia protrude from the leading edge of many motile cells, including fibroblasts and nerve growth cones (reviewed in Refs. [143–145]). During cell and tissue motility, filopodia are believed to explore the cell's surroundings, such as adhesive surfaces and sense soluble cues, to determine the direction of cell locomotion. These functions are particularly important for the guidance of neuronal growth cones and angiogenic blood vessels. Viruses often bind at filopodial tips and are transported back to the cell body before being internalized. Leading-edge filopodia contain actin–fascin bundles. Some studies suggest that the individual filaments span the entire length of the filopodium and are parallel to the filopodium axis [145], while other studies suggest that the filopodium backbone consists of filaments that are shorter than the entire filopodium and aligned in parallel or obliquely to the filopodium axis [146]. The number and length of the filaments at the filopodial tip are important for force generation.

During protrusion, actin subunits are incorporated at the filopodial tips and are released at the rear of filopodia in a treadmilling process. Actin filament elongation in filopodia is assisted by barbed end-binding proteins, mDia2 (formin) and Ena/VASP proteins. One of the challenges is to explore how Dia-related formins and Ena/VASP proteins cooperate in this process. Previous in vivo and in vitro studies showed that the interaction between the two proteins is crucial for efficient elongation of actin filaments at the tips of protruding filopodia. The combination of both protein activities ensures the formation of appropriate cytoskeletal architecture, formation of filopodial protrusions, and the generation of the force required to push the membrane outward [52,147]. After assembling, leading-edge filopodia elongate at rates between 1 and 5 µm/min in multiple cell types. However, the elongation dynamics vary, even during the lifetime of a single filopodium, with periods of elongation and retraction [148]. In growth-cone filopodia, the elongation rate depends on the balance between monomer addition at the tip and retrograde flow combined with depolymerization at the base [149]. In contrast to the overall consensus regarding the mechanisms of actin filament elongation and turnover in leading-edge filopodia, the site and mechanisms of filopodial initiation are currently subject to debate [144,145]. One model proposes that filopodia assemble by *convergent elongation* of lamellipodial filaments, through reorganization of the Arp2/3 complex-assembled dendritic network into bundles [128,136,150]. The *convergent elongation* mechanism was recently reconstituted in vitro [127,128,136] (Figure 3.3), where it was suggested that the structural properties of the branched actin network (mesh, filament length, etc.) mediate the transition and the bundling efficiency [128,136]. The *tip nucleation* model of filopodial initiation proposes that filopodial filaments are directly nucleated by a cluster of formins on the plasma membrane, which would instantly produce a bunch of elongating actin filaments that become bundled by fascin [151,152]. The two proposed models of filopodial initiation, convergent elongation and tip nucleation, are both theoretically possible. They are not mutually exclusive and can easily coexist, especially considering the diversity and variability of broadly defined filopodial protrusions.

3.1.2.1.1 *Reconstitution of Processes Driven by the Assembly/Disassembly Dynamics of Cytoskeletal Filaments Coupled to Surfaces*

In vitro studies using cell-free extracts and purified proteins have been shown useful for reconstituting cellular motility and for the investigation of the mechanism of force generation by actin polymerization [153–155]. Since biological activities and reactions are confined to surfaces (e.g., fluid membranes or bacterial surface), reconstitution of spatially organized processes was done by restricting biochemical activities to surfaces, such as micrometer-sized polystyrene beads or lipid bilayers.

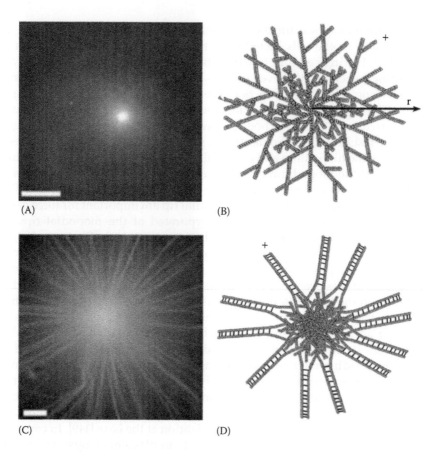

(A)   (B)

(C)   (D)

FIGURE 3.3

Transition from Arp2/3 branched actin network (asters) to filopodium-like bundles (stars). (A, B) Network structure is driven by autocatalytic process of branch formation and filament growth mediated by Arp2/3 complex. Time growth of the cluster that becomes isotropic, resulting in an aster of roughly spherical symmetry. The growth of the network advances with the barbed ends of the actin filaments (+) pointing outward. (C, D) The transition of an aster into a star is driven by the addition of the bundling protein fascin and is initiated by the reorganization of the network structure into bundles, which further elongate by continuous actin polymerization with the filament barbed ends pointing outward (+). During the growth process, the density of the star core continues to increase because of the continuous nucleation of new Arp2/3-actin branches. (A and C) Shown are experimental data—actin is fluorescently labeled; bars are 10 µm. (A and C: Adapted from Ideses, Y., Brill-Karniely, Y., Haviv, L., Ben-Shaul, A., and Bernheim-Groswasser, A., Arp2/3 branched actin network mediates filopodia-like bundles formation in vitro, *PloS One*, 3, e3297. Copyright 2008 National Academy of Sciences, USA; B and D: Adapted from Haviv, L., Brill-Karniely, Y., Mahaffy, R., Backouche, F., Ben-Shaul, A., Pollard, T.D., and Bernheim-Groswasser, A., Reconstitution of the transition from lamellipodium to filopodium in a membrane-free system, *Proc. Natl. Acad. Sci. USA*, 103, 4906–4911. Copyright 2006 National Academy of Sciences, USA)

3.1.2.1.1.1 Reconstitution on Rigid Surfaces   Reconstitution of crawling cellular motility began with studies of the pathogenic bacterium *L. monocytogenes. Listeria* constitutes a simple model system for studying movement induced by actin polymerization; it expresses an actin NPF, ActA, which exploits the cellular machinery of the host (infected) cell to generate its own movement [156]. *Listeria* induces the assembly of a branched actin tail at its surface by recruiting actin monomers and associated proteins from the pool of the infected cell. The forces generated by this site-regulated polymerization process enable the *Listeria* to move within the cell cytoplasm and spread from cell to cell [157]. *Listeria* has been instrumental in determining the biochemistry and biophysics of the actin machinery of cells. Initially, *Listeria* motility was reconstituted in *Xenopus* eggs cell [158]; these studies demonstrated that the recruitment of eukaryotic proteins is necessary for the motility of the *Listeria*. A major breakthrough was obtained in 1999, with the discovery of the essential set of proteins necessary and sufficient to reconstitute the motility of *Listeria* in vitro [57] (Figure 3.4A a). These experiments provided strong evidence that polymerization forces suffice to push a load in the absence of motor proteins.

In the past decade, additional model systems were developed. *Listeria* was replaced with rigid microspheres that were used to study the polymerization forces generated by growing cytoskeletal networks in cell extracts [159,160] (Figure 3.4A b) and purified protein solutions [129] (Figure 3.4A c–f). Actin nucleation is induced from the bead surface and, in the presence of CPs, which block filament elongation from the fast-growing end, forms a dense branched network that can drive the beads forward. Several studies investigated the effect of accessory proteins on movement generation of beads coated with branched [129,130,159,161–165] or unbranched actin nucleators [166,167] (e.g., mDia2 formin, Figure 3.4A c). Figure 3.4B describes schematically two distinct mechanisms by which actin polymerization is transformed into motion, that is, for a branched and unbranched actin nucleator. The physical mechanism of force generation by actin polymerization has been described theoretically using microscopic theories, which describe actin polymerization as a Brownian ratchet [140,168], and macroscopic models that consider elastic stresses developing in the entire gel [169,170]. The elastic modulus of the gel was measured for *Listeria* in cell extracts by using optical tweezers to buckle comet fragments [169] and was shown to range between 10^3 and 10^4 Pa. The same values were measured in a solution of purified proteins using micropipette micromanipulation [131].

An early step in actin-based motility of spherical objects in cells, such as lysosomes and endocytic vesicles [171,172], is the breakage of the symmetrical actin gel that surrounds the bead [129,173]. The time it takes for symmetry breaking to take place was shown to depend linearly on the size of the beads [129]. This dependence was explained to result from the spherical geometry of the system; on curved surfaces, the growing actin gel is stretched that results in the development of a lateral stress that is maximal at the outer surface of the gel. The time of symmetry breaking is thus associated to the time it takes for the gel to fracture [129,173]. Other models propose that symmetry breaking results from the depolymerization of the gel [174], initiated at the outer surface, where the lateral stresses are maximal. These two scenarios recover the experimentally observed linear dependence of the time of symmetry breaking with the size of the beads. After symmetry is broken, the beads start to move. It has been shown that changes in the surface density of actin nucleators (WASP-VCA) or microsphere diameter markedly affect the velocity regime, shifting it from a continuous to a periodic (saltatory) movement [129,130] (Figure 3.4A d–f), resembling that of the mutated *hopping Listeria* [175]. These results highlight how simple physical parameters such as surface geometry and protein density directly affect spatially controlled actin polymerization and play a fundamental role in actin-dependent movement.

3.1.2.1.1.2 Reconstitution on Soft Surfaces Despite the ease of manipulation, NPFs that are directly attached to the bead surface may adapt nonphysiological conformation and are not free to diffuse along the surface; therefore, their own dynamics cannot be studied. To reconstitute membrane-based phenomena, oil droplets [176–178] (Figure 3.4C a), lipid vesicles [179–181] (Figure 3.4C b), and supported membranes [182,183] (Figure 3.4D) replaced the rigid microspheres, thus enabling the coupling of 2D diffusion within the

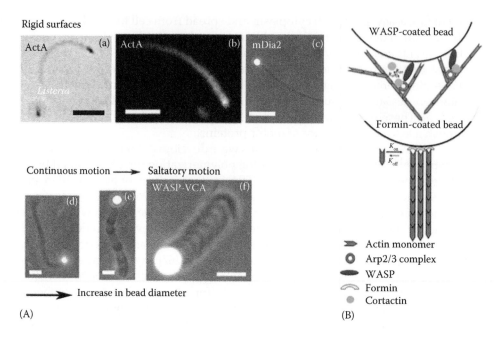

FIGURE 3.4

Reconstitution of actin-based cellular processes. (A) Rigid surfaces. (a) Phase-contrast image of the movement of the *L. monocytogenes* in pure actin and accessory proteins solution. Scale bar is 10 μm. (Reprinted by permission from Macmillan Publishers from *Nature*, Loisel, T.P., Boujemaa, R., Pantaloni, D., and Carlier, M.F., Reconstitution of actin-based motility of *Listeria* and *Shigella* using pure proteins, 401, 613–616. Copyright 1999.) (b) Reconstitution of actin-based movement of a 0.5 μm carboxylated microsphere coated with His-ActA immersed in *Xenopus* egg cytoplasmic extract supplemented by fluorescent actin. Scale bar is 5 μm. (Cameron, L.A., Footer, M.J., Van Oudenaarden, A., and Theriot, J.A., Motility of ActA protein-coated microspheres driven by actin polymerization, *Proc. Natl. Acad. Sci. USA*, 96, 4908–4913. Copyright 1999 National Academy of Sciences, USA) (c) Phase-contrast image of a 2 μm mDia2-coated bead propelled by a thin actin bundles (A.B.G unpublished data). (d–f) Three main regimes of motion of beads coated with WASP-VCA as a function of the bead diameter. (d) Small beads (<2.5 μm diameter) with a comet of constant gel density (gray level) exhibit continuous motion. (e) Medium-size beads of 3 μm diameter move in an intermittent manner, as manifested by the irregular gel density of the comet. Large beads (4.5 μm diameter) progress in periodic fashion, as reflected by the alternating comet gray level. Scale bars are (d, e) 5 μm and (f) 10 μm. (Reprinted by permission from Macmillan Publishers Ltd *Nature*, Bernheim-Groswasser, A., Wiesner, S., Golsteyn, R.M., Carlier, M.F., and Sykes, C., The dynamics of actin-based motility depend on surface parameters, 417, 308–311. Copyright 2002.) (B) Schematic diagram showing actin-based motility by two different actin nucleators localized to a bead surface. (Top) A bead coated with WASP molecules polymerizing a branch actin network at the rear of the bead through the activation of Arp2/3 complex. WASP-Arp2/3 complex associates with the side of the mother filaments during filament polymerization. At a second stage, WASP is released from the branch point, a step that is triggered by the binding of cortactin [165]. (Bottom) Formin mediates bead propulsion by polymerizing unbranched actin bundles at the rear of the bead. *(Continued)*

fluid interface with actin network reorganization. The deformability of lipid bilayer vesicles made them an ideal tool for studying the biophysical effects of force generation by actin and MT. Giant vesicles (GVs) can be composed of a single lipid bilayer (unilamellar) or multiple bilayers (multilamellar). They support the diffusion of molecules in plane and morphological deformations out of plane, such as protrusions or invaginations. Yet, making a large number of GVs with defined size and composition is a challenging mission. Liu and Fletcher produced giant unilamellar vesicles (GUVs) containing PIP2 (initially homogeneously distributed) and mixed them with purified proteins required for actin

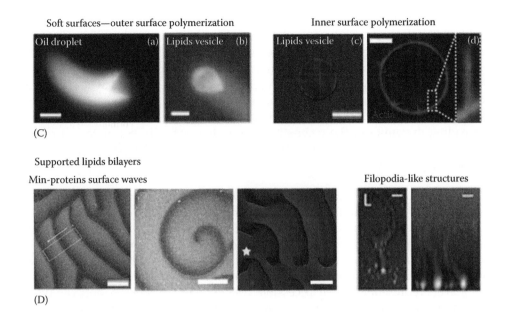

FIGURE 3.4 (Continued)
Reconstitution of actin-based cellular processes. (C) Soft surfaces. (a and b) Polymerization from the exterior and (c and d) polymerization from the interior. (a) An oil droplet coated on its outer surface with a branched actin nucleator is propelled by a comet tail; the forces generated by the growing gel induce droplet deformation into a teardrop shape. Scale bar is 4 µm. (Reprinted with permission from Boukellal, H., Campás, O., Joanny, J.F., Prost, J., and Sykes, C., Soft *Listeria*: Actin-based propulsion of liquid drops, *Phys. Rev. E.*, 69, 061906. Copyright 2004 American Physical Society.) (b) Teardrop shape of a vesicle. Scale bar is 3 µm. (From Giardini, P.A., Fletcher, D.A., and Theriot, J.A., Compression forces generated by actin comet tails on lipid vesicles, *Proc. Natl. Acad. Sci. USA*, 100, 6493–6498. Copyright 2003 National Academy of Sciences, USA) (c) Filopodium-like protrusions emerge from a dendritic actin network that is polymerized at the interior of a GUV. Overlay of the fluorescence images of actin (red) and the membrane (green) confirms that the membrane protrusions are supported by actin filaments. Scale bar is 5 µm. (Reprinted by permission from Macmillan Publishers Ltd. *Nat. Phys.* Liu, A.P., Richmond, D.L., Maibaum, L., Pronk, S., Geissler, P.L., and Fletcher, D.A., Membrane-induced bundling of actin filaments, 4, 789–793. Copyright 2008.) (d) Cortical-like actin network is polymerized at the membrane surface from the interior of a vesicle. Scale bar is 10 µm. (Reprinted from *Biophys. J.*, 96, Pontani, L., Van der Gucht, J., Salbreux, G., Heuvingh, J., Joanny, J.F., and Sykes, C., Reconstitution of an actin cortex inside a liposome, 192. Copyright 2009 with permission from Elsevier.) (D) Supported lipid bilayers. (left) Min protein waves on a lipid membrane (MinD [green] and MinE [red]). Surface waves and double spirals formed by Min proteins. All scale bars are 50 µm. (Loose, M., Fischer-Friedrich, E., Ries, J., Kruse, K., and Schwille, P., Spatial regulators for bacterial cell division self-organize into surface waves in vitro, *Science*, 320, 789–792, 2008. Reprinted with permission from AAAS.) (right) Reconstitution of filopodium-like structures on supported lipid bilayers with *Xenopus* egg extracts. The reaction is started with Alexa 647-actin (red) and chased by Alexa 488-actin (green). Scale bars are 2 µm. (Lee, K., Gallop, J.L., Rambani, K., and Kirschner, M.W., Self-assembly of filopodia-like structures on supported lipid bilayers, *Science*, 329, 1341–1345, 2010. Reprinted with permission from AAAS.)

assembly of branched actin networks on the outer surface of the vesicles [184]. At the end of the process, PIP2 was no more homogenously distributed but rather localized to regions where the branched actin network polymerized [184]. Recently, filopodium-like structures emanating from a branched actin network were reconstituted on GUVs in the absence of a tip complex (e.g., formin or VASP) and bundling proteins [185] (Figure 3.4C c). This study indicates that the elastic interaction between the membrane and the actin cytoskeleton is sufficient to induce structural transitions from a branched actin network into parallel filaments without the need of specific proteins.

In the aim to reconstitute artificial cells and reconstitute actin–membrane interactions that control the cell shape, mechanics, and motility, pioneer works encapsulated actin filaments/networks inside vesicles. The attachment of the actin filaments to the membrane was done using specific anchoring proteins such as spectrin/ankyrin complex [186], or via site-directed actin polymerization, as recently done in the group of Sykes [187] (Figure 3.4C d). Using this system, the characterization of the physical and geometrical constraints controlling the thickness of the actin network that grew from the surface was established. The spreading dynamics of the reconstituted vesicle upon adhesion onto a surface was also investigated, which turned out to be a function of the mechanics and density of the cortical actin layer [188].

3.1.2.1.1.3 Reconstitution on Supported Lipid Bilayer Supported lipid bilayers have been used, in both curved [189] and planar configurations. Planar-supported lipid bilayers are used as substrate in model systems for biological membranes [182,183]. These can be made by fusing small unilamellar vesicles (SUVs) made of lipids, on clean glass coverslips [190]. Planar-supported membranes are used to study lipid diffusion and are particularly suited for investigating processes that occur at the plasma membrane. In contrast to vesicles, supported bilayers represent an easily accessible system where the components can be incorporated in the system gradually. This enables to control the experimental condition and study the role of functional membrane proteins, such as actin-anchoring proteins present at the surface. High resolution of surface-based imaging and manipulation, including total internal reflection fluorescence microscopy (TIRFM), fluorescence correlation spectroscopy (FCS), surface plasmon resonance (SPR), and atomic force microscopy (AFM), can be applied on membranes that are coupled to the planar surface of the support. Using these technologies, several groups were able to anchor actin networks to supported lipid bilayers by using recombinant proteins, such as ponticulin, a transmembrane protein that anchors the actin network to the cell membrane and serves as a nucleation site for actin assembly [191]. In the presence of ponticulin, a thin layer of actin network bound to the supported bilayer was formed [192]. More recently, reconstitution of filopodium-like structures on planar-supported lipid bilayers has been carried out [193] (Figure 3.4D, right). In the first step, mixture of proteins consisting of Arp2/3 complex, Cdc42, N-WASP, and toca-1 was added to supported bilayer and led to the formation of a thin layer of actin on the surface. In the second step, cell extracts were added, leading to the growth of filopodium-like structures from the surface. This system allowed following the recruitment of filopodial proteins to their sites of formation.

Reconstitution on planar bilayer was also used to study the self-organization and dynamics of the MinD/MinE system into traveling waves on supported lipid bilayers in vitro [133] (Figure 3.4D, left). It was shown that the Min proteins MinD and MinE (proteins that oscillate between the cell poles to select the cell center as division site of the bacterium *Escherichia coli*) in the presence of ATP spontaneously generated propagating waves on supported bilayer [133]. Actin-based traveling waves also appear in vivo [194,195] and were predicted theoretically [196,197]. Yet, their reconstitution in vitro was not established.

3.1.2.2 Motor–Filament-Based Processes

3.1.2.2.1 Cooperative Transport: Directional versus Bidirectional Motion

3.1.2.2.1.1 Cooperative Transport by a Few Motor Proteins While transport of cargoes in the cell can be accomplished by single processive motors, there are many examples where a few motors cooperate in transport. In *Xenopus melanophores*, 1–2 kinesin II motors transport melanosomes on the MT plus direction, while 1–3 centrosomal dyneins transport melanosomes in the opposite direction [198]. Despite the identical cargo, the groups of opposing motors do not compete. Kinesins and dyneins are also involved in the transport of peroxisomes in *Drosophila* cells [199]. Up to 11 of both kinds of motors drive a shared cargo. In vivo observations of bidirectional cargo transport involving plus- and minus-end motors were also frequently observed [200,201]. In vitro, the motion of MTs on a bed of a mixed population of plus-end (kinesin-5 KLP61F) and minus-end (Ncd) driven motors was also examined. It was shown to exhibit dynamics whose directionality depends on the ratio of the two motor species, including bidirectional movement over a narrow range of relative concentrations around the *balance point* [202]. Bidirectional motion was also observed in a motility assay consisting of kinesin-1 motors interacting with antiparallel MT doublets [203].

3.1.2.2.1.2 Cooperative Behavior of Many Motor Proteins While some processes, such as the transport of cargoes, are achieved mainly by the action of a few motors (as discussed earlier), other processes, such as cell motility [117,204] and cytokinesis, require the cooperative work of many motors. Muscle contraction, for instance, involves the simultaneous action of hundreds of myosin II motors pulling on attached actin filaments and causing them to slide against each other [205]. Similarly, groups of nonmuscle myosin II motors participate in the assembly and contraction of the contractile ring during cytokinesis [4,206,207] and play an essential role in the formation and maintenance of stress fibers and focal adhesion [208]. In certain biological systems, cooperative behavior of molecular motors produces oscillatory motion. In some insects, for instance, autonomous oscillations are generated within the flight muscle [209]. Spontaneous oscillations have also been observed in single myofibrils in vitro [210]. Finally, dynein motors are responsible for the oscillatory motion of axonemal cilia and flagella [211–213].

3.1.2.2.1.3 Reconstitution of Bidirectionality In Vitro The directionality of individual motors stems from interactions between different parts of the motor (e.g., the neck and the motor domain in kinesins) and from interactions between the motor and the track filament [89,90,214]. The direction of motion of a large collection of motors may also be influenced by their cooperative mode of action. Specifically, in several experiments, the ability of motors to cooperatively induce bidirectional motion has been demonstrated. In one such experiment, unidirectional motion of actin filaments due to the action of myosin II motors was transformed into bidirectional motion by the application of an external stalling electric field [215] (Figure 3.5A and B). Under such conditions, the external forces acting on the actin filament nearly balance the forces generated by the motors such that a fluctuation in the number of attached motor can reverse the motion direction. The authors interpret this instability as a dynamical phase transition resulting from a collective effect of many motors attached to a single filament. This type of transition has been predicted theoretically several years before by Jülicher and Prost [216]. Electric field was also used to bias the direction of motion in kinesin–MT systems [217]. In another experiment, bidirectional motion of MTs was observed when subjected to the action of an ensemble of NK11 motors. These motors are a mutant form of the kinesin-related Ncd, which individually exhibits random motion with no preferred directionality [218]. Bidirectional motion was also observed when apolar actin bundles glided over a bed of myosin II motors [135] (Figure 3.5C and D). The experimental results show that the *reversal time*, which is the characteristic

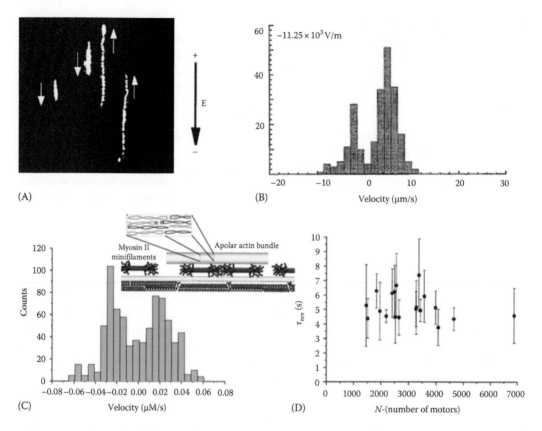

FIGURE 3.5
(A, B) Bidirectional cooperative motion of actin filaments gliding over a bed of myosin II motors in the presence of an electric field E; arrows indicate direction of motion. (B) Velocity histogram of the actin filaments for a strong opposing electric field shows clear bimodal distribution. (A and B: Reprinted from Riveline, D. et al., *Eur. Biophys. J.*, 27, 403, 1998.) (C, D) Bidirectional cooperative motion of actin tracks with randomly alternating polarities moving over a bed of myosin II motors cluster. (C) Velocity histogram of apolar bundles exhibiting a clear bimodal distribution. Inset: schematic diagram of the experimental assay where myosin II clusters are the multiheaded brown objects and apolar bundle are thick yellow tube. The internal structure of such a bundle, consisting of individual actin filaments with randomly oriented polarities, is also given. (D) The characteristic reversal time, t_{rev}, as a function of the number of working motors N is independent of system size N. (C and D: Gilboa, B., Gillo, D., Farago, O., and Bernheim-Groswasser, A., Bidirectional cooperative motion of myosin-II motors on actin tracks with randomly alternating polarities, *Soft Matter*, 5, 2223–2231, 2009. Reproduced by permission of The Royal Society of Chemistry.)

time in which the filament undergoes direction reversal, t_{rev} does not depend on the size of the system (i.e., on the number of motors, N) [135] (Figure 3.5D). Interestingly, these apolar bundles could exhibit dynamic transition from bidirectional to directional and then fuse with a nearby bundle due to interaction with that bundle [219]. The transition is size dependent and the magnitude of the attracting forces was shown to increase as the separation between the two bundles decreases. Two kinds of forces were proposed to exist: (1) external long-ranged forces (of elastic and/or hydrodynamic origin) and (2) internal short-ranged forces (resulting from interactions between individual actin filaments and motors from the two bundles). In the only experiment to date that studied collective behavior using an optical trap, Plaçais et al. have shown that spontaneous oscillations occur when a group of heavy meromyosin motors pulls on a filament attached to a bead in the optical trap which acts as an elastic load [220]. In their paper, the authors show that the typical frequency of oscillations is related to the stiffness of the trap.

3.1.2.2.1.4 Theoretical Aspects of Bidirectionality Several aspects of cooperativity in molecular motor systems have been addressed using different theoretical approaches [216,221–230]. These studies focus on systems consisting of large group of motors interacting with the cytoskeletal tracks, thus mimicking the conditions that are found in muscles or in the contractile ring. However, in order to explain the transport of a cargo by a small number motors, a different approach was needed. Müller et al. have described the behavior of motors having opposite polarity as a tug-of-war (TOW) between opposing molecular motors [231]. In the TOW model, the cargo moves in the direction of the motor party that exerts the larger force. The balance of power is shifting between the two parties as a result of stochastic events of binding and unbinding of motors. The main feature of the TOW model lies in the fact that the unbinding rates depend exponentially on the force load experienced by the motors, which itself depends on the number of attached plus-end-directed and minus-end-directed motors. This leads to a very rich dynamic behavior that is very sensitive to the model parameters (which include the stall force, detachment force, unbinding and binding rates, forward velocity, and superstall velocity amplitude). Specifically, for certain sets of parameter values, the motion is bidirectional, that is, switches between periods of plus-directed and minus-directed movements. Interestingly, during these periods of unidirectional motion, the motors that win the contest cause the detachment of all the motors of the other type. Recent experiments, which have carefully analyzed the bidirectional transport of vesicles along MTs, concluded that the dynamics is indeed consistent with the TOW mechanism [201,232].

Bidirectional motion does not necessarily require the existence of two types of motors but may be also observed when one group of motors is driving the motion of filaments and bundles with mixed polarities [135,203] (Figure 3.5C). Hexner and Kafri have recently analyzed the TOW model in the large number of motor (N) limit and found two patterns of bidirectional motion: the first one is of a rapid oscillating-like motion, with microscopic reversal times of the order of the ATPase cycle (i.e., the typical attachment time of a single motor to the filament) [233]. The second one is bidirectional motion with macroscopically large reversal times t_{rev} that grow exponentially with N. Badoual et al. [138] used a two-state ratchet model to explain the emergence of bidirectionality in a system composed of a large group of motors interacting with a given track. The model presented in Ref. [138] demonstrates the ability of a large group of motors working cooperatively to induce bidirectional motion, even when individually the motors do not show preferential directionality. According to this model and similarly to the results obtained in Ref. [233], the characteristic reversal time increases exponentially with the number of motors, N.

By using apolar actin bundles of different sizes, we were able to measure the dependence of the reversal times t_{rev} on N and show that while N varies over half an order of magnitude, the corresponding t_{rev} are similar to each other ($3 < t_{rev} < 10$ s) and show no apparent correlation with N [135] (Figure 3.5D). Using a modified version of Ref. [138] model, we argue that the origin of this behavior can be attributed to the tension developed in the actin track due to the action of the attached motors, which in turn increase the detachment rate of the motors. This strong effect, which is an indirect manifestation of cooperativity between the motors, recovered the independency of t_{rev} on N, in accord with the experimental observation [134,135,234].

3.1.2.2.2 Collective Behavior: Active Gels

Besides being involved in intracellular transport, motor proteins play a pivotal role in cell cytoskeleton remodeling and force generation. The cell cytoskeleton is an active, ATP-driven network that constantly remodels to fit the processes required by the cell. These cytoskeletal networks are viscoelastic materials driven away from equilibrium by ATP hydrolysis [235]. They resemble in many aspects of polymer solutions or gels. The main difference from usual passive polymer solutions is their intrinsic activity resulting from filament polymerization/depolymerization kinetics and from the action of molecular motors that act as active cross-linkers, constantly breaking and reforming [83,84]. The ability of molecular motors to exert forces on cytoskeletal polymers (e.g., actin filaments) introduces a new class of materials called active gels. Active gels can deform themselves by transforming enzymatic activity (ATP hydrolysis) to mechanical forces. There are two main mechanisms that govern the remodeling of the network: force production by motors and network turnover. These processes are by no means independent of each other, as motors were demonstrated to have a huge impact on the depolymerization of the network [69,84,117,236], while network turnover can change the network structure, and thus the activity, of motors [84].

The characteristic network turnover time is affected by the type of passive cross-linkers that participate, as well as by the activity/concentration of molecular motors. Therefore, on long timescales, cytoskeletal networks behave like a viscous fluid because actin turnover and dissociation of cross-linkers dissipate stresses and enable the polymer network to remodel [2]. Actin turnover can be modulated by molecular motor activity as is the case of the cleavage furrow of dividing cells, in which actin turnover is accelerated by myosin II activity [119,121]. In the cortex, the dynamics of actin cross-linkers and myosin II motors is usually 5–10 times faster than actin turnover, which suggests that the relaxation of the cortex is dominated by the turnover of cross-links rather than the actin turnover [73,74,78,121]. Cortical actin turnover is slowed by overexpression of passive cross-linkers (α-actinin) that hinders actin disassembly [74]. The same effect is also observed in stereocilia, where overexpression of espin generates longer stereocilia, by reducing the rate of actin disassembly at their pointed end [73,237].

From a theoretical perspective, the study of an active system such as the cytoskeleton cannot rely on a thermodynamic approach based on a free energy minimization; it must thus rely on a dynamic theory reflecting the local force balance in the system. Different approaches were used to describe the dynamics of such systems. Several models were based on the introduction of continuum mean field kinetic equations to describe the dynamics of filaments moving relatively to each other due to the presence of cross-linking motors [230,238–240]. Another approach that is being proven useful to study the physics of liquid crystals is to treat filament–motor systems as a viscoelastic polar active gel [241,242]. In these generalized hydrodynamic theories, the dynamics is inferred from symmetry considerations or by coarse-graining the mesoscopic kinetic equations. Several inhomogeneous structures have been identified as steady-state solutions of the macroscopic equations, including asters, vortices, and spirals [243–246].

3.1.2.2.2.1 Cellular Reconstitution of Active Gels

3.1.2.2.2.1.1 Self-Organization and Pattern Formation   Reconstituted in vitro systems composed of molecular motors, protein cross-linkers, and cytoskeletal polymers have shed light on the complex behavior of active gels. The first to reconstitute active patterns in motor–filaments systems were Nedelec and coworkers [247,248] (Figure 3.6, blue rectangle).

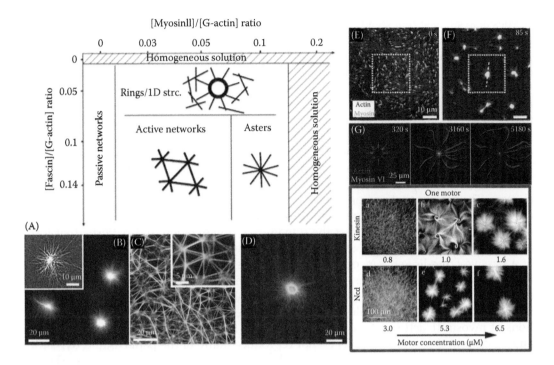

FIGURE 3.6

(A) Schematic phase diagram exhibiting the four essential types of patterns (i.e., *passive networks* and *rings and 1D curve structures* embedded in a surrounding network, *asters*, and *active networks*) formed in vitro for small/intermediate motor clusters as a function of [F]/[A] ([fascin]/[actin]) and [M]/[A] ([myosin II]/[actin] ratios. Patterns do not form at high concentrations of motors or in the absence of fascin (*homogeneous solutions*). For clarity, schematic drawings of the patterns appear in the appropriate regions of the phase diagram where the lines mark the approximate boundaries between the phases. (B–D) Typical patterns visualized by fluorescence microscopy: (B) asters, (C) tensile active networks, (D) and rings (A–D: Adapted from Backouche, F., Haviv, L., Groswasser, D., and Bernheim-Groswasser, A. (2006) Active gels: dynamics of patterning and self-organization. *Physical biology.* 3, 264) (E–F) Time sequence of actin networks confined to a supported lipid bilayer reorganizing into asters in the presence of myosin II and the passive cross-linker α-actinin. (E–F: From Murrell, M.P. and Gardel, M.L., *Proc. Natl. Acad. Sci. USA,* 109, 20820, 2012.) (G) Polarized actin networks grown on patterned surfaces. Time series of actin network reorganization (contraction and disassembly) of an eight-branch radial array in the presence of myosin VI motors. (Reymann, A., Boujemaa-Paterski, R., Martiel, J., Guérin, C., Cao, W., Chin, H.F., Enrique, M., Théry, M., and Blanchoin, L., Actin network architecture can determine myosin motor activity, *Science,* 336, 1310–1314, 2012. Reprinted with permission of AAAS.) (Blue rectangle) Pattern formation of MTs in the presence of kinesin or dynein motor complexes. In the presence of increasing amounts of kinesin complexes, (A) networks, (B) spirals, and (C) asters are formed, where in the presence of dynein complexes, (A) networks and (B–C) asters form. (Surrey, T., Nédélec, F., Leibler, S., and Karsenti, E., Physical properties determining self-organization of motors and microtubules, *Science,* 292, 1167–1171, 2001. Reprinted with permission of AAAS.)

It was shown that small processive clusters of kinesins and/or dyneins can organize MT into asters, vortices, or bundles, depending on the motor concentration [247–250]. The same patterns form whether tubulin or taxol-stabilized MTs are used, suggesting that kinesin clusters do not function in the dynamics of polymerization/depolymerization of MT but serve only as active reorganizing centers for the MT. The development of asters, vortices, and bundles is also observed numerically in numerous theoretical models and computer simulations studying motor–filament systems [241,243–247,251–255]. These theoretical studies also suggest that steady-state patterns are generic and therefore should be experimentally observable in any motor/filament system.

Reconstitution of actomyosin network dynamics requires the use of actin monomers (G-actin) as a starting point for in vitro network formation [83] in order to reflect faithfully the tight interplay between actin polymerization/depolymerization dynamics, myosin II motors' contractile and reorganization activity, and their coupling. Ideses et al. provide first evidence for myosin II multifunctionality, which ranges from a network conucleator, through an active organizer of the network structure, to a severing and regulating agent of actin turnover. The authors show that during the initial process of network formation, myosin II motors become embedded within the network structure and take part in its formation [84]. This process is further enhanced when myosin II is in the form of motor clusters. The motors that are embedded within the network are acting as internal active cross-links that apply pinching forces at the molecular level. The size and concentration of myosin II clusters highly affect the dynamic of organization and patterns that form at steady state [84]. In contrast to the MT/kinesin/dynein systems [247–250], the addition of a passive cross-linker is necessary for actomyosin networks to form [69,83,84]; in its absence, the severing and disassembly activity of myosin II dominates, and no networks form [69,83,84] (Figure 3.2A; Figure 3.6A shows a phase diagram for small/intermediate myosin II motor clusters). While asters (Figure 3.6A and B) and bundles do form in solutions of myosin II–actin systems, or when coupled to a flat lipid membrane (Figure 3.6E and F), current theoretical models do not reproduce many of the patterns and dynamics generated in the experiments [83,84,256], such as the formation of tensile networks (Figure 3.6C) and rings (Figure 3.6D), or the effect of myosin II cluster size on networks reorganization dynamics and patterning [84]. Finally, while vortices form in MT/kinesin systems [247], they have never been observed in myosin II–actin systems, regardless of the myosin II cluster size used [84]. This discrepancy may originate from the nature of the myosin II motor itself (highly nonprocessive), from the size of the myosin cluster (which must be relatively large due to the low processivity of individual myosin II motors), from the coupling between actin filaments' disassembly dynamics and myosin II activity [69,122], or from the difference in mechanical properties of MT and actin filaments. In comparison to MT, actin filaments have a highly asymmetric load response, that is, to support large tensions but buckle easily under piconewton (pN) compressive loads [256–258].

While most studies investigate the reorganization properties of actin solutions in the bulk, another approach was recently undertaken. Using micropatterning methods, polarized actin networks were grown onto a patterned surface [122]. Using this type of approach, myosin II activity was shown to correlate with network architecture and filament organization/polarity (Figure 3.6G). Patterns can also emerge in motility assays where dense populations of cytoskeletal filaments/bundles are driven on a bed of molecular motors. These systems exhibit complex collective motion, such as vortices, clusters, bands, and other nematic structures [259–261]. Such systems are of great importance as they can mimic, in a controlled manner, the collective motion and pattern formation observed in other systems such as animal flocks [262].

3.1.2.2.2.1.2 Contractility and Generation of Flows Being actively driven, active gels can spontaneously contract, flow and deform, and exhibit autonomous motility. Gradients of contractility (induced by biochemical cues [263,264] or boundary conditions [114]) can lead to actomyosin flows. Actomyosin flows appear in cell division [265,266], wound healing, morphogenesis, and many contexts of development [111–115]. Gradients of contractility produce a mechanical imbalance leading to actomyosin flows toward the strongest pulling area. This mechanism may also involve positive feedbacks through the redistribution of myosin II by the flow that could amplify the gradient of contractility. Thus, actomyosin flows may only require a small initial bias in contractility distribution and could be amplified by biomechanical feedbacks. Recent theoretical studies predict that steady-state flows can emerge spontaneously in the absence of any bias [267–269]. A very general prediction emerging from these studies is that a nonmoving state is unstable and the gel flows if the gel thickness is increased beyond a critical value or, equivalently, if motor activity is increased at constant thickness above a critical value.

Contractile systems were reconstituted in vitro, in 1D [270] and in 3D [84,271,272] (Figure 3.7A and B) configurations. It has recently been shown that actomyosin networks undergo massive contraction only when large clusters are used (apply forces > tens of pN) and when their density is sufficiently low, such that they are sparsely distributed, resulting in local force inhomogeneities within the network structure [84]. Contraction is initiated at the periphery and propagates inward (Figure 3.7B). Macroscopic contraction occurs in a certain myosin II concentration range (Figure 3.7C). Below it, no contraction is observed (cyan hexagons, Figure 3.7C), whereas above it, disruption of the networks occurs (green rectangle, Figure 3.7C). At very large motor concentrations, the severing and disassembly activity of myosin II dominates and no networks form (*homogenous solution*, in Figure 3.7C). Contractility seems to be a generic property of myosin II motor/filament systems and has to do with the fact that the motors within a motor aggregate can interact with multiple filaments simultaneously. Those filaments, which can be of opposite orientations/polarity, lead to the formation of *contractile units* within the network. Global contraction is induced by the collective action of those randomly distributed contractile units, as long as they apply sufficiently large forces and are sparsely distributed. The formation of contractile elements was also used to explain the contractility of reconstituted actomyosin bundles (1D structure) in vitro [270]. Kruse and Jülicher predicted theoretically the spontaneous contraction of bundles of polar filaments by molecular motors [273]. Membrane-based actomyosin contractility was recently reconstituted [257,274,275] and was shown to depend on actin–membrane anchoring and network connectivity (Figure 3.7D). The spontaneous generation of F-actin flows was also reconstituted using cell extracts [276,277]. Quiet surprising, the spatial localization of F-actin nucleators and actin turnover play a decisive role in generating flow, but inhibition of myosin II activity did not perturbed the flows [277].

In recent years, there has been a growing interest in the collective behavior of molecular motors, which is ubiquitous in biology and physiology. Much progress has been achieved experimentally using biomimetic systems with new assays and through a variety of theoretical models that have been proposed to interpret the experimental results. Our understanding is though still relatively limited, as the activity leads to spectacular and unexpected behaviors that are still not fully understood. This is specifically true in the case of actomyosin networks, where coupling between motor activity and actin turnover exists. The implementation of such a coupling in theoretical model is nontrivial nor the description of the activity and its dependence on motor properties and concentration.

3.1.3 Concluding Remarks

This chapter provides an overview of the current achievements of cellular reconstitutions. Although these systems contain a relatively small number of components, the richness of behaviors that emerge is spectacular and enables to gain better understanding of the complex behaviors and phenomena observed in biological systems. We believe that understanding the principles governing the multiscale, multicomponent, and self-organization

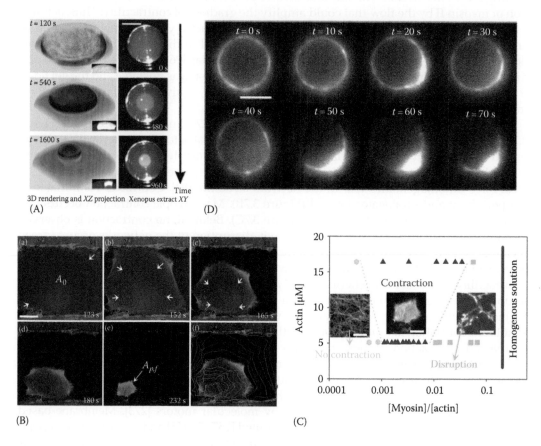

FIGURE 3.7
Contractility driven by myosin II activity. (A) Left—3D rendering of *XY*-confocal slices of the fluorescently labeled F-actin network (orange) contracting within the water droplet (blue) in the presence of α-actinin. (Inset) Corresponding *XZ* projections through the gel. Right—dark-field images of a contracting *Xenopus* extract that is placed within a layer of mineral oil. Scale bar, 400 µm. (Reprinted from *Biophys. J.*, 94, Bendix, P.M., Koenderink, G.H., Cuvelier, D., Dogic, Z., Koeleman, B.N., Brieher, W.M., Field, C.M., Mahadevan, L., and Weitz, D.A., A quantitative analysis of contractility in active cytoskeletal protein networks, 3126–3136. Copyright 2008 with permission from Elsevier.) (B) Contractile network dynamics confined in a chamber. (a) Network formation typically takes 1–2 min; A_0 is the initial area of the network. (b–e) Global contraction of the network as a function of time. Contraction begins at the periphery (white arrows show the direction of contraction). At the end of contraction, the final area is a fraction (typically 5%) of the initial network area A_0. (e) A_{pf} is the final network area. (f) Stages of contraction are followed using polygons that mark the network perimeter at different times. (C) Phase diagram of a system consisting of large myosin II motor clusters (~150 molecules/cluster) in the presence of fascin ([fascin]/[actin] = 1/18) and 5 or 16 µM of actin. The lines mark the approximate limits between the regions in the phase diagram. Below a certain motor concentration, the system does not contract (cyan hexagons—the picture shows an entangled actin network); at intermediate concentration, the network contracts macroscopically (blue triangles); at higher concentrations, the network undergoes enhanced disruption (green squares); and at very large concentration, no patterns form (*homogenous solution*). (D) Time lapse of the contraction of an actin network localized to the outer surface of a lipid vesicle after injection of myosin II motors. Scale bar, 5 µm. (From Carvalho, K., Lemiere, J., Faqir, F., Manzi, J., Blanchoin, L., Plastino, J., Betz, T., and Sykes, C. (2013) Actin polymerization or myosin contraction: Two ways to build up cortical tension for symmetry breaking. *Philos. Trans. R. Soc. Lond. B. Biol. Sci.* 368, 20130005. With permission.)

of a cell into a functional robust machine is expected to open new directions for the engineering of biomimetic active materials from microscopic components that consume energy to generate continuous motion. Being actively driven, these materials can spontaneously flow and deform and exhibit autonomous motility that can be used to develop self-propelled machines that are capable of executing mechanical work.

References

1. Bray, D. (2001) *Cell Movements: From Molecules to Motility*, Taylor & Francis Group, Boca Raton, FL.
2. Howard, J. (2001) *Mechanics of Motor Proteins and the Cytoskeleton*, Sinauer Associates Inc., Sunderland, MA.
3. Wu, J.Q. and Pollard, T.D. (2005) Counting cytokinesis proteins globally and locally in fission yeast. *Sci. Signal.* 310, 310.
4. Carvalho, A., Desai, A., and Oegema, K. (2009) Structural memory in the contractile ring makes the duration of cytokinesis independent of cell size. *Cell.* 137, 926–937.
5. Ma, X., Kovács, M., Conti, M.A., Wang, A., Zhang, Y., Sellers, J.R., and Adelstein, R.S. (2012) Nonmuscle myosin II exerts tension but does not translocate actin in vertebrate cytokinesis. *Proc. Natl. Acad. Sci. USA* 109, 4509–4514.
6. Pollard, T.D. and Cooper, J.A. (2009) Actin, a central player in cell shape and movement. *Science.* 326, 1208–1212.
7. Vicente-Manzanares, M., Ma, X., Adelstein, R.S., and Horwitz, A.R. (2009) Non-muscle myosin II takes centre stage in cell adhesion and migration. *Nat. Rev. Mol. Cell Biol.* 10, 778–790.
8. Wang, Y. (1985) Exchange of actin subunits at the leading edge of living fibroblasts: Possible role of treadmilling. *J. Cell Biol.* 101, 597–602.
9. Dean, S.O., Rogers, S.L., Stuurman, N., Vale, R.D., and Spudich, J.A. (2005) Distinct pathways control recruitment and maintenance of myosin II at the cleavage furrow during cytokinesis. *Proc. Natl. Acad. Sci. USA* 102, 13473–13478.
10. Aebi, U., Millonig, R., Salvo, H., and Engel, A. (1986) The three-dimensional structure of the actin filament revisited. *Ann. N.Y. Acad. Sci.* 483, 100–119.
11. Yanagida, T., Nakase, M., Nishiyama, K., and Oosawa, F. (1984) Direct observation of motion of single F-actin filaments in the presence of myosin. *Nature.* 307, 58–60.
12. Gittes, F., Mickey, B., Nettleton, J., and Howard, J. (1993) Flexural rigidity of microtubules and actin filaments measured from thermal fluctuations in shape. *J. Cell Biol.* 120, 923–934.
13. Chhabra, E.S. and Higgs, H.N. (2007) The many faces of actin: Matching assembly factors with cellular structures. *Nat. Cell Biol.* 9, 1110–1121.
14. Pollard, T.D. and Borisy, G.G. (2003) Cellular motility driven by assembly and disassembly of actin filaments. *Cell.* 112, 453–465.
15. Pantaloni, D., Le Clainche, C., and Carlier, M.F. (2001) Mechanism of actin-based motility. *Science.* 292, 1502–1506.
16. Pollard, T.D. (2007) Regulation of actin filament assembly by Arp2/3 complex and formins. *Annu. Rev. Biophys. Biomol. Struct.* 36, 451–477.
17. Goode, B.L. and Eck, M.J. (2007) Mechanism and function of formins in the control of actin assembly. *Annu. Rev. Biochem.* 76, 593–627.
18. Faix, J. and Grosse, R. (2006) Staying in shape with formins. *Dev. Cell.* 10, 693–706.
19. Evangelista, M., Zigmond, S., and Boone, C. (2003) Formins: Signaling effectors for assembly and polarization of actin filaments. *J. Cell. Sci.* 116, 2603–2611.
20. Kerkhoff, E. (2011) Actin dynamics at intracellular membranes: The Spir/formin nucleator complex. *Eur. J. Cell Biol.* 90, 922–925.

21. Firat-Karalar, E.N. and Welch, M.D. (2011) New mechanisms and functions of actin nucleation. *Curr. Opin. Cell Biol.* 23, 4–13.
22. Qualmann, B. and Kessels, M.M. (2009) New players in actin polymerization–WH2-domain-containing actin nucleators. *Trends Cell Biol.* 19, 276–285.
23. Chesarone, M.A. and Goode, B.L. (2009) Actin nucleation and elongation factors: Mechanisms and interplay. *Curr. Opin. Cell Biol.* 21, 28–37.
24. Didry, D., Cantrelle, F.X., Husson, C., Roblin, P., Moorthy, A.M.E., Perez, J., Le Clainche, C., Hertzog, M., Guittet, E., and Carlier, M.F. (2011) How a single residue in individual β-thymosin/WH2 domains controls their functions in actin assembly. *EMBO J.* 31, 1000–1013.
25. Trichet, L., Sykes, C., and Plastino, J. (2008) Relaxing the actin cytoskeleton for adhesion and movement with Ena/VASP. *J. Cell Biol.* 181, 19–25.
26. Bear, J.E. and Gertler, F.B. (2009) Ena/VASP: Towards resolving a pointed controversy at the barbed end. *J. Cell. Sci.* 122, 1947–1953.
27. Domann, E., Wehland, J., Rohde, M., Pistor, S., Hartl, M., Goebel, W., Leimeister-Wächter, M., Wuenscher, M., and Chakraborty, T. (1992) A novel bacterial virulence gene in *Listeria monocytogenes* required for host cell microfilament interaction with homology to the proline-rich region of vinculin. *EMBO J.* 11, 1981.
28. Rottner, K., Hänisch, J., and Campellone, K.G. (2010) WASH, WHAMM and JMY: Regulation of Arp2/3 complex and beyond. *Trends Cell Biol.* 20, 650–661.
29. Cosen-Binker, L.I. and Kapus, A. (2006) Cortactin: The gray eminence of the cytoskeleton. *Physiology.* 21, 352–361.
30. Daly, R.J. (2004) Cortactin signalling and dynamic actin networks. *Biochem. J.* 382, 13.
31. Ammer, A.G. and Weed, S.A. (2008) Cortactin branches out: Roles in regulating protrusive actin dynamics. *Cell Motil. Cytoskeleton.* 65, 687–707.
32. Uruno, T., Liu, J., Li, Y., Smith, N., and Zhan, X. (2003) Sequential interaction of actin-related proteins 2 and 3 (Arp2/3) complex with neural Wiscott-Aldrich syndrome protein (N-WASP) and cortactin during branched actin filament network formation. *J. Biol. Chem.* 278, 26086–26093.
33. Dominguez, R. (2010) Structural insights into *de novo* actin polymerization. *Curr. Opin. Struct. Biol.* 20, 217–225.
34. Svitkina, T.M. and Borisy, G.G. (1999) Arp2/3 complex and actin depolymerizing factor/cofilin in dendritic organization and treadmilling of actin filament array in lamellipodia. *J. Cell Biol.* 145, 1009–1026.
35. Zuchero, J.B., Coutts, A.S., Quinlan, M.E., La Thangue, N.B., and Mullins, R.D. (2009) p53-cofactor JMY is a multifunctional actin nucleation factor. *Nat. Cell Biol.* 11, 451–459.
36. Watanabe, N. and Higashida, C. (2004) Formins: Processive cappers of growing actin filaments. *Exp. Cell Res.* 301, 16–22.
37. Imamura, H., Tanaka, K., Hihara, T., Umikawa, M., Kamei, T., Takahashi, K., Sasaki, T., and Takai, Y. (1997) Bni1p and Bnr1p: Downstream targets of the Rho family small G-proteins which interact with profilin and regulate actin cytoskeleton in *Saccharomyces cerevisiae*. *EMBO J.* 16, 2745–2755.
38. Sagot, I., Klee, S.K., and Pellman, D. (2001) Yeast formins regulate cell polarity by controlling the assembly of actin cables. *Nat. Cell Biol.* 4, 42–50.
39. Koka, S., Neudauer, C.L., Li, X., Lewis, R.E., McCarthy, J.B., and Westendorf, J.J. (2003) The formin-homology-domain-containing protein FHOD1 enhances cell migration. *J. Cell. Sci.* 116, 1745–1755.
40. Ingouff, M., Gerald, J.N.F., Guérin, C., Robert, H., Sørensen, M.B., Van Damme, D., Geelen, D., Blanchoin, L., and Berger, F. (2005) Plant formin AtFH5 is an evolutionarily conserved actin nucleator involved in cytokinesis. *Nat. Cell Biol.* 7, 374–380.
41. Feierbach, B. and Chang, F. (2001) Roles of the fission yeast formin for3p in cell polarity, actin cable formation and symmetric cell division. *Curr. Biol.* 11, 1656–1665.
42. Li, F. and Higgs, H.N. (2005) Dissecting requirements for auto-inhibition of actin nucleation by the formin, mDia1. *J. Biol. Chem.* 280, 6986–6992.

43. Gasman, S., Kalaidzidis, Y., and Zerial, M. (2003) RhoD regulates endosome dynamics through Diaphanous-related Formin and Src tyrosine kinase. *Nat. Cell Biol.* 5, 195–204.
44. Habas, R., Kato, Y., and He, X. (2001) Wnt/Frizzled activation of Rho regulates vertebrate gastrulation and requires a novel Formin homology protein Daam1. *Cell.* 107, 843–854.
45. Pellegrin, S. and Mellor, H. (2005) The Rho family GTPase Rif induces filopodia through mDia2. *Curr. Biol.* 15, 129–133.
46. Schirenbeck, A., Arasada, R., Bretschneider, T., Schleicher, M., and Faix, J. (2005) Formins and VASPs may co-operate in the formation of filopodia. *Biochem. Soc. Trans.* 33, 1256–1259.
47. Schirenbeck, A., Bretschneider, T., Arasada, R., Schleicher, M., and Faix, J. (2005) The Diaphanous-related formin dDia2 is required for the formation and maintenance of filopodia. *Nat. Cell Biol.* 7, 619–625.
48. Higashida, C., Miyoshi, T., Fujita, A., Oceguera-Yanez, F., Monypenny, J., Andou, Y., Narumiya, S., and Watanabe, N. (2004) Actin polymerization-driven molecular movement of mDia1 in living cells. *Sci. Signall.* 303, 2007.
49. Kovar, D.R., Harris, E.S., Mahaffy, R., Higgs, H.N., and Pollard, T.D. (2006) Control of the assembly of ATP-and ADP-actin by formins and profilin. *Cell.* 124, 423–435.
50. Kovar, D.R. and Pollard, T.D. (2004) Insertional assembly of actin filament barbed ends in association with formins produces piconewton forces. *Proc. Natl. Acad. Sci. USA* 101, 14725–14730.
51. Zigmond, S.H., Evangelista, M., Boone, C., Yang, C., Dar, A.C., Sicheri, F., Forkey, J., and Pring, M. (2003) Formin leaky cap allows elongation in the presence of tight capping proteins. *Curr. Biol.* 13, 1820–1823.
52. Drees, F. and Gertler, F.B. (2008) Ena/VASP: Proteins at the tip of the nervous system. *Curr. Opin. Neurobiol.* 18, 53–59.
53. Krause, M., Bear, J.E., Loureiro, J.J., and Gertler, F.B. (2002) The Ena/VASP enigma. *J. Cell. Sci.* 115, 4721–4726.
54. Krause, M., Dent, E.W., Bear, J.E., Loureiro, J.J., and Gertler, F.B. (2003) Ena/VASP proteins: Regulators of the actin cytoskeleton and cell migration. *Annu. Rev. Cell Dev. Biol.* 19, 541–564.
55. Chakraborty, T., Ebel, F., Domann, E., Niebuhr, K., Gerstel, B., Pistor, S., Temm-Grove, C., Jockusch, B., Reinhard, M., and Walter, U. (1995) A focal adhesion factor directly linking intracellularly motile *Listeria monocytogenes* and *Listeria ivanovii* to the actin-based cytoskeleton of mammalian cells. *EMBO J.* 14, 1314.
56. Laurent, V., Loisel, T.P., Harbeck, B., Wehman, A., Gröbe, L., Jockusch, B.M., Wehland, J., Gertler, F.B., and Carlier, M.F. (1999) Role of proteins of the Ena/VASP family in actin-based motility of *Listeria monocytogenes*. *J. Cell Biol.* 144, 1245–1258.
57. Loisel, T.P., Boujemaa, R., Pantaloni, D., and Carlier, M.F. (1999) Reconstitution of actin-based motility of *Listeria* and *Shigella* using pure proteins. *Nature.* 401, 613–616.
58. Bear, J.E., Loureiro, J.J., Libova, I., Fässler, R., Wehland, J., and Gertler, F.B. (2000) Negative regulation of fibroblast motility by Ena/VASP proteins. *Cell.* 101, 717–728.
59. Lanier, L.M., Gates, M.A., Witke, W., Menzies, A.S., Wehman, A.M., Macklis, J.D., Kwiatkowski, D., Soriano, P., and Gertler, F.B. (1999) Mena is required for neurulation and commissure formation. *Neuron.* 22, 313–325.
60. Wills, Z., Marr, L., Zinn, K., Goodman, C.S., and Van Vactor, D. (1999) Profilin and the Abl tyrosine kinase are required for motor axon outgrowth in the *Drosophila* embryo. *Neuron.* 22, 291–299.
61. Lacayo, C.I., Pincus, Z., VanDuijn, M.M., Wilson, C.A., Fletcher, D.A., Gertler, F.B., Mogilner, A., and Theriot, J.A. (2007) Emergence of large-scale cell morphology and movement from local actin filament growth dynamics. *PLoS Biol.* 5, e233.
62. Schirenbeck, A., Arasada, R., Bretschneider, T., Stradal, T.E.B., Schleicher, M., and Faix, J. (2006) The bundling activity of vasodilator-stimulated phosphoprotein is required for filopodium formation. *Proc. Natl. Acad. Sci. USA* 103, 7694–7699.
63. Barzik, M., Kotova, T.I., Higgs, H.N., Hazelwood, L., Hanein, D., Gertler, F.B., and Schafer, D.A. (2005) Ena/VASP proteins enhance actin polymerization in the presence of barbed end capping proteins. *J. Biol. Chem.* 280, 28653–28662.

64. Skoble, J., Auerbuch, V., Goley, E.D., Welch, M.D., and Portnoy, D.A. (2001) Pivotal role of VASP in Arp2/3 complex–mediated actin nucleation, actin branch-formation, and *Listeria monocytogenes* motility. *J. Cell Biol.* 155, 89–100.

65. Samarin, S., Romero, S., Kocks, C., Didry, D., Pantaloni, D., and Carlier, M.F. (2003) How VASP enhances actin-based motility. *J. Cell Biol.* 163, 131–142.

66. Korenbaum, E. and Rivero, F. (2002) Calponin homology domains at a glance. *J. Cell. Sci.* 115, 3543–3545.

67. Jansen, S., Collins, A., Yang, C., Rebowski, G., Svitkina, T., and Dominguez, R. (2011) Mechanism of actin filament bundling by fascin. *J. Biol. Chem.* 286, 30087–30096.

68. Haviv, L., Gov, N., Ideses, Y., and Bernheim-Groswasser, A. (2008) Thickness distribution of actin bundles in vitro. *Eur. Biophys. J.* 37, 447–454.

69. Haviv, L., Gillo, D., Backouche, F., and Bernheim-Groswasser, A. (2008) A cytoskeletal demolition worker: Myosin II acts as an actin depolymerization agent. *J. Mol. Biol.* 375, 325–330.

70. Courson, D.S. and Rock, R.S. (2010) Actin cross-link assembly and disassembly mechanics for α-actinin and fascin. *J. Biol. Chem.* 285, 26350–26357.

71. Stossel, T.P., Condeelis, J., Cooley, L., Hartwig, J.H., Noegel, A., Schleicher, M., and Shapiro, S.S. (2001) Filamins as integrators of cell mechanics and signalling. *Nat. Rev. Mol. Cell Biol.* 2, 138–145.

72. Vignjevic, D., Kojima, S., Aratyn, Y., Danciu, O., Svitkina, T., and Borisy, G.G. (2006) Role of fascin in filopodial protrusion. *J. Cell Biol.* 174, 863–875.

73. Rzadzinska, A., Schneider, M., Noben-Trauth, K., Bartles, J.R., and Kachar, B. (2005) Balanced levels of espin are critical for stereociliary growth and length maintenance. *Cell Motil. Cytoskeleton.* 62, 157–165.

74. Mukhina, S., Wang, Y., and Murata-Hori, M. (2007) α-Actinin is required for tightly regulated remodeling of the actin cortical network during cytokinesis. *Dev. Cell.* 13, 554–565.

75. Laporte, D., Ojkic, N., Vavylonis, D., and Wu, J. (2012) α-Actinin and fimbrin cooperate with myosin II to organize actomyosin bundles during contractile-ring assembly. *Mol. Biol. Cell.* 23, 3094–3110.

76. Salbreux, G., Charras, G., and Paluch, E. (2012) Actin cortex mechanics and cellular morphogenesis. *Trends Cell Biol.* 22, 536–545.

77. Feng, Y. and Walsh, C.A. (2004) The many faces of filamin: A versatile molecular scaffold for cell motility and signalling. *Nat. Cell Biol.* 6, 1034–1038.

78. Reichl, E.M., Ren, Y., Morphew, M.K., Delannoy, M., Effler, J.C., Girard, K.D., Divi, S., Iglesias, P.A., Kuo, S.C., and Robinson, D.N. (2008) Interactions between myosin and actin crosslinkers control cytokinesis contractility dynamics and mechanics. *Curr. Biol.* 18, 471–480.

79. MacKintosh, F., Käs, J., and Janmey, P. (1995) Elasticity of semiflexible biopolymer networks. *Phys. Rev. Lett.* 75, 4425.

80. Lieleg, O., Claessens, M.M., Heussinger, C., Frey, E., and Bausch, A.R. (2007) Mechanics of bundled semiflexible polymer networks. *Phys. Rev. Lett.* 99, 088102.

81. Gardel, M., Shin, J., MacKintosh, F., Mahadevan, L., Matsudaira, P., and Weitz, D. (2004) Elastic behavior of cross-linked and bundled actin networks. *Science.* 304, 1301–1305.

82. Gardel, M., Nakamura, F., Hartwig, J., Crocker, J., Stossel, T., and Weitz, D. (2006) Prestressed F-actin networks cross-linked by hinged filamins replicate mechanical properties of cells. *Proc. Natl. Acad. Sci. USA* 103, 1762–1767.

83. Backouche, F., Haviv, L., Groswasser, D., and Bernheim-Groswasser, A. (2006) Active gels: Dynamics of patterning and self-organization. *Phys. Biol.* 3, 264.

84. Ideses, Y., Sonn-Segev, A., Roichman, Y., and Bernheim, A. (2013) Myosin II does it all: Assembly, remodeling, and disassembly of actin networks are governed by myosin II activity. *Soft Matter.* 9, 7127–7137.

85. Koenderink, G.H., Dogic, Z., Nakamura, F., Bendix, P.M., MacKintosh, F.C., Hartwig, J.H., Stossel, T.P., and Weitz, D.A. (2009) An active biopolymer network controlled by molecular motors. *Proc. Natl. Acad. Sci. USA* 106, 15192–15197.

86. Vale, R.D. (2003) The molecular motor toolbox for intracellular transport. *Cell.* 112, 467–480.

87. Rock, R.S., Rice, S.E., Wells, A.L., Purcell, T.J., Spudich, J.A., and Sweeney, H.L. (2001) Myosin VI is a processive motor with a large step size. *Proc. Natl. Acad. Sci. USA* 98, 13655–13659.

88. Vale, R.D. and Milligan, R.A. (2000) The way things move: Looking under the hood of molecular motor proteins. *Science.* 288, 88–95.

89. Endow, S.A. (1999) Determinants of molecular motor directionality. *Nat. Cell Biol.* 1, E163–E167.

90. Ménétrey, J., Bahloul, A., Wells, A.L., Yengo, C.M., Morris, C.A., Sweeney, H.L., and Houdusse, A. (2005) The structure of the myosin VI motor reveals the mechanism of directionality reversal. *Nature.* 435, 779–785.

91. Rüegg, C., Veigel, C., Molloy, J.E., Schmitz, S., Sparrow, J.C., and Fink, R.H. (2002) Molecular motors: Force and movement generated by single myosin II molecules. *Physiology.* 17, 213–218.

92. Tyska, M., Dupuis, D., Guilford, W., Patlak, J., Waller, G., Trybus, K., Warshaw, D., and Lowey, S. (1999) Two heads of myosin are better than one for generating force and motion. *Proc. Natl. Acad. Sci. USA* 96, 4402–4407.

93. Ricca, B.L. and Rock, R.S. (2010) The stepping pattern of myosin X is adapted for processive motility on bundled actin. *Biophys. J.* 99, 1818–1826.

94. Svoboda, K., Schmidt, C.F., Schnapp, B.J., and Block, S.M. (1993) Direct observation of kinesin stepping by optical trapping interferometry. *Nature.* 365, 721–727.

95. Mallik, R., Carter, B.C., Lex, S.A., King, S.J., and Gross, S.P. (2004) Cytoplasmic dynein functions as a gear in response to load. *Nature.* 427, 649–652.

96. Schmitz, K.A., Holcomb-Wygle, D.L., Oberski, D.J., and Lindemann, C.B. (2000) Measurement of the force produced by an intact bull sperm flagellum in isometric arrest and estimation of the dynein stall force. *Biophys. J.* 79, 468–478.

97. Wrobel, L.K., Fray, T.R., Molloy, J.E., Adams, J.J., Armitage, M.P., and Sparrow, J.C. (2002) Contractility of single human dermal myofibroblasts and fibroblasts. *Cell Motil. Cytoskeleton.* 52, 82–90.

98. Smith, P.G., Roy, C., Fisher, S., Huang, Q., and Brozovich, F. (2000) Selected contribution: Mechanical strain increases force production and calcium sensitivity in cultured airway smooth muscle cells. *J. Appl. Physiol.* 89, 2092–2098.

99. Higuchi, H. and Endow, S.A. (2002) Directionality and processivity of molecular motors. *Curr. Opin. Cell Biol.* 14, 50–57.

100. Toprak, E., Yildiz, A., Hoffman, M.T., Rosenfeld, S.S., and Selvin, P.R. (2009) Why kinesin is so processive. *Proc. Natl. Acad. Sci. USA* 106, 12717–12722.

101. Walker, M.L., Burgess, S.A., Sellers, J.R., Wang, F., Hammer, J.A., Trinick, J., and Knight, P.J. (2000) Two-headed binding of a processive myosin to F-actin. *Nature.* 405, 804–807.

102. Kambara, T. and Ikebe, M. (2006) A unique ATP hydrolysis mechanism of single-headed processive myosin, myosin IX. *J. Biol. Chem.* 281, 4949–4957.

103. Brunello, E., Reconditi, M., Elangovan, R., Linari, M., Sun, Y., Narayanan, T., Panine, P., Piazzesi, G., Irving, M., and Lombardi, V. (2007) Skeletal muscle resists stretch by rapid binding of the second motor domain of myosin to actin. *Proc. Natl. Acad. Sci. USA* 104, 20114–20119.

104. Piazzesi, G., Reconditi, M., Linari, M., Lucii, L., Bianco, P., Brunello, E., Decostre, V., Stewart, A., Gore, D.B., and Irving, T.C. (2007) Skeletal muscle performance determined by modulation of number of myosin motors rather than motor force or stroke size. *Cell.* 131, 784–795.

105. Harris, D., Work, S., Wright, R., Alpert, N., and Warshaw, D. (1994) Smooth, cardiac and skeletal muscle myosin force and motion generation assessed by cross-bridge mechanical interactions in vitro. *J. Muscle Res. Cell Motil.* 15, 11–19.

106. Borejdo, J. and Burlacu, S. (1992) Velocity of movement of actin filaments in in vitro motility assay. Measured by fluorescence correlation spectroscopy. *Biophys. J.* 61, 1267–1280.

107. Howard, J. (1997) Molecular motors: Structural adaptations to cellular functions. *Nature.* 389, 561–567.

108. Pierobon, P., Achouri, S., Courty, S., Dunn, A.R., Spudich, J.A., Dahan, M., and Cappello, G. (2009) Velocity, processivity, and individual steps of single myosin V molecules in live cells. *Biophys. J.* 96, 4268–4275.

109. Tsaturyan, A.K., Bershitsky, S.Y., Koubassova, N.A., Fernandez, M., Narayanan, T., and Ferenczi, M.A. (2011) The fraction of myosin motors that participate in isometric contraction of rabbit muscle fibers at near-physiological temperature. *Biophys. J.* 101, 404–410.

110. Kovács, M., Thirumurugan, K., Knight, P.J., and Sellers, J.R. (2007) Load-dependent mechanism of nonmuscle myosin 2. *Proc. Natl. Acad. Sci. USA* 104, 9994–9999.

111. Bray, D. and White, J. (1988) Cortical flow in animal cells. *Science.* 239, 883–888.

112. Fernandez-Gonzalez, R., Simoes, S.d.M., Röper, J., Eaton, S., and Zallen, J.A. (2009) Myosin II dynamics are regulated by tension in intercalating cells. *Dev. Cell.* 17, 736–743.

113. Martin, A.C., Gelbart, M., Fernandez-Gonzalez, R., Kaschube, M., and Wieschaus, E.F. (2010) Integration of contractile forces during tissue invagination. *J. Cell Biol.* 188, 735–749.

114. Rauzi, M., Verant, P., Lecuit, T., and Lenne, P. (2008) Nature and anisotropy of cortical forces orienting *Drosophila* tissue morphogenesis. *Nat. Cell Biol.* 10, 1401–1410.

115. Martin, A.C. (2010) Pulsation and stabilization: Contractile forces that underlie morphogenesis. *Dev. Biol.* 341, 114–125.

116. Stewart, M. and Kensler, R.W. (1986) Arrangement of myosin heads in relaxed thick filaments from frog skeletal muscle. *J. Mol. Biol.* 192, 831–851.

117. Wilson, C.A., Tsuchida, M.A., Allen, G.M., Barnhart, E.L., Applegate, K.T., Yam, P.T., Ji, L., Keren, K., Danuser, G., and Theriot, J.A. (2010) Myosin II contributes to cell-scale actin network treadmilling through network disassembly. *Nature.* 465, 373–377.

118. Pelham, R.J. and Chang, F. (2002) Actin dynamics in the contractile ring during cytokinesis in fission yeast. *Nature.* 419, 82–86.

119. Guha, M., Zhou, M., and Wang, Y. (2005) Cortical actin turnover during cytokinesis requires myosin II. *Curr. Biol.* 15, 732–736.

120. Burgess, D.R. (2005) Cytokinesis: New roles for myosin. *Curr. Biol.* 15, R310–R311.

121. Murthy, K. and Wadsworth, P. (2005) Myosin-II-dependent localization and dynamics of F-actin during cytokinesis. *Curr. Biol.* 15, 724–731.

122. Reymann, A., Boujemaa-Paterski, R., Martiel, J., Guérin, C., Cao, W., Chin, H.F., Enrique, M., Théry, M., and Blanchoin, L. (2012) Actin network architecture can determine myosin motor activity. *Science.* 336, 1310–1314.

123. Helenius, J., Brouhard, G., Kalaidzidis, Y., Diez, S., and Howard, J. (2006) The depolymerizing kinesin MCAK uses lattice diffusion to rapidly target microtubule ends. *Nature.* 441, 115–119.

124. Veksler, A. and Gov, N.S. (2007) Phase transitions of the coupled membrane-cytoskeleton modify cellular shape. *Biophys. J.* 93, 3798–3810.

125. Shlomovitz, R. and Gov, N. (2007) Membrane waves driven by actin and myosin. *Phys. Rev. Lett.* 98, 168103.

126. Gov, N.S. and Gopinathan, A. (2006) Dynamics of membranes driven by actin polymerization. *Biophys. J.* 90, 454–469.

127. Haviv, L., Brill-Karniely, Y., Mahaffy, R., Backouche, F., Ben-Shaul, A., Pollard, T.D., and Bernheim-Groswasser, A. (2006) Reconstitution of the transition from lamellipodium to filopodium in a membrane-free system. *Proc. Natl. Acad. Sci. USA* 103, 4906–4911.

128. Ideses, Y., Brill-Karniely, Y., Haviv, L., Ben-Shaul, A., and Bernheim-Groswasser, A. (2008) Arp2/3 branched actin network mediates filopodia-like bundles formation in vitro. *PloS One.* 3, e3297.

129. Bernheim-Groswasser, A., Wiesner, S., Golsteyn, R.M., Carlier, M.F., and Sykes, C. (2002) The dynamics of actin-based motility depend on surface parameters. *Nature.* 417, 308–311.

130. Bernheim-Groswasser, A., Prost, J., and Sykes, C. (2005) Mechanism of actin-based motility: A dynamic state diagram. *Biophys. J.* 89, 1411–1419.

131. Marcy, Y., Prost, J., Carlier, M., and Sykes, C. (2004) Forces generated during actin-based propulsion: A direct measurement by micromanipulation. *Proc. Natl. Acad. Sci. USA* 101, 5992–5997.

132. Paluch, E., van der Gucht, J., Joanny, J.F., and Sykes, C. (2006) Deformations in actin comets from rocketing beads. *Biophys. J.* 91, 3113–3122.
133. Loose, M., Fischer-Friedrich, E., Ries, J., Kruse, K., and Schwille, P. (2008) Spatial regulators for bacterial cell division self-organize into surface waves in vitro. *Science.* 320, 789–792.
134. Gillo, D., Gur, B., Bernheim-Groswasser, A., and Farago, O. (2009) Cooperative molecular motors moving back and forth. *Phys. Rev. E.* 80, 021929.
135. Gilboa, B., Gillo, D., Farago, O., and Bernheim-Groswasser, A. (2009) Bidirectional cooperative motion of myosin-II motors on actin tracks with randomly alternating polarities. *Soft Matter.* 5, 2223–2231.
136. Brill-Karniely, Y., Ideses, Y., Bernheim-Groswasser, A., and Ben-Shaul, A. (2009) From branched networks of actin filaments to bundles. *Chem Phys Chem.* 10, 2818–2827.
137. Karsenti, E., Nédélec, F., and Surrey, T. (2006) Modelling microtubule patterns. *Nat. Cell Biol.* 8, 1204–1211.
138. Badoual, M., Jülicher, F., and Prost, J. (2002) Bidirectional cooperative motion of molecular motors. *Proc. Natl. Acad. Sci. USA* 99, 6696–6701.
139. Klein, G.A., Kruse, K., Cuniberti, G., and Jülicher, F. (2005) Filament depolymerization by motor molecules. *Phys. Rev. Lett.* 94, 108102.
140. Mogilner, A. and Oster, G. (2003) Force generation by actin polymerization II: The elastic ratchet and tethered filaments. *Biophys. J.* 84, 1591–1605.
141. Lenz, M., Thoresen, T., Gardel, M.L., and Dinner, A.R. (2012) Contractile units in disordered actomyosin bundles arise from F-actin buckling. *Phys. Rev. Lett.* 108, 238107.
142. Rougerie, P., Miskolci, V., and Cox, D. (2013) Generation of membrane structures during phagocytosis and chemotaxis of macrophages: Role and regulation of the actin cytoskeleton. *Immunol. Rev.* 256, 222–239.
143. Mattila, P.K. and Lappalainen, P. (2008) Filopodia: Molecular architecture and cellular functions. *Nat. Rev. Mol. Cell Biol.* 9, 446–454.
144. Faix, J., Breitsprecher, D., Stradal, T.E.B., and Rottner, K. (2009) Filopodia: Complex models for simple rods. *Int. J. Biochem. Cell Biol.* 41, 1656–1664.
145. Yang, C. and Svitkina, T. (2011) Filopodia initiation: Focus on the Arp2/3 complex and formins. *Cell Adh. Migr.* 5, 402–408.
146. Medalia, O., Beck, M., Ecke, M., Weber, I., Neujahr, R., Baumeister, W., and Gerisch, G. (2007) Organization of actin networks in intact filopodia. *Curr. Biol.* 17, 79–84.
147. Applewhite, D.A., Barzik, M., Kojima, S., Svitkina, T.M., Gertler, F.B., and Borisy, G.G. (2007) Ena/VASP proteins have an anti-capping independent function in filopodia formation. *Mol. Biol. Cell.* 18, 2579–2591.
148. Mallavarapu, A. and Mitchison, T. (1999) Regulated actin cytoskeleton assembly at filopodium tips controls their extension and retraction. *J. Cell Biol.* 146, 1097–1106.
149. Dent, E.W. and Gertler, F.B. (2003) Cytoskeletal dynamics and transport in growth cone motility and axon guidance. *Neuron.* 40, 209–227.
150. Svitkina, T.M., Bulanova, E.A., Chaga, O.Y., Vignjevic, D.M., Kojima, S., Vasiliev, J.M., and Borisy, G.G. (2003) Mechanism of filopodia initiation by reorganization of a dendritic network. *J. Cell Biol.* 160, 409–421.
151. Block, J., Stradal, T., Hänisch, J., Geffers, R., Köstler, S., Urban, E., Small, J., Rottner, K., and Faix, J. (2008) Filopodia formation induced by active mDia2/Drf3. *J. Microsc.* 231, 506–517.
152. Steffen, A., Faix, J., Resch, G.P., Linkner, J., Wehland, J., Small, J.V., Rottner, K., and Stradal, T.E.B. (2006) Filopodia formation in the absence of functional WAVE-and Arp2/3-complexes. *Mol. Biol. Cell.* 17, 2581–2591.
153. Liu, A.P. and Fletcher, D.A. (2009) Biology under construction: In vitro reconstitution of cellular function. *Nat. Rev. Mol. Cell Biol.* 10, 644–650.
154. Vignaud, T., Blanchoin, L., and Théry, M. (2012) Directed cytoskeleton self-organization. *Trends Cell Biol.* 22, 671–682.
155. Vogel, S.K. and Schwille, P. (2012) Minimal systems to study membrane–cytoskeleton interactions. *Curr. Opin. Biotechnol.* 23, 758–765.

156. Pistor, S., Chakraborty, T., Niebuhr, K., Domann, E., and Wehland, J. (1994) The ActA protein of *Listeria monocytogenes* acts as a nucleator inducing reorganization of the actin cytoskeleton. *EMBO J.* 13, 758.

157. Tilney, L.G. and Portnoy, D.A. (1989) Actin filaments and the growth, movement, and spread of the intracellular bacterial parasite, *Listeria monocytogenes*. *J. Cell Biol.* 109, 1597–1608.

158. Theriot, J.A., Rosenblatt, J., Portnoy, D.A., Goldschmidt-Clermont, P.J., and Mitchison, T.J. (1994) Involvement of profilin in the actin-based motility of *L. monocytogenes* in cells and in cell-free extracts. *Cell.* 76, 505–517.

159. Cameron, L.A., Footer, M.J., Van Oudenaarden, A., and Theriot, J.A. (1999) Motility of ActA protein-coated microspheres driven by actin polymerization. *Proc. Natl. Acad. Sci. USA* 96, 4908–4913.

160. Noireaux, V., Golsteyn, R., Friederich, E., Prost, J., Antony, C., Louvard, D., and Sykes, C. (2000) Growing an actin gel on spherical surfaces. *Biophys. J.* 78, 1643–1654.

161. Akin, O. and Mullins, R.D. (2008) Capping protein increases the rate of actin-based motility by promoting filament nucleation by the Arp2/3 complex. *Cell.* 133, 841–851.

162. Delatour, V., Shekhar, S., Reymann, A.C., Didry, D., Lê, K.H.D., Romet-Lemonne, G., Helfer, E., and Carlier, M.F. (2008) Actin-based propulsion of functionalized hard versus fluid spherical objects. *New J. Phys.* 10, 025001.

163. Plastino, J., Olivier, S., and Sykes, C. (2004) Actin filaments align into hollow comets for rapid VASP-mediated propulsion. *Curr. Biol.* 14, 1766–1771.

164. Plastino, J., Lelidis, I., Prost, J., and Sykes, C. (2004) The effect of diffusion, depolymerization and nucleation promoting factors on actin gel growth. *Eur. Biophys. J.* 33, 310–320.

165. Siton, O., Ideses, Y., Albeck, S., Unger, T., Bershadsky, A.D., Gov, N.S., and Bernheim-Groswasser, A. (2011) Cortactin releases the brakes in actin-based motility by enhancing WASP-VCA detachment from Arp2/3 branches. *Curr. Biol.* 21, 2092–2097.

166. Breitsprecher, D., Kiesewetter, A.K., Linkner, J., Urbanke, C., Resch, G.P., Small, J.V., and Faix, J. (2008) Clustering of VASP actively drives processive, WH2 domain-mediated actin filament elongation. *EMBO J.* 27, 2943–2954.

167. Michelot, A., Berro, J., Guérin, C., Boujemaa-Paterski, R., Staiger, C.J., Martiel, J.L., and Blanchoin, L. (2007) Actin-filament stochastic dynamics mediated by ADF/cofilin. *Curr. Biol.* 17, 825–833.

168. Mogilner, A. and Oster, G. (1996) Cell motility driven by actin polymerization. *Biophys. J.* 71, 3030–3045.

169. Gerbal, F., Chaikin, P., Rabin, Y., and Prost, J. (2000) An elastic analysis of *Listeria monocytogenes* propulsion. *Biophys. J.* 79, 2259–2275.

170. Gerbal, F., Noireaux, V., Sykes, C., Jülicher, F., Chaikin, P., Ott, A., Prost, J., Golsteyn, R., Friederich, E., and Louvard, D. (1999) On the 'listeria' propulsion mechanism. *Pramana.* 53, 155–170.

171. Merrifield, C.J., Moss, S.E., Ballestrem, C., Imhof, B.A., Giese, G., Wunderlich, I., and Almers, W. (1999) Endocytic vesicles move at the tips of actin tails in cultured mast cells. *Nat. Cell Biol.* 1, 72–74.

172. Taunton, J., Rowning, B.A., Coughlin, M.L., Wu, M., Moon, R.T., Mitchison, T.J., and Larabell, C.A. (2000) Actin-dependent propulsion of endosomes and lysosomes by recruitment of N-WASP. *J. Cell Biol.* 148, 519–530.

173. Van Der Gucht, J., Paluch, E., Plastino, J., and Sykes, C. (2005) Stress release drives symmetry breaking for actin-based movement. *Proc. Natl. Acad. Sci. USA* 102, 7847–7852.

174. Sekimoto, K., Prost, J., Jülicher, F., Boukellal, H., and Bernheim-Groswasser, A. (2004) Role of tensile stress in actin gels and a symmetry-breaking instability. *Eur. Phys. J. E.* 13, 247–259.

175. Lasa, I., Gouin, E., Goethals, M., Vancompernolle, K., David, V., Vandekerckhove, J., and Cossart, P. (1997) Identification of two regions in the N-terminal domain of ActA involved in the actin comet tail formation by *Listeria monocytogenes*. *EMBO J.* 16, 1531–1540.

176. Trichet, L., Campàs, O., Sykes, C., and Plastino, J. (2007) VASP governs actin dynamics by modulating filament anchoring. *Biophys. J.* 92, 1081–1089.

177. Boukellal, H., Campás, O., Joanny, J.F., Prost, J., and Sykes, C. (2004) Soft *Listeria*: Actin-based propulsion of liquid drops. *Phys. Rev. E.* 69, 061906.

178. Plastino, J. and Sykes, C. (2005) The actin slingshot. *Curr. Opin. Cell Biol.* 17, 62–66.

179. Cortese, J.D., Schwab, B., Frieden, C., and Elson, E.L. (1989) Actin polymerization induces a shape change in actin-containing vesicles. *Proc. Natl. Acad. Sci. USA* 86, 5773–5777.

180. Giardini, P.A., Fletcher, D.A., and Theriot, J.A. (2003) Compression forces generated by actin comet tails on lipid vesicles. *Proc. Natl. Acad. Sci. USA* 100, 6493–6498.

181. Upadhyaya, A., Chabot, J.R., Andreeva, A., Samadani, A., and Van Oudenaarden, A. (2003) Probing polymerization forces by using actin-propelled lipid vesicles. *Proc. Natl. Acad. Sci. USA* 100, 4521–4526.

182. Sackmann, E. (1996) Supported membranes: Scientific and practical applications. *Science-AAAS-Weekly Paper Edition.* 271, 43–48.

183. Sackmann, E. and Tanaka, M. (2000) Supported membranes on soft polymer cushions: Fabrication, characterization and applications. *Trends Biotechnol.* 18, 58–64.

184. Liu, A.P. and Fletcher, D.A. (2006) Actin polymerization serves as a membrane domain switch in model lipid bilayers. *Biophys. J.* 91, 4064–4070.

185. Liu, A.P., Richmond, D.L., Maibaum, L., Pronk, S., Geissler, P.L., and Fletcher, D.A. (2008) Membrane-induced bundling of actin filaments. *Nat. Phys.* 4, 789–793.

186. Merkle, D., Kahya, N., and Schwille, P. (2008) Reconstitution and anchoring of cytoskeleton inside giant unilamellar vesicles. *ChemBioChem.* 9, 2673–2681.

187. Pontani, L., Van der Gucht, J., Salbreux, G., Heuvingh, J., Joanny, J.F., and Sykes, C. (2009) Reconstitution of an actin cortex inside a liposome. *Biophys. J.* 96, 192–198.

188. Murrell, M., Pontani, L., Guevorkian, K., Cuvelier, D., Nassoy, P., and Sykes, C. (2011) Spreading dynamics of biomimetic actin cortices. *Biophys. J.* 100, 1400–1409.

189. Co, C., Wong, D.T., Gierke, S., Chang, V., and Taunton, J. (2007) Mechanism of actin network attachment to moving membranes: Barbed end capture by N-WASP WH2 domains. *Cell.* 128, 901–913.

190. Richter, R.P., Bérat, R., and Brisson, A.R. (2006) Formation of solid-supported lipid bilayers: An integrated view. *Langmuir.* 22, 3497–3505.

191. Hitt, A.L., Hartwig, J.H., and Luna, E.J. (1994) Ponticulin is the major high affinity link between the plasma membrane and the cortical actin network in *Dictyostelium. J. Cell Biol.* 126, 1433–1444.

192. Barfoot, R.J., Sheikh, K.H., Johnson, B.R., Colyer, J., Miles, R.E., Jeuken, L.J., Bushby, R.J., and Evans, S.D. (2008) Minimal F-actin cytoskeletal system for planar supported phospholipid bilayers. *Langmuir.* 24, 6827–6836.

193. Lee, K., Gallop, J.L., Rambani, K., and Kirschner, M.W. (2010) Self-assembly of filopodia-like structures on supported lipid bilayers. *Science.* 329, 1341–1345.

194. Giannone, G., Dubin-Thaler, B.J., Döbereiner, H., Kieffer, N., Bresnick, A.R., and Sheetz, M.P. (2004) Periodic lamellipodial contractions correlate with rearward actin waves. *Cell.* 116, 431–443.

195. Weiner, O.D., Marganski, W.A., Wu, L.F., Altschuler, S.J., and Kirschner, M.W. (2007) An actin-based wave generator organizes cell motility. *PLoS Biol.* 5, e221.

196. Doubrovinski, K. and Kruse, K. (2008) Cytoskeletal waves in the absence of molecular motors. *Europhys. Lett.* 83, 18003.

197. Doubrovinski, K. and Kruse, K. (2011) Cell motility resulting from spontaneous polymerization waves. *Phys. Rev. Lett.* 107, 258103.

198. Levi, V., Serpinskaya, A.S., Gratton, E., and Gelfand, V. (2006) Organelle transport along microtubules in *Xenopus* melanophores: Evidence for cooperation between multiple motors. *Biophys. J.* 90, 318–327.

199. Kural, C., Kim, H., Syed, S., Goshima, G., Gelfand, V.I., and Selvin, P.R. (2005) Kinesin and dynein move a peroxisome in vivo: A tug-of-war or coordinated movement? *Science.* 308, 1469–1472.

200. Morris, R.L. and Hollenbeck, P.J. (1993) The regulation of bidirectional mitochondrial transport is coordinated with axonal outgrowth. *J. Cell. Sci.* 104, 917–927.

201. Soppina, V., Rai, A.K., Ramaiya, A.J., Barak, P., and Mallik, R. (2009) Tug-of-war between dissimilar teams of microtubule motors regulates transport and fission of endosomes. *Proc. Natl. Acad. Sci. USA* 106, 19381–19386.

202. Tao, L., Mogilner, A., Civelekoglu-Scholey, G., Wollman, R., Evans, J., Stahlberg, H., and Scholey, J.M. (2006) A homotetrameric kinesin-5, KLP61F, bundles microtubules and antagonizes Ncd in motility assays. *Curr. Biol.* 16, 2293–2302.

203. Leduc, C., Pavin, N., Jülicher, F., and Diez, S. (2010) Collective behavior of antagonistically acting kinesin-1 motors. *Phys. Rev. Lett.* 105, 128103.

204. Meili, R., Alonso-Latorre, B., del Álamo, J.C., Firtel, R.A., and Lasheras, J.C. (2010) Myosin II is essential for the spatiotemporal organization of traction forces during cell motility. *Mol. Biol. Cell.* 21, 405–417.

205. Geeves, M.A. and Holmes, K.C. (1999) Structural mechanism of muscle contraction. *Annu. Rev. Biochem.* 68, 687–728.

206. Calvert, M.E., Wright, G.D., Leong, F.Y., Chiam, K., Chen, Y., Jedd, G., and Balasubramanian, M.K. (2011) Myosin concentration underlies cell size–dependent scalability of actomyosin ring constriction. *J. Cell Biol.* 195, 799–813.

207. Pollard, T.D. (2010) Mechanics of cytokinesis in eukaryotes. *Curr. Opin. Cell Biol.* 22, 50–56.

208. Geiger, B., Spatz, J.P., and Bershadsky, A.D. (2009) Environmental sensing through focal adhesions. *Nat. Rev. Mol. Cell Biol.* 10, 21–33.

209. Machin, K. and Pringle, J. (1959) The physiology of insect fibrillar muscle. II. Mechanical properties of a beetle flight muscle. *Proc. R. Soc. Lond. Series B. Biol. Sci.* 151, 204–225.

210. Yasuda, K., Shindo, Y., and Ishiwata, S. (1996) Synchronous behavior of spontaneous oscillations of sarcomeres in skeletal myofibrils under isotonic conditions. *Biophys. J.* 70, 1823–1829.

211. Brokaw, C.J. (1975) Molecular mechanism for oscillation in flagella and muscle. *Proc. Natl. Acad. Sci. USA* 72, 3102–3106.

212. Camalet, S. and Jülicher, F. (2000) Generic aspects of axonemal beating. *New J. Phys.* 2, 24.

213. Hayashi, S. and Shingyoji, C. (2008) Mechanism of flagellar oscillation–bending-induced switching of dynein activity in elastase-treated axonemes of sea urchin sperm. *J. Cell. Sci.* 121, 2833–2843.

214. Liao, J., Elting, M.W., Delp, S.L., Spudich, J.A., and Bryant, Z. (2009) Engineered myosin VI motors reveal minimal structural determinants of directionality and processivity. *J. Mol. Biol.* 392, 862–867.

215. Riveline, D., Ott, A., Jülicher, F., Winkelmann, D.A., Cardoso, O., Lacapère, J., Magnúsdóttir, S., Viovy, J., Gorre-Talini, L., and Prost, J. (1998) Acting on actin: The electric motility assay. *Eur. Biophys. J.* 27, 403–408.

216. Jülicher, F. and Prost, J. (1995) Cooperative molecular motors. *Phys. Rev. Lett.* 75, 2618.

217. Van den Heuvel, Martin G.L., De Graaff, M.P., and Dekker, C. (2006) Molecular sorting by electrical steering of microtubules in kinesin-coated channels. *Science.* 312, 910–914.

218. Endow, S.A. and Higuchi, H. (2000) A mutant of the motor protein kinesin that moves in both directions on microtubules. *Nature.* 406, 913–916.

219. Gillo, D., Gilboa, B., Gurka, R., and Bernheim-Groswasser, A. (2009) The fusion of actin bundles driven by interacting motor proteins. *Phys. Biol.* 6, 036003.

220. Plaçais, P., Balland, M., Guérin, T., Joanny, J.F., and Martin, P. (2009) Spontaneous oscillations of a minimal actomyosin system under elastic loading. *Phys. Rev. Lett.* 103, 158102.

221. Brokaw, C.J. (2005) Computer simulation of flagellar movement IX. Oscillation and symmetry breaking in a model for short flagella and nodal cilia. *Cell Motil. Cytoskeleton.* 60, 35–47.

222. DeVille, R.L. and Vanden-Eijnden, E. (2008) Regularity and synchrony in motor proteins. *Bull. Math. Biol.* 70, 484–516.

223. Duke, T. (1999) Molecular model of muscle contraction. *Proc. Natl. Acad. Sci. USA* 96, 2770–2775.

224. Guérin, T., Prost, J., Martin, P., and Joanny, J.F. (2010) Coordination and collective properties of molecular motors: Theory. *Curr. Opin. Cell Biol.* 22, 14–20.

225. Jülicher, F. and Prost, J. (1997) Spontaneous oscillations of collective molecular motors. *Phys. Rev. Lett.* 78, 4510.

226. Lan, G. and Sun, S.X. (2005) Dynamics of myosin-driven skeletal muscle contraction: I. Steady-state force generation. *Biophys. J.* 88, 4107–4117.

227. Shu, Y. and Shi, H. (2004) Cooperative effects on the kinetics of ATP hydrolysis in collective molecular motors. *Phys. Rev. E.* 69, 021912.

228. Smith, D. and Geeves, M. (1995) Strain-dependent cross-bridge cycle for muscle. *Biophys. J.* 69, 524–537.

229. Touya, C., Schwalger, T., and Lindner, B. (2011) Relation between cooperative molecular motors and active Brownian particles. *Phys. Rev. E.* 83, 051913.

230. Zemel, A. and Mogilner, A. (2009) Motor-induced sliding of microtubule and actin bundles. *Phys. Chem. Chem. Phys.* 11, 4821–4833.

231. Müller, M.J., Klumpp, S., and Lipowsky, R. (2010) Bidirectional transport by molecular motors: Enhanced processivity and response to external forces. *Biophys. J.* 98, 2610–2618.

232. Hendricks, A.G., Perlson, E., Ross, J.L., Schroeder III, H.W., Tokito, M., and Holzbaur, E.L. (2010) Motor coordination via a tug-of-war mechanism drives bidirectional vesicle transport. *Curr. Biol.* 20, 697–702.

233. Hexner, D. and Kafri, Y. (2009) Tug of war in motility assay experiments. *Phys. Biol.* 6, 036016.

234. Farago, O. and Bernheim-Groswasser, A. (2011) Crosstalk between non-processive myosin motors mediated by the actin filament elasticity. *Soft Matter.* 7, 3066–3073.

235. Mizuno, D., Tardin, C., Schmidt, C., and MacKintosh, F. (2007) Nonequilibrium mechanics of active cytoskeletal networks. *Science.* 315, 370–373.

236. Mendes Pinto, I., Rubinstein, B., Kucharavy, A., Unruh, J.R., and Li, R. (2012) Actin depolymerization drives actomyosin ring contraction during budding yeast cytokinesis. *Dev. Cell.* 22, 1247–1260.

237. Prost, J., Barbetta, C., and Joanny, J.F. (2007) Dynamical control of the shape and size of stereocilia and microvilli. *Biophys. J.* 93, 1124–1133.

238. Nakazawa, H. and Sekimoto, K. (1996) Polarity sorting in a bundle of actin filaments by two-headed myosins. *J. Phys. Soc. Jpn.* 65, 2404.

239. Kruse, K., Camalet, S., and Jülicher, F. (2001) Self-propagating patterns in active filament bundles. *Phys. Rev. Lett.* 87, 138101.

240. Liverpool, T.B. and Marchetti, M.C. (2002) Organization and instabilities of entangled active polar filaments. arXiv preprint cond-mat/0207320.

241. Jülicher, F., Kruse, K., Prost, J., and Joanny, J.F. (2007) Active behavior of the cytoskeleton. *Phys. Rep.* 449, 3–28.

242. Joanny, J.F. and Prost, J. (2009) Active gels as a description of the actin-myosin cytoskeleton. *HFSP J.* 3, 94–104.

243. Bassetti, B., Lagomarsino, M.C., and Jona, P. (2000) A model for the self-organization of microtubules driven by molecular motors. *Eur. Phys. J. B-Condens. Matter Complex Syst.* 15, 483–492.

244. Kruse, K., Joanny, J.F., Jülicher, F., Prost, J., and Sekimoto, K. (2004) Asters, vortices, and rotating spirals in active gels of polar filaments. *Phys. Rev. Lett.* 92, 078101.

245. Aranson, I.S. and Tsimring, L.S. (2005) Pattern formation of microtubules and motors: Inelastic interaction of polar rods. *Phys. Rev. E.* 71, 050901.

246. Ziebert, F. and Zimmermann, W. (2005) Nonlinear competition between asters and stripes in filament-motor systems. *Eur. Phys. J. E.* 18, 41–54.

247. Surrey, T., Nédélec, F., Leibler, S., and Karsenti, E. (2001) Physical properties determining self-organization of motors and microtubules. *Science.* 292, 1167–1171.

248. Nédélec, F., Surrey, T., Maggs, A.C., and Leibler, S. (1997) Self-organization of microtubules and motors. *Nature.* 389, 305–308.

249. Hentrich, C. and Surrey, T. (2010) Microtubule organization by the antagonistic mitotic motors kinesin-5 and kinesin-14. *J. Cell Biol.* 189, 465–480.

250. Urrutia, R., McNiven, M.A., Albanesi, J.P., Murphy, D.B., and Kachar, B. (1991) Purified kinesin promotes vesicle motility and induces active sliding between microtubules in vitro. *Proc. Natl. Acad. Sci. USA* 88, 6701–6705.

251. Sankararaman, S., Menon, G.I., and Kumar, P.S. (2004) Self-organized pattern formation in motor-microtubule mixtures. *Phys. Rev. E.* 70, 031905.

252. Liverpool, T.B. (2006) Active gels: Where polymer physics meets cytoskeletal dynamics. *Philos. Trans. R. Soc. A: Math., Phys. Eng. Sci.* 364, 3335–3355.

253. Nédélec, F. (2002) Computer simulations reveal motor properties generating stable antiparallel microtubule interactions. *J. Cell Biol.* 158, 1005–1015.

254. Gordon, D., Bernheim-Groswasser, A., Keasar, C., and Farago, O. (2012) Hierarchical self-organization of cytoskeletal active networks. *Phys. Biol.* 9, 026005.

255. Lee, H.Y. and Kardar, M. (2001) Macroscopic equations for pattern formation in mixtures of microtubules and molecular motors. *Phys. Rev. E.* 64, 056113.

256. e Silva, M.S., Depken, M., Stuhrmann, B., Korsten, M., MacKintosh, F.C., and Koenderink, G.H. (2011) Active multistage coarsening of actin networks driven by myosin motors. *Proc. Natl. Acad. Sci. USA* 108, 9408–9413.

257. Murrell, M.P. and Gardel, M.L. (2012) F-actin buckling coordinates contractility and severing in a biomimetic actomyosin cortex. *Proc. Natl. Acad. Sci. USA* 109, 20820–20825.

258. Footer, M.J., Kerssemakers, J.W., Theriot, J.A., and Dogterom, M. (2007) Direct measurement of force generation by actin filament polymerization using an optical trap. *Proc. Natl. Acad. Sci. USA* 104, 2181–2186.

259. Sanchez, T., Chen, D.T., DeCamp, S.J., Heymann, M., and Dogic, Z. (2012) Spontaneous motion in hierarchically assembled active matter. *Nature.* 491, 431–434.

260. Schaller, V., Weber, C., Semmrich, C., Frey, E., and Bausch, A.R. (2010) Polar patterns of driven filaments. *Nature.* 467, 73–77.

261. Sumino, Y., Nagai, K.H., Shitaka, Y., Tanaka, D., Yoshikawa, K., Chaté, H., and Oiwa, K. (2012) Large-scale vortex lattice emerging from collectively moving microtubules. *Nature.* 483, 448–452.

262. Marchetti, M., Joanny, J.F., Ramaswamy, S., Liverpool, T., Prost, J., Rao, M., and Simha, R.A. (2013) Hydrodynamics of soft active matter. *Rev. Mod. Phys.* 85, 1143.

263. Mayer, M., Depken, M., Bois, J.S., Jülicher, F., and Grill, S.W. (2010) Anisotropies in cortical tension reveal the physical basis of polarizing cortical flows. *Nature.* 467, 617–621.

264. Munro, E., Nance, J., and Priess, J.R. (2004) Cortical flows powered by asymmetrical contraction transport PAR proteins to establish and maintain anterior-posterior polarity in the early *C. elegans* embryo. *Dev. Cell.* 7, 413–424.

265. Zhou, M. and Wang, Y. (2008) Distinct pathways for the early recruitment of myosin II and actin to the cytokinetic furrow. *Mol. Biol. Cell.* 19, 318–326.

266. Mori, M., Monnier, N., Daigle, N., Bathe, M., Ellenberg, J., and Lénárt, P. (2011) Intracellular transport by an anchored homogeneously contracting F-actin meshwork. *Curr. Biol.* 21, 606–611.

267. Hawkins, R.J., Poincloux, R., Bénichou, O., Piel, M., Chavrier, P., and Voituriez, R. (2011) Spontaneous contractility-mediated cortical flow generates cell migration in three-dimensional environments. *Biophys. J.* 101, 1041–1045.

268. Bois, J.S., Jülicher, F., and Grill, S.W. (2011) Pattern formation in active fluids. *Phys. Rev. Lett.* 106, 028103.

269. Voituriez, R., Joanny, J.F., and Prost, J. (2005) Spontaneous flow transition in active polar gels. *Europhys. Lett.* 70, 404.

270. Thoresen, T., Lenz, M., and Gardel, M.L. (2011) Reconstitution of contractile actomyosin bundles. *Biophys. J.* 100, 2698–2705.

271. Bendix, P.M., Koenderink, G.H., Cuvelier, D., Dogic, Z., Koeleman, B.N., Brieher, W.M., Field, C.M., Mahadevan, L., and Weitz, D.A. (2008) A quantitative analysis of contractility in active cytoskeletal protein networks. *Biophys. J.* 94, 3126–3136.

272. Köhler, S. and Bausch, A.R. (2012) Contraction mechanisms in composite active actin networks. *PloS One.* 7, e39869.

273. Kruse, K. and Jülicher, F. (2000) Actively contracting bundles of polar filaments. *Phys. Rev. Lett.* 85, 1778.

274. Carvalho, K., Tsai, F.C., Lees, E., Voituriez, R., Koenderink, G.H., and Sykes, C. (2013) Cell-sized liposomes reveal how actomyosin cortical tension drives shape change. *Proc. Natl. Acad. Sci. USA* 110, 16456–16461.

275. Carvalho, K., Lemiere, J., Faqir, F., Manzi, J., Blanchoin, L., Plastino, J., Betz, T., and Sykes, C. (2013) Actin polymerization or myosin contraction: Two ways to build up cortical tension for symmetry breaking. *Philos. Trans. R. Soc. Lond. B. Biol. Sci.* 368, 20130005.

276. Field, C.M., Wühr, M., Anderson, G.A., Kueh, H.Y., Strickland, D., and Mitchison, T.J. (2011) Actin behavior in bulk cytoplasm is cell cycle regulated in early vertebrate embryos. *J. Cell. Sci.* 124, 2086–2095.

277. Pinot, M., Steiner, V., Dehapiot, B., Yoo, B., Chesnel, F., Blanchoin, L., Kervrann, C., and Gueroui, Z. (2012) Confinement induces actin flow in a meiotic cytoplasm. *Proc. Natl. Acad. Sci. USA,* 109, 11705–11710.

4

Structural and Dynamical Hierarchy of Fibrillar Collagen

Xiaojing Teng and Dr. Wonmuk Hwang

CONTENTS

4.1 Introduction

Collagen is the major component of the extracellular matrix (ECM), forming key structures for most parts of our body such as skin, bone, tendon, and blood vessel. Besides its prominent mechanical property to provide strength and stability to the tissue (Fratzl, 2003), collagen is also an important scaffold with numerous ligand-binding sites for cells to attach and migrate, as well as serve as ligands for intracellular signaling (Sweeney et al., 2008; Leitinger, 2011). Collagen thus plays key roles in numerous pathophysiological processes such as development, wound healing, arthritis, cancer, and atherosclerosis (Myllyharju and Kivirikko, 2001; Sternlicht and Werb, 2001; Fields, 2013).

The remarkable properties of collagen as a building block of living tissues are reflected in the conditions characterizing *hierarchical materials systems in biology*, which include recurrent use of molecular constituents such that widely variable properties are attained from apparently similar elementary units, controlled orientation of structural elements, durable interfaces between hard and soft materials, sensitivity to the presence of water, properties that vary in response to performance requirements, fatigue resistance and resiliency, controlled and often complex shapes, and capacity for self-repair (Tirrell et al., 1994; Cowin and Doty, 2007). One question that naturally arises is, "What are the molecular properties of collagen that are responsible for making it almost an ideal material for building animal tissues?" Apart from its chemical and biological actions, the sheer ability of collagen to self-assemble into hierarchically ordered structures over several orders of magnitude in length scale already makes it distinct from most other proteins. By comparison, other structural proteins forming the ECM, such as elastin and fibrillin (Kielty et al., 2002; Vakonakis and Campbell, 2007), are less ordered. Another class of structural proteins, the cytoskeleton, is suited for the dynamical behavior of cells that can change shape and move in timescales much shorter than those of tissue remodeling. Only collagen appears to possess robust mechanical properties to keep the integrity of the ECM and yet remains sufficiently dynamical to self-assemble and allow tissue remodeling.

To date, 28 types of collagens have been identified (Ricard-Blum, 2011). Among them, types I, II, III, V, XI, XXIV, and XXVII are fibrillar collagens (Kadler et al., 2007; Muiznieks and Keeley, 2013), the main focus of this chapter. Types I, II, and III take up 80%–90% of collagens in human body, where type I is the most abundant (Lodish et al., 2012). In this chapter, we give an overview of molecular structures of fibrillar collagens and their assemblies (herein, we call fibrillar collagen simply as collagen, unless noted otherwise). Since the first characterization of the collagen triple helical structure (Ramachandran and Kartha, 1954, 1955; Rich and Crick, 1955, 1961), many review papers have been written describing collagen structures (Bhattacharjee and Bansal, 2005; Brodsky and Persikov, 2005; Wess, 2005; Kadler et al., 2007; Shoulders and Raines, 2009). But they tend to focus on a single length scale, mostly at the molecular level, or do not explain physicochemical interactions underlying self-assembly of collagen. Here, we start from the analysis of the amino acid sequence (primary structure) up to the fibril level and also discuss dynamical aspects that may affect the assembly process. Using a recent 3D structure of a collagen microfibrillar bundle (Orgel et al., 2006), we revisit intermolecular interactions, for which studies at the amino acid sequence level have been the predominant basis for explaining the D-period packing of collagen molecules within a fibril (Petruska and Hodge, 1964; Hulmes et al., 1973). The ability of collagen to serve as a mechanically robust building block of tissues while allowing dynamical assembly and turnover is manifested across multiple levels of organization, where the underlying *design principle* (by design we mean natural selection through evolution) is a delicate balance between semicrystalline order and disorder in relevant interactions. Through such properties of fibrillar collagen that permit *disorder* and *additives*, diversity in the structure and composition of the ECM in different tissue types can be achieved. Also, anomaly in the assembly and turnover of fibrillar collagen is implicated in a number of diseases, including osteogenesis imperfecta (also known as brittle bone disease, characterized by low bone mass, bone fragility, and susceptibility to fracture) and Ehlers–Danlos syndrome (involves severe skin extensibility, tissue fragility, and joint hyperlaxity).

4.2 Primary Structure

The amino acid sequence of collagen is characterized by the repeating motif of Gly-X-Y, where Gly is glycine and X and Y are other residues that are often imino acids, proline (Pro; P) and hydroxyproline (Hyp; O) (Figure 4.1a). The positioning of Gly in every third residue is necessary for forming a triple helix (see the next section). Each polypeptide chain forming a collagen triple helix is called an α chain (note that this is different from the α-helix, which is another protein secondary structure). A *collagen molecule* typically refers to the triple helix formed by three α chains and we denote each with the corresponding collagen type in parentheses. For example, a type I collagen is a heterotrimer that consists of two α1 (I) and one α2 (I) chains, whereas a type III collagen is a homotrimer of three α1 (III) chains.

Table 4.1 shows the amino acid composition of major human collagens. The α2(I) chain has the least number of imino acids that stabilize the collagen triple helical structure

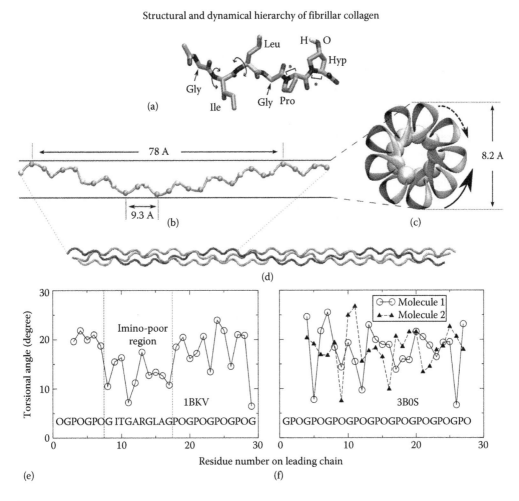

Structural and dynamical hierarchy of fibrillar collagen

FIGURE 4.1

Structure of a collagen triple helix. (a) Molecular structure of an α chain segment (Gly-Ile-Leu-Gly-Pro-Hyp). Note that glycine does not have a side chain, making it suitable for the inner side of the triple helix, as in panel (c). Hydrogen atoms are not shown except for that of the hydroxyl group in the Hyp ring. While the backbone of non–imino acids such as Ile and Leu can rotate (marked by circular arrows), imino rings of Pro and Hyp prevent backbone rotation (marked by *). Structures in (b–d) are for the GPO (Gly-Pro-Hyp) repeat constructed using the THeBuScr script (From Rainey, J.K. and Goh, M.C., *Bioinformatics*, 20(15), 2458, 2004). (b) Lateral and (c) axial views of a single α chain. C_α atoms of glycine are denoted as white spheres, which are located closer to the central axis of the triple helix than those of X and Y. In (c), dotted arrow denotes the left-handed turn of polyproline II helix (out of page). Solid arrow denotes the right-handed super-helical twist of the α chain within a triple helix. The α chain backbones of the triple helix span a cylinder ~8.2 Å in diameter. Including side chains and the hydration shell, the diameter becomes ~14–15 Å. (From Ravikumar, K.M. and Hwang, W., *Proteins: Struct. Funct. Bioinf.*, 72(4), 1320, 2008.) (d) Overview of a collagen triple helix, revealing staggering of the α chains. Molecular structures are rendered using the VMD program. (From Humphrey, W. et al., *J. Mol. Graph.*, 14(1), 33, 1996.) (e, f) Torsional angles between successive triads (Ravikumar et al., 2007; Ravikumar and Hwang, 2008) in two Protein Data Bank (PDB) structures. (e) PDB 1BKV that contains the imino-poor region in the middle (Kramer et al., 1999) and (f) PDB 3B0S, which is (GPO)₉. (From Okuyama, K. et al., *Biopolymers*, 97(8), 607, 2012.) The corresponding sequence of the leading chain is given in the figure. PDB 3B0S contains two molecules, both of which are analyzed. Overall, the imino-poor region in 1BKV is less wound than the GPO regions. However, even for the same GPO triplet sequence, the conformation varies. Large variations in torsional angles near termini are due to the end effect.

TABLE 4.1

Amino Acid Composition of Human Collagen Types I, II, and III

Amino Acid	α1 (I)	α2 (I)	α1 (II)	α1 (III)
Gly	341	341	342	362
Pro	236	203	225	239
Asn	11	24	12	23
Gln	27	21	37	24
Ser	34	32	28	42
Thr	17	19	21	15
Arg	51	54	52	48
His	2	12	2	7
Lys	36	30	37	38
Asp	31	20	29	26
Glu	47	45	53	48
Ala	117	107	102	90
Cys	0	0	0	1
Ile	6	18	10	14
Leu	19	33	25	22
Met	7	5	7	9
Phe	12	10	13	8
Tyr	0	1	1	2
Val	20	39	18	11
UniProt ID	P02452	P08123	P02458	P02461
Net charge	+11	+31	+9	+19
Large nonpolar residues	64	106	74	66

Type I collagen is a heterotrimer of two α1 (I) and one α2 (I) chains, whereas types II and III collagens are homotrimers of α1 (II) and α1 (III) chains, respectively. Only the triple helix part was considered (residue 179–1192 for α1 (I); 91–1104 for α2 (I); 201–1214 for α1 (II); 168–1196 for α1 (III)). Regions outside of this part are nonhelical telopeptides, which do not play any major role in the recognition between collagens (Kuznetsova, N. and Leikin, S., *J. Biol. Chem.*, 274(51), 36083, 1999). For counting large nonpolar residues (bottom row), residues Ile, Leu, Met, Phe, Tyr, and Val were used. The Amino Acid Calculator from the MacCoss lab (http://proteome.gs.washington.edu/cgi-bin/aa_calc.pl) was used to count residues.

(see the next section). This is consistent with the increased stability of the homotrimer made of three α1 (I) chains compared to the wild-type type I collagen that consists of two α1 (I) and one α2 (I) chains (Miles et al., 2002; Han et al., 2010). The α2 (I) chain also has the highest net charge and the greatest number of large nonpolar residues (Ile, Leu, Met, Phe, Tyr, and Val). As discussed in the following, since hydrophobic as well as hydration forces are major drivers at different stages of collagen assembly, α2 (I) chain plays an important role for the assembly of type I collagen, where homotrimers of three α1 (I) chains assemble less efficiently and also are less ordered in fibrils compared to wild-type heterotrimers (McBride Jr. et al., 1992, 1997). The α1 (I) homotrimer is found in fetal tissues and diseased tissues, such as fibrotic tissue and carcinomas (Jimenez et al., 1977; Han et al., 2010; Makareeva et al., 2010) as well as in osteogenesis imperfecta (Chipman et al., 1993). It is also known to be resistant to cleavage by matrix metalloproteinases (MMPs) (Makareeva et al., 2010). Likewise, the reduced net charge and nonpolar residues of type II collagen may be related to its slow assembly kinetics (Dong et al., 2007) and inability to form extended fibrils (Birk and Brückner, 2011). Type II collagen-rich fibrils are also less ordered and heterogeneous, such as in cartilage (Wess, 2005).

As is the case for nearly all proteins, knowledge of the amino acid sequence of collagen is a necessary, but not a sufficient condition for understanding its structure and function. Still, the sensitivity of the collagen structure to its sequence is manifested by point mutations that lead to various types of diseases (Myllyharju and Kivirikko, 2001). Out of ~340 Gly-X-Y triplets, replacing a single Gly with other residues can destabilize the entire protein or alter its assembly behavior, which are observed in diseases such as osteogenesis imperfecta types 1 to 4 (involves mutations of type I collagen) (Marini et al., 2007) and Ehlers–Danlos syndrome type IV (involves mutations of type III collagen) (Bodian and Klein, 2009). Severity of the effect depends on the location of the mutation and the replacing residue. For example, Gly → Ala (small nonpolar side chain) mutation leads to a mild phenotype, in contrast to Gly → Glu (negatively charged), which is too severe and thus is not even viable in most cases (Persikov et al., 2004). Considering the requirement of a glycine in every third residue for forming a triple helix (cf., Figure 4.1), it is expected that its mutation severely destabilizes the structure (Long et al., 1993; Lee et al., 2011) or disrupts supramolecular assembly (Myllyharju and Kivirikko, 2001). In addition, there are mutations on residues other than glycines, which cause diseases including Ehlers–Danlos syndrome type I (Nuytinck et al., 2000) and various types of osteogenesis imperfecta (Pollitt et al., 2006).

4.3 Secondary and Tertiary Structures

In a collage triple helix, each α-chain forms a left-handed *polyproline type II* helix (Figure 4.1b). The axial rise per residue is about 3.1 Å, making 9.3 Å per helical turn that consists of a Gly-X-Y triplet. Such conformation is most easily taken by imino acids due to the constraint on the peptide's backbone dihedral angles imposed by the imino rings (Figure 4.1a). By comparison, a polyproline type I is a more tightly wound, *right-handed* helix that occurs more rarely in nature (Beausoleil and Lubell, 2000).

Three α chains twist together into a *right-handed* triple helix. In the axial view of a single α chain within a triple helix, glycines line up along the inner side (Figure 4.1c, white spheres), which is why its presence in every third residue is necessary for forming a triple helix—only glycine that does not have a bulky side chain can be accommodated in this inner position. The pitch of the right-handed turn is about 78 Å. The degree of the triple helical twist varies depending on residues in the X and Y position. A customary way of describing the helical twist is using the number of Gly-X-Y triplets per given number of turns around the triple helix, such as 7/2 (7 triplet units in 2 turns around a triple helix) or 10/3 (10 units in 3 turns) (Kramer et al., 1999; Brodsky and Persikov, 2005; Okuyama, 2008). However, these descriptors are two idealized limits and do not adequately capture the twist of the actual molecule (Bella, 2010). We thus developed a triad-based description to quantitate the collagen conformation in greater detail (Ravikumar et al., 2007; Ravikumar and Hwang, 2008). A triad is formed by three C_α atoms in respective α chains of a triple helix. By measuring Euler angles between two successive triads, one can describe torsional as well as bending motion of a filament. As Figure 4.1e and f shows, the region of collagen that is devoid of imino acids is less tightly wound than the imino-rich GPO regions, although the latter also exhibits substantial variation in local torsional angles.

The imino-poor region is also called the *labile* domain (Kramer et al., 1999; Miles and Bailey, 2001), which is considered to provide flexibility to the molecule and also serves as binding sites for various collagen receptors (Miles and Bailey, 2001; Sweeney et al., 2008).

A collagen triple helix is stabilized by extensive hydrogen bond networks (Figure 4.2a). The most common ones are between the amide hydrogen of Gly and the carbonyl oxygen of the residue in the X position (N-H···O=C; Figure 4.2a, solid line). Since imino acids do not have amide hydrogen, one interchain backbone hydrogen bond is formed per Gly-X-Y triplet (Figure 4.2b and c). Additional stabilization of the triple helix depends on the presence of imino acids on the X and Y positions. In the imino-rich domain, the constraint imposed on the backbone dihedral angle by the imino ring itself provides stability to the triple helix. Hydroxylation of proline in the Y position to hydroxyproline (which occurs via posttranslational modifications in the cell) further stabilizes the triple helix. The main stabilizing mechanism by Hyp is suggested to be inductive or stereoelectronic effect that favors the exo-puckered conformation of the imino ring (Shoulders and Raines, 2009), although stabilization via water bridges involving the OH group in Hyp may also play a role (Brodsky and Persikov, 2005). By contrast, peptides with Hyp on the X position do not form triple helix, indicating that the location of hydroxylation sensitively affects the triple helix stability (Jenkins et al., 2003; Mizuno et al., 2004). Overall, the Gly-Pro-Hyp triplet makes the most stable structural motif forming the triple helix.

In the imino-poor domain, even if the amide hydrogens on the X or Y positions are present, they are positioned radially outside (cf., Figure 4.2c) so direct backbone-to-backbone hydrogen bonds cannot form. Instead, water-mediated bridges form between the carbonyl oxygen of glycine and amide hydrogen of the residue on the X position (Figure 4.2a, dashed line). Additional water-mediated hydrogen bonds including those involving amino acid side chains are found in crystal structures of collagen mimetic peptides (Bella et al., 1995; Kramer et al., 2001; Brodsky and Persikov, 2005). But they likely play secondary roles in stabilizing the collagen triple helix compared to the more direct contacts and water bridges described earlier. It should be noted that water bridges, despite being present in crystal structures, are highly dynamic, with lifetime on the order of only a few picoseconds (Ravikumar and Hwang, 2008). Surface-bound water molecules in general have low entropic cost of binding (Dunitz, 1994). Water molecules thus stabilize the structure by rapidly forming and breaking bridges. This is likely responsible for the high-temperature sensitivity of the conformation of the imino-poor labile domain. Gradual reduction in the lifetime of water bridges with temperature leads to unwinding of the labile domain at temperatures as low as 300 K (Ravikumar and Hwang, 2008). This may contribute to the instability of isolated type I collagen molecules at body temperature (Leikina et al., 2002). The *micro-unfolding* of the labile domain is considered as a requirement for collagen cleavage by MMPs (Fields, 1991; Chung et al., 2004). Collagen is known to be extremely resistant to proteolytic cleavage, highlighting its importance as a robust building block of tissues. Native collagens (not unfolded) are cleaved almost exclusively by MMPs. However, since the catalytic cleft of an MMP is too narrow to accommodate a collagen triple helix, unwinding appears to be necessary (Chung et al., 2004). Since the MMP cleavage site of collagen (about 3/4 along the length of the molecule; Figure 4.3a) is in a labile domain, its spontaneous thermal unwinding may be harnessed by MMPs for cleavage. Suppression of its unwinding activity by mechanical load may be a basis for the load-dependent collagen cleavage by MMP, although the issue is still under substantial debate (Flynn et al., 2010; Camp et al., 2011; Adhikari et al., 2011, 2012).

Since type I collagen is a heterotrimer, there could be three isomers depending on whether the α2 is in the leading, middle, or trailing position of the triple helix

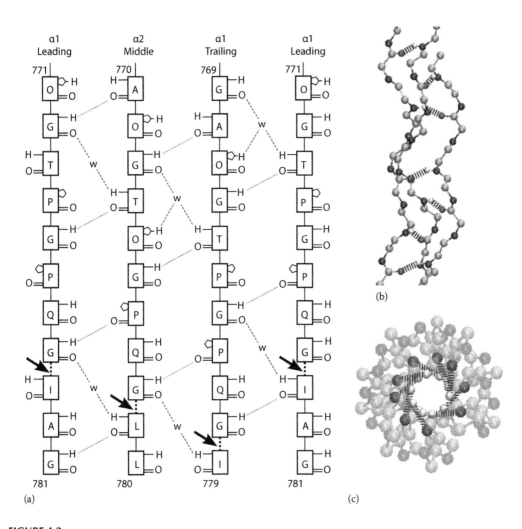

(a)

(b)

(c)

FIGURE 4.2

An example of the hydrogen bonding network in a collagen triple helix. (a) Schematic model of the hydrogen bond network around the MMP cleavage site of human type I collagen. Scissile bonds are marked by arrows. Similar hydrogen bond networks are observed in x-ray structures of collagen mimetic peptides, in particular PDB 1BKV that contains a labile domain in the middle (Figure 4.1e). (From Kramer, R.Z. et al., *Nat. Struct. Biol.*, 6(5), 454, 1999.) Due to the staggered arrangement, α chains are labeled leading (α1), middle (α2), and trailing (α1) on top. Residue numbers of the first and the last amino acids are shown. Direct hydrogen bonds between backbone atoms are denoted by solid lines, and water bridges are denoted by dashed lines joined by the letter *w*. Only bridges involving a single water molecule between α chain backbones are shown. When the X position is occupied by a proline, the water bridge cannot form due to the lack of an amide hydrogen. (b) Hydrogen bonds between α chain backbones in a collagen triple helix (dotted lines). (c) Axial view of panel (b) revealing a circular arrangement of the backbone hydrogen bonds along the inner portion of the triple helix. For clarity, other backbone atoms are rendered transparent. In (b, c), the triple helix model was GPO repeats built using the THeBuScr program. (From Rainey, J.K. and Goh, M.C., *Bioinformatics*, 20(15), 2458, 2004.)

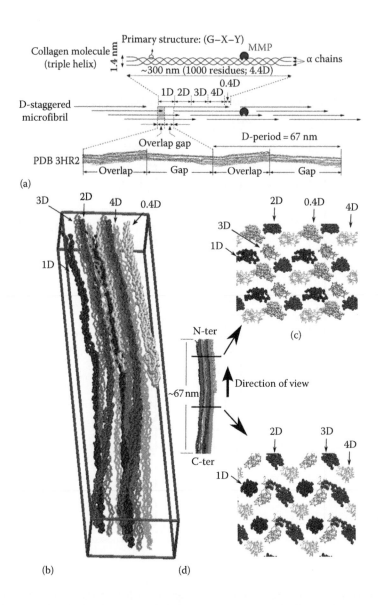

FIGURE 4.3

Organization of collagen molecules within a fibrillar bundle. (a) Schema of the collagen hierarchy. In the Hodge–Petruska model, five collagen molecules staggered by ~234 residues possess a D-periodic band with overlap and gap regions. (From Petruska, J.A. and Hodge, A.J., *Proc. Natl Acad. Sci. USA*, 51(5), 871, 1964). The MMP cleavage site is at about 3/4 along the length of a molecule. (b–d) In situ structure of a D-periodic bundle of type I collagen from rat tail tendon obtained via x-ray fiber diffraction (PDB 3HR2, 5.16 Å resolution). (From Orgel, J.P.R.O. et al., *Structure*, 8(2), 137, 2000; Orgel, J.P.R.O. et al., *Proc. Natl. Acad. Sci. USA*, 103, 9001, 2006.) (b) Overview of the bundle. Different rendering methods were used to distinguish between D-period sections of individual monomers. Since the aspect ratio of the unit cell is large ($4.0 \times 2.7 \times 67.8$ nm³), the bundle is shown in a slanted manner. The enclosing box is a visual guide indicating tilting of the long bundle (small panel). (c, d) Axial views of the (c) overlap and (d) gap regions, taken from the horizontal lines on the left. The view is from the C- to the N-terminal direction (vertical arrow in the small panel). There are only C_α atoms in the crystal structure, and the missing atoms were built using the CHARMM program. (From Brooks, B.R. et al., *J. Comput. Chem.*, 4, 187–217, 1983; Brooks, B.R. et al., *J. Comput. Chem.*, 30(10), 1545, 2009.) Molecules are packed in quasi-hexagonal manner in the overlap region while the absence of the 0.4D section renders molecules in the gap region less well packed.

(Bella et al., 1994). Figure 4.2a shows the case where α2 is in the middle. Chain registry within a heterotrimer of short collagen mimetic peptides is known to have strong influence on the stability of the molecule (Saccà et al., 2002; Fallas et al., 2010; Russell et al., 2010; Li et al., 2011). It is thus expected that the three isomers of the heterotrimeric type I collagen behave differently. The registry of α chains in a heterotrimer (such as type I) is likely important for collagen catabolism since it will affect how MMP binds to the molecule and cleaves it. A nuclear magnetic resonance (NMR)-based study suggested that MMP interacts predominantly with the leading and middle chains (Bertini et al., 2011). However, the recently obtained first x-ray structure of the MMP-1–collagen complex (PDB 4AUO) indicates that all three chains interact with MMP-1, which is possible due to the flexibility between the catalytic and hemopexin domains of MMP-1 in its dumbbell geometry (Manka et al., 2012). However, PDB 4AUO was obtained with a homotrimeric collagen mimetic peptide, also the structure is considered not to represent the active cleavage-prone state. Further studies are required to determine the registry of α chains within native collagen heterotrimers and its impact on the binding mode of collagen-associated enzymes.

4.4 Quaternary Structure and Heterogeneity in Higher-Order Organization

One of the remarkable properties of fibrillar collagen is its ability to self-assemble into ordered fibrils. Although collagen assembly in living tissues is assisted by cells (e.g., by the organelle *fibripositor* in embryonic fibroblasts [Canty et al., 2004]) and the existing ECM (e.g., by proteoglycans [PGs]) (Kalamajski and Oldberg, 2010), under appropriate buffer condition, collagen molecules can readily assemble into fibrils possessing the characteristic 67 nm D-period (Williams et al., 1978; Gelman et al., 1979; Jiang et al., 2004; Leow and Hwang, 2011; Fang et al., 2013). This suggests that the native-like ordering of collagen molecules in a fibril is at an energy minimum state that is independent of the influence by cells.

During early stages of in vitro collagen assembly, fibrils appear without any D-period or with a larger value (e.g., 88 nm), which later reduces to 67 nm (Leow and Hwang, 2011). This indicates that the assembly of a collagen fibril is multistep (Gelman et al., 1979), where molecules initially form a rather loose bundle, after which they perform axial diffusion and find the native-like D-periodic order. Hydration force is a major driver for the attraction between collagen molecules (Leikin et al., 1995, 1997), which is also the case for other filamentous molecules including DNA and stiff polysaccharides (Leikin et al., 1993, 1994). A detailed computational analysis reveals that the attraction is mediated by the interaction between hydration shells that ubiquitously cover the surfaces of the molecules (Ravikumar and Hwang, 2011), further supporting that the initial attraction does not depend sensitively on the amino acid sequence of collagen.

Since hydration shell is an integral part of a collagen molecule (even crystal structures of collagen peptides possess hydration shells) (Ravikumar and Hwang, 2008, 2011), after initial bundling, hydration shells likely serve as lubrication layers that assist with axial diffusion. A D-period is formed when neighboring collagen molecules have about 234-residue stagger (length of a one 1 D-period; Figure 4.3a). Since a single collagen molecule is about 1030 residues (with nonhelical region), it spans $1030/234 = 4.4$ D-periods.

When five collagen molecules assemble, this leaves a *gap* region, 0.6 D in size (Figure 4.3a). The minimal 5-collagen unit that exhibits the 0.4 D overlap and 0.6 D gap regions is called the microfibril. Although a collagen microfibril has not been observed in an isolated form, ~5 nm microfibrillar structures were seen on surfaces of collagen fibrils via atomic force microscopy (AFM) (Baselt et al., 1993), and in situ AFM also shows that collagen layers on mica widen in discrete 4 nm steps, which approximately corresponds to the thickness of a microfibril (Cisneros et al., 2006). Although collagen assembly in this study was assisted by the mica surface, it is possible that, when a collagen molecule adds to an existing filament, locally finding the D-periodic order within a microfibrillar unit is relatively rapid, whereas formation of a D-period over the entire fibril (tens to hundreds of nanometers in diameter) may take a longer time. After all, the molecule needs to axially diffuse at most its own length (~300 nm) to find a minimum-energy state. Another possibility is that a small number of collagen molecules loosely associate and then together deposit onto existing filament, after which higher ordering can occur.

Unlike the initial bundling, formation of a D-period involves more specific interactions between amino acids. So far, this picture has been based largely on the pioneering analysis made by Hulmes and coworkers in 1973 (Hulmes et al., 1973). They placed two α1 (I) chains and slid one relative to the other (at that time, the α2 (I) sequence was not known completely) while scoring electrostatic and hydrophobic interactions. Peaks in the number of interactions were found when the molecules were staggered by every D-period. This provided a rationale for the classic Hodge–Petruska model of collagen D-period (Petruska and Hodge, 1964). Similar result for the 234-residue stagger forming the D-period was also obtained for type II collagen (Ortolani et al., 2000). However, these studies were performed at the sequence level, and the Hodge–Petruska model of the collagen D-period is essentially 2-dimensional (Figure 4.3a). Only recently has the 3-dimensional arrangement of collagen molecules within the microfibrillar bundle been resolved via x-ray fiber diffraction (Orgel et al., 2000, 2006) (PDB 3HR2; Figure 4.3). Overall, a single collagen molecule has a right-handed super-twist along the axis of the fibril. Within the overlap region of a D-periodic bundle, a D-section of a collagen molecule (1D, 2D, 3D, 4D, or 0.4D) is surrounded by other sections of the neighboring molecules in quasi-hexagonal manner (Figure 4.3c). In the gap region, the 0.4D section is absent and the molecules are less tightly packed, so lateral positions are less clearly resolved compared to those in the overlap region (Figure 4.3d) (Orgel et al., 2006).

With the availability of a 3D microfibrillar structure (Orgel et al., 2006), it is possible to examine the intermolecular contacts in greater details. We counted nonpolar (hydrophobic) and charged (electrostatic) contacts along the length of a collagen molecule in a bundle. Given the modest resolution of the PDB 3HR2 structure (5.16 Å), instead of finding clear-cut atomic contacts, we used certain cutoff distances between C_α atoms of residues in counting the number of contacts. Besides, it is believed that contacts between neighboring collagen molecules are dynamic, switching between different residues, rather than making unique combinations along the entire length of the molecule (Brodsky and Persikov, 2005). Since the intercollagen distance is about 15 Å (Leikin et al., 1995; Ravikumar and Hwang, 2008, 2011), we used 10 or 15 Å as cutoff distances for counting hydrophobic and electrostatic contacts. Figure 4.4 shows the result of our calculation. Electrostatic contacts are spread out along the molecule, which was also found in Hulmes' sequence-based linear model (Hulmes et al., 1973). Its relatively uniform distribution indicates that electrostatic interaction is unlikely to have a strong influence on forming a D-periodic stagger. Hydrophobic interaction displays a more punctate distribution (Figure 4.4a and c). When the numbers of contacts are aligned based on the position of each amino acid relative to its

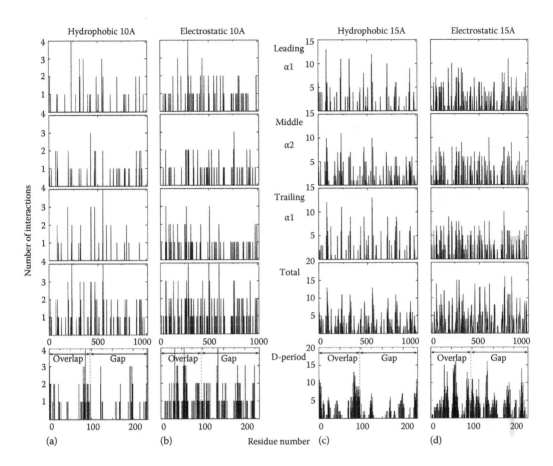

FIGURE 4.4
Hydrophobic and electrostatic contacts within a microfibrillar bundle along each α chain of type I collagen (PDB 3HR2). For hydrophobic contacts, we considered Leu, Ile, Phe, Met, Val, and Tyr. Cutoff distances used for counting contacts are (a, b) 10 Å and (c, d) 15 Å. The first three rows are the numbers of contacts with surrounding molecules that each α chain makes. Intramolecular contacts between α chains within the same triple helix were excluded from our calculation. The fourth row (*total*) adds counts in the first three rows. Due to the staggering of α chains, the sum is not necessarily larger than counts for individual chains. Residue numbering is based on the leading chain. Bottom row: the fourth row rearranged according to the 234-residue D-periodic stagger (Figure 4.3a).

position within the corresponding D-period, more pronounced clustering of hydrophobic contacts can be seen at the beginning and end of the overlap region (Figure 4.4 bottom row). There are also more electrostatic contacts in the overlap region, although the distribution is more uniform compared to hydrophobic contacts. The MMP cleavage site (G778~I779; Figure 4.2a) is located in the overlap region, about 30 residues (~90 Å) N-terminal from the boundary between the overlap and gap regions (Sweeney et al., 2008). A previous theoretical study suggested that the scissile bond is located at the point where the collagen triple helix changes from well-folded imino-rich region to a more flexible imino-poor region (Fields, 1991). Our contact analysis indicates that a similar conformational environment for the MMP cleavage site exists at the fibrillar level as well: The marginally stable, *transition* zone between overlap and gap regions (neither too stable nor too disordered) would be the best place where collagen cleavage can be sensitively and dynamically controlled, such as by mechanical load. In addition, packing of collagen molecules within a fibril may have a regulatory role, as neighboring molecules may block access of MMP to its cleavage site (Perumal et al., 2008). However, this picture is based on static x-ray structure of the collagen microfibrillar bundle, and little is known regarding whether or to what extent such steric hindrance can contribute to collagen regulation when molecules undergo thermal motion.

The basic structural organization of fibrillar collagen described earlier serves as a *chassis* that can be further modified and combined to give the extreme structural and functional diversity seen in different tissues or in different stages of development. The first modification we consider is collagen cross-linking (Eyre and Wu, 2005). Lysyl oxidase is the main enzyme carrying out cross-linking between the nonhelical telopeptides at either the N- or C-terminal ends of the collagen molecule and the triple helix region of neighboring molecules. Covalent cross-links are made through lysines or hydroxylysines. Uncontrolled, nonenzymatic cross-links can also form with age, which are typically deleterious (Robins, 2007). Covalent cross-links are necessary for the mechanical integrity of collagen fibrils and networks, which is also an essential aspect in collagen remodeling. This may be a reason why lysyl oxidase is involved in cancer cell invasion (Kirschmann et al., 2002). Since its upregulation is associated with poor prognosis in cancer, lysyl oxidase can be a viable therapeutic target (Nishioka et al., 2012).

Another variation to the basic chassis is the heterotypic nature of collagen. For example, depending on the tissue type, a type I collagen fibril contains up to about 20% of type III or type V collagen (Wess, 2005; Bruckner, 2010). Cross-links between different collagen types exist, indicating specific interactions (Eyre and Wu, 2005; Wess, 2005). Heterotypic collagen fibrils also vary greatly in diameter, for example, as small as 20 nm in cornea (Birk and Brückner, 2005) to 50–250 nm in the case of tendon fibrils (Wess, 2005). Even greater complexity and diversity of fibrils is achieved via surface decoration. Many different types of collagens protrude out from the fibril's surface, including type V collagen (Birk and Brückner, 2005) and a whole family of fibril-associated collagen with interrupted triple helices (FACITs) (Birk and Brückner, 2005, 2011). PG is another large set that decorates the surface (Yoon and Halper, 2005; Kalamajski and Oldberg, 2010). It is found that there are bridges formed by PGs between collagen fibrils periodically (Scott, 1992), which means PGs are the connectors between collagen fibrils. The possible binding sites on collagen has been reported (Sweeney et al., 2008). It seems that there is electrostatic interaction between the core protein and collagen molecule, and also the extensive hydrogen bonding and hydrophobic interactions further consolidate the binding (Orgel et al., 2009; Kalamajski and Oldberg, 2010), which guarantees the stability of collagen fibers and hence the ECM integrity. In cornea, PGs separate collagen fibrils and arrange them in hexagonal manner, ensuring corneal transparency (Lewis et al., 2010).

In addition to organizing the fibril network, the large variety of PGs is also important in regulating the structure of collagen fibril (Kalamajski and Oldberg, 2010). For example, competition between different types of PGs controls fibril development. Asporin and decorin compete for the same binding site on collagen, and they differ in collagen affinity and functional role (Kalamajski et al., 2009; Kalamajski and Oldberg, 2010). Asporin is a fibril nucleator and induces bone mineralization during early stages of fibril development, which can later be regulated by decorin as it has a higher affinity for collagen. In this way, fibril diameter can be controlled by decorin. Fibromodulin and lumican are another pair of PGs that employ competitive binding and time-dependent expression to control tendon development (Svensson et al., 1999; Kalamajski and Oldberg, 2010).

The heterotypic and surface-decorated collagen fibrils further organize and assemble into collagen fibers and networks, as well as interact with other matrix molecules and cells to serve as the backbone of the ECM milieu. Since the basic structural building blocks of collagen fibrils are similar, heterogeneity in fibril composition and their higher-order organization at the network level may define tissue-specific characteristics. Microscopy images and schematic illustrations of the different types of higher-level organization can be found in Ushiki (2002) and Birk and Brückner (2005, 2011). Bruckner aptly coins the term *biological alloy*, to emphasize the importance of such heterogeneity at the suprastructural level (Bruckner, 2010). Nevertheless, as evidenced by the organism-level phenotypes caused by single point mutations in collagen, understanding and controlling the mechanobiology of collagen requires knowledge of individual proteins.

4.5 Conclusion

A main theme in the structural hierarchy of collagens is the balance between crystallinity and disorder (Wess, 2005). As the major load-bearing constituent of the ECM, fibrillar collagens must be able to build strong and precisely ordered structures, for which atomistic-level specificity and semicrystalline packing is required. On the other hand, given the diversity in mechanical loads experienced by the ECM, as well as the different functional requirements of tissues including the ability to remodel, collagens must possess *loose parts* that allow controlled deformation and cleavage. Such a dual nature of collagen also poses a great challenge in mechanistic understanding of its dynamic behavior. Common properties of the basic structural chassis and properties that arise from the heterotypic organization must be delineated, where theoretical and computational analysis employing the principles of physics and chemistry will be as important as experimental and clinical studies on collagen.

References

Adhikari, A. S., Chai, J., and Dunn, A. R. (2011). Mechanical load induces a 100-fold increase in the rate of collagen proteolysis by MMP-1. *J. Am. Chem. Soc.*, 133(6):1686–1689.

Adhikari, A. S., Glassey, E., and Dunn, A. R. (2012). Conformational dynamics accompanying the proteolytic degradation of trimeric collagen I by collagenases. *J. Am. Chem. Soc.*, 134(32):13259–13265.

Baselt, D. R., Revel, J.-P., and Baldeschwieler, J. D. (1993). Subfibrillar structure of type I collagen observed by atomic force microscopy. *Biophys. J.*, 65(6):2644–2655.

Beausoleil, E. and Lubell, W. D. (2000). An examination of the steric effects of 5-tert-butylproline on the conformation of polyproline and the cooperative nature of type II to type I helical interconversion. *Biopolymers*, 53(3):249–256.

Bella, J. (2010). A new method for describing the helical conformation of collagen: Dependence of the triple helical twist on amino acid sequence. *J. Struct. Biol.*, 170(2):377–391.

Bella, J., Brodsky, B., and Berman, H. M. (1995). Hydration structure of a collagen peptide. *Structure*, 3:893–906.

Bella, J., Eaton, M., Brodsky, B., and Berman, H. M. (1994). Crystal and molecular structure of a collagen-like peptide at 1.9 Å resolution. *Science*, 266:75–81.

Bertini, I., Fragai, M., Luchinat, C., Melikian, M., Toccafondi, M., Lauer, J. L., and Fields, G. B. (2011). Structural basis for matrix metalloproteinase 1-catalyzed collagenolysis. *J. Am. Chem. Soc.*, 134:2100–2110.

Bhattacharjee, A. and Bansal, M. (2005). Collagen structure: The Madras triple helix and the current scenario. *IUBMB Life*, 57(3):161–172.

Birk, D. E. and Brückner, P. (2005). Collagen suprastructures. In Brinckmann, J., Notbohm, H., and Müller, P. K., eds., *Collagen*, pp. 185–205. Springer, New York.

Birk, D. E. and Brückner, P. (2011). Collagens, suprastructures, and collagen fibril assembly. In Mecham, R. P., ed., *The Extracellular Matrix: An Overview*, pp. 77–115. Springer, New York.

Bodian, D. L. and Klein, T. E. (2009). COLdb, a database linking genetic data to molecular function in fibrillar collagens. *Hum. Mutat.*, 30(6):946–951.

Brodsky, B. and Persikov, A. V. (2005). Molecular structure of the collagen triple helix. *Adv. Protein Chem.*, 70:301–339.

Brooks, B. R., Brooks III, C. L., Mackerell Jr., A. D., Nilsson, L., Petrella, R. J., Roux, B., Won, Y. et al. (2009). CHARMM: The biomolecular simulation program. *J. Comput. Chem.*, 30(10):1545–1614.

Brooks, B. R., Bruccoleri, R. E., Olafson, B. D., States, D. J., Swaminathan, S., and Karplus, M. (1983). CHARMM: A program for macromolecular energy, minimization, and dynamics calculations. *J. Comput. Chem.*, 4:187–217.

Bruckner, P. (2010). Suprastructures of extracellular matrices: Paradigms of functions controlled by aggregates rather than molecules. *Cell Tissue Res.*, 339(1):7–18.

Camp, R. J., Liles, M., Beale, J., Saeidi, N., Flynn, B., Moore, E., Murthy, S. K., and Ruberti, J. W. (2011). Molecular mechanochemistry: Low force switch slows enzymatic cleavage of human type I collagen monomer. *J. Am. Chem. Soc.*, 133:4073–4078.

Canty, E. G., Lu, Y., Meadows, R. S., Shaw, M. K., Holmes, D. F., and Kadler, K. E. (2004). Coalignment of plasma membrane channels and protrusions (fibripositors) specifies the parallelism of tendon. *J. Cell Biol.*, 165(4):553–563.

Chipman, S. D., Sweet, H. O., McBride Jr., D. J., Davisson, M. T., Marks Jr., S. C., Shuldiner, A. R., Wenstrup, R. J., Rowe, D. W., and Shapiro, J. R. (1993). Defective pro α 2(I) collagen synthesis in a recessive mutation in mice: A model of human osteogenesis imperfecta. *Proc. Natl Acad. Sci. USA*, 90:1701–1705.

Chung, L., Dinakarpandian, D., Yoshida, N., Lauer-Fields, J. L., Fields, G. B., Visse, R., and Nagase, H. (2004). Collagenase unwinds triple-helical collagen prior to peptide bond hydrolysis. *EMBO J.*, 23(15):3020–3030.

Cisneros, D. A., Hung, C., Franz, C. M., and Müller, D. J. (2006). Observing growth steps of collagen self-assembly by time-lapse high-resolution atomic force microscopy. *J. Struct. Biol.*, 154(3):232–245.

Cowin, S. C. and Doty, S. B. (2007). *Tissue Mechanics*. Springer, New York.

Dong, M., Xu, S., Bünger, M., Birkedal, H., and Besenbacher, F. (2007). Temporal assembly of collagen type II studied by atomic force microscopy. *Adv. Eng. Mater.*, 9(12):1129–1133.

Dunitz, J. (1994). The entropic cost of bound water in crystals and biomolecules. *Science*, 264:670.

Eyre, D. R. and Wu, J.-J. (2005). Collagen cross-links. *Top. Curr. Chem.*, 247:207–229.

Fallas, J. A., O'Leary, L. E., and Hartgerink, J. D. (2010). Synthetic collagen mimics: Self-assembly of homotrimers, heterotrimers and higher order structures. *Chem. Soc. Rev.*, 39(9):3510–3527.

Fang, M., Goldstein, E. L., Matich, E. K., Orr, B. G., and Banaszak Holl, M. M. (2013). Type I collagen self-assembly: The roles of substrate and concentration. *Langmuir*, 29(7):2330–2338.

Fields, G. B. (1991). A model for interstitial collagen catabolism by mammalian collagenases. *J. Theor. Biol.*, 153(4):585–602.

Fields, G. B. (2013). Interstitial collagen catabolism. *J. Biol. Chem.*, 288(13):8785–8793.

Flynn, B. P., Bhole, A. P., Saeidi, N., Liles, M., DiMarzio, C. A., and Ruberti, J. W. (2010). Mechanical strain stabilizes reconstituted collagen fibrils against enzymatic degradation by mammalian collagenase matrix metalloproteinase 8 (MMP-8). *PLoS One*, 5(8):e12337.

Fratzl, P. (2003). Cellulose and collagen: From fibres to tissues. *Curr. Opin. Colloid Interface Sci.*, 8(1):32–39.

Gelman, R. A., Williams, B. R., and Piez, K. A. (1979). Collagen fibril formation. Evidence for a multistep process. *J. Biol. Chem.*, 254(1):180–186.

Han, S., Makareeva, E., Kuznetsova, N. V., DeRidder, A. M., Sutter, M. B., Losert, W., Phillips, C. L., Visse, R., Nagase, H., and Leikin, S. (2010). Molecular mechanism of type I collagen homotrimer resistance to mammalian collagenases. *J. Cell Biol.*, 285(29):22276–22281.

Hulmes, D. J. S., Miller, A., Parry, D. A. D., Piez, K. A., and Woodhead-Galloway, J. (1973). Analysis of the primary structure of collagen for the origins of molecular packing. *J. Mol. Biol.*, 79:137–148.

Humphrey, W., Dalke, A., and Schulten, K. (1996). VMD: Visual molecular dynamics. *J. Mol. Graph.*, 14(1):33–38.

Jenkins, C. L., Bretscher, L. E., Guzei, I. A., and Raines, R. T. (2003). Effect of 3-hydroxyproline residues on collagen stability. *J. Am. Chem. Soc.*, 125(21):6422–6427.

Jiang, F., Hörber, H., Howard, J., and Müller, D. J. (2004). Assembly of collagen into microribbons: Effects of pH and electrolytes. *J. Struct. Biol.*, 148(3):268–278.

Jimenez, S. A., Bashey, R. I., Benditt, M., and R, Y. (1977). Identification of collagen alpha1(I) trimer in embryonic chick tendons and calvaria. *Biochem. Biophys. Res. Commun.*, 78:1354–1361.

Kadler, K. E., Baldock, C., Bella, J., and Boot-Handford, R. P. (2007). Collagens at a glance. *J. Cell Sci.*, 120(12):1955–1958.

Kalamajski, S., Aspberg, A., Lindblom, K., Heinegard, D., and Oldberg, A. (2009). Asporin competes with decorin for collagen binding, binds calcium and promotes osteoblast collagen mineralization. *Biochem. J.*, 423:53–59.

Kalamajski, S. and Oldberg, A. (2010). The role of small leucine-rich proteoglycans in collagen fibrillogenesis. *Matrix Biol.*, 29:248–253.

Kielty, C. M., Sherratt, M. J., and Shuttleworth, C. A. (2002). Elastic fibres. *J. Cell Sci.*, 115(14):2817–2828.

Kirschmann, D. A., Seftor, E. A., Fong, S. F., Nieva, D. R., Sullivan, C. M., Edwards, E. M., Sommer, P., Csiszar, K., and Hendrix, M. J. (2002). A molecular role for lysyl oxidase in breast cancer invasion. *Cancer Res.*, 62(15):4478–4483.

Kramer, R. Z., Bella, J., Brodsky, B., and Berman, H. M. (2001). The crystal and molecular structure of a collagen-like peptide with a biologically relevant sequence. *J. Mol. Biol.*, 311:131–147.

Kramer, R. Z., Bella, J., Mayville, P., Brodsky, B., and Berman, H. M. (1999). Sequence dependent conformational variations of collagen triple-helical structure. *Nat. Struct. Biol.*, 6(5):454–457.

Kuznetsova, N. and Leikin, S. (1999). Does the triple helical domain of type I collagen encode molecular recognition and fiber assembly while telopeptides serve as catalytic domains? Effect of proteolytic cleavage on fibrillogenesis and on collagen-collagen interaction in fibers. *J. Biol. Chem.*, 274(51):36083–36088.

Lee, K.-H., Kuczera, K., and Holl, M. M. B. (2011). Effect of osteogenesis imperfecta mutations on free energy of collagen model peptides: A molecular dynamics simulation. *Biophys. Chem.*, 156(2):146–152.

Leikin, S., Parsegian, V. A., Rau, D. C., and Rand, R. P. (1993). Hydration forces. *Annu. Rev. Phys. Chem.*, 44(1):369–395.

Leikin, S., Parsegian, V. A., Yang, W., and Walrafen, G. E. (1997). Raman spectral evidence for hydration forces between collagen triple helices. *Proc. Natl Acad. Sci. USA*, 94:11312–11317.

Leikin, S., Rau, D. C., and Parsegian, V. A. (1994). Direct measurement of forces between self-assembled proteins: Temperature-dependent exponential forces between collagen triple helices. *Proc. Natl Acad. Sci. USA*, 91:276–280.

Leikin, S., Rau, D. C., and Parsegian, V. A. (1995). Temperature-favoured assembly of collagen is driven by hydrophilic not hydrophobic interactions. *Nat. Struct. Biol.*, 2:205–210.

Leikina, E., Mertts, M. V., Kuznetsova, N., and Leikin, S. (2002). Type I collagen is thermally unstable at body temperature. *Proc. Natl Acad. Sci. USA*, 99:1314–1318.

Leitinger, B. (2011). Transmembrane collagen receptors. *Annu. Rev. Cell Dev. Biol.*, 27:265–290.

Leow, W. W. and Hwang, W. (2011). Epitaxially guided assembly of collagen layers on mica surfaces. *Langmuir*, 27:10907–10913.

Lewis, P. N., Pinali, C., Young, R. D., Meek, K. M., Quantock, A. J., and Knupp, C. (2010). Structural interactions between collagen and proteoglycans are elucidated by three-dimensional electron tomography of bovine cornea. *Structure*, 18(2):239–245.

Li, Y., Mo, X., Kim, D., and Yu, S. M. (2011). Template-tethered collagen mimetic peptides for studying heterotrimeric triple-helical interactions. *Biopolymers*, 95(2):94–104.

Lodish, H., Berk, A., Kaiser, C. A., Krieger, M., Bretscher, A., Ploegh, H., Amon, A., and Scott, M. P. (2012). *Molecular Cell Biology*, 7th edn. W. H. Freeman, Gordonsville, VA.

Long, C. G., Braswell, E., Zhu, D., Apigo, J., Baum, J., and Brodsky, B. (1993). Characterization of collagen-like peptides containing interruptions in the repeating Gly-X-Y sequence. *Biochemistry*, 32(43):11688–11695.

Makareeva, E., Han, S., Vera, J. C., Sackett, D. L., Holmbeck, K., Phillips, C. L., Visse, R., Nagase, H., and Leikin, S. (2010). Carcinomas contain a matrix metalloproteinase–resistant isoform of type I collagen exerting selective support to invasion. *Cancer Res.*, 70(11):4366–4374.

Manka, S. W., Carafoli, F., Visse, R., Bihan, D., Raynal, N., Farndale, R. W., Murphy, G., Enghild, J. J., Hohenester, E., and Nagase, H. (2012). Structural insights into triple-helical collagen cleavage by matrix metalloproteinase 1. *Proc. Natl Acad. Sci. USA*, 109(31):12461–12466.

Marini, J. C., Forlino, A., Cabral, W. A., Barnes, A. M., San Antonio, J. D., Milgrom, S., Hyland, J. C. et al. (2007). Consortium for osteogenesis imperfecta mutations in the helical domain of type I collagen: Regions rich in lethal mutations align with collagen binding sites for integrins and proteoglycans. *Hum. Mutat.*, 28(3):209–221.

McBride Jr., D. J., Choe, V., Shapiro, J. R., and Brodsky, B. (1997). Altered collagen structure in mouse tail tendon lacking the α2(I) chain. *J. Mol. Biol.*, 170:275–284.

McBride Jr., D. J., Kadler, K. E., Hojima, Y., and Prockop, D. J. (1992). Self-assembly into fibrils of a homotrimer of type I collagen. *Matrix*, 12(4):256–263.

Miles, C. A. and Bailey, A. J. (2001). Thermally labile domains in the collagen molecule. *Micron*, 32(3):325–332.

Miles, C. A., Sims, T. J., Camacho, N. P., and Bailey, A. J. (2002). The role of the α 2 chain in the stabilization of the collagen type I heterotrimer: A study of the type I homotrimer in *oim* mouse tissues. *J. Mol. Biol.*, 321:797–805.

Mizuno, K., Hayashi, T., Peyton, D. H., and Bächinger, H. P. (2004). The peptides acetyl-(Gly-3(*S*) Hyp-4(*R*)Hyp)$_{10}$-NH$_2$ and acetyl-(Gly-Pro-3(*S*)Hyp)$_{10}$-NH$_2$ do not form a collagen triple helix. *J. Biol. Chem.*, 279(1):282–287.

Muiznieks, L. D. and Keeley, F. W. (2013). Molecular assembly and mechanical properties of the extracellular matrix: A fibrous protein perspective. *Biochim. Biophys. Acta, Mol. Basis Dis.*, 1832:866–875.

Myllyharju, J. and Kivirikko, K. I. (2001). Collagens and collagen-related diseases. *Ann. Med.*, 33(1):7–21.

Nishioka, T., Eustace, A., and West, C. (2012). Lysyl oxidase: From basic science to future cancer treatment. *Cell Struct. Funct.*, 37(1):75–80.

Nuytinck, L., Freund, M., Lagae, L., Pierard, G. E., Hermanns-Le, T., and De Paepe, A. (2000). Classical Ehlers-Danlos syndrome caused by a mutation in type I collagen. *Am. J. Hum. Genet.*, 66(4):1398–1402.

Okuyama, K. (2008). Revisiting the molecular structure of collagen. *Connect. Tissue Res.*, 49(5):299–310.

Okuyama, K., Miyama, K., Mizuno, K., and Bächinger, H. P. (2012). Crystal structure of (Gly-Pro-Hyp)$_9$: Implications for the collagen molecular model. *Biopolymers*, 97(8):607–616.

Orgel, J. P. R. O., Eid, A., Antipova, O., Bella, J., and Scott, J. E. (2009). Decorin core protein (decoron) shape complements collagen fibril surface structure and mediates its binding. *PLoS ONE*, 4:e7028.

Orgel, J. P. R. O., Irving, T. C., Miller, A., and Wess, T. J. (2006). Microfibrillar structure of type I collagen *in situ*. *Proc. Natl Acad. Sci. USA*, 103:9001–9005.

Orgel, J. P. R. O., Wess, T. J., and Miller, A. (2000). The in situ conformation and axial location of the intermolecular cross-linked non-helical telopeptides of type I collagen. *Structure*, 8(2):137–142.

Ortolani, F., Giordano, M., and Marchini, M. (2000). A model for type II collagen fibrils: Distinctive D-band patterns in native and reconstituted fibrils compared with sequence data for helix and telopeptide domains. *Biopolymers*, 54(6):448–463.

Persikov, A. V., Pillitteri, R. J., Amin, P., Schwarze, U., Byers, P. H., and Brodsky, B. (2004). Stability related bias in residues replacing glycines within the collagen triple helix (Gly-Xaa-Yaa) in inherited connective tissue disorders. *Hum. Mutat.*, 24(4):330–337.

Perumal, S., Antipova, O., and Orgel, J. P. R. O. (2008). Collagen fibril architecture, domain organization, and triple-helical conformation govern its proteolysis. *Proc. Natl Acad. Sci. USA*, 105:2824–2829.

Petruska, J. A. and Hodge, A. J. (1964). A subunit model for the tropocollagen macromolecule. *Proc. Natl Acad. Sci. USA*, 51(5):871–876.

Pollitt, R., McMahon, R., Nunn, J., Bamford, R., Afifi, A., Bishop, N., and Dalton, A. (2006). Mutation analysis of COL1A1 and COL1A2 in patients diagnosed with osteogenesis imperfecta type I-IV. *Hum. Mutat.*, 27(7):716.

Rainey, J. K. and Goh, M. C. (2004). An interactive triple-helical collagen builder. *Bioinformatics*, 20(15):2458–2459.

Ramachandran, G. N. and Kartha, G. (1954). Structure of collagen. *Nature*, 174:269–270.

Ramachandran, G. N. and Kartha, G. (1955). Structure of collagen. *Nature*, 176:593–595.

Ravikumar, K. M., Humphrey, J. D., and Hwang, W. (2007). Spontaneous unwinding of a labile domain in a collagen triple helix. *J. Mech. Mater. Struct.*, 2(6):999–1010.

Ravikumar, K. M. and Hwang, W. (2008). Region-specific role of water in collagen unwinding and assembly. *Proteins: Struct. Funct. Bioinf.*, 72(4):1320–1332.

Ravikumar, K. M. and Hwang, W. (2011). Role of hydration force in the self-assembly of collagens and amyloid steric zipper filaments. *J. Am. Chem. Soc.*, 133:11766–11773.

Ricard-Blum, S. (2011). The collagen family. *Cold Spring Harb. Perspect. Biol.*, 3(1):a004978.

Rich, A. and Crick, F. H. C. (1955). The structure of collagen. *Nature*, 176(4489):915.

Rich, A. and Crick, F. H. C. (1961). The molecular structure of collagen. *J. Mol. Biol.*, 3(5):483–506.

Robins, S. (2007). Biochemistry and functional significance of collagen cross-linking. *Biochem. Soc. Trans.*, 35(Pt 5):849.

Russell, L. E., Fallas, J. A., and Hartgerink, J. D. (2010). Selective assembly of a high stability AAB collagen heterotrimer. *J. Am. Chem. Soc.*, 132(10):3242–3243.

Saccà, B., Renner, C., and Moroder, L. (2002). The chain register in heterotrimeric collagen peptides affects triple helix stability and folding kinetics. *J. Mol. Biol.*, 324(2):309–318.

Scott, J. E. (1992). Supramolecular organization of extracellular matrix glycosaminoglycans, in vitro and in the tissue. *FASEB J.*, 6:2639–2645.

Shoulders, M. D. and Raines, R. T. (2009). Collagen structure and stability. *Annu. Rev. Biochem.*, 78:929–958.

Sternlicht, M. D. and Werb, Z. (2001). How matrix metalloproteinases regulate cell behavior. *Annu. Rev. Cell Dev. Biol.*, 17:463–516.

Svensson, L., Aszodi, A., Reinholt, F. P., Fassler, R., Heinegard, D., and Oldberg, A. (1999). Fibromodulin-null mice have abnormal collagen fibrils, tissue organization, and altered lumican deposition in tendon. *J. Biol. Chem.*, 274:9636–9647.

Sweeney, S. M., Orgel, J. P. R. O., Fertala, A., McAuliffe, J. D., Turner, K. R., Di Lullo, G. A., Chen, S. et al. (2008). Candidate cell and matrix interaction domains on the collagen fibril, the predominant protein of vertebrates. *J. Biol. Chem.*, 283(30):21187–21197.

Tirrell, D. A., Aksay, I., Baer, E., Calvert, P. D., Cappello, J., Dimarzio, E. A., Evans, E. A., and Fessler, J. (1994). *Hierarchical Structures in Biology as a Guide for New Materials Technology*. National Academy of Sciences, Washington, DC.

Ushiki, T. (2002). Collagen fibers, reticular fibers and elastic fibers. A comprehensive understanding from a morphological viewpoint. *Arch. Histol. Cytol.*, 65(2):109–126.

Vakonakis, I. and Campbell, I. D. (2007). Extracellular matrix: From atomic resolution to ultrastructure. *Curr. Opin. Cell Biol.*, 19:578–583.

Wess, T. J. (2005). Collagen fibril form and function. *Adv. Protein Chem.*, 70:341–374.

Williams, B. R., Gelman, R. A., Poppke, D. C., and Piez, K. A. (1978). Collagen fibril formation. Optimal in vitro conditions and preliminary kinetic results. *J. Biol. Chem.*, 253(18):6578–6585.

Yoon, J. H. and Halper, J. (2005). Tendon proteoglycans: Biochemistry and function. *J. Musculoskelet. Neuronal Interact.*, 5(1):22–34.

5

Cell–Matrix and Cell–Cell Mechanical Interactions

Assaf Zemel and Ralf Kemkemer

CONTENTS

Cell adhesion to the extracellular matrix (ECM) is a vital process of normal tissue cells such as fibroblasts, endothelial cells, and stem cells. Normal cells that fail to anchor to a solid environment and spread would not grow and divide and eventually enter apoptosis (programmed cell death) (Folkman and Greenspan, 1975; Discher et al., 2005). Furthermore, loss of anchorage dependence and the ability of cells to proliferate and survive in very soft environments (e.g., agar gel) or suspension is one of the long recognized signatures of cancerous transformation (Stoker et al., 1968; Wittelsberger et al., 1981; Vasiliev, 1984). The mechanisms governing cell behavior following its engagement with the ECM have been the focus of extensive research in recent years. Many intriguing new effects of the surrounding mechanics on cell behavior and fate have been discovered, and much has been learned about the mechanical and biochemical mechanisms involved, both on the cellular and molecular levels.

The mechanical interplay between the cell and the environment in the early stages of cell adhesion plays a critical role in the regulation of cell behavior and fate. During cell adhesion, cells establish their morphology and adhesion pattern to the ECM, and forces that develop in this process orchestrate global changes in cytoskeleton organization as well as in nucleus positioning and form. The surroundings present a variety of mechanical and geometrical cues, such as the ECM topography, rigidity, or ligand density, as well as local stresses and strains that play an important part in establishing the mechanical state of the cell, including its shape and orientation. These mechanical processes are accompanied and regulated by force-sensitive biochemical cascades that not only feedback on the mechanical response but also impinge on other processes such as cell cycle control and differentiation.

In addition, a growing number of studies highlight the importance of ECM mechanics in the control of collective behaviors of cells and cell–cell interactions in processes such as self-assembly and morphogenesis (Folkman and Greenspan, 1975; Harris et al., 1980; Oster et al., 1983; Ingber and Folkman, 1989; Paszek et al., 2005; Califano and Reinhart-King, 2008; Yu et al., 2011; Calvo et al., 2013).

In this chapter, we review two major aspects of cell mechanosensitivity. The first part concerns the process of cell adhesion. The effects of surrounding mechanics and cell morphology on cell behavior are reviewed, and a discussion of the mechanisms involved is presented. This includes research on the dynamics and mechanics of cell adhesion and spreading and the accompanying remodeling processes of the cytoskeleton. Our focus here is on the mechanical processes associated with cell adhesion; the reader is referred to other texts for in-depth discussions of the biochemical processes involved (Geiger et al., 2009). In the second part of the chapter, we discuss two distinct routes of cell–cell mechanical interactions and the role they play in cellular self-assembly and morphogenesis.

5.1 Mechanical Cues in the Cellular Environment and Their Effect on Cell Behavior and Fate

5.1.1 Topographic Determinants of Cell Shape and Function

One of the first recognized mechanical factors shown to guide cells in their environment is topographical cues—small ridges and grooves and fibers in the ECM that lead to cell elongation and orientation in preferred directions (see Figure 5.1). Dating back to the beginning of the previous century, experiments by Harrison and others have shown that solid mechanical structures, in this case small fibers from a spider web, can guide cells and cause them to elongate along the fiber direction (Harrison, 1914). Weiss has later demonstrated that the outgrowth of nerves can be guided by small elongated features on the substrate (Weiss, 1934) supporting the hypothesis of *contact guidance* that provided a complementary mechanism (to chemical guidance) for directing cell orientation.

With improved technologies for fabricating micro- or nanotopographies on surfaces (Curtis and Wilkinson, 1997; Whitesides et al., 2001), cell responses to submicron or nano-sized cues have been extensively examined in the past decade. It has been demonstrated that even surface features with heights in the range of a few *nanometers* or small variation in adhesive ligand distance on the scale of nanometers can induce changes in cell shape and other cellular responses (Cavalcanti-Adam et al., 2007; Kim et al., 2012).

The extent to which cells adapt their morphology to surface topography depends on its geometrical characteristics and the cell type. A comparative study demonstrated that there are differences in the topography sensitivity between human fibroblasts, smooth muscle cells, and endothelial cells (Biela et al., 2009). Also the width and depth of grooves have been shown to affect cell shape (Karuri et al., 2008). To separate the effects of groove size and curvature, Mathur et al. developed a method to fabricate grooved substrates with different edge radii keeping other dimensions constant. Mouse fibroblast increased their

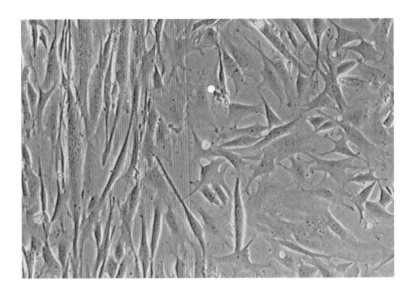

FIGURE 5.1
Fibroblast cells on a flat elastomer surface (right side) and adhering on the surface with a small topography given by grooves with 200 nm height and 2 μm ridge width (left part of the image). The adaptation of cell shape and the alignment of the cells with respect to the topography is well visible.

spreading area but reduced their polarization with increasing radius of curvature (Mathur et al., 2012). An interesting study on the effects of surface topography on nerve cell branching has been discussed by Baranes et al. (2012a,b). Photolithography was used to fabricate substrates with repeatable line-pattern ridges of nanoscale heights. Neuronal processes were shown to sense ridges of height as low as 10 nm. The interaction between the neuronal process and the ridge leads to a deflection of growth direction and a preferred alignment with the ridges. The incoming angle between the neuronal process and the ridge was shown to have an effect on the growth direction. This study demonstrates the sensitivity of growing neurites to nanoscale cues and opens a new avenue of research for predesigned neuronal growth and circuitry.

Despite the growing interest in the effects of surface topography, there are still a limited number of quantitative studies determining how cell elongation or other morphological characteristics depend on specific aspects of the topography. A quantitative analysis of the orientation response of human melanocytes cultured on rectangular grooves with heights between 25 and 200 nm and spatial frequencies between 100 and 500 mm^{-1} measured a dose–response-like relationship (Kemkemer et al., 2006). To quantitatively predict the dose–response characteristics of the cell behavior, a simple system analysis related to engineering control principles has been used, assuming that the orientation response of the cells is regulated by a kind of automatic controller. Although this approach is solely phenomenological, it provided some insight into the nature of the response without knowing the molecular steps involved. It demonstrated that cell orientation to the microgroove topography depends on the square of the product of groove height and spatial frequency. Further studies may correlate these quantitative dependencies with molecular details and thereby provide insight into the molecular details of cellular topography sensing.

One of the important findings of recent research is that surface topography not only affects cell morphology and orientation but also influences important cell functions such as gene expression, proliferation, or migration. Among the striking demonstrations is the impact of surface topography on in vitro stem cell behavior. In their recent study, Moe et al. (2012) have shown that anisotropic topographies enhance neuronal differentiation, while isotropic topographies enhance glial differentiation of neural progenitor cells. In another study, it has been demonstrated that nanostructured surfaces can prolong the stem cell phenotype of adult mesenchymal stem cells in culture (McMurray et al., 2011). An additional impressive example is the reprograming of somatic cells into pluripotent stem cells. A simple topography of the culture substrate, in the form of parallel microgrooves, can substitute the effects of epigenetic modifiers and improves the reprogramming efficiency (Downing et al., 2013). This study suggests that simple changes in the morphology of cells and their nucleus, caused by the alignment along simple grooves in the culture surface, may modulate the reprogramming process substantially. This not only highlights the importance of mechanical signals in directing the native cell behavior but also suggests the use of mechanical signals for artificially controlling cells. The premise to control cell behavior by specific topographical features has significant implications in biomaterial research, tissue engineering, and regenerative medicine (Harvey et al., 2013). Another interesting implication of cellular responsiveness to surface topography follows from the fact that the cell sensitivity might be altered by disease-related mutations, as shown, for example, in Booth-Gauthier et al. (2012) and Kaufmann et al. (2012). These studies hint toward the use of mechanical surface signals in cellular diagnostics.

5.1.2 Cell Shape and the Regulation of Cell Behavior and Fate

Surface topography has seen to induce changes in cell orientation and shape. The hypothesis that the cell morphology might play an important role in the regulation of gene expression and therefore in cell behavior and morphogenesis has been raised already in the 1970s by cell biologists such as Folkman and coworkers (Folkman and Greenspan, 1975; Folkman and Moscona, 1978; Ingber and Folkman, 1989). By changing the thickness of adhesive layer (poly-2-hydroxyethyl methacrylate) on a surface, Folkman and Moscona (1978) could systematically control the flattening of cells on the surface and showed that the degree of DNA synthesis and cell proliferation are tightly coupled to the cell spreading area. Similar variations in the levels of DNA synthesis were found upon a change in the plating density of cells that restrict cell spreading. Extensive research along these lines confirmed the importance of spreading area and cell morphology in the mechanics of cell division and cell cycle control (Chen et al., 1997; Huang et al., 1998; Théry and Bornens, 2006), stem cell differentiation (McBeath et al., 2004; Gao et al., 2010; Kilian et al., 2010), and cancer transformation (Folkman and Greenspan, 1975; Paszek et al., 2005), and researchers are currently investigating the subcellular and molecular mechanisms underlying these responses including the effects on cytoskeleton organization, adhesion pattern, nucleus deformation, and gene expression regulation.

The development of microlithography techniques provided an exquisite tool to manipulate cell morphology and adhesion characteristics (adhesive pattern and ligand type), essentially at will, and to decouple effects of cell shape and adhesion properties from other factors such as the surroundings elasticity, dimensionality, and cell density; for a review, see Ruiz and Chen (2007). Accordingly, cell-adhesive proteins such as fibronectin and collagen are micropatterned in restricted islands of any desired shape on a surface to precisely control (in submicron resolution) cell spreading area and shape (Singhvi et al., 1994; Chen et al., 1997; Kilian et al., 2010). Most studies of this sort have been carried out on rigid (glass or plastic) substrates; however, more recently, efforts have been made to pattern cells on soft elastic foundations (Grevesse et al., 2013; Versaevel et al., 2014). These efforts have the premise of elucidating the relative importance of ECM rigidity and topography in controlling cell form and function. The micropatterning technology enabled to study the effects of cell shape and size on the establishment of cytoskeleton structure (Théry et al., 2006a,b), focal adhesion redistribution (Chen et al., 2003), nuclear positioning and morphology (Mazumder and Shivashankar, 2010; Versaevel et al., 2012), and the effects on cell survival and differentiation.

Ingber and coworkers (Chen et al., 1997; Huang et al., 1998) have used the micropatterning technique to investigate the effect of cell spreading area on DNA synthesis and apoptosis. Capillary endothelial cells were allowed to adhere to differently sized (~10–50 μm in diameter) fibronectin islands on a surface. Levels of DNA synthesis were shown to increase with cell spreading area, while the percentage of cells that entered apoptosis was significantly higher on the smaller-sized islands. This was consistent with effects of plating density on cell proliferation and the response of these capillary cells to damaged/fragmented ECM in vivo. It was then questioned whether it is the amount of ligand binding (that increases with the contact area between the cell and the substrate) or the spreading area per se, which is responsible for the observed effects on cell fate. By fabricating adhesive patches with tunable (small) spacings between them, the authors could generate islands that enclosed different areas but whose total available (fibronectin) adhesive area was the same. Both DNA synthesis and apoptosis rates were shown to be sensitive to actual projected cell area rather than to the total adhesive area. This finding suggested that a mechanical (elastic in origin), rather than a chemical, mechanism is responsible to the apparent geometrical response, namely, cells with larger projected areas died less and proliferated more due to their elastically more stretched configuration rather than their larger contact area with ligands on the surface. This hypothesis gained additional support in experiments in which internal cellular tension has been modulated via variations in Rho signaling and consequent cytoskeleton contractility (Welsh et al., 2001; Assoian and Klein, 2008) and changes in ECM rigidity.

McBeath et al. (2004) used a similar approach to examine the effects of stem cell morphology on differentiation. When human mesenchymal stem cells (hMSCs) were grown in a medium with mixed adipogenic (fat cell inducing) and osteogenic (bone cell inducing) cocktails, small adhesive areas promoted adipogenic fate, while large adhesive areas promoted the osteogenic fate. These results were consistent with experiments in which the plating cell density was varied. Cells plated at low densities (thus assuming well-spread morphologies) preferentially differentiated into bone cells, while at high densities, adipogenic (fat cell) fate was promoted. Variations in cellular tension, governed by the extent of cell spreading, have been suggested to underlay these responses via the effect on integrin activation and downstream RhoA signaling (McBeath et al., 2004). A follow-up, more detailed analysis of the effect of cell shape on mesenchymal stem cell differentiation has been reported by Kilian et al. (2010). These authors have demonstrated that even remarkably fine variations in the cell shape may have significant influences on cell differentiation. In comparing cells with pentagonal symmetry with minor differences in the sharpness of their corners, "starlike" shapes promoted osteogenesis, while "flowerlike" shapes promoted adipogenesis. As in previous studies, internal cellular tension has been suggested to underlay this prominent shape sensitivity. Consistently, interference with actomyosin contractility, using specially designated drugs, eliminated the observed shape sensitivity. In line with this study, when adipogenesis was studied in cells with small adhesive areas (Song et al., 2011), no significant shape effects have been observed; triangle, square, hexagon, and circular cell shapes were examined and yielded similar results. In contrast, osteogenesis has been shown to depend on the details of the cell shape (Kilian et al., 2010) and the underlying substrate topography (Seo et al., 2011).

The study by Kilian et al. showed that even minute changes in cell morphology might have profound effects on cell fate. Cell shape is likely to exert its regulatory effects via the role it plays in the determination of the stress distribution in the cell. For instance, stresses tend to concentrate at sharp cell corners, and this results in recruitment of adhesion receptors and stress fiber formation at these sites. However, how these changes transduce to specific effects on gene expression is still not fully understood; for texts concerning these mechanisms, see Geiger et al. (2009), Assoian and Klein (2008), and Wang et al. (2009).

5.1.3 Effects of Extracellular Matrix Rigidity

The factor that perhaps most directly controls the force balance between the cell and its environment is the elastic rigidity of the ECM. The ECM rigidity is implicated in the determination of many morphologic and mechanical characteristics of cells, including the cell size and shape, internal structure, adhesion pattern, cytoskeleton stiffness, force generation, and locomotion (Yeung et al., 2005). It has also been shown to have appreciable effects on cell proliferation (Wang et al., 2001; Klein et al., 2009; Ulrich et al., 2009; Tilghman et al., 2010; Mih et al., 2011, 2012), stem cell differentiation (Engler et al., 2006), cancer transformation (Wang et al., 2001; Paszek et al., 2005; Ulrich et al., 2009; Tilghman et al., 2010; DuFort et al., 2011; Dupont et al., 2011), cell–cell interactions, and morphogenesis (Guo et al., 2006; Califano and Reinhart-King, 2009; Bajaj et al., 2010; Tang et al., 2011). Moreover, a growing body of evidence support, the paradigm that cellular processes have evolved to optimally function in a certain range of stiffness (Discher et al., 2005) and conversely that an improper rigidity of the surroundings contributes to the onset of pathology and disease (Ingber, 2003; Park et al., 2007; Janmey and Miller, 2011).

One of the striking reported effects of substrate rigidity on cell behavior has been the demonstration by Engler et al. (2006) that the lineage specification of hMSCs could be directed by the elastic rigidity of the substrate on which the cells are grown. Substrates of stiffness similar to that of brain, muscle, or bone tissues caused optimal expression of neurogenic, myogenic, or osteogenic differentiation, respectively. Many subsequent studies showed the potential to manipulate stem cell fates by suitably choosing the stiffness of the environment; for reviews, see Discher et al. (2009), Guilak et al. (2009), and Dado et al. (2012). Osteogenic and chondrogenic differentiation markers were generally more prominently expressed in more rigid environments, whereas neurogenic and adipogenic markers were more pronounced in soft environments. Recent studies have shown that the elastic properties of the stem cell niche may play a role in supporting their multi/pluripotency and self-renewal capabilities (Winer et al., 2008; Gilbert et al., 2010). Winer et al. have shown that hMSCs remain quiescent (i.e., do not proliferate or differentiate) when sparsely grown on soft substrates (0.25 kPa) that mimic bone marrow conditions, but still preserve their capacity to differentiate upon exposure to appropriate mechanical or chemical cues (Winer et al., 2008). Similarly, embryonic stem cells were shown to maintain their pluripotency and self-renewal capabilities when grown in a soft environment that is similar to their own stiffness (Wang et al., 2009). What maintains the cells in a quiescent state is not fully understood. However, evidences from a variety of experiments suggest that the entry to S (DNA synthesis) phase of the cell cycle requires the buildup of tension in the cytoskeleton (Chen et al., 1997; McBeath et al., 2004; Assoian and Klein, 2008; Klein et al., 2009); this is only possible in a sufficiently rigid matrix that can sustain the stress in the cell. Similarly, the muscle microenvironment (niche) enables muscle stem cells to contribute to skeletal muscle regeneration, but when grown on standard tissue culture plastic, they lose "stemness" yielding progenitors with greatly diminished regenerative potential. When grown in hydrogels with rigidity similar to that of muscle tissue, these cells self-renewed and featured enhanced potential to repair damaged muscle tissue after transplantation (Gilbert et al., 2010).

ECM rigidity also plays an important role in the regulation of cell proliferation (Assoian and Klein, 2008) and has recently been identified as an important factor in breast cancer development (Paszek et al., 2005; Provenzano et al., 2009) and metastasis (Levental et al., 2009; Calvo et al., 2013); for reviews, see DuFort et al. (2011) and Bissell and Hines (2011). Cell proliferation is seen to robustly increase with substrate rigidity across a wide range of cell types including both normal and transformed (cancerous) cell types (Chen et al., 1997; Wang et al., 2001; Klein et al., 2009; Tilghman et al., 2010; Mih et al., 2011, 2012). The stimulating effect of matrix rigidity on cell growth has been attributed to the higher tension, resulting with better cell spreading. Elevated tension induces integrin clustering and focal adhesion formation and consequently activates the Rho and Erk signaling pathways that promote cell division (Assoian and Klein, 2008). Remarkably, in some transformed cells, this regulatory pathway is attenuated, thereby leading to uncontrolled cell proliferation. Furthermore, recent studies have shown that an abnormal increase in the elastic rigidity of the cellular surroundings may induce a malignant phenotype of self-assembling, normal mammary endothelial cells (Paszek et al., 2005). When grown in 3D collagen gels with normal tissue rigidity, the cells nicely develop functional acinar structures and show growth arrest; however, when the rigidity is slightly increased beyond normal, the cells show increased proliferation, loss of internal polarity, increase in cellular forces, as well as increased Rho and Erk expressions, all of which are the hallmarks of the malignant phenotype.

5.2 Mechanics of Cell Adhesion and the Acquisition of Cell Morphology and Internal Structure

The studies surveyed earlier have demonstrated the importance of the surrounding geometry and mechanics in the regulation of cellular processes. While these are generally long-term phenomena, many central mechanical processes already initiate with the first engagement of the cell with the ECM. Thus, this period plays a crucial role in the life of adherent cells. During the process of cell adhesion, the cell establishes its morphology, size, and adhesive pattern to the ECM, and remodeling processes take place in the cytoskeleton that dictate the cells' internal structure and polarity. Cell shape and matrix rigidity govern the remodeling processes in the cytoskeleton and the establishment of cell polarity. This raises several fundamental questions: By what mechanisms do cells regulate their size, shape, and internal structure? What is the origin of the coupling between the cell shape and the spatial organization of the cytoskeleton, and how does the rigidity of the surroundings or forces in the ECM govern this coupling to ensure optimal functioning of the cell? Experiments have also shown that the stiffness of cells is actively tuned in response to the rigidity of the environment (Solon et al., 2007; Tee et al., 2011) and that rigidity gradients orient cells and trigger their locomotion from soft to more rigid areas (Tse and Engler, 2011). The observation that all these structural and

mechanical characteristics of cells depend on the elastic rigidity of the surroundings indicates that these properties are mechanically regulated by forces. Indeed, upon binding to
the ECM, adherent cells generate and maintain elastic tension (often known as prestress)
in their cytoskeleton. This is achieved by the distribution of myosin II molecular motors
that apply contractile forces on actin filaments and locally compress the cytoskeleton.
Actin polymerization at the cell front also contributes to the development of cellular tension (Nisenholz et al., 2014b). Since these elastic stresses generally propagate throughout
the cytoskeleton, they constitute the means for globally regulating geometric features and
processes in the cell. For example, the magnitude, orientation, and spatial distribution of
cytoskeletal stresses govern the size of the cell, the strength, and *pattern* of cell adhesion
to the substrate and dictate the spatial architecture and polarity of cytoskeletal filaments
among other effects to be elaborated in the following. In this section, we touch upon the
physics underlaying these processes, starting with the dynamics and mechanics of cell
spreading on a substrate, a process that leads to establishment of cell morphology. We
then discuss the accompanying processes that take place in the cytoskeleton and their
regulation by internal forces.

5.2.1 Mechanism of Cell Spreading; Dynamics and Mechanics

The dynamics and mechanics of cell spreading and shape acquisition are subjects of
major current interest. As shown earlier, topographical cues in the environment such as
thick fibers, ridges, and grooves regulate the spreading process and thereby modulate
the alignment and shapes that the cells adopt; other factors include the ECM rigidity
and ligand density. Interestingly, while the dynamics of spreading and the cell shape
generally vary from one cell type to the other, establishment of cell area is often reached
prior to final shape acquisition, and the steady-state area shows a generic dependence on
rigidity (Vasiliev, 1984; Prager-Khoutorsky et al., 2011; Rehfeldt et al., 2012) (see Figure 5.2).
When sparsely grown on isotropic soft substrates, cells assume round and small morphologies, and as the rigidity increases, a variety of shapes develop, but the spreading

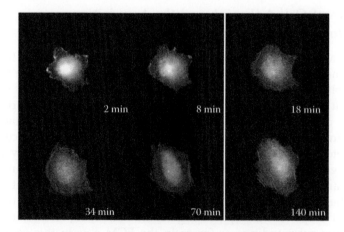

FIGURE 5.2
Time-lapse fluorescence microscopy sequence demonstrating the active spreading of a mouse fibroblast
(NIH3T3) on a flat elastomer surface covered with fibronectin. Actin was visualized by using GFP-Lifeact. Note
the two distinct stages of spreading: initial isotropic spreading followed by stress fiber formation and concurrent shape acquisition. Time after seeding the cell is given for each image in minutes.

area often increases according to a simple empirical relation (Engler et al., 2004a,b; Zemel et al., 2010b; Rehfeldt et al., 2012):

$$\frac{A - A_{\min}}{A_{\max} - A_{\min}} = \frac{E_m}{E_m + E_0} \tag{5.1}$$

where

$A_{\min}$ and $A_{\max}$ are the mean projected cell areas in the limit of infinitely soft and hard matrices, respectively

E_m is the Young's modulus of the substrate

E_0 is a parameter associated with an effective Young's modulus of the cell (Zemel et al., 2010b)

The observation that cell spreading area depends and generally increases with substrate rigidity suggests that there is a mechanical contribution to the determination of cell size and that cell size is dictated by a balance of forces between the cell and the environment. Indeed, experiments show that cell spreading proceeds with a continuous development of elastic tension in the cytoskeleton (Reinhart-King et al., 2005; Cai et al., 2010; Nisenholz et al., 2014b), and a similar scaling of total cell force and cell area on substrate rigidity is obtained (Paszek et al., 2005; Reinhart-King et al., 2005; Ghibaudo et al., 2008; Fu et al., 2010). Interestingly, Equation 5.1 may be deduced from simple elastic considerations where the cell and matrix elasticities are represented by two effective springs that are connected in series and where the area difference, $A_{\max} - A_{\min}$, reflects an inherent (limiting) length scale born by the dynamics of the machinery that drives cell spreading (Zemel et al., 2010b; Nisenholz et al., 2014b). The intrinsic cellular property, E_0, plays an interesting biological role since it governs the range of cell sensitivity to substrate stiffness; cells with small E_0 will show a sharp area dependence on rigidity, reaching saturation already on soft matrices, while cells with higher E_0 would show a broader range of rigidity dependence.

The establishment of cell morphology and the reorganization of the cytoskeleton and adhesion pattern to the ECM often continue after cell area has reached a steady state (Vasiliev, 1984; Prager-Khoutorsky et al., 2011; Rehfeldt et al., 2012). For some cells (e.g., hMSCs (Rehfeldt et al., 2012)), it has been reported that a breaking of symmetry in cell shape (e.g., cell elongation) develops continuously from initial spreading, but the rate of change is slower than the rate of area change. In contrast, the spreading of fibroblasts on rigid substrates has been described to proceed in two distinct phases (Prager-Khoutorsky et al., 2011): an isotropic phase, where cell area establishes first, and then a second phase where cell elongation and cytoskeleton polarization develop with minor change in cell area. A demonstration of these two distinct stages can be seen in Figure 5.2, where the first three images show isotropic spreading and the last three images show the formation of actin bundles and the concurrent shape changes. During the initial phase, microtubules and intermediate filaments develop from the perinuclear regime into the lamellar cytoplasm. At the same time, bundles of actomyosin stress fibers develop circumferentially at the cell periphery. As the cell begins to elongate, these actomyosin bundles reorganize in parallel to the long axis of the cell. Time-resolved microscopy in live cells revealed that focal adhesion redistribution and orientation at the cell poles precede the elongation processes, suggesting a role of these adhesions in controlling cell alignment (Prager-Khoutorsky et al., 2011). Unlike the fibroblasts, the spreading of many epithelial and endothelial cells often (yet not always) lacks the second elongation stage; thus, these cells spread more isotropically acquiring more symmetric morphologies (Vasiliev, 1984).

For some cell types, the level of cell elongation and actomyosin polarization were shown to attain a maximum for intermediate level of substrate rigidity (Zemel et al., 2010a; Chopra et al., 2011; Rehfeldt et al., 2012). In contrast, myoblasts (Engler et al., 2004b) and smooth muscle cells (Engler et al., 2006) showed no significant dependence of cell aspect ratio on substrate rigidity. Furthermore, on soft fibronectin-coated polyacrylamide gels (5 kPa), fibroblasts failed to polarize in the second stage and remained round in morphology (Prager-Khoutorsky et al., 2011), and the mean cell aspect ratio was shown to increase monotonically with substrate rigidity. How substrate rigidity functions in these different behaviors is still not known.

5.2.1.1 Subcellular Mechanisms of Cell Spreading Dynamics

Much attention has been devoted in recent years to study the molecular processes that take place at the leading edge of the cell and that are responsible for powering cell spreading and locomotion. Several lines of evidence suggest that the first few minutes of cell substrate interaction is driven by passive adhesion (of membrane receptors to ligands in the ECM), resembling the wetting phenomenon of liquid droplets on a solid substrate (Cuvelier et al., 2007). During this phase, cells often show recurrent attempts to explore adhesive domains by dynamically extruding and retracting thin filopodia protrusions in arbitrary directions. As the cell begins to flatten on the surface, wider lamellipodia protrusions filled by a dense network of actin filaments establish at the leading edge of the cell and actively propel the cell front forward (Döbereiner et al., 2004; Dubin-Thaler et al., 2004, 2008; Fardin et al., 2010). This marks the second phase of spreading—an active phase driven by actin polymerization.

Cuvilier et al. (2007) have found that the initial phase of cell spreading shows a universal power law dependence of the cell radius on time, $R(t) \sim t^{1/2}$ (equivalently the area grows linearly in time, $A(t) \sim t$). Earlier studies of the dynamics and metabolitic requirements of cell spreading showed that inhibitors of energy metabolism have no affect on early cell attachment and deformation (Bereiter-Hahn et al., 1990; Pierres et al., 2003). These evidences suggested that the early phase of cell spreading is a passive phase, driven by adhesion energy of cell membrane receptors to ligands in the ECM. To reconcile the universal power law of spreading dynamics with that of passive wetting, Cuvelier et al. suggested that the viscous dissipation during spreading is restricted to a thin cortex layer adjacent to the cell membrane, rather than the whole cell volume.

The $R(t) \sim t^{1/2}$ scaling law appears to extend beyond the purely passive regime of spreading where actin polymerization is known to be an essential driving force. There thus seems to be an overlap period where both active and passive forces drive cell spreading. Fardin et al. (2010) have developed an alternative hydrodynamic model of early cell spreading in which actin polymerization provides the driving force for cell spreading, while cytoskeleton viscosity is the time dictating factor, assuming volume conservation leads them to a similar scaling law $R(t) \sim t^{1/2}$ of fast spreading.

Lamellipodia protrusions establishing at the leading edges of cells and driving the active phase of spreading comprise a highly dynamic and dense network of actin filaments that constantly polymerize at the cell front and depolymerize few microns behind, forming a treadmilling front that advances forward relative to the underlying substrate. Interestingly, early investigations of actin dynamics at the cell leading edge have found that membrane protrusion is accompanied by a centripetal (retrograde) flow of the actin network. There has been much interest in this phenomenon because it was thought to reflect the mechanism by which the cell applies forces on the substrate to propel itself forward (Harris, 1994; Cramer, 1997). The net forward protrusion speed, v_{prot}, of an extending membrane protrusion is therefore dictated by the sum of two oppositely directed motions, the actin polymerization speed, v_{pol}, and the retrograde flow speed, v_{ret}:

$$v_{prot} = v_{pol} - v_{ret} \tag{5.2}$$

A number of experiments have demonstrated that the sum of protrusion speed and retrograde flow, $v_{prot} + v_{ret}$, remains constant during cell spreading. The retrograde flow and protrusion speed develop in opposite directions during spreading; while the retrograde flow increases with time, the protrusion speed decelerates until a steady state is reached (Lin and Forscher, 1995; Giannone et al., 2004; Jurado et al., 2005; Szewczyk et al., 2013). These observations suggest that the radial polymerization speed is slowly varying during spreading. Insight into this behavior comes from measurements of the force–velocity relation associated with the polymerizing actin network (Prass et al., 2006). An "autocatalytic" mechanism (Carlsson, 2004) suggests that as the forces opposing the network growth increase in magnitude, more branching of the network occurs, thus reducing the actual load per filament and allowing the polymerization to proceed with fixed radial speed.

The relation between retrograde flow and protrusion rate speed has been suggested to be mediated by "molecular clutches" that establish between the moving cytoskeleton and the substrate (Chan and Odde, 2008). Because these clutches are only transiently engaged (having a finite lifetime), they collectively give rise to cell–substrate friction (Walcott and Sun, 2010). This cell–substrate interfacial friction is what supports the outward polymerization force against the cell membrane and enables the cell periphery to advance forward. The frictional slippage of the polymerizing actin network relative to the substrate is monitored as a retrograde flow of this network (Giannone et al., 2004; Hu et al., 2007; Alexandrova et al., 2008). For a more detailed discussion about these frictional interactions, the reader is referred to several recent theoretical investigations of this subject (Sabass and Schwarz, 2010; Walcott and Sun, 2010; Bangasser et al., 2013; Sens, 2013; Nisenholz et al., 2014b; Schweitzer et al., 2014).

Closer investigation of the actin-based machinery at the leading edge of the cell revealed two distinct, yet overlapping, networks extending 2–4 μm away from the leading edge: these are the so-called lamella and lamellipodium networks (Ponti et al., 2004). The two networks differ in structure, density, molecular composition, as well as in their turnover properties and the retrograde flow speed. Moreover, the mechanism driving the retrograde flow of the two networks appears to be different and sensitive to different drug treatments (Hu et al., 2007; Alexandrova et al., 2008; Gardel et al., 2008). The lamellipodium network is a dense dendritic network, rich in the Arp 2/3 complex that nucleates new actin branches and cofilin that catalyzes F-actin disassembly. Actin polymerization at the cell front is thought to drive rapid retrograde flow of the lamellipodium network. Indeed, inhibition of actin polymerization with cytochalasin D selectively slows down flow in the lamellipodium; however, due to the low myosin density in this network, its retrograde flow is insensitive to myosin inhibition.

Engagement of these moving networks to the substrate results in transmission of traction forces. Gardel et al. focused on the relation between the retrograde flow across adhesion complexes and the traction forces exerted onto the substrate (Gardel et al., 2008). Good correlation has been found between the local directions of the retrograde flow and the traction forces exerted into the matrix. Surprisingly, however, their experiments revealed a robust biphasic relationship between the F-actin speed and magnitude of the traction force. In the lamella, where F-actin flow was relatively slow (2–10 nm/s), traction force was found to be an increasing function of the retrograde flow. In contrast, at more peripheral locations in the lamellipodium, traction force was a decreasing function of the retrograde flow speed. Some insight about these different behaviors comes from theoretical studies of the effects of force on the binding kinetics of the actin network to the ECM that constitute the cell–substrate interfacial friction (Sabass and Schwarz, 2010; Walcott and Sun, 2010; Bangasser et al., 2013; Sens, 2013; Nisenholz et al., 2014b; Schweitzer et al., 2014).

Toward late stages of spreading, cell spreading decelerates until the cell area stabilizes around an asymptotic value. Monitoring the temporal changes in the local speed along the cell front revealed an interesting sharp crossover between two different behaviors (dynamical phases) (Döbereiner et al., 2004; Dubin-Thaler et al., 2008b). During the "fast spreading" phase where the cell radius increases as $\sim t^{1/2}$, the cell front shows continuous forward development; however, beyond a certain threshold time, local oscillations of fast extensions and retractions decorate the recordings of cell front velocity. These oscillations continue during the stall phase where the cell font ceases to grow and reflect the stick–slip interaction between the cell and the substrate (Chan and Odde, 2008).

A number of theoretical models of different levels of sophistication have been developed to predict the evolution of cell spreading (Frisch and Thoumine, 2002; Cuvelier et al., 2007; Fardin et al., 2010; Zeng and Li, 2011; Vernerey and Farsad, 2014; Nisenholz et al., 2014b). A particularly simple theory has recently been developed by Nisenholz et al. In their theory, the cell is idealized as a flat homogeneous and isotropic elastic disk that is actively pulled by lamellipodia protrusions at the cell front; some known details of the motor unit at the cell front, including actin polymerization, interfacial friction, and the retrograde flow, are taken into account. The theory predicts closed-form expressions for the simultaneous evolution of cell size and force and the dependence on matrix rigidity and ligand density on a substrate (Nisenholz et al., 2014b). The authors verify their predictions against simultaneous measurements of cell area and force. To unravel some of the features discussed previously of cell spreading dynamics, more elaborate treatments of cell spreading dynamics would have to account for concurrent shape changes, cytoskeletal, and adhesion pattern rearrangements that typically occur during spreading.

5.2.2 Active Remodeling of Stress Fibers in the Cytoskeleton

In this section, we discuss the remodeling processes that take place in the cytoskeleton during, or following, cell spreading. Of particular interest is the development of long and thick actomyosin fibers, known as stress fibers, tens of minutes to hours after cell spreading (Hall, 1998; Ren et al., 1999; Wang et al., 2002; Théry et al., 2006a). The formation and remodeling of stress fibers in the cytoskeleton play an essential role in the acquisition of cell shape (Tluasty et al., 1999; Bischofs et al., 2008, 2009; Paul et al., 2008). The tension produced by actomyosin contraction is transmitted through the cytoskeleton to the focal adhesions where they activate integrin receptors and control the strength and pattern of cell adhesion to the matrix (Burridge and Chrzanowska-Wodnicka, 1996; Bershadsky et al., 2006).

A clue to understanding the development of stress fibers in cells comes from a variety of experiments that demonstrated that focal adhesion and stress fiber formation depend on the presence of tension in the cytoskeleton and that their spatial distribution correlates with the direction of principal stresses in the cell. Experiments by Balaban et al. (2001) have shown that focal adhesions grow in the direction of applied force and that the focal adhesion size increases linearly with the magnitude of the force. Similarly, when tension is applied to a localized region of the cell surface, a bundle of actin filaments is induced immediately adjacent to the site of applied tension (Kolega, 1986; Choquet et al., 1997). When placed on a rigid substrate or incorporated in a gel with fixed boundaries, more prominent stress fibers develop in the cell as compared with softer systems (i.e., soft substrates or floating gels with free boundaries) (Discher et al., 2005). Conversely, when cells are detached from the surfaces they adhere to and their internal tension is released, or when myosin is inhibited, focal adhesions and stress fibers disassemble in the cell (Burridge and Chrzanowska-Wodnicka, 1996). Moreover, experiments show that stress fiber assembly is in general a global cell phenomenon that is governed by the overall shape of the cell and elastic properties of the cell and its environment (Wang et al., 2002; Discher et al., 2005; Yeung et al., 2005; Théry et al., 2006a). Using microlithography techniques to systematically manipulate the spreading area and shape of cells, Wang et al. (2002) and later on Théry et al. (2006a,b) were able to control the patterning of stress fibers and other cytoskeleton components in the cell (see also Parker et al., 2002). Typically one finds that the density of stress fibers is higher in cells with larger spreading areas and more numerous stress fibers are found along the perimeter and corners of the cell; this is consistent with measurements of the spatial distribution of forces that are exerted by the cell (Wang et al., 2002).

Another generic observation is that stress fibers generally orient parallel to the long axis of cells (Curtis et al., 2006). However, an interesting recent observation reported for hMSCs (Zemel et al., 2010a; Rehfeldt et al., 2012) and cardiomyocytes (Chopra et al., 2011) has been that the level of cell elongation (aspect ratio) and the orientational order of stress fibers in these cells depended nonmonotonically on the rigidity of the substrate on which the cells were grown. On soft substrates, cells appeared round and small, and the stress fibers were sparse and disordered in the cytoskeleton. As the matrix rigidity increased, cells obtained the characteristic elongated structure of muscle cells, and stress fiber alignment attained a maximal level and was found lower in cells grown on glass surfaces. Interestingly, even for cells of given aspect ratio, stress fiber alignment was found to be highest in the cells grown on substrates of intermediate rigidity ~10 kPa. An explanation to these observations was proposed based on elastic calculation of the stress anisotropy that develops in the cell as function of ECM rigidity (Zemel et al., 2010a). It was shown that even if myosin motors exert isotropic forces, the stresses that develop in the cytoskeleton may be anisotropic due to the anisotropy of the cell shape. Furthermore, the calculations showed that the breaking of symmetry (i.e., the difference between the stresses produced along the long and short axes of the cell) is maximal for an intermediate range of the substrate rigidity, suggesting an explanation for the nonmonotonic dependence of stress fiber alignment on the substrate rigidity.

A number of theoretical models have been developed to predict the spontaneous alignment of stress fibers and focal adhesion redistribution in cells. Computer simulations of actin filaments interacting with myosin motors (Kim et al., 2009; Walcott and Sun, 2010; Borau et al., 2012), spring networks (Paul et al., 2008; Torres et al., 2012), finite element calculations (Deshpande et al., 2006, 2008; Pathak et al., 2008), and continuum models (Zemel et al., 2010a; Nisenholz et al., 2014a) have shed light on how cell shape and boundary conditions may dictate the spatial distribution of cellular stresses and how these may induce the bundling of filaments along principal stress directions.

In addition to elastic forces, excluded volume interactions between actin filaments were also suggested to drive cytoskeleton alignment in confined cell geometries (Silva et al., 2011). Dense actin networks were grown in cell-sized microchambers to investigate the combined effects of filament packing constraints and global spatial confinement. The authors showed that the filaments spontaneously formed dense, bundle-like structures above a threshold actin concentration, in contrast to unconfined networks, which are homogeneous and undergo a bulk isotropic-to-nematic phase transition at a higher actin concentration. A more elaborate discussion of this mechanism can be found in Chapter 1 by Alvarado and Koenderink.

It is becoming clear that cellular shape acquisition, focal adhesion redistribution, and stress fiber assembly are coupled phenomena that influence each other and proceed on similar timescales. It is also known that microtubules, other cytoskeleton elements, and the nucleus adjust to these changes and there are works investigating the role of nucleus deformation in cellular mechanosensing and gene expression regulation. For reviews of these topics, see Théry and Bornens (2006), Wang et al. (2009), and Shivashankar (2011). Furthermore, we refer the reader to the chapter by Roland Kaunas for a discussion on the effects of applied loads on stress fiber assembly and reorganization in the cell.

5.3 Mechanical Interactions of Cells

5.3.1 Excluded Volume Interactions

The interaction of cells with each other and with the medium they live in may lead to correlated collective motion and formation of ordered patterns in cell ensembles. For many cell biologists, a well-known example is the observation of an apparent emerging orientational order in nearly confluent cultures of some cell types such as human fibroblasts. In these patterns, the cells are generally elongated, and their alignment is set by their shape anisotropy. A typical example of such an ordered arrangement is given in Figure 5.3.

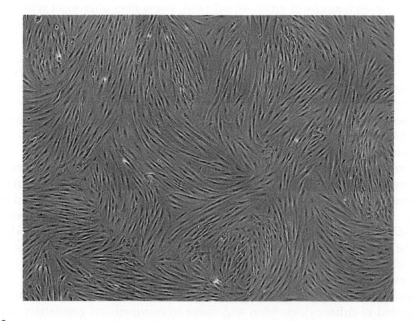

FIGURE 5.3
Human skin fibroblasts cultured on a collagen-coated rigid cell culture surface. Over several days, cells grow to this confluent monolayer with a high orientational order. Similar patterns can often be observed in cell cultures of other cell types. The entire area is approximately 1200 μm × 1200 μm. Defects in the pattern are well visible and can provide additional information about the structure.

The cells have no particular positional order but align by taking on an elongated shape and arranging their main cell axis in an ordered way with respect to each other. Such collective arrangements are commonly found in assemblies of self-propelled organisms but also with synthetic particles, the later frequently termed *active fluids*. Biological examples of such systems, ranging over several orders of magnitude in size, are mixtures of microtubules and associated motor proteins, the whole cytoskeleton of living cells, and animal flocks such as bird or fish flocks (Surrey et al., 2001; Katz et al., 2011; Schaller and Bausch, 2013). Nonliving active fluids with a high degree of order can arise in layers of granular rods, colloidal, or nanoscale particles (Narayan et al., 2007; Aranson et al., 2008). Because of the nonequilibrium nature, such active pattern formation has attracted enormous attention in the broad scientific community, and there are various theoretical approaches to describe such systems (D'Orsogna et al., 2006; Peruani et al., 2006, 2012; Farrell et al., 2012; Romanczuk et al., 2012). In classical physics, the interactions between the building blocks of such fluids are suggested to consist of short-range repulsive (steric) interactions and other long-range interaction due to forces such as the van der Waals force. The relative importance of short-range and long-range interactions to disorder–order phase transitions has been extensively studied by using idealized colloidal systems. In particular, Onsager demonstrated in his pioneering work of 1947 that hard-core repulsion between highly anisotropic cylinders is sufficient to produce a disorder–order transition and thus the formation of a stable liquid-crystalline phase (Onsager, 1949). Ordering in cellular systems may be more complex in nature, but short-range excluded volume interactions are likely to contribute to phenomena seen in Figure 5.3.

Dynamic and ordered structures in cultured cell layers can serve as interesting model systems to study emerging collective behaviors (Stopak and Harris, 1982). Ordered patterns in some 2D cell layers show a remarkable similarity with nematic phases in soft matter systems such as liquid crystals. Considering the symmetry of the cells, there are two distinct cases: elongated cells can be polar entities with distinct fronts and ends, which can cooperatively order in condensed structures described later. The second class of frequently observed ordered pattern consists of cells that have rather an apolar symmetry with no distinct leading front and rear part. Such symmetry can be observed, for example, in cultured human melanocytes, fibroblasts, or osteoblasts. These cells have typically an elongated shape but no distinct polar symmetry and show commonly a nondirected motion. In particular, cultured human melanocytes can form an ideal bipolar shape mode with two opposing dendrite-like extensions and a nucleus region in the center of the cell, thus resembling rods. At low cell densities in a 2D layer, there is no apparent order, and the elongated melanocytes are randomly oriented. However, at higher cell densities, the dendrites interact with other cells, and the elongated melanocytes get increasingly anisotropically oriented and finally form a nematic state with structural similarities to nematic fluids (Kemkemer et al., 2000b). A very similar observation can be made in fibroblast cultures (Figure 5.4). The cells in this example show no apparent order at low densities, but with increasing cell number, they form ordered patches, finally yielding a pattern with long-range orientational order as seen in Figure 5.4 at t_4. The fibroblasts show initially a cell shape that is typical for mesenchymal-like cells but get very elongated with increasing cell density. Such density-dependent order–disorder transitions in cell cultures can be observed on very rigid substrates such as on glass or typical cell culture plates made from hard polymers. The proposed mechanism is, therefore, based on excluded volume interactions occurring at increasing cell density. Elastic interactions (discussed in the following) arising from the cell-generated deformations of the substrate, for example, could modify such behavior but may be no prerequisite for the

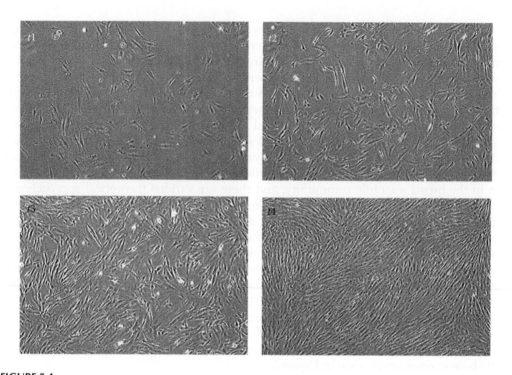

FIGURE 5.4

A culture of human skin fibroblasts at four different time points in a time course of 7 days. At $t1$ (day 1) and $t2$ (day 2), cells are at a low density and hardly interfere by contact with each other directly (no steric interaction). No apparent order is visible. At $t3$ (day 4), cells density exceeded a threshold, and cells interact directly. They take on an elongated shape and align in a common direction, resulting in a pattern with high orientational order at day 7 ($t4$).

formation of a nematic state at sufficiently dense cultures. To elucidate the contribution of elastic interactions to the ordering phenomena, experiments with soft substrates in the regime of kPa are required.

One way to describe the formation of such nematic states is to approximate the cell–cell interaction by a mean-field approach similar to the description of nematics in liquid crystal physics. Although strictly being a nonequilibrium system, the excluded volume interactions between cells may be quantified by effective free energy functionals, which, of course, need different interpretations. Other distinct differences between the two systems need to be considered. Instead of passive motion and thermal processes ($k_B T$), one needs to take into account the active motion of cells and stochastic processes taking place in the biochemistry of the cells and in their interaction responses. Furthermore, the formation of ordered patterns does not depend on a constant rodlike shape. In case of fibroblasts, the variable shape depending on the density needs to be considered (Figure 5.4). However, there are phenomenological approaches assuming that the cells interact with the mean field produced by all the surrounding cells, which can describe various aspects of the pattern formation (Edelstein-Keshet and Ermentrout, 1990; Gruler et al., 1999; Kemkemer et al., 2000b). In soft matter physics, there are essentially two mean-field approaches for nematics formed by elongated molecules: the Onsager theory and the Maier–Saupe theory. The Onsager theory is based on the steric repulsion of long rigid rods that cannot penetrate each other (volume exclusion). The formation of a nematic state is predicted for rather high rod concentrations. The second theory (Maier–Saupe) assumes a soft anisotropic interaction potential and considers flexible elongated molecules and their steric interaction. Taking the latter into account and following the approach given in Kemkemer et al. (2000b), the experimentally observed density-dependent state transition from a randomly oriented cell ensemble to pattern with highly ordered orientation of cells (Figure 5.4) could be well described and quantitatively predicted.

First, the orientation angle ψ of self-propelled elongated particles such as the cells is described by the stochastic differential equation,

$$\frac{d\psi}{dt} = kE(s)\sin 2\psi + \Gamma(t) \tag{5.3}$$

An elongated cell regulates its orientation angle ψ with respect to an extracellular guiding signal, $E(signal)$, which in this case is made up by surrounding cells. The entire response system of the cell is characterized by k. Random processes leading to a random orientation in the absence of an external field ($E(s) = 0$) are described by a stochastic source, $\Gamma(t)$, which can in the first approximation be quantified by an additive white noise source with strength q. For low cell densities with hardly interacting cells, this term will lead to the observed random orientation as seen in Figure 5.4 (t_0). From Equation 5.3, a further equation can be derived, which describes the temporal variations of the angle density distribution function, $f(\psi,t)$, for a cell ensemble. The solutions of this so-called Fokker–Planck equation predict the behavior of the cell ensemble with respect to the orientation angle of the cells (Kemkemer et al., 2006). The distribution $f(\psi)$ can be experimentally measured and compared with the theoretical prediction. However, the unknown interaction of the cells needs to be approximated. To achieve this, a mean-field procedure similar to the theory of Maier and Saupe is used in order to approximate the interaction of the amoeboid cells. That approach leads to the assumption that $E(signal)$ is proportional to the orientational order parameter, $S = \langle\cos 2\psi\rangle$, of the cells (see Equation 5.6) and their density ρ. If the surrounding cells are highly orientated in a specific direction, there will be a strong guiding signal. In addition, a large guiding signal is expected for strongly interacting cells (high cell density = small cell–cell distance) and a weak guiding signal for weakly interacting cells (low cell density = large cell–cell distance): the strength of the anisotropic cell–cell interaction is quantified by the coefficient a, which is a cell-type-specific quantity that can be derived from the experimental data. In summary, the guiding field is approximated by

$$E(signal) = aS\rho \tag{5.4}$$

This assumption has been used to describe the orientation behavior of melanocytes. Cells with variable cell shape, respectively, shape-polarizable cell types like fibroblasts or osteoblasts, require further modification of the coefficients k and a since these quantities are expected to be guiding signal dependent. Generally, in steady state, the predicted density distribution function has the characteristic of a Boltzmann distribution and is given by

$$f(\psi) = f_0 \exp\left[\frac{kE(signal)\cos 2\psi}{q}\right] \tag{5.5}$$

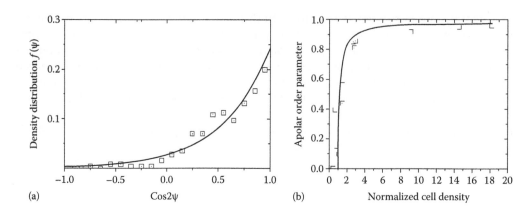

FIGURE 5.5

(a) An example for an experimentally determined distribution of the orientation angle, given in cos2ψ. The orientation angle was measured for human melanocytes cells, which were cultured on a standard culture dish. The cells were grown to an intermediate cell density at which an order effect is visible. The data are fitted with Equation 5.5. (b) Experimental values for the apolar order parameter ⟨cos2ψ⟩ for various melanocyte cell densities, normalized to the cell density at which ordering occurs $\rho_0 = 120/mm^2$. The line represents the prediction of the described model assuming an excluded volume interaction. The only fitting parameter is ρ_0.

This distribution could be experimentally verified by measuring the orientation angle of cells in images such as Figure 5.1 and calculating the frequency of ψ in given intervals $\Delta\psi_i$. An example for a typical distribution is given in Figure 5.5 demonstrating the Boltzmann-like characteristics.

To evaluate the disorder–order transition, one can also calculate an apolar order parameter ⟨cos2ψ⟩ by

$$\langle \cos 2\psi \rangle = \int f(\psi) \cos 2\psi \, d\psi \tag{5.6}$$

which quantifies the mean orientational order of the cell system in dependence of the cell density ρ. The predicted behavior of that order parameter can be compared with experimental data as demonstrated in Figure 5.5. The only fitting parameter is the threshold cell density ρ_0 at which ordering starts to occur. For densities below ρ_0, cells have a random orientation. For increasing cell densities, a transition to a nematic state is observed for fibroblasts with $\rho > \rho_0 \approx 250$ cells/mm² or human melanocytes with $\rho > \rho_0 \approx 100$ cells/mm² (Kemkemer et al., 2000b). In particular, for melanocytes, which are rather stiff, rodlike cells, this density corresponds to a mean distance between cells in the size of a typical cell length emphasizing the effect of steric interactions. Similar transitions to ordered states have also been observed with cells embedded in collagen gels (Fernandez and Bausch, 2009).

Such consideration from soft matter physics is not only useful to describe the self-organized order–disorder state transition. Similarly, elastic properties of these nematic structures could be quantified in analogy to that of classical nematic liquid crystals by investigating the structural defects of the patterns. An orientational elastic energy could be derived, and the appearance of disclinations, orientational defects, in the nematic pattern was used to approximate their elastic constants. Interestingly, in cultures of human fibroblasts and melanocytes, only half-numbered disclinations are experimentally observed, indicating that the nematic structure has an apolar symmetry as

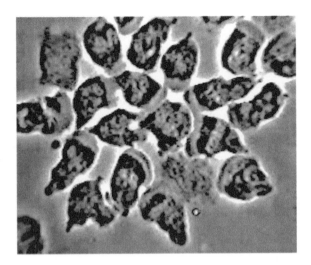

FIGURE 5.6
Single migrating neutrophil granulocytes start to form a polar cluster of densely packed cells if calcium in the surrounding media is depleted. A monocyte close to the center of the droplet acts as nucleation center for the migrating granulocytes. The polar order parameter of this cluster is approximately 0.82. (After Kemkemer, R. et al., *Eur. Phys. J. E*, 1(2–3), 215, 2000.)

initially assumed. From the analysis of the defect structures, the relation of the splay and bend elastic constant in this planar cell culture pattern could be estimated (Kemkemer et al., 2000a).

Considering the symmetry of the cells, there are, in contrast to the nonpolar cells treated so far, also elongated cells that form polar entities with distinct fronts and ends. A typical example for such cells is keratinocytes or leukocytes such as neutrophil granulocytes. The latter can cooperatively order in condensed structure with some common features to the order–disorder transitions discussed previously. An example is given by the in vitro assembly of human granulocytes as shown in Figure 5.6. The cells form a condensed state similar to a fluid with no positional order of the cells but with their fronts pointing toward the center of the cluster. This orientational order can be quantified by a polar order parameter and described in analogy to the mean-field approach previously. Instead of excluded volume interactions, here, other short-range interaction may contribute to this ordering effect. A polar cluster may form if: (1) the cells are polar, i.e., their front and back are distinct, and (2) if there is an attractive signal between them (Kemkemer et al., 2000a). The cellular response under such conditions can be directed migration, for example, when calcium is depleted from the surrounding cell culture media. The actively migrating granulocytes attract each other and form a cluster as demonstrated in Figure 5.6. This process is reversible, and the polar ordered cell cluster disappears if the calcium concentration is increased.

These results show that applying concepts from soft matter physics to simple cell culture experiments is very helpful in getting descriptions for such complex systems in which disorder–order transitions occur. Although phenomenological, such models are usually easy to interpret and may be suitable for general predictions and a precise quantification of experimental observations. The process by which migrating cells exchange information and use them to form pattern with long-range order constitutes an intriguing area in cell biology. Steric interactions, in analogy to nonliving systems, may lead to a condensed state of cell ensembles as described previously. From the biologist's perspective, it may be an intriguing question, why such ordered pattern cannot be observed in cultures of all cell types. Cells from epithelial or endothelial origin, for example, having a different cell shape and morphology, show usually no ordered nematic-like pattern in culture but rather a paving-stone appearance. Even considering fibroblasts from human skin samples, there are variations in the degree of pattern formation. Cells taken from patients with specific diseases such as neurofibromatosis type 1 (NF1) show a less pronounced transition to an ordered system than fibroblasts from healthy donors. NF1 is a common monogenic tumor predisposition disorder caused by germ line mutations in the neurofibromin gene (NF1+/−) (Kaufmann et al., 2012). One of the typical consequences is the development of multiple benign tumors, called neurofibromas. Experiments showed that cells taken from benign tumors of such patients build in vitro distinctively different patterns than cells from healthy donors. That observation suggests that an increased understanding about the formation of ordered patterns in culture may also be a helpful analytical tool and may contribute to an understanding of physiological or pathophysiological in vivo situations.

5.3.2 Elastic Interactions of Cells

We now consider a second route of cell–cell interactions that naturally arise from cell contractility. When a population of dispersed cells applies forces on their environment, they elastically interact via the deformations they impose on their surrounding matrix. Among the first to notice this effect and to point out its implications for morphogenesis were Albert K. Harris and coworkers who studied the regular patterns that formed in a collagen gel by cell contraction (Harris et al., 1981, 1984; Stopak and Harris, 1982). The forces exerted by the cells were seen to align collagen fibers along principle stress directions. The formed fibrils provided tracks for cells to migrate in the matrix. Between explants of fibroblasts, traction forces were seen to align collagen fibers into linear tracts as much as 4 cm long. The findings of Harris were quantitatively formulated by an elastic diffusion–advection theory by Oster et al. (1983). This theory takes into account the effects of cell traction on the local density of the ECM, which in turn provides a cue for cells to migrate in the matrix, thus affecting the flux of cells down the (produced) adhesive gradient. Simulations carried out by these authors enabled them to reproduce the spatial patterns found in experiments. More recent use of these simulations was able to reproduce vascular network structures formed by endothelial cells in vitro (Manoussaki et al., 1996; Murray, 2003).

The ability of cells to align ECM fibers and to migrate, via contact guidance, along the fibers' direction allows the cells to interact over long distances, several orders of magnitude longer than the cell size. However, the existence of elastic interaction of cells depends on the ability of cells to respond to local deformations or force and does not necessarily require that cells would locally align the ECM. ECM remodeling is a *significant* secondary effect that generally enhances the response. Indeed, similar, but less profound, behaviors of cell interactions were reported with cells placed on top of linearly elastic inert polyacrylamide gels that cells are unable to remodel (Guo et al., 2006; Califano and Reinhart-King, 2008, 2009, Reinhart-King et al., 2008). By plating endothelial cells on collagen-coated polyacrylamide gels of varying rigidities, Califano et al. have noticed that only on sufficiently soft gels (see Figure 5.7) would the cells be able to form organized network structures (Califano and Reinhart-King, 2008, 2009). On stiffer gels, no patterns were visible. The alignment of cell-secreted fibronectin has been shown to be essential for network formation (Califano and Reinhart-King, 2008), thus demonstrating the necessity and limitation

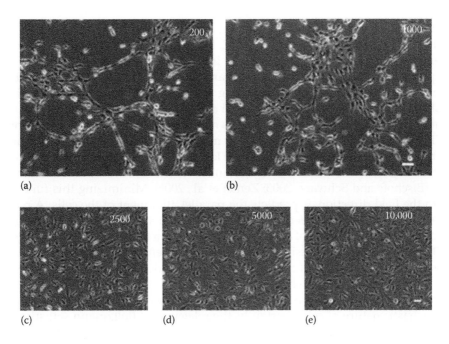

FIGURE 5.7
Compliant substrates promote bovine aortic endothelial cell (BAEC) network assembly. (a and b) Phase contrast images of BAECs on PA gels derivatized with 100 μg/mL of collagen I assemble into networks on 200 and 1000 Pa substrates, respectively. This phenotype was characterized by cords of cells and ringlike morphologies. (c–e) This organization was not present when substrate stiffness was increased to 2500, 5000, and 10,000 Pa, respectively. Bar = 50 μm. (Reproduced from Califano, J.P. and Reinhart-King, C.A., *Cell. Mol. Bioeng.*, 1(2–3), 122, 2008. With permission.)

of the pure elastic response. A recent study by Saif and coworkers has shown that elastic interactions between beating heart cells can tune the synchrony of cell beating (Tang et al., 2011). Consistent observations of the tendency of cells to disperse on stiff gels and to merge on soft gels have been previously reported by Yu-Li Wang and coworkers (Guo et al., 2006). A similar phenomenon has also been shown with the wetting of spherical cell aggregates or rigid but not on soft substrates (Douezan et al., 2011). Because the force exerted by cells generally increases with the rigidity of the surroundings, whereas the deformations decrease with rigidity, elastic interactions of cells are likely to be most significant for some intermediate range of elasticity (Zemel and Safran, 2007; Zemel et al., 2010a; Friedrich et al., 2011). While there are some evidences for this optimization principle (Engler et al., 2004b), this issue still needs to be explored more systematically.

A quantity that proved useful for modeling the mechanical activity of cells and for studying the interaction of cells with local or applied elastic fields is the (anisotropic) cellular point force-dipole tensor defined by $p_{ij} = f_i l_j = p n_i n_j$ (Schwarz and Safran, 2002; Bischofs and Schwarz, 2003, 2006; Bischofs et al., 2004; Zemel et al., 2006; De et al., 2007; Zemel and Safran, 2007; Paul and Schwarz, 2010). In this coarse-grained picture, the entire mechanical activity of the cell is modeled by two oppositely oriented forces, $\vec{f}/2$ and $-\vec{f}/2$, separated by a short distance $\vec{l}$, where $\vec{f} = f\hat{n}$ and $\vec{l} = l\hat{n}$ are parallel to each other and $p = lf \sim 10\,\mu m \times 10\,nN$ is the magnitude of the cellular dipole and has units of energy (force times distance). Each cell's force dipole creates an elastic strain field around it, and one may calculate its interaction energy with the local strain, u_{ij}^{loc}, produced by other neighboring cells or by externally applied forces. This interaction energy is given by $W = p_{ij}u_{ij}^{loc}$, where summation of repeated indices is implied (Schwarz and Safran, 2002; Bischofs and Schwarz, 2003; Bischofs et al., 2004; Zemel et al., 2006). Minimization of the interaction energy has shown to be a useful tool for predicting cell orientation under a variety of boundary conditions (Bischofs and Schwarz, 2003; Bischofs et al., 2004), including the parallel alignment of cells in the direction of static/quasistatic load, the head-to-tail orientation of cells in string structures, and the parallel (perpendicular) alignment of cells next to a boundary with a softer (more rigid) matrix. This property was used to predict the macroscopic susceptibility and orientational distribution of a population of cells to applied loads. For a cell interacting with a static, uniaxial tensile stress, T, one finds $W \sim -(|p|T/2\mu)\cos^2\theta$, where μ is the shear modulus of the medium (Bischofs and Schwarz, 2003; Zemel et al., 2006). Minimizing this function with respect to the field direction, θ, predicts the parallel alignment of the cells. A mean-field theory, inspired by Onsager's theory for polar liquids (Onsager, 1936), has been carried out to predict the mean alignment of cells in the presence of a large population of cells (Zemel et al., 2006). A modified expression for the interaction energy, W, has been derived that takes into account the density of cells in the medium. To devise the theory, Boltzmann-like statistics, $P(\theta) \sim \exp[-\beta W(\theta)]$, were postulated based on experimental observations cited in the previous section; here, β represents an effective inverse temperature that accounts for the stochastic nature of cells. Certainly, the relevance of Boltzmann statistics to cellular systems and systematic investigations of the noise associated with cell behavior need still to be studied more thoroughly in the experiment. One of the interesting predictions of this theory is that cell interactions generally screen each other's effect, as well as the entire response of cells to applied loads—similar to the way that dielectrics screen the electrostatic interaction between charged particles. The screening effect of cells manifests itself in the enhanced rigidity of a cell containing gel, $\tilde{\mu} = (1+\chi)\mu$, where $\chi = \chi(\rho)$ is the calculated active susceptibility of the cells in the medium and ρ is their density. An alternative mean-field theory of cell–cell interactions, which also predicts the nematic ordering of cells by elastic interactions, has been developed by Friedrich and Safran (2011). The energy functional concept has also been used in Monte Carlo simulations of cell interactions and predicted different structural phases of cells in culture (Bischofs and Schwarz, 2006). Such concepts and systematic experiments of cell–cell interactions are valuable for understanding the mechanisms and consequences of these less familiar routes of cell–cell interactions, whose implications and implementations in medical and biological science are intensively being discovered in recent years. For additional reviews of this topic, see Zemel et al. (2011), Schwarz and Safran (2013), and chapters 10–14 of this book.

5.4 Conclusion

The newly developed technologies to manipulate cell shape and adhesion pattern down to the submicron and nanometer scale and the ability to systematically control the mechanical and topographical properties of the extracellular environment provided exquisite tools to investigate the mechanisms underlying cellular mechanosensitivity. Many unexpected manifestations of cellular mechanosensitivity have extensively been discovered, for example, the sensitivity of stem cell differentiation to the details of cell shape and matrix rigidity. The elastic stresses that develop in the cell during cell adhesion seem to be at the core of cellular mechanosensitivity. These seem to be essential for globally regulating many concurrent processes in the cell, including the establishment of cell morphology, adhesion pattern, cytoskeleton structure, nucleus deformation, and biochemical signaling. Current research focuses on measuring the forces that cells exert into the environment at high spatial and temporal resolution while simultaneously imaging the molecular rearrangements that take place in the cell. Moreover, considerable efforts are aimed at elucidating the mechanisms of mechanotransduction, namely, how the various mechanical processes in the cell eventually lead to regulation of gene expression. Among the differently proposed routes are integrin signaling, stretch-activated ionic channels, and the nucleus deformation. A better understanding of these processes and of the mechanical mechanisms that mediate cell–cell interactions (elastic, excluded volume, and others) will open up new avenues in the development of biological and medical science.

References

Alexandrova, A., Arnold, K., Schaub, S., Vasiliev, J., Meister, J., Bershadsky, A., and Verkhovsky, A. (2008). Comparative dynamics of retrograde actin flow and focal adhesions: Formation of nascent adhesions triggers transition from fast to slow flow. *PLoS One*, 3(9):e3234.

Aranson, I. S., Snezhko, A., Olafsen, J. S., and Urbach, J. S. (2008). Comment on "Long-lived giant number fluctuations in a swarming granular nematic." *Science*, 320(5876):612c.

Assoian, R. and Klein, E. (2008). Growth control by intracellular tension and extracellular stiffness. *Trends Cell Biol.*, 18(7):347–352.

Bajaj, P., Tang, X., Saif, T. A., and Bashir, R. (2010). Stiffness of the substrate influences the phenotype of embryonic chicken cardiac myocytes. *J. Biomed. Mater. Res. A*, 95(4):1261–1269.

Balaban, N. Q., Schwarz, U. S., Riveline, D., Goichberg, P., Tzur, G., Sabanay, I., Mahalu, D., Safran, S., Bershadsky, A., Addadi, L., and Geiger, B. (2001). Force and focal adhesion assembly: A close relationship studied using elastic micropatterned substrates. *Nat. Cell Biol.*, 3(5):466–472.

Bangasser, B. L., Rosenfeld, S. S., and Odde, D. J. (2013). Determinants of maximal force transmission in a motor-clutch model of cell traction in a compliant microenvironment. *Biophys. J.*, 105(3):581–592.

Baranes, K., Chejanovsky, N., Alon, N., Sharoni, A., and Shefi, O. (2012a). Topographic cues of nanoscale height direct neuronal growth pattern. *Biotechnol. Bioeng.*, 109(7):1791–1797.

Baranes, K., Kollmar, D., Chejanovsky, N., Sharoni, A., and Shefi, O. (2012b). Interactions of neurons with topographic nano cues affect branching morphology mimicking neuron–neuron interactions. *J. Mol. Histol.*, 43(4):437–447.

Bereiter-Hahn, J., Luck, M., Miebach, T., Stelzer, H., and Voth, M. (1990). Spreading of trypsinized cells: Cytoskeletal dynamics and energy requirements. *J. Cell Sci.*, 96(1):171–188.

Bershadsky, A., Kozlov, M., and Geiger, B. (2006). Adhesion-mediated mechanosensitivity: A time to experiment, and a time to theorize. *Curr. Opin. Cell Biol.*, 18:472–481.

Biela, S. A., Su, Y., Spatz, J. P., and Kemkemer, R. (2009). Different sensitivity of human endothelial cells, smooth muscle cells and fibroblasts to topography in the nano–micro range. *Acta Biomater.*, 5(7):2460–2466.

Bischofs, I. B., Klein, F., Lehnert, D., Bastmeyer, M., and Schwarz, U. S. (2008). Filamentous network mechanics and active contractility determine cell and tissue shape. *Biophys. J.*, 95:3488–3496.

Bischofs, I. B., Safran, S. A., and Schwarz, U. S. (2004). Elastic interactions of active cells with soft materials. *Phys. Rev. E*, 69(2):021911.

Bischofs, I. B., Schmidt, S. S., and Schwarz, U. S. (2009). Effect of adhesion geometry and rigidity on cellular force distributions. *Phys. Rev. Lett.*, 103:048101.

Bischofs, I. B. and Schwarz, U. S. (2003). Cell organization in soft media due to active mechanosensing. *Proc. Natl. Acad. Sci. USA*, 100(16):9274–9279.

Bischofs, I. B. and Schwarz, U. S. (2006). Collective effects in cellular structure formation mediated by compliant environments: A Monte Carlo study. *Condens. Matter.*, 2:0510391.

Bissell, M. J. and Hines, W. C. (2011). Why don't we get more cancer? A proposed role of the microenvironment in restraining cancer progression. *Nat. Med.*, 17(3):320–329.

Booth-Gauthier, E. A., Alcoser, T. A., Yang, G., and Dahl, K. N. (2012). Force-induced changes in subnuclear movement and rheology. *Biophys. J.*, 103(12):2423–2431.

Borau, C., Kim, T., Bidone, T., García-Aznar, J. M., and Kamm, R. D. (2012). Dynamic mechanisms of cell rigidity sensing: Insights from a computational model of actomyosin networks. *PloS One*, 7(11):e49174.

Burridge, K. and Chrzanowska-Wodnicka, M. (1996). Focal adhesions, contractility, and signaling. *Annu. Rev. Cell Develop. Biol.*, 12:463–519.

Cai, Y., Rossier, O., Gauthier, N. C., Biais, N., Fardin, M. A., Zhang, X., Miller, L. W., Ladoux, B., Cornish, V. W., and Sheetz, M. P. (2010). Cytoskeletal coherence requires myosin-IIA contractility. *J. Cell Sci.*, 123:413–423.

Califano, J. P. and Reinhart-King, C. A. (2008). A balance of substrate mechanics and matrix chemistry regulates endothelial cell network assembly. *Cell. Mol. Bioeng.*, 1(2–3):122–132.

Califano, J. P. and Reinhart-King, C. A. (2009). The effects of substrate elasticity on endothelial cell network formation and traction force generation. In *Engineering in Medicine and Biology Society, 2009. EMBC 2009. Annual International Conference of the IEEE*, pp. 3343–3345. IEEE, Washington, DC.

Calvo, F., Ege, N., Grande-Garcia, A., Hooper, S., Jenkins, R. P., Chaudhry, S. I., Harrington, K., Williamson, P., Moeendarbary, E., Charras, G., Sahai, E. (2013). Mechanotransduction and yap-dependent matrix remodelling is required for the generation and maintenance of cancer-associated fibroblasts. *Nat. Cell Biol.*, 15(6):637–646.

Carlsson, A. (2004). Structure of autocatalytically branched actin solutions. *Phys. Rev. Lett.*, 92(23):238102.

Cavalcanti-Adam, E. A., Volberg, T., Micoulet, A., Kessler, H., Geiger, B., and Spatz, J. P. (2007). Cell spreading and focal adhesion dynamics are regulated by spacing of integrin ligands. *Biophys. J.*, 92(8):2964–2974.

Chan, C. E. and Odde, D. J. (2008). Traction dynamics of filopodia on compliant substrates. *Science*, 322:16871691.

Chen, C. S., Alonso, J. L., Ostuni, E., Whitesides, G. M., and Ingber, D. E. (2003). Cell shape provides global control of focal adhesion assembly. *Biochem. Biophys. Res. Commun.*, 307(2):355–361.

Chen, C. S., Mrksich, M., Huang, S., Whitesides, G. M., and Ingber, D. E. (1997). Geometric control of cell life and death. *Science*, 276(5317):1425–1428.

Chopra, A., Tabdanov, E., Patel, H., Janmey, P., and Kresh, J. (2011). Cardiac myocyte remodeling mediated by n-cadherin-dependent mechanosensing. *Am. J. Physiol-Heart Circ. Physiol.*, 300(4):H1252.

Choquet, D., Felsenfeld, D. P., and Sheetz, M. P. (1997). Extracellular matrix rigidity causes strengthening of integrin-cytoskeleton linkages. *Cell*, 88:39–48.

Cramer, L. P. (1997). Molecular mechanism of actin-dependent retrograde flow in lamellipodia of motile cells. *Front. Biosci.*, 2:d260–d270.

Curtis, A., Aitchison, G., and Tsapikouni, T. (2006). Orthogonal (transverse) arrangements of actin in endothelia and fibroblasts. *J. R. Soc. Interface*, 3:753–756.

Curtis, A. and Wilkinson, C. (1997). Topographical control of cells. *Biomaterials*, 18(24):1573–1583.

Cuvelier, D., Théry, M., Chu, Y., Dufour, S., Thiéry, J., Bornens, M., Nassoy, P., and Mahadevan, L. (2007). The universal dynamics of cell spreading. *Curr. Biol.*, 17(8):694–699.

Dado, D., Sagi, M., Levenberg, S., and Zemel, A. (2012). Mechanical control of stem cell differentiation. *Regen. Med.*, 7(1):101–116.

De, R., Zemel, A., and Safran, S. A. (2007). Dynamics of cell orientation. *Nat. Phys.*, 3:655–659.

Deshpande, V. S., McMeeking, R. M., and Evans, A. G. (2006). A bio-chemo-mechanical model for cell contractility. *Proc. Natl. Acad. Sci. USA*, 103:14015–14020.

Deshpande, V. S., Mrksich, M., McMeeking, R. M., and Evans, A. G. (2008). A bio-mechanical model for coupling cell contractility with focal adhesion formation. *J. Mech. Phys Solids*, 56:1484–1510.

Discher, D. E., Janmey, P., and Wang, Y. (2005). Tissue cells feel and respond to the stiffness of their substrate. *Science*, 310:1139–1143.

Discher, D. E., Mooney, D. J., and Zandstra, P. W. (2009). Growth factors, matrices, and forces combine and control stem cells. *Science*, 324:1673–1677.

Döbereiner, H. G., Dubin-Thaler, B., Giannone, G., Xenias, H. S., and Sheetz, M. P. (2004). Dynamic phase transitions in cell spreading. *Phys. Rev. Lett.*, 93:108105.

D'Orsogna, M., Chuang, Y., Bertozzi, A., and Chayes, L. (2006). Self-propelled particles with soft-core interactions: Patterns, stability, and collapse. *Phys. Rev. Lett.*, 96(10):104302.

Douezan, S., Guevorkian, K., Naouar, R., Dufour, S., Cuvelier, D., and Brochard-Wyart, F. (2011). Spreading dynamics and wetting transition of cellular aggregates. *Proc. Natl. Acad. Sci. USA*, 108(18):7315–7320.

Downing, T. L., Soto, J., Morez, C., Houssin, T., Fritz, A., Yuan, F., Chu, J., Patel, S., Schaffer, D. V., and Li, S. (2013). Biophysical regulation of epigenetic state and cell reprogramming. *Nat. Mater.*, 12(12):1154–1162.

Dubin-Thaler, B. J., Giannone, G., Döbereiner, H.-G., and Sheetz, M. P. (2004). Nanometer analysis of cell spreading on matrix-coated surfaces reveals two distinct cell states and steps. *Biophys. J.*, 86(3):1794–1806.

Dubin-Thaler, B. J., Hofman, J. M., Cai, Y., Xenias, H., Spielman, I., Shneidman, A. V., David, L. A., Döbereiner, H.-G., Wiggins, C. H., and Sheetz, M. P. (2008). Quantification of cell edge velocities and traction forces reveals distinct motility modules during cell spreading. *PLoS One*, 3(11):e3735.

DuFort, C. C., Paszek, M. J., and Weaver, V. M. (2011). Balancing forces: Architectural control of mechanotransduction. *Nat. Rev. Mol. Cell Biol.*, 12(5):308–319.

Dupont, S., Morsut, L., Aragona, M., Enzo, E., Giulitti, S., Cordenonsi, M., Zanconato, F. et al. (2011). Role of YAP/TAZ in mechanotransduction. *Nature*, 474(7350):179–183.

Edelstein-Keshet, L. and Ermentrout, G. B. (1990). Contact response of cells can mediate morphogenetic pattern formation. *Differentiation*, 45(3):147–159.

Engler, A., Bacakova, L., Newman, C., Hategan, A., Griffin, M., and Discher, D. (2004a). Substrate compliance versus ligand density in cell on gel responses. *Biophys. J.*, 86(1):617–628.

Engler, A. J., Griffin, M. A., Sen, S., Bönnemann, C. G., Sweeney, H. L., and Discher, D. E. (2004b). Myotubes differentiate optimally on substrates with tissue-like stiffness: Pathological implications for soft or stiff microenvironments. *J. Cell Biol.*, 166:877–887.

Engler, A. J., Sen, S., Sweeney, H. L., and Discher, D. E. (2006). Matrix elasticity directs stem cell lineage specification. *Cell*, 126:677–689.

Fardin, M., Rossier, O., Rangamani, P., Avigan, P., Gauthier, N., Vonnegut, W., Mathur, A., Hone, J., Iyengar, R., and Sheetz, M. (2010). Cell spreading as a hydrodynamic process. *Soft Matter*, 6(19):4788–4799.

Farrell, F., Marchetti, M., Marenduzzo, D., and Tailleur, J. (2012). Pattern formation in self-propelled particles with density-dependent motility. *Phys. Rev. Lett.*, 108(24):248101.

Fernandez, P. and Bausch, A. R. (2009). The compaction of gels by cells: A case of collective mechanical activity. *Integr. Biol.*, 1:252–259.

Folkman, J. and Greenspan, H. P. (1975). Influence of geometry on control of cell growth. *Biochim. Biophys. Acta*, 417:211–236.

Folkman, J. and Moscona, A. (1978). Role of cell shape in growth control. *Nature*, 273:345–349.

Friedrich, B. M., Buxboim, A., Discher, D. E., and Safran, S. A. (2011). Striated acto-myosin fibers can reorganize and register in response to elastic interactions with the matrix. *Biophys. J.*, 100(11):2706–2715.

Friedrich, B. M. and Safran, S. A. (2011). Nematic order by elastic interactions and cellular rigidity sensing. *Europhys. Lett.*, 93(2):28007.

Frisch, T. and Thoumine, O. (2002). Predicting the kinetics of cell spreading. *J. Biomech.*, 35(8): 1137–1141.

Fu, J., Wang, Y. K., Yang, M. T., Desai, R. A., Yu, X., Liu, Z., and Chen, C. S. (2010). Mechanical regulation of cell function with geometrically modulated elastomeric substrates. *Nat. Methods*, 7:733–736.

Gao, L., McBeath, R., and Chen, C. (2010). Stem cell shape regulates a chondrogenic versus myogenic fate through Rac1 and N-cadherin. *Stem Cells*, 28(3):564–572.

Gardel, M. L., Sabass, B., Ji, L., Danuser, G., Schwarz, U. S., and Waterman, C. M. (2008). Traction stress in focal adhesions correlates biphasically with actin retrograde flow speed. *J. Cell Biol.*, 183(6):999–1005.

Geiger, B., Spatz, J. P., and Bershadsky, A. D. (2009). Environmental sensing through focal adhesions. *Nat. Rev. Mol. Cell Biol.*, 10(1):21–33.

Ghibaudo, M., Saez, A., Trichet, L., Xayaphoummine, A., Browaeys, J., Silberzan, P., Buguinb, A., and Ladoux, B. (2008). Traction forces and rigidity sensing regulate cell functions. *Soft Matter*, 4:1836–1843.

Giannone, G., Dubin-Thaler, B. J., Döbereiner, H.-G., Kieffer, N., Bresnick, A. R., and Sheetz, M. P. (2004). Periodic lamellipodial contractions correlate with rearward actin waves. *Cell*, 116(3):431–443.

Gilbert, P., Havenstrite, K., Magnusson, K., Sacco, A., Leonardi, N., Kraft, P., Nguyen, N., Thrun, S., Lutolf, M., and Blau, H. (2010). Substrate elasticity regulates skeletal muscle stem cell self-renewal in culture. *Science*, 329(5995):1078.

Grevesse, T., Versaevel, M., Circelli, G., Desprez, S., and Gabriele, S. (2013). A simple route to functionalize polyacrylamide hydrogels for the independent tuning of mechanotransduction cues. *Lab on a Chip*, 13(5):777–780.

Gruler, H., Dewald, U., and Eberhardt, M. (1999). Nematic liquid crystals formed by living amoeboid cells. *Eur. Phys. J. B-Condens. Matter Complex Syst.*, 11(1):187–192.

Guilak, F., Cohen, D., Estes, B., Gimble, J., Liedtke, W., and Chen, C. (2009). Control of stem cell fate by physical interactions with the extracellular matrix. *Cell Stem Cell*, 5(1):17–26.

Guo, W. H., Frey, M. T., Burnham, N. A., and Wang, Y. L. (2006). Substrate rigidity regulates the formation and maintenance of tissues. *Biophys. J.*, 90:2213–2220.

Hall, A. (1998). Rho GTPases and the actin cytoskeleton. *Science*, 23:509–514.

Harris, A. K. (1994). Locomotion of tissue culture cells considered in relation to ameboid locomotion. *Int. Rev. Cytol.*, 150:35–68.

Harris, A. K., Stopak, D., and Wild, P. (1981). Fibroblast traction as a mechanism for collagen morphogenesis. *Nature*, 290:249–251.

Harris, A. K., Warner, P., and Stopak, D. (1984). Generation of spatially periodic patterns by a mechanical instability: A mechanical alternative to the turing model. *J. Embryol. Exp. Morphol.*, 80(1):1–20.

Harris, A. K., Wild, P., and Stopak, D. (1980). *Science*, 208:177–179.

Harrison, R. G. (1914). The reaction of embryonic cells to solid structures. *J. Exp. Zool.*, 17(4):521–544.

Harvey, A. G., Hill, E. W., and Bayat, A. (2013). Designing implant surface topography for improved biocompatibility. *Expert Rev. Med. Dev.*, 10(2):257–267.

Hu, K., Ji, L., Applegate, K. T., Danuser, G., and Waterman-Storer, C. M. (2007). Differential transmission of actin motion within focal adhesions. *Science*, 315(5808):111–115.

Huang, S., Chen, C. S., and Ingber, D. E. (1998). Control of cyclin D1, p27Kip1, and cell cycle progression in human capillary endothelial cells by cell shape and cytoskeletal tension. *Mol. Biol. Cell*, 9(11):3179–3193.

Ingber, D. E. (2003). Mechanobiology and diseases of mechanotransduction. *Ann. Med.*, 35:1–14.

Ingber, D. E. and Folkman, J. (1989). Mechanochemical switching between growth and differentiation during fibroblast growth factor-stimulated angiogenesis in vitro: Role of extracellular matrix. *J. Cell Biol.*, 109(1):317–330.

Janmey, P. and Miller, R. (2011). Mechanisms of mechanical signaling in development and disease. *J. Cell Sci.*, 124(1):9.

Jurado, C., Haserick, J. R., and Lee, J. (2005). Slipping or gripping? Fluorescent speckle microscopy in fish keratocytes reveals two different mechanisms for generating a retrograde flow of actin. *Mol. Biol. Cell*, 16(2):507–518.

Karuri, N. W., Nealey, P. F., Murphy, C. J., and Albrecht, R. M. (2008). Structural organization of the cytoskeleton in SV40 human corneal epithelial cells cultured on nano- and microscale grooves. *Scanning*, 30(5):405–413.

Katz, Y., Tunstrøm, K., Ioannou, C. C., Huepe, C., and Couzin, I. D. (2011). Inferring the structure and dynamics of interactions in schooling fish. *Proc. Natl. Acad. Sci. USA*, 108(46):18720–18725.

Kaufmann, D., Hoesch, J., Su, Y., Deeg, L., Mellert, K., Spatz, J. P., and Kemkemer, R. (2012). Partial blindness to submicron topography in NF1 haploinsufficient cultured fibroblasts indicates a new function of neurofibromin in regulation of mechanosensoric. *Mol. Syndromol*, 3(4):169–179.

Kemkemer, R., Jungbauer, S., Kaufmann, D., and Gruler, H. (2006). Cell orientation by a micro-grooved substrate can be predicted by automatic control theory. *Biophys. J.*, 90(12):4701–4711.

Kemkemer, R., Kling, D., Kaufmann, D., and Gruler, H. (2000a). Elastic properties of nematoid arrangements formed by amoeboid cells. *Eur. Phys. J. E*, 1(2–3):215–225.

Kemkemer, R., Teichgräber, V., Schrank-Kaufmann, S., Kaufmann, D., and Gruler, H. (2000b). Nematic order-disorder state transition in a liquid crystal analogue formed by oriented and migrating amoeboid cells. *Eur. Phys. J. E*, 3(2):101–110.

Kilian, K., Bugarija, B., Lahn, B., and Mrksich, M. (2010). Geometric cues for directing the differentiation of mesenchymal stem cells. *Proc. Natl. Acad. Sci. USA*, 107(11):4872.

Kim, D. H., Provenzano, P. P., Smith, C. L., and Levchenko, A. (2012). Matrix nanotopography as a regulator of cell function. *J. Cell Biol.*, 197(3):351–360.

Kim, T., Hwang, W., and Kamm, R. (2009). Computational analysis of a cross-linked actin-like network. *Exp. Mech.*, 49(1):91–104.

Klein, E., Yin, L., Kothapalli, D., Castagnino, P., Byfield, F., Xu, T., Levental, I., Hawthorne, E., Janmey, P., and Assoian, R. (2009). Cell-cycle control by physiological matrix elasticity and in vivo tissue stiffening. *Curr. Biol.*, 19(18):1511–1518.

Kolega, J. (1986). Effects of mechanical tension on protrusive activity and microfilament and intermediate filament organization in an epidermal epithelium moving in culture. *J. Cell Biol.*, 102:1400–1411.

Levental, K. R., Yu, H., Kass, L., Lakins, J. N., Egeblad, M., Erler, J. T., Fong, S. F. et al. (2009). Matrix crosslinking forces tumor progression by enhancing integrin signaling. *Cell*, 139(5):891–906.

Lin, C.-H. and Forscher, P. (1995). Growth cone advance is inversely proportional to retrograde F-actin flow. *Neuron*, 14(4):763–771.

Manoussaki, D., Lubkin, S., Vemon, R., and Murray, J. (1996). A mechanical model for the formation of vascular networks in vitro. *Acta Biotheor.*, 44(3–4):271–282.

Mathur, A., Moore, S. W., Sheetz, M. P., and Hone, J. (2012). The role of feature curvature in contact guidance. *Acta Biomater.*, 8(7):2595–2601.

Mazumder, A. and Shivashankar, G. V. (2010). Emergence of a prestressed eukaryotic nucleus during cellular differentiation and development. *J. R. Soc. Interface*, 7:S321–S330.

McBeath, R., Pirone, D. M., Nelson, C. M., Bhadriraju, K., and Chen, C. S. (2004). Cell shape, cytoskeleton tension, and RhoA regulate stem cell lineage commitment. *Dev. Cell*, 6:483–495.

McMurray, R. J., Gadegaard, N., Tsimbouri, P. M., Burgess, K. V., McNamara, L. E., Tare, R., Murawski, K., Kingham, E., Oreffo, R. O. C., and Dalby, M. J. (2011). Nanoscale surfaces for the long-term maintenance of mesenchymal stem cell phenotype and multipotency. *Nat. Mater.*, 10(8):637–644.

Mih, J. D., Marinkovic, A., Liu, F., Sharif, A. S., and Tschumperlin, D. J. (2012). Matrix stiffness reverses the effect of actomyosin tension on cell proliferation. *J. Cell Sci.*, 125(24):5974–5983.

Mih, J. D., Sharif, A. S., Liu, F., Marinkovic, A., Symer, M. M., and Tschumperlin, D. J. (2011). A multi-well platform for studying stiffness-dependent cell biology. *PloS One*, 6(5):e19929.

Moe, A. A. K., Suryana, M., Marcy, G., Lim, S. K., Ankam, S., Goh, J. Z. W., Jin, J. et al. (2012). Microarray with micro- and nano-topographies enables identification of the optimal topography for directing the differentiation of primary murine neural progenitor cells. *Small*, 8(19):3050–3061.

Murray, J. D. (2003). On the mechanochemical theory of biological pattern formation with application to vasculogenesis. *C. R. Biol.*, 326(2):239–252.

Narayan, V., Ramaswamy, S., and Menon, N. (2007). Long-lived giant number fluctuations in a swarming granular nematic. *Science*, 317(5834):105–108.

Nisenholz, N., Botton, M., and Zemel, A. (2014a). Early-time dynamics of actomyosin polarization in cells of confined shape in elastic matrices. *Soft Matter*, 14:2453–2462.

Nisenholz, N., Rajendran, K., Dang, Q., Chen, H., Kemkemer, R., Krishnan, R., and Zemel, A. (2014b). Active dynamics and mechanics of cell spreading on elastic substrates. *Soft Matter*, in press.

Onsager, L. (1936). Electric moments of molecules in liquids. *J. Am. Chem. Soc.*, 58:1486–1493.

Onsager, L. (1949). The effects of shape on the interaction of colloidal particles. *Ann. NY Acad. Sci.*, 51(4):627–659.

Oster, G. F., Murray, J. D., and Harris, A. (1983). Mechanical aspects of mesenchymal morphogenesis. *J. Embryol. Exp. Morphol.*, 78(1):83–125.

Park, J., Huang, N., Kurpinski, K., Patel, S., Hsu, S., and Li, S. (2007). Mechanobiology of mesenchymal stem cells and their use in cardiovascular repair. *Front. Biosci.*, 12:5098–5116.

Parker, K. K., Brock, A. L., Brangwyne, C., Mannix, R. J., Wang, N., Ostuni, E., Geisse, N. A., Adams, J. C., Whitesides, G. M., and Ingber, D. E. (2002). Directional control of lamellipodia extension by constraining cell shape and orienting cell tractional forces. *FASEB J.*, 16:1195–1204.

Paszek, M. J., Zahir, N., Johnson, K. R., Lakins, J. N., Rozenberg, G. I., Gefen, A., Reinhart-King, C. A. et al. (2005). Tensional homeostasis and the malignant phenotype. *Cancer Cell*, 8:241–254.

Pathak, A., Deshpande, V. S., McMeeking, R. M., and Evans, A. G. (2008). The simulation of stress fibre and focal adhesion development in cells on patterned substrates. *J. R. Soc. Interface*, 5:507–524.

Paul, R., Heil, P., Spatz, J. P., and Schwarz, U. S. (2008). Propagation of mechanical stress through the actin cytoskeleton toward focal adhesions: Model and experiment. *Biophys. J.*, 94:1470–1482.

Paul, R. and Schwarz, U. S. (2010). Pattern formation and force generation by cell ensembles in a filamentous matrix. *IUTAM Bookseries*, 16:203–213.

Peruani, F., Deutsch, A., and Bär, M. (2006). Nonequilibrium clustering of self-propelled rods. *Phys. Rev. E*, 74(3):030904.

Peruani, F., Starruß, J., Jakovljevic, V., Søgaard-Andersen, L., Deutsch, A., and Bär, M. (2012). Collective motion and nonequilibrium cluster formation in colonies of gliding bacteria. *Phys. Rev. Lett.*, 108(9):098102.

Pierres, A., Eymeric, P., Baloche, E., Touchard, D., Benoliel, A.-M., and Bongrand, P. (2003). Cell membrane alignment along adhesive surfaces: Contribution of active and passive cell processes. *Biophys. J.*, 84(3):2058–2070.

Ponti, A., Machacek, M., Gupton, S., Waterman-Storer, C., and Danuser, G. (2004). Two distinct actin networks drive the protrusion of migrating cells. *Science*, 305(5691):1782–1786.

Prager-Khoutorsky, M., Lichtenstein, A., Krishnan, R., Rajendran, K., Mayo, A., Kam, Z., Geiger, B., and Bershadsky, A. D. (2011). Fibroblast polarization is a matrix-rigidity-dependent process controlled by focal adhesion mechanosensing. *Nat. Cell Biol.*, 13(12):1457–1465.

Prass, M., Jacobson, K., Mogilner, A., and Radmacher, M. (2006). Direct measurement of the lamellipodial protrusive force in a migrating cell. *J. Cell Biol.*, 174(6):767–772.

Provenzano, P. P., Inman, D. R., Eliceiri, K. W., and Keely, P. J. (2009). Matrix density-induced mechanoregulation of breast cell phenotype, signaling and gene expression through a FAK–ERK linkage. *Oncogene*, 28(49):4326–4343.

Rehfeldt, F., Brown, A. E., Raab, M., Cai, S., Zajac, A. L., Zemel, A., and Discher, D. E. (2012). Hyaluronic acid matrices show matrix stiffness in 2D and 3D dictates cytoskeletal order and myosin-ii phosphorylation within stem cells. *Integr. Biol.*, 4(4):422–430.

Reinhart-King, C., Dembo, M., and Hammer, D. (2005). The dynamics and mechanics of endothelial cell spreading. *Biophys. J.*, 89(1):676–689.

Reinhart-King, C. A., Dembo, M., and Hammer, D. A. (2008). Cell-cell mechanical communication through compliant substrates. *Biophys. J.*, 95(12):6044–6051.

Ren, X. D., Kiosses, W. B., and Schwartz, M. A. (1999). Regulation of the small GTP-binding protein Rho by cell adhesion and the cytoskeleton. *EMBO J.*, 18:578–585.

Romanczuk, P., Bär, M., Ebeling, W., Lindner, B., and Schimansky-Geier, L. (2012). Active brownian particles. *Eur. Phys. J. Spec. Top.*, 202(1):1–162.

Ruiz, S. A. and Chen, C. S. (2007). Microcontact printing: A tool to pattern. *Soft Matter*, 3(2):168–177.

Sabass, B. and Schwarz, U. S. (2010). Modeling cytoskeletal flow over adhesion sites: Competition between stochastic bond dynamics and intracellular relaxation. *J. Phys. Condens. Matter*, 22(19):194112.

Schaller, V. and Bausch, A. R. (2013). Topological defects and density fluctuations in collectively moving systems. *Proc. Natl. Acad. Sci. USA*, 110(12):4488–4493.

Schwarz, U. S. and Safran, S. A. (2002). Elastic interactions of cells. *Phys. Rev. Lett.*, 88(4):048102.

Schwarz, U. S. and Safran, S. A. (2013). Physics of adherent cells. *Rev. Mod. Phys.*, 85(3):1327.

Schweitzer, Y., Lieber, A. D., Keren, K., and Kozlov, M. M. (2014). Theoretical analysis of membrane tension in moving cells. *Biophys. J.*, 106(1):84–92.

Sens, P. (2013). Rigidity sensing by stochastic sliding friction. *Europhys. Lett.*, 104(3):38003.

Seo, C., Furukawa, K., Suzuki, Y., Kasagi, N., Ichiki, T., and Ushida, T. (2011). A topographically optimized substrate with well-ordered lattice micropatterns for enhancing the osteogenic differentiation of murine mesenchymal stem cells. *Macromol. Biosci.*, 11(7):938–945.

Shivashankar, G. (2011). Mechanosignaling to cell nucleus and gene regulation. *Annu. Rev. Biophys.*, 40:361–378.

Silva, M. S., Alvarado, J., Nguyen, J., Georgoulia, N., Mulder, B. M., and Koenderink, G. H. (2011). Self-organized patterns of actin filaments in cell-sized confinement. *Soft Matter*, 7(22):10631–10641.

Singhvi, R., Kumar, A., Lopez, G. P., Stephanopoulos, G. N., Wang, D., Whitesides, G. M., and Ingber, D. E. (1994). Engineering cell shape and function. *Science*, 264(5159):696–698.

Solon, J., Levental, I., Sengupta, K., Georges, P. C., and Janmey, P. A. (2007). Fibroblast adaptation and stiffness matching to soft elastic substrates. *Biophys. J.*, 93:4453–4461.

Song, W., Lu, H., Kawazoe, N., and Chen, G. (2011). Adipogenic differentiation of individual mesenchymal stem cell on different geometric micropatterns. *Langmuir*, 27(10):6155–6162.

Stoker, M., O'Neill, C., Berryman, S., and Waxman, V. (1968). Anchorage and growth regulation in normal and virus-transformed cells. *Int. J. Cancer*, 3(5):683–693.

Stopak, D. and Harris, A. K. (1982). Connective tissue morphogenesis by fibroblast traction: I. Tissue culture observations. *Dev. Biol.*, 90(2):383–398.

Surrey, T., Nedelec, F., Leibler, S., and Karsenti, E. (2001). Physical properties determining self-organization of motors and microtubules. *Science*, 292(5519):1167–1171.

Szewczyk, D., Yamamoto, T., and Riveline, D. (2013). Non-monotonic relationships between cell adhesion and protrusions. *New J. Phys.*, 15(3):035031.

Tang, X., Bajaj, P., Bashir, R., and Saif, T. A. (2011). How far cardiac cells can see each other mechanically. *Soft Matter*, 7(13):6151–6158.

Tee, S., Fu, J., Chen, C., and Janmey, P. (2011). Cell shape and substrate rigidity both regulate cell stiffness. *Biophys. J.*, 100(5):L25–L27.

Théry, M. and Bornens, M. (2006). Cell shape and cell division. *Curr. Opin. Cell Biol.*, 18(6):648–657.

Théry, M., Pépin, A., Dressaire, E., Chen, Y., and Bornens, M. (2006a). Cell distribution of stress fibres in response to the geometry of the adhesive environment. *Cell Motil. Cytoskeleton.*, 63:341–355.

Théry, M., Racine, V., Piel, M., Pépin, A., Dimitrov, A., Chen, Y., Sibarita, J. B., and Bornens, M. (2006b). Anisotropy of cell adhesive microenvironment governs cell internal organization and orientation of polarity. *Proc. Natl. Acad. Sci. USA*, 103:19771–19776.

Tilghman, R. W., Cowan, C. R., Mih, J. D., Koryakina, Y., Gioeli, D., Slack-Davis, J. K., Blackman, B. R., Tschumperlin, D. J., and Parsons, J. T. (2010). Matrix rigidity regulates cancer cell growth and cellular phenotype. *PLoS One*, 5(9):e12905.

Tlusty, R. B.-Z. T., Moses, E., Safran, S. A., and Bershadsky, A. (1999). Pearling in cells: A clue to understanding cell shape. *Proc. Natl. Acad. Sci. USA*, 96:10140–10145.

Torres, P. G., Bischofs, I., and Schwarz, U. (2012). Contractile network models for adherent cells. *Phys. Rev. E*, 85(1):011913.

Tse, J. and Engler, A. (2011). Stiffness gradients mimicking in vivo tissue variation regulate mesenchymal stem cell fate. *PLoS One*, 6(1):e15978.

Ulrich, T. A., de Juan Pardo, E. M., and Kumar, S. (2009). The mechanical rigidity of the extracellular matrix regulates the structure, motility, and proliferation of glioma cells. *Cancer Res.*, 69(10):4167–4174.

Vasiliev, J. M. (1984). Spreading of non-transformed and transformed cells. *Biochim. Biophys. Acta*, 780(1):21–65.

Vernerey, F. J. and Farsad, M. (2014). A mathematical model of the coupled mechanisms of cell adhesion, contraction and spreading. *J. Math. Biol.*, 68(4):989–1022.

Versaevel, M., Grevesse, T., and Gabriele, S. (2012). Spatial coordination between cell and nuclear shape within micropatterned endothelial cells. *Nat. Commun.*, 3:671.

Versaevel, M., Grevesse, T., Riaz, M., Lantoine, J., and Gabriele, S. (2014). Micropatterning hydroxy-PAAm hydrogels and Sylgard 184 silicone elastomers with tunable elastic moduli. *Methods Cell Biol.*, 121:33–48.

Walcott, S. and Sun, S. (2010). A mechanical model of actin stress fiber formation and substrate elasticity sensing in adherent cells. *Proc. Natl. Acad. Sci. USA*, 107(17):7757.

Wang, J. H. C., Goldschmidt-Clermont, P., Wille, J., and Yin, F. (2001). Specificity of endothelial cell reorientation in response to cyclic mechanical stretching. *J. Biomech.*, 34(12):1563–1572.

Wang, N., Ostuni, E., Whitesides, G. M., and Ingber, D. E. (2002). Micropatterning tractional forces in living cells. *Cell Motil. Cytoskeleton*, 52:97–106.

Wang, N., Tytell, J. D., and Ingber, D. E. (2009). Mechanotransduction at a distance: Mechanically coupling the extracellular matrix with the nucleus. *Nat. Rev. Mol. Cell Biol.*, 10:75–82.

Weiss, P. (1934). In vitro experiments on the factors determining the course of the outgrowing nerve fiber. *J. Exp. Zool.*, 68(3):393–448.

Welsh, C. F., Roovers, K., Villanueva, J., Liu, Y., Schwartz, M. A., and Assoian, R. K. (2001). Timing of cyclin D1 expression within G1 phase is controlled by Rho. *Nat. Cell Biol.*, 3(11):950–957.

Whitesides, G. M., Ostuni, E., Takayama, S., Jiang, X., and Ingber, D. E. (2001). Soft lithography in biology and biochemistry. *Annu. Rev. Biomed. Eng.*, 3:335–373.

Winer, J., Janmey, P., McCormick, M., and Funaki, M. (2008). Bone marrow-derived human mesenchymal stem cells become quiescent on soft substrates but remain responsive to chemical or mechanical stimuli. *Tissue Eng. A*, 15(1):147–154.

Wittelsberger, S. C., Kleene, K., and Penman, S. (1981). Progressive loss of shape-responsive metabolic controls in cells with increasingly transformed phenotype. *Cell*, 24(3):859–866.

Yeung, T., Georges, P. C., Flanagan, L. A., Marg, B., Ortiz, M., Funaki, M., Zahir, N., Ming, W., Weaver, V., and Janmey, P. A. (2005). Effects of substrate stiffness on cell morphology, cytoskeletal structure, and adhesion. *Cell Motil. Cytoskeleton*, 60:24–34.

Yu, H., Mouw, J. K., and Weaver, V. M. (2011). Forcing form and function: Biomechanical regulation of tumor evolution. *Trends Cell Biol.*, 21(1):47–56.

Zemel, A., Bischofs, I. B., and Safran, S. A. (2006). Active elasticity of gels with contractile cells. *Phys. Rev. Lett.*, 97:128103.

Zemel, A., De, R., and Safran, S. (2011). Mechanical consequences of cellular force generation. *Curr. Opin. Solid State Mater. Sci.*, 15(5):169–176.

Zemel, A., Rehfeldt, F., Brown, A., Discher, D., and Safran, S. (2010a). Optimal matrix rigidity for stress-fibre polarization in stem cells. *Nat. Phys.*, 6(6):468–473.

Zemel, A., Rehfeldt, F., Brown, A. E. X., Discher, D. E., and Safran, S. A. (2010b). Cell shape, spreading symmetry and the polarization of stress-fibers in cells. *J. Phys. Condens. Matter.*, 22:194110.

Zemel, A. and Safran, S. A. (2007). Active self-polarization of contractile cells in asymmetrically shaped domains. *Phys. Rev. E.*, 76:021905.

Zeng, X. and Li, S. (2011). Multiscale modeling and simulation of soft adhesion and contact of stem cells. *J. Mech. Behav. Biomed. Mater.*, 4(2):180–189.

6

Dynamic Stress Fiber Reorganization on Stretched Matrices

Roland Kaunas

CONTENTS

6.1 Introduction

Cells in various tissues throughout the body sense and respond to deformation of their surrounding extracellular matrix (ECM). Some examples include cyclic loading of endothelial and smooth muscle cells in the arterial wall, chronic loading of fibroblasts during expansion-induced skin growth,[1] and transient loading of fibroblasts in tendons during muscle contractions. In each of these examples, tensile forces acting at a macroscopic level are transmitted to the local pericellular environment to regulate a number of cellular events, including proliferation, differentiation, and gene expression. These cellular responses are necessary to facilitate the adaptation of tissues to these loads to contribute to tissue homeostasis or dysfunction.

Cells also generate contractile forces that are transmitted to the ECM. These forces are necessary for cell functions, such as migration and spreading, and also contribute to the mechanical properties of the tissue. For example, highly contractile myofibroblasts contribute to the process of wound closure. Interestingly, cells also sense the stiffness of the ECM, which in turn not only regulates the contractile state of the cell but can also contribute to cell differentiation.

Three aspects of the study of cell responses to tissue deformation will be discussed in this chapter, with a focus on remodeling of actin stress fibers (SFs). First, we will describe actin SFs and establish the mechanical state of SFs in adherent cells on static substrates. Next, techniques for applying tensile loads to cells cultured on or in ECM will be reviewed along with observations from experiments. Finally, mathematical models that provide insight into how cells sense and respond to matrix deformation will be described. The reader may refer to Chapters 1 and 3 for discussions regarding the mechanics of other cytoskeletal constituents (i.e., microtubules and intermediate filaments) and in-depth descriptions of actin cytoskeletal structures other than SFs.

6.2 Actin Stress Fibers

Most images of cells outside the body have been obtained from cells adhered onto stiff substrates such as glass slides or tissue culture plastic. Under these conditions, most cells will develop SFs and associated focal adhesions (FAs) to an extent that depends on the contractile state of the cell. For example, highly contractile cells, such as myofibroblasts, contain pronounced SFs, while leukocytes are largely devoid of these structures. Cell contractility was nicely demonstrated in the pioneering experiments of Albert Harris[2] in the 1980s (see Chapter 10 for details) and later quantified using flexible hydrogels[3] and micropost arrays.[4] SFs are also observed in vivo in situations where cells are subjected to high mechanical stresses such as in contracting wound granulation tissue[5] and in arterial endothelial cells in regions of high fluid shear stress[6] and elevated wall stress.[7] The following sections summarize the composition of SFs and factors that contribute to their assembly and disassembly. The reader may also refer to Chapter 3 for additional details on SFs.

Although the precise molecular architecture remains to be clarified, SFs are thought to resemble myofibrils for several reasons. First, SFs contain many of the same proteins found in myofibrils, including actin, non-muscle myosin II motor protein, and α-actinin.[8,9] Second, electron micrographs and fluorescent labeling experiments demonstrate a striated organization of myosin II and α-actinin reminiscent of myofibrils, with the level of organization dependent on the size of the SF.[10,11] Third, these sarcomere-like structures can individually contract to result in the overall contraction of the SF.[12]

Contractile force is regulated by myosin II that physically interacts and moves actin filaments. As described in Chapter 1, myosin activity is driven by ATP hydrolysis to cause conformational changes in myosin necessary for myosin to move relative to the associated actin filament. The rate of ATP hydrolysis depends on the phosphorylation status of myosin light chain (MLC), which is regulated by MLC kinases (MLCKs) and MLC phosphatases. Specifically, Rho-associated kinase (Rho-kinase/ROCK/ROK) mediates MLC phosphorylation downstream of Rho small GTPase, while MLCK phosphorylates MLC through a calcium-dependent pathway.[13] Interestingly, Rho-kinase and MLCK regulate separate SF populations: peripheral SFs are MLCK dependent, while central SFs are Rho-kinase dependent.[13–15]

6.2.1 Mechanical Properties

Unique methods have been devised to characterize the mechanical properties of SFs. Deguchi et al.[16] isolated intact SFs and obtained the force–strain relationship in the absence of ATP using microcantilevers. In the absence of ATP, the myosin heads are expected to be firmly attached to actin filaments (so-called *rigor state*); hence, these measurements represent the passive mechanical properties of the SF. To study functional SFs in intact living cells, Yin et al.[17,18] imaged the buckling of SFs in cells that were suddenly uniaxially shortened on prestretched silicone rubber sheets. These studies indicated that SFs are under a level of tension equivalent to applying 10%–35% prestretch depending on the contractile activity in the cell. Kumar et al.[19] severed individual SFs using laser scissors and observed that the severed ends retracted at a rate that was fit to a simple viscoelastic model. Using SFs expressing α-actinin labeled with Green Fluorescent Protein (GFP), Russell et al.[12] demonstrated that the apparent viscoelastic behavior was caused by the individual sarcomeres shortening at constant rates until reaching their minimum size. Based on this model, the total shortening length of individual sarcomeres is distributed; hence, the rate of SF retraction slows over time as fewer sarcomeres participate in the overall shortening of the SF.

The behavior of SFs is quite interesting and complex. For example, Peterson et al.[20] showed simultaneous shortening of peripheral sarcomeres and lengthening of central sarcomeres within single contracting SFs. This behavior was described in a model by Colombelli et al.[21] and Stachowiak and O'Shaughnessy[22] using viscoelastic contractile elements. Russell et al.[23] demonstrated that nascent sarcomeres appear to originate at FAs and move centripetally, while other sarcomeres disappear at discrete *sinks* along the SF—often at junctions between adjacent SFs.

6.2.2 Assembly

SFs form in response to serum factors that activate Rho GTPase through the generation of force by myosin motors.[24,25] Little actin polymerization occurs during the formation of ventral SFs in quiescent cells in which RhoA had been activated,[26] suggesting that assembly occurs through bundling of preexisting actin filaments. SF assembly has been observed in permeabilized cells after the addition of ATP and Ca^{2+} and involves centripetal movement of actin filaments.[27] Interestingly, an applied force could also induce SF assembly in the absence of ATPase activity.[27] These results suggest SFs form through force-mediated bundling of actin filaments rather than actin polymerization.

SF assembly is also important for the assembly of associated FAs. Contractility drives centripetal flow of actin filaments that are captured at nascent cell adhesions and converted by Dia1 and α-actinin into dense actin bundles.[28] These actin bundles then serve as a scaffold for recruitment of additional FA proteins. Smith et al.[29] demonstrated a key role for the FA protein zyxin in promoting actin polymerization at FAs and reinforcing SF integrity in response to applied force.

6.2.3 Disassembly

While SF assembly has been studied extensively, relatively little is known about the mechanisms regulating SF disassembly. SFs disassemble when contractility is deactivated, such as when elevated cAMP and consequent activation of cAMP-dependent protein kinase (PKA) results in phosphorylation and inhibition of RhoA and MLCK.[30,31]

Costa's[17] observation of rapid SF disassembly following rapid cell shorting, if the cells were not immediately fixed, nicely illustrates the importance of tension in maintaining the stability of SFs. Interestingly, laser severing of individual SFs does not cause SF disassembly, but instead, the SFs stabilized and formed new FAs near the site of severing.[19,21] Further, zyxin was lost from the original FAs and SF and then redistributed to distinct regions along the new SF fragments. Colombelli et al.[21] attributed this behavior to integrin-mediated attachments to the substratum along the length of the SF. In general, loss of tension of many SFs results in SF disassembly while loss of tension in single SFs does not.

A role for myosin II in promoting actin filament disassembly has recently emerged (see Chapter 3 for a detailed discussion). We have also reported a role for myosin II in promoting SF disassembly using a preparation in which cells were *de-roofed* so that only SFs remained attached to the substratum.[32] Addition of ATP to drive actomyosin crossbridge cycling resulted in contraction and disassembly, of the SFs depending on the concentration of ATP—low concentrations (<2 mM ATP) induced slow contraction of the SFs, while high concentrations (>4 mM ATP) induced rapid disassembly before contraction could occur. Only at an intermediate concentration (2–4 mM ATP) was there rapid contraction that only resulted in disassembly after the SFs shortening to ~25% of their original lengths. Interestingly, this intermediate range of intracellular [ATP] is within the physiological range. This suggests that rapid actomyosin crossbridge cycling may promote SF disassembly in intact cells, likely through a decrease in the number of actomyosin crossbridges.[32]

6.3 Stretching Cells on 2D Substrates

Most stretch devices stretch cells cultured on flat, elastomeric silicone rubber sheets. To support cell adhesion, the sheets are typically coated with ECM protein, including fibronectin and collagen type I, or short synthetic peptides containing the integrin-binding sequences of these matrix proteins (i.e., RGD).[33]

In 1985, Banes et al.[34] introduced and later commercialized the Flexcell® system that applies vacuum suction to conventional six-well plates containing silicone rubber bottoms. The current version of the device includes a loading post that increases the uniformity of the strain field over the post and allowing generation of equibiaxial and uniaxial strains. Variations on this approach have also been developed.[35]

As an alternative to the vacuum approach, the loading post can be pushed against the elastomer sheet to produce a homogenous equibiaxial strain,[36–38] as well as uniaxial strain by modifying the post and chamber geometries.[39,40] A DC motor-cam is typically used to drive the cyclic load, which allows control of both stretch magnitude and frequency, but the stretch waveform is limited to sinusoidal function,[38] though a stepper motor can be used to generate other waveforms.[40]

Simple uniaxial strain has often been used to stretch rectangular elastomeric sheets with a motorized actuator.[41] When a material is stretched in one direction, it tends to contract in the other two directions perpendicular to the direction of stretch. This phenomenon, termed Poisson's effect, results in cell and SF alignment at an oblique angle relative to the direction of stretch. Various stretch devices have been developed that limit the extent of lateral contraction,[17,42,43] including the loading post systems described above and the commercially available STREX® system.[44]

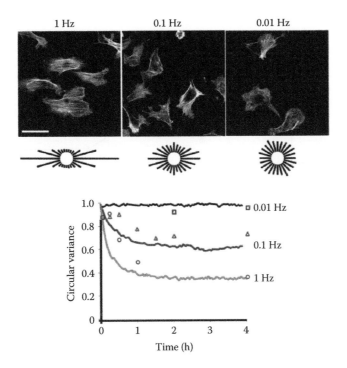

FIGURE 6.1
The extent of SF alignment depends on the frequency of cyclic uniaxial stretch. Representative images and circular histograms are shown of sparsely seeded bovine aortic endothelial cells (ECs) that were subjected to 4 h of 10% cyclic uniaxial stretch at frequencies of 1–0.01 Hz, fixed, and stained for F-actin. The direction of stretch is vertical with respect to the page. Circular variances of the SF distributions are plotted over time along with predictions of the sarcomere-based model described in Section 6.6.3. (Adapted from Hsu, H.J. et al., *PLoS One*, 4, e4853, 2009; Kaunas, R., et al., *Cell Health Cytoskelet.*, 3, 13–22, 2011. With permission.)

Cell remodeling has been shown to be exquisitely sensitive to various spatiotemporal characteristics of cyclic strain, including strain magnitude, rate, and direction. There is a greater tendency for alignment as the magnitude of stretch increases from 1% to 20%[39,45] and as the frequency of stretching increases from 0.01 to 1 Hz[46,47] (Figure 6.1). The extent of SF alignment can be reported as a circular variance quantifying the uniformity in the orientations by vectorially summing each orientation vector component, normalizing the result by the total number of vectors (N), and subtracting the value from unity.[46]

$$\text{Circular Variance} = 1 - \frac{1}{N}\sqrt{\left(\sum_{i=1}^{N}\sin 2\theta^i\right)^2 + \left(\sum_{i=1}^{N}\cos 2\theta^i\right)^2} \qquad (6.1)$$

The values range from zero (perfect alignment) to one (random distribution). For endothelial cells subjected to cyclic stretch at 1 Hz, there is an apparent minimum stretch of 3% necessary to promote alignment, but this is lowered to 1% when the cells express constitutively active RhoA GTPase[39] (Figure 6.2). Lowering RhoA activity using C3 exoenzyme or pharmacological inhibition of the downstream effector Rho-kinase, on the other hand, changes the direction of SF alignment to be parallel to the stretch direction.[15,39]

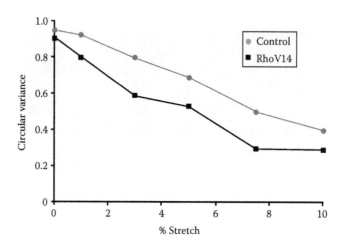

FIGURE 6.2
The extent of SF alignment depends on the magnitude of cyclic stretch. Circular variances of the SF distributions achieved after subjecting bovine aortic endothelial cells to 6 h of 1 Hz cyclic uniaxial stretch are plotted as a function of stretch magnitude. The cells were either transfected with GFP alone (control) or co-transfected with GFP and constitutively active RhoV14.

The minimum stretch frequency needed to induce alignment is ~0.01 Hz.[46,47] Further, the rate of cell alignment decreases as either the amplitude or frequency of stretching decreases.[47] Typical stretch waveforms are symmetric, that is, the lengthening and shortening phases are applied at similar rates. Using asymmetric waveforms in which either the rate of elongation or shortening of a stretch cycle is much faster than the other, Tondon et al.[48] demonstrated that the alignment response is much more sensitive to the rate of elongation than the rate of shortening.

Fluorescent imaging studies have provided important insight into how cells and their actin cytoskeleton reorganize over time in response to stretch. Katsumi et al.[49] demonstrated that a step increase in equibiaxial stretch causes a transient, uniform retraction of lamellipodia, while uniaxial stretch only causes retraction of lamellipodia located along the sides of the cell perpendicular to the direction of stretching. Using a fluorescence resonance energy transfer (FRET) biosensor for Rac GTPase activation, these authors showed that Rac activity is transiently decreased at the edges that are subjected to increased tangential tension because of the uniaxial stretch.[49] Time-lapse imaging of cells expressing GFP-labeled actin subjected to cyclic equibiaxial stretching shows sustained lamellipodial retraction applied at 1 Hz, but not at 0.01 Hz.[50] These results suggest that cells can adapt to a transient or slow stretch, but not to rapid cyclic stretch. In response to cyclic uniaxial stretch, time-lapse images illustrate that SFs in cells expressing fluorescently labeled actin gradually align perpendicular to the stretch direction through SF turnover.[15] Time-lapse studies reported by the Kemkemer group demonstrate that realignment can also occur through FA sliding and SF rotation.[51,52]

6.4 Stretching Cells in 3D Matrices

Cells in 3D matrices are exposed to a much different environment than cells attached to flat substrates.[53] Unlike cells on flat substrates, cells in 3D matrices have no apical–basolateral

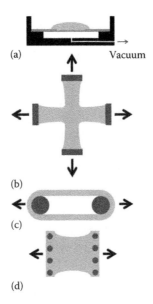

FIGURE 6.3
Diagrams illustrating methods for stretching 3D collagen gels. (a) The Flexcell Tissue Train culture system anchors the gels at two ends with stems, and a vacuum applied from below stretches the silicone rubber membrane downward to pull on the gel. (b) A biaxial system anchors the crucifix-shaped gel from each end and is stretched symmetrically along two axes. (c) A ring-shaped collagen gel is pulled along one axis using two hooks (cross section is shown only). (d) A collagen gel is anchored to silicone rubber pillars that are used to pull the gel.

polarity, interact with soft networks of discrete fibrils, and may be sterically hindered from spreading.[54] Furthermore, cells in 3D collagen matrices align parallel, not perpendicular, to the direction of stretch, as do the collagen fibrils themselves.[55,56] These observations are consistent with the circumferential alignment of smooth muscle cells and collagen fibrils parallel to the principal direction of cyclic stretch in arteries. Thus, when studying the effects of stretch on cells that reside within tissues, it is appropriate to use a system in which the cells are cultured in a 3D matrix. More details regarding cell behavior in mechanically loaded 3D collagen matrices can be found in Chapter 7.

Fibrillar collagen hydrogels are commonly used for stretching cells in a 3D matrix. Various approaches have been devised to handle these very soft gels. A rigid material (e.g., hydrophilic porous polyethylene) can be used as an interface between the gel and the actuators used to stretch the gel by allowing the gel to intercalate within the pores before gelation.[57] One example is the Flexcell Tissue Train® culture system (Figure 6.3a), which applies a uniaxial strain to the matrix to result in alignment of F-actin, fibroblasts, and collagen fibrils parallel to the direction of stretch.[55,58]

Uniaxial systems typically align collagen fibrils and cells in the direction of stretch in a manner that also involves lateral contraction.[59] A biaxial configuration (Figure 6.3b) that provides control over strains in the x- and y-directions has been used to demonstrate that the extent of collagen fibril alignment depends on the ratio of strain applied in the x- and y-directions.[57] The incorporation of force transducers along each axis allows simultaneous measurement of strains and forces to potentially identify the roles of these two factors on cell-induced reorganization of matrix architecture.

Collagen gels can be cast in a hoop shape and pulled in opposite directions using a pair of hooks[56] (Figure 6.3c). This system has been used to characterize the active contribution of cells to the mechanical properties of these constructs[60] and the dynamic reorganization of the actin cytoskeleton upon changes in mechanical loading of the constructs.[56] With regard to cytoskeletal reorganization, two kinds of responses were observed: retraction and reinforcement. Retraction responses consisted of SF fragmentation and retraction of filopodia-like cellular protrusions. These responses were observed for SFs and filopodia-like cellular protrusions oriented in all directions relative to the stretch direction. Reinforcement responses consisted of an increase of the size and number of SFs and extension of cellular protrusions, both mainly in the direction of the applied stretch. Furthermore, it appeared that F-actin was transferred from fragmenting SFs during retraction responses to *reservoirs* that diminished during reinforcement responses, presumably due to transfer of F-actin to growing SFs.

Collagen gels containing cells can also be stretched using silicone rubber sheets with vertical posts to constrain the gels along the edges orthogonal to the direction of stretching (Figure 6.3d). Using this system, Foolens et al.[93] observed that F-actin alignment perpendicular to the direction of uniaxial stretching at gel surfaces and not in the core. Perpendicular alignment was observed in the core under conditions in which collagen fibrils were suppressed, such as stretching prior to full collagen polymerization or under conditions of high-collagen proteolysis. Thus, like in the 2D environment, cells and their SFs can align perpendicular to the direction of stretching in 3D collagen gels in the absence of the strong influence of collagen fibrils.

6.5 Stretching Cells on Soft 2.5D Matrices

Substrate rigidity regulates a number of cell functions, including cell–cell adhesion,[61,62] cell–substrate adhesion,[63] and cell differentiation.[64] Given that cells behave differently on soft and rigid substrates, it is expected that cells will also respond differently to stretch on soft substrates. These soft substrates are typically 3D hydrogels with the cells confined to the apical surface. Since this environment has characteristics of both 2D and 3D environments, it is sometimes referred to as 2.5D.

Trepat et al.[65] cultured human airway smooth muscle cells on polyacrylamide gels coated with matrix protein and demonstrated that cell stiffness (quantified by magnetic twisting cytometry) rapidly decreased and slowly recovered after a transient isotropic biaxial stretch–unstretch maneuver of 4 s duration. These results were interpreted as stretch-induced fluidization of the cell cytoskeleton (in a manner comparable to the effect of shear on soft materials such as colloidal glasses, emulsions, and pastes[66,67]), followed by resolidification. Krishnan et al.[68] later measured cell contractile forces in human umbilical vein endothelial cells that were uniaxially stretched repeatedly for 3 s every 49 s on soft polyacrylamide gels. The periodic stretching resulted in an initial drop in contractile force within 50 min and the subsequent recovery of force after 100 min once the cell had aligned perpendicular to the direction of stretching. Further, the contractile force development was anisotropic, with the cell contractility only recovering perpendicular to the stretch direction. Interestingly, the reorientation of the contractile moment preceded the realignment of the cell body. While a transient stretch–unstretch regimen results in a drop in cell stiffness, a transient compression–uncompression of the same magnitude and timing did not,[69] this observation led the authors to propose that the extent of the fluidization response was localized to the unstretch phase of the stretch–unstretch maneuver in a manner suggesting cytoskeletal catch bonds. This contrasts with the results by Tondon et al.[48] that demonstrated cells are more responsive to the rate of stretching than the rate of unstretching in response to cyclic uniaxial stretching with asymmetric waveforms. Also, fluidization is not observed in cells expressing fluorescently labeled SFs subjected to cyclic stretch.[15,50,51]

Throm Quinlan et al.[70] stretched aortic valve interstitial cells on polyacrylamide gels of a range of stiffnesses covalently attached to silicone rubber sheets. Cyclic equibiaxial stretching on soft (0.5 kPa) gels promoted cell spreading and suppressed cell alignment perpendicular to the direction of cyclic uniaxial stretch. Using soft polydimethylsiloxane (PDMS) stretch chambers, Faust et al.[45] reported that cell and SF alignment in the direction of zero strain in response to cyclic stretching at mHz frequencies was dependent on substrate stiffness, with significant alignment observed on relatively stiff substrates (50 kPa), an intermediate response when cells were stretched on less stiff substrates (11 kPa), and no alignment on soft substrates (≤3 kPa).

Tondon and Kaunas[71] observed that SFs in human mesenchymal stem cells subjected to cyclic uniaxial stretch at 1 Hz on soft collagen gels (on the order of tens of Pa[72]) align parallel to the stretch direction, in contrast to the perpendicular alignment observed on collagen-coated silicone rubber sheets cells, and similar results were observed using U2OS osteosarcoma cells. Parallel alignment was also induced in response to a step stretch on soft collagen gels, but not on stiff collagen-coated silicone rubber. Interestingly, alignment was not observed on either substrate when the cyclic or step stretches were applied at a slow rate (e.g., 0.01 Hz cyclic stretch).

6.6 Mathematical Modeling

As can be seen in the previous section, the effect of stretch depends on many factors, including the type of strain applied, the contractile state of the cells, and the organization and mechanical properties of the ECM. Mathematical models provide tools to interpret seemingly diverse experimental observations under a unifying framework. A number of such models have been developed over the years, some of which are discussed in the following sections. Given that these models all describe stretch-induced cell and/or SF alignment, it is perhaps not surprising that there are several similarities between them. Differences mainly lie in the mathematical treatments, as shall be summarized below. The review begins with coarse-grained models that are motivated by the molecular mechanisms regulating cell and SF reorganization, but lack specific molecular details. Next, models that mathematically describe cytoskeletal turnover are discussed. Finally, models that address FA remodeling are described.

6.6.1 Coarse-Grained Modeling

De et al.[73] employed a contractile force dipole model of a cell to predict the dynamics of stretch-induced cell alignment in soft, elastic materials subjected to cyclic loading. A force dipole is defined by two equal and oppositely directed forces, $\vec{P}/2$ and $-\vec{P}/2$, separated by a short distance, $\vec{l}$, which is directed at an angle θ relative to the direction of loading

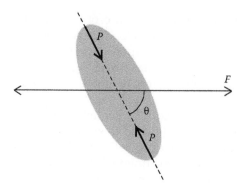

FIGURE 6.4
Coarse-grained dipole model. The model consists of a contractile force dipole P acting along the cell axis oriented at an angle θ relative to the direction of an external force field F.

(Figure 6.4). The two point forces represent the entire contractile machinery of the cell, in a coarse-grained fashion. Along the axis of the force dipole, there is local compression of the substrate between the two forces and expansion of the substrate just outside of the dipole. Under substrate loading, the local stress field depends on the interaction of the applied stress field, elastic restoring forces in the substrate, and the contractile force dipole. A free energy functional was used to predict the forces that act on the cellular force dipole in response to application of external load. The free energy includes two terms. The first term measures the difference between the actual magnitude of the cellular dipole and an optimal (homeostatic) level of stress in the matrix. This contribution is motivated by the observed tendency for cells to lower the magnitude of their forces when subjected to applied tensile stress.[59] The second term in the free energy is the interaction (strain) energy between the force dipole and the applied load. This term reflects the passive forces in the matrix that tend to orient the cell in the direction of applied load. To describe the dynamic behavior of the system, the rate of change in orientation and magnitude of the dipole was assumed to be linearly dependent on the variation of the effective free energy of the system, which allows the cells to relax perturbations in matrix stress through both changes in orientation and contractile dipole magnitude. By incorporating dynamics, the model predicts that cells orient parallel to the direction of slowly applied cyclic stress, but that at high stretch frequencies, the inability for the contractile moment to relax quickly enough forces the dipoles to orient nearly perpendicular to the applied stress. For high stretch frequencies, De et al.[74] predicted that the experimentally observed oblique alignment of cells on substrates that contract in the direction orthogonal to the direction of loading is due to cell sensitivity to strain rather than stress. If the cells were instead sensitive to stress, the model would predict that the cells would align perpendicular to the direction of loading (exerting their optimal stress) regardless of orthogonal contraction (however, see Ref. [74] for an alternative explanation for oblique alignment in case of relatively soft substrates). Further, the model provides a tool for predicting the dependence of the response on substrate rigidity since the magnitude of contractile dipole increases with increasing rigidity.[73] To explain the inherent disorder in SF alignment, random *noise* modeled as an effective temperature was added to the model to dissipate the effects of stretch at low stretch frequencies. It is worth noting that this model is based on the response to applied stresses. This is a distinction from the models discussed below that impose applied strains. While stresses and strains are clearly related, they are not interchangeable.

6.6.2 Modeling Based on Cytoskeletal Fluidization and Resolidification

Motivated by the experiments reported by Krishnan et al.,[75] Stamenovic et al.[76] developed a model incorporating the phenomena of cytoskeletal fluidization and resolidification in response to cyclic stretching. Individual SFs are treated as prestressed linearly elastic line elements anchored to an elastic substrate via elastic FAs. Stretch-induced SF fluidization decreases fiber stress to zero, resulting in fiber disassembly and subsequent reassembly in a direction where the fiber's total potential energy (Π) is a global minimum. The total potential energy is defined as the sum of the elastic potential energy stored in the SF and the force potential, which the authors show reduces to

$$\Pi = -\frac{\sigma^2 V}{4E} \tag{6.2}$$

at equilibrium, where σ is the stress acting on the SF, V is its volume, and E is its elastic modulus. The stress is taken as the sum of the prestress (σ_0) and the additional stress generated by stretching the SF (i.e., $\sigma = \sigma_0 + E\varepsilon$). For pure uniaxial strain, the linear approximation of strain on an SF oriented at an angle θ relative to the stretch direction is taken to be

$$\varepsilon = \frac{\alpha \varepsilon_s}{2}(1 + \cos 2\theta), \tag{6.3}$$

where
  ε_s is the substrate strain in the direction of stretching
  α represents the fraction of substrate strain transmitted to the SF

Since the physical origins of fluidization are unknown, two possible mechanisms were considered[76]: (1) fluidization results in ablation of stress without changing the elastic modulus of the SF and (2) fluidization reduces the elastic modulus. Further, the authors considered two cyclic stretch–unstretch patterns. First, square-wave stretching consisting of stretch and unstretch intervals of equal duration was used to consider the situation where there is insufficient time for reinforcement following fluidization. In comparison, pulse-wave stretching employed unstretch intervals much longer than the stretch interval, thus allowing SF reinforcement. Numerical simulations of square-wave stretching predict that the direction of minimal energy evolves over the first several stretch cycles, eventually leading to stable SFs oriented perpendicular to the direction of stretch, though there are differences in the time course for the two cases. In contrast, pulse-wave stretching simulations predict that SFs align parallel to the direction of stretch.

6.6.3 Modeling Based on Sarcomere Mechanics

To account for the dynamic behavior of individual SFs, Kaunas et al. developed a model consisting of a population of SFs capable of responding to stretch through both tension relaxation and SF turnover.[77] Figure 6.5 illustrates a simplified representation of an individual SF, consisting of a series of identical sarcomeres, adhered to an elastic substrate at each end through FAs. Each sarcomere consists of several actomyosin complexes in parallel, which can be represented mechanically as a single spring in series with a contractile element (Figure 6.5b). Since the sarcomeres are in series, the force acting on each sarcomere is identical to the force acting on the entire SF. The passive and active behavior of the sarcomere, which reflects that of the entire SF, is modeled using a linearized version of the

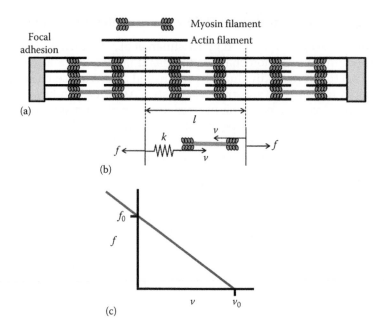

(a)

(b)

(c)

FIGURE 6.5

Sarcomere-based model. (a) An SF is modeled as a series of sarcomeric bands of length l, each composed of a bundle of N individual actomyosin subunits. (b) Each actomyosin subunit is modeled mechanically as a composite elastic–contractile element. The velocity v of myosin sliding depends on the force f acting on a subunit. The boundary conditions are determined by the relative displacements of the FAs during the deformation of the underlying substrate. (c) A linearized Hill model describes the relationship between force and the velocity of shortening of each end of a myosin filament in an actomyosin subunit. (Reprinted from Kaunas, R. et al., *Cell Health Cytoskelet.*, 3, 13–22, 2011. With permission.)

relationship between force and shortening velocity (Figure 6.5c) proposed by Hill.[78] The force acting on each sarcomere is described by the following:

$$f(t) = k\frac{l(t) - l_{ref}(t)}{l_{ref}(t)},$$
(6.4)

where k is the effective spring constant of an element. The lengths of the sarcomere in the unloaded (i.e., when $f = 0$) and loaded states are l_{ref} and l, respectively, and each may change over time. At rest, this model predicts that myosin motors adjust l_{ref} until the stall force (f_0) is attained as determined by Equation 6.4. Changing the length of the SF by some value (e.g., 10%) by stretching the cell will change the length of the sarcomere proportionally. Changing the SF length will also change the tension in a manner that depends on the time history of the change in l. The reason is that a perturbation in force from f_0 causes the myosin motors to move, hence changing l_{ref} as described by the following:

$$\frac{dl_{ref}}{dt} = -2v$$
(6.5)

where the factor of two arises from the lengthening of the element in both directions at a rate v that depends on the current tension (Figure 6.5c). Differentiation of Equation 6.4

and substitution of Equation 6.5 into the result give the governing equation describing the dynamic mechanical response of the element:

$$\frac{df}{dt} = \frac{k}{l_{ref}}\frac{dl}{dt} + \frac{2klv}{l_{ref}^2}$$

(6.6)

The first term on the right side of the equation represents the elastic response. The second term represents the rate of force relaxation toward f_0.

To illustrate the behavior of Equation 6.6, consider the response to a 10% step increase in length in an SF (Figure 6.6a). There is an initial instantaneous increase in tension due to the elastic response of the element. Thereafter, the first term on the right side of Equation 6.6 is zero, and the tension decreases since l_{ref} increases (i.e., $v<0$). The rate of relaxation decreases as f approaches f_0, ultimately resulting in equilibrium with a new value for l_{ref}. Returning the SF length to the original value results in a similar response, though now the tension drops below f_0 and gradually returns to f_0 as the element contracts (i.e., $v>0$) and l_{ref} returns to the original value.

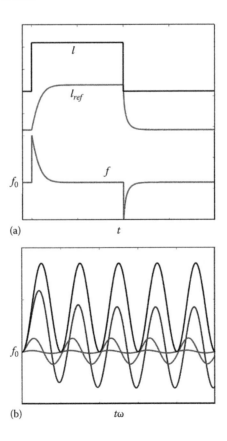

(a)

(b)

FIGURE 6.6
The response of an individual actomyosin subunit to transient and cyclic stretch patterns. (a) Transient changes in actual and reference length of an actomyosin subunit result in transient changes in tension. (b) The time course of force generation is shown for an actomyosin subunit in an SF subject to 10% sinusoidal strain at frequencies of 1–0.01 Hz. The black curve indicates the purely elastic response (i.e., v always zero). The time axis is normalized by frequency ω to facilitate comparisons.

Now, consider the response to cyclic stretching, which is less intuitive. Figure 6.6b illustrates numerical simulations of the behavior of Equation 6.6 for 10% cyclic sinusoidal stretching at frequencies ranging from 0.01 to 1 Hz. The model parameters were estimated from the literature and are listed in the original report.[77] The model predicts that the amplitude of SF tension exhibits two characteristic regimes. Above 1 Hz, the first term in Equation 6.6 is much larger than the second term, and hence, the amplitude in tension is proportional to the amplitude in SF stretch. At lower frequencies, the two terms become comparable and of opposite sign, resulting in less and less perturbation in tension from the stall force as the strain rate decreases. Specifically, the amplitude of SF tension increases with a power law as the frequency increases toward 1 Hz and then saturates to a constant value above 1 Hz. This behavior is similar to that of a simple linear viscoelastic model consisting of a dashpot in series with spring, except that the equilibrium force response is nonzero.

With the mechanical behavior of individual SFs now established, we next consider the turnover of SFs in a population. Rapid myosin sliding[10,32] and excessive SF loading[18] can each accelerate SF disassembly. To account for SF disassembly and reassembly, a stochastic approach was employed in which the probability of each individual SF disassembling over a time interval Δt depended on force and the velocity of myosin motor movement:

$$P(\text{disassembly of the } i\text{th SF}) = \alpha\exp\left(-\frac{\gamma}{|v|}\right) + \beta\exp\left(\frac{f}{f_\beta}\right) \tag{6.7}$$

where α, β, γ, and f_β are constants. The first term on the right side represents the decrease in fraction of bound myosin heads as actomyosin sliding velocity increases.[79] The second term is the well-known slip bond expression by Bell[80] describing the increase in dissociation rate of protein–protein interactions due to force. The bonds being broken can be taken to be those between myosin and actin or between integrins and their cognate receptors on the ECM. Equation 6.7 predicts that the probability of disassembly increases much faster for positive perturbations in tension than negative perturbations. Physically, this predicts that SFs are more likely to disassemble in response to rapid lengthening than rapid shortening, which was observed experimentally when subjecting cells to asymmetric cyclic stretch waveforms.[48]

6.6.4 Polymerization-Based Models

Deshpande et al.[81] developed a model of stretch-induced SF reorganization motivated by tension-dependent stimulation of actin polymerization and contraction. The model simulation begins with an imposed activation signal for stimulating contractile force in the cell (e.g., increased intracellular calcium concentration) that decays exponentially over time with a characteristic time τ_a. Subsequently, the effect of the activation signal for the level of polymerized actin oriented in the θ direction, $\eta(\theta)$, varies as a function of the stress acting in that direction, $\sigma(\theta)$:

$$\frac{d\eta(\theta)}{dt} = \left[1 - \eta(\theta)\right]\frac{\exp(-t/\tau_a)k_p}{\tau_a} - \left[1 - \frac{\sigma(\theta)}{\eta(\theta)\sigma_{\max}}\right]\eta(\theta)\frac{k_d}{\tau_a} \tag{6.8}$$

where
  $\sigma_{\max}$ represents the tensile stress exerted by an SF at maximum activation (i.e., when $\eta(\theta)$ equals one)
  k_p and k_d are dimensionless actin polymerization and depolymerization rate constants

The first term on the right represents the rate of increase in activation that depends on the decaying activation signal. The second term describes the rate of decrease in activation that depends on the magnitude of stress. The value of σ depends on the rate of shortening in the θ direction using a modified linear Hill relationship relating stress and strain rate:

$$\frac{\sigma(\theta)}{\eta(\theta)\sigma_{max}} = \begin{cases} 0 & \frac{\dot{\varepsilon}}{\dot{\varepsilon}_0} < -\frac{\eta(\theta)}{k_v} \\ 1 + \frac{k_v}{\eta(\theta)}\frac{\dot{\varepsilon}}{\dot{\varepsilon}_0} & -\frac{\eta(\theta)}{k_v} \le \frac{\dot{\varepsilon}}{\dot{\varepsilon}_0} \le 0 \\ 1 & \frac{\dot{\varepsilon}}{\dot{\varepsilon}_0} > 0 \end{cases} \tag{6.9}$$

where
$\dot{\varepsilon}_0$ is the strain rate at zero stress
k_v is a constant

The intermediate filament network is also assumed to contribute to the stress response as an isotropic, linear elastic material. The average Cauchy stress is then taken as the linear superposition of the active actin response and the passive intermediate filament response to stretch.

Starting with a cell lacking any actin filaments, the Deshpande model predicts that actin polymerizes preferentially in the direction of lowest shortening rate. In the case of cyclic pure uniaxial stretch, the distribution of filaments is greatest perpendicular to the stretch direction with some dispersion. The dispersion decreases as strain amplitude and frequency increase. Consistent with experiments,[82] a 90° change in stretching direction results in concomitant reorientation of the actin distribution.

Obbink-Huizer et al.[83] proposed a model incorporating some of the concepts from Desphande's model capable of predicting the effects of both stretch and matrix stiffness on cytoskeletal alignment. In this model, the stress in SFs depends on both strain and strain rate:

$$\sigma_\theta^p = \sigma_{max} f_\varepsilon(\varepsilon_\theta) f_{\dot{\varepsilon}}(\dot{\varepsilon}_\theta) \tag{6.10}$$

where
σ_θ^p is the stress in SFs oriented in the θ direction
σ_{max} is a constant representing the maximum SF stress

The length–tension relationship $f_\varepsilon(\varepsilon_\theta)$, taken from the model of Vernerey and Farsad,[84] consists of active ($f_{\varepsilon,a}$) and passive ($f_{\varepsilon,p}$) components:

$$f_\varepsilon(\varepsilon_\theta) = f_{\varepsilon,a} + f_{\varepsilon,p} = \exp\left[-\left(\frac{\varepsilon_\theta}{\varepsilon_0}\right)^2\right] + \left(\frac{\varepsilon_\theta}{\varepsilon_1}\right)^2 \tag{6.11}$$

where ε_0 and ε_1 are scaling factors for the active and passive terms, respectively. Further, the passive term is taken to be zero for compressive strains, that is, when $\varepsilon_\theta < 0$, since SFs buckle under any compression.[17] The strain rate-dependent component $f_{\dot{\varepsilon}}(\dot{\varepsilon}_\theta)$, motivated by Hill's model, is a smooth function also taken from the model of Vernerey and Farsad[84]:

$$f_{\dot{\varepsilon}}(\dot{\varepsilon}_\theta) = \frac{1}{1+\left(2/\sqrt{5}\right)}\left[1+\frac{2+k_v\dot{\varepsilon}_\theta}{\sqrt{1+\left(2+k_v\dot{\varepsilon}_\theta\right)^2}}\right] \tag{6.12}$$

where k_v is a scaling factor describing the sensitivity to shortening rate. Thus, $f_{\dot{\varepsilon}}(\dot{\varepsilon}_\theta)$ transitions from zero at high shortening rates to a constant at high lengthening rates with an intermediate value of unity when shortening rate is zero.

The model accounts for conservation of total actin mass (φ^{tot}):

$$\varphi^{tot} = \varphi^m + \frac{1}{N}\sum_\theta \varphi_\theta^p \tag{6.13}$$

where

   N is the total number of directions for SFs to orient
   φ^m and φ_θ^p represent the mass volume fractions of monomeric actin and polymerized actin in the θ direction, respectively

Similar to the models of Deshpande and of Vernerey and Farsad,[84] the rate of actin polymerization depends on stress, except only the active component of stress affects SF assembly:

$$\frac{d\varphi_\theta^p}{dt} = \left[k_0^f + k_1^f\sigma_{max}f_{\varepsilon,a}(\varepsilon_\theta)f_{\dot{\varepsilon}}(\dot{\varepsilon}_\theta)\right]\varphi^m - k_d\varphi_\theta^p \tag{6.14}$$

where k_0^f, k_1^f, and k_d are rate constants describing the basal rate of assembly, stress-dependent assembly, and disassembly of SFs, respectively. Unlike the model by Deshpande et al., actin polymerization does not require an activation signal to initiate (cf. Equation 6.8).

The cell stress is expressed as the summed contributions of SFs in all directions attached to a substrate modeled as an isotropic Neo-Hookean material that can be stretched. The model is capable of several interesting predictions. On a material with anisotropic stiffness, SFs are predicted to align parallel to the direction of greatest stiffness, but only when the stiffness is within a range that cells can sense (<100 kPa). These results are consistent with experimental observations.[85] On stiff substrates, the model predicts that increasing stretch frequency and amplitude each increases the extent of SF alignment perpendicular to the direction of cyclic uniaxial stretch. On a soft substrate with stiffness comparable to that of cells, the model predicts SF alignment parallel to the direction of stretch along with obvious lateral constriction of the cell-tissue construct since the tissue Poisson's ratio was taken to be 0.4. These results are consistent with experimental observations reported by the Baaijens[83] and Kaunas[71] groups.

6.6.5 Models Motivated by Focal Adhesion Dynamics

Kong et al.[86] modeled the behavior of populations of SFs in a cell and their associated FAs in response to cyclic stretch. The probability of the dissociation of the ith FA was taken to be proportional to the deviation of the force F_i acting on the FA relative to an optimal value F_0:

$$P(\text{disassembly of the } i\text{th FA}) = \begin{cases} k_0 + k_1 \left(\dfrac{F_i - F_0}{F_0} \right)^2 & F_i < F_0 \\[3mm] k_0 + k_2 \left(\dfrac{F_i - F_0}{F_0} \right)^2 & F_i \geq F_0 \end{cases} \tag{6.15}$$

where k_0, k_1, and k_2 are rate constants. The difference in behaviors at low and high forces is based on the authors' previous study.[87] When $F_i F_0$, the integrins cannot be activated, resulting in small, unstable FAs. When $F_i \geq F_0$, disassembly is driven by bond rupture. The force acting on the FA depends on the viscoelastic response of SFs to stretch:

$$F = EA \left(\frac{l}{l_{\text{ref}}} - 1 \right) + \mu \frac{d\Delta l}{dt} \tag{6.16}$$

where
E and A are the elastic modulus and cross-sectional area of the SF
μ is the effective viscosity
Δl equals the current length of the SF (l) minus the reference length (l_{ref}) of the SF

In addition to FA disassembly, the SFs disassemble at a rate v based on the following expression:

$$v = v_{\max} \left[1 - \exp \left(-\frac{|F_i - F_0|}{F^*} \right) \right] \tag{6.17}$$

where
$v_{\max}$ is the maximum rate
F^* is a scaling factor

Equations 6.15 and 6.17 resemble the expressions used previously by the Kaunas group to model stretch-induced SF alignment[46,50] and similarly predict frequency-dependent alignment of SFs perpendicular to the direction of stretch. Consistent with experimental observations,[47] the Kong model also predicts that SF reorientation time exhibits two characteristic regimes. For frequencies below 1 Hz, the time decreases with a power law as the frequency increases, while it saturates at a constant value for frequencies above 1 Hz.

Gao et al.[51] developed a model inspired by the observation that cyclic stretch-induced SF alignment involves FA sliding and SF rotation, rather than SF disassembly and reassembly.[51,52] Similar to the model used by Kaunas et al.,[77] a sarcomere unit in an SF consists of a spring in series with a contractile element (cf. Figure 6.4b). Periodic stretching of the sarcomere with an initial length l_0 (i.e., $l = l_0 + \Delta l/2(1 - \cos\omega t)$) produces a force variation in the sarcomere described by the following governing equation:

$$\frac{df}{dt} = kv + k\frac{dl}{dt} = kv + \frac{k\Delta l}{2}\sin\omega t \tag{6.18}$$

where
v_0 is the no-load shortening velocity
f_0 is the stall force (cf. Figure 6.4c)

This is essentially the same form as Equation 6.6 and predicts that the amplitude of SF tension increases with a power law as the frequency increases toward a characteristic value $\omega_h = kv_0/f_0$ and then saturates to a constant value at higher frequencies.

Motivated by the observation that contracting SFs are seen to shorten near the periphery of the cell and lengthen in the center,[20] the model assumes that at low stretch frequencies, a single sarcomere located at the cell periphery is stretched, in which case the spring constant is reduced by a factor N equal to the number of sarcomere units in the SF ($k_l = k/N$). Now the characteristic frequency decreases to $\omega_l = k_l v_0/f_0 = \omega_h/N$ representing the threshold frequency below which there is no stretch-induced cell reorientation.

Rather than complete SF turnover, the SFs are assumed to reorient through force-induced breaking of cell–matrix adhesions at a rate given by

$$k_t = k_1 \left(\frac{f_t - f_0}{f_b} \right)^2 + k_m \tag{6.19}$$

where
  f_t is the force acting on an individual bond
  f_t is the optimal level of force
  k_m is a basal rate constant
  k_1 and f_b are scaling factors

Thus, the bond acts as a catch bond for forces lower than f_0 and transition into a slip bond when force exceeds f_0. This equation predicts that adhesions associated with SFs oriented in the direction of least amplitude in stretching are most stable. This is consistent with predictions from SF-based models that employ a dissociation kinetics expression similar to Equation 6.19.[46,50,88] Instead of assuming disassembly and reassembly of SFs caused by bond dissociation, however, the authors postulated that SFs with less stable adhesions slide to result in SF rotation toward orientations that result in more stable adhesions. The rate of fiber rotation was expressed as the angular variation in the amplitude of force acting on the adhesions and predicts that the rotation rate decreases with decreasing strain rate and increasing SF length.

To describe SF reorientation in experiments by Tondon et al. applying asymmetric stretch waveforms,[48] a modified Hill model was employed that assumed the slope (df/dv) of the curve for $f > f_0$ is 10-fold smaller than when $f_0 > f > 0$.[89] The relatively slow response of the contractile element during the lengthening part of a stretch waveform results in greater force generation and subsequently greater bond breaking in the cell–matrix adhesions.

6.7 Conclusions

The mounting data quantifying cytoskeletal reorganization in cells subjected to applied stresses and strains has provided the information needed to generate and test mathematical models of the process of reorganization. It is now evident that stretch-induced cytoskeletal remodeling is complex, involving force-induced conformational changes of proteins within SFs, cell adhesions, and likely within other intracellular structures. For example, cytoskeletal connections to the nucleus via nesprins have recently been demonstrated to be important for transmitting forces and for cyclic stretch-induced cell alignment.[90]

Comparisons of the models used to describe stretch-induced cytoskeletal reorganization illustrate that although they are often motivated by different phenomena (e.g., actomyosin sliding vs. integrin–matrix bond detachment), the equations used to describe the behavior are not dissimilar (e.g., Equations 6.6 and 6.18). In general, these models are all based on the premise that cytoskeletal structures are destabilized when perturbed by applied forces. In some cases, such as when cells treated with inhibitors of cell contractility[39] or adhering to soft substrates,[71] applied forces appear to stabilize these structures. Consequently, these models need to be extended to address these new observations.

The cellular response to stretch depends on not only the magnitude, rate, and waveform of stretching but also cell type, substrate composition, architecture, and mechanical properties. High-throughput systems will facilitate future studies to elucidate the interactive effects of these different factors on cell responses to stretch.[91] Most studies have employed simple scaffolds, usually elastomeric sheets. As described in the next chapter, cells stretched in 3D collagen gels exhibit complicated behavior that provides unique challenges to modeling.[92] To understand how cells respond to stretch in a physiological environment, careful studies are still needed to elucidate how macroscopically applied forces applied to heterogeneous materials such as 3D tissues and engineered scaffolds are transmitted into local stresses and strains within the cells' extracellular niche.

References

1. Zollner, A. M., Holland, M. A., Honda, K. S., Gosain, A. K., and Kuhl, E. Growth on demand: Reviewing the mechanobiology of stretched skin. *J Mech Behav Biomed Mater* **28**, 495–509 (2013).
2. Harris, A. K., Wild, P., and Stopak, D. Silicone rubber substrata: A new wrinkle in the study of cell locomotion. *Science* **208**, 177–179 (1980).
3. Dembo, M. and Wang, Y. L. Stresses at the cell-to-substrate interface during locomotion of fibroblasts. *Biophys J* **76**, 2307–2316 (1999).
4. Tan, J. L. et al. Cells lying on a bed of microneedles: An approach to isolate mechanical force. *Proc Natl Acad Sci USA* **100**, 1484–1489 (2003).
5. Tomasek, J. J., Gabbiani, G., Hinz, B., Chaponnier, C., and Brown, R. A. Myofibroblasts and mechano-regulation of connective tissue remodelling. *Nat Rev Mol Cell Biol* **3**, 349–363 (2002).
6. Wong, A. J., Pollard, T. D., and Herman, I. M. Actin filament stress fibers in vascular endothelial cells in vivo. *Science* **219**, 867–869 (1983).
7. White, G. E., Gimbrone, M. A., Jr., and Fujiwara, K. Factors influencing the expression of stress fibers in vascular endothelial cells in situ. *J Cell Biol* **97**, 416–424 (1983).
8. Cramer, L. P., Siebert, M., and Mitchison, T. J. Identification of novel graded polarity actin filament bundles in locomoting heart fibroblasts: Implications for the generation of motile force. *J Cell Biol* **136**, 1287–1305 (1997).
9. Lazarides, E. and Burridge, K. Alpha-actinin: Immunofluorescent localization of a muscle structural protein in nonmuscle cells. *Cell* **6**, 289–298 (1975).
10. Matsui, T. S., Kaunas, R., Kanzaki, M., Sato, M., and Deguchi, S. Non-muscle myosin II induces disassembly of actin stress fibres independently of myosin light chain dephosphorylation. *Interf Focus* **1**, 754–766 (2011).
11. Langanger, G. et al. The molecular organization of myosin in stress fibers of cultured cells. *J Cell Biol* **102**, 200–209 (1986).
12. Russell, R. J., Xia, S. L., Dickinson, R. B., and Lele, T. P. Sarcomere mechanics in capillary endothelial cells. *Biophys J* **97**, 1578–1585 (2009).

13. Katoh, K., Kano, Y., and Ookawara, S. Rho-kinase dependent organization of stress fibers and focal adhesions in cultured fibroblasts. *Genes Cells* **12**, 623–638 (2007).
14. Totsukawa, G. et al. Distinct roles of ROCK (Rho-kinase) and MLCK in spatial regulation of MLC phosphorylation for assembly of stress fibers and focal adhesions in 3T3 fibroblasts. *J Cell Biol* **150**, 797–806 (2000).
15. Lee, C. F., Haase, C., Deguchi, S., and Kaunas, R. Cyclic stretch-induced stress fiber dynamics – dependence on strain rate, Rho-kinase and MLCK. *Biochem Biophy Res Commun* **401**, 344–349 (2010).
16. Deguchi, S., Ohashi, T., and Sato, M. Tensile properties of single stress fibers isolated from cultured vascular smooth muscle cells. *J Biomech* **39**, 2603–2610 (2006).
17. Costa, K. D., Hucker, W. J., and Yin, F. C. Buckling of actin stress fibers: A new wrinkle in the cytoskeletal tapestry. *Cell Motil Cytoskeleton* **52**, 266–274 (2002).
18. Lu, L. et al. Actin stress fiber pre-extension in human aortic endothelial cells. *Cell Motil Cytoskeleton* **65**, 281–294 (2008).
19. Kumar, S. et al. Viscoelastic retraction of single living stress fibers and its impact on cell shape, cytoskeletal organization, and extracellular matrix mechanics. *Biophys J* **90**, 3762–3773 (2006).
20. Peterson, L. J. et al. Simultaneous stretching and contraction of stress fibers in vivo. *Mol Biol Cell* **15**, 3497–3508 (2004).
21. Colombelli, J. et al. Mechanosensing in actin stress fibers revealed by a close correlation between force and protein localization. *J Cell Sci* **122**, 1665–1679 (2009).
22. Stachowiak, M. R. and O'Shaughnessy, B. Recoil after severing reveals stress fiber contraction mechanisms. *Biophys J* **97**, 462–471 (2009).
23. Russell, R. J. et al. Sarcomere length fluctuations and flow in capillary endothelial cells. *Cytoskeleton (Hoboken)* **68**, 150–156 (2011).
24. Ridley, A. J. and Hall, A. The small GTP-binding protein rho regulates the assembly of focal adhesions and actin stress fibers in response to growth factors. *Cell* **70**, 389–399 (1992).
25. Chrzanowska-Wodnicka, M. and Burridge, K. Rho-stimulated contractility drives the formation of stress fibers and focal adhesions. *J Cell Biol* **133**, 1403–1415 (1996).
26. Machesky, L. M. and Hall, A. Role of actin polymerization and adhesion to extracellular matrix in Rac- and Rho-induced cytoskeletal reorganization. *J Cell Biol* **138**, 913–926 (1997).
27. Hirata, H., Tatsumi, H., and Sokabe, M. Dynamics of actin filaments during tension-dependent formation of actin bundles. *Biochim Biophys Acta* **1770**, 1115–1127 (2007).
28. Oakes, P. W., Beckham, Y., Stricker, J., and Gardel, M. L. Tension is required but not sufficient for focal adhesion maturation without a stress fiber template. *Cell Biol* **196**, 363–374 (2012).
29. Smith, M. A. et al. A zyxin-mediated mechanism for actin stress fiber maintenance and repair. *Dev Cell* **19**, 365–376 (2010).
30. Lamb, N. J. et al. Regulation of actin microfilament integrity in living nonmuscle cells by the cAMP-dependent protein kinase and the myosin light chain kinase. *J Cell Biol* **106**, 1955–1971 (1988).
31. Lang, P. et al. Protein kinase A phosphorylation of RhoA mediates the morphological and functional effects of cyclic AMP in cytotoxic lymphocytes. *EMBO J* **15**, 510–519 (1996).
32. Matsui, T. S., Ito, K., Kaunas, R., Sato, M., and Deguchi, S. Actin stress fibers are at a tipping point between conventional shortening and rapid disassembly at physiological levels of MgATP. *Biochem Biophy Res Commun* **395**, 301–306 (2010).
33. Boateng, S. Y. et al. RGD and YIGSR synthetic peptides facilitate cellular adhesion identical to that of laminin and fibronectin but alter the physiology of neonatal cardiac myocytes. *Am J Physiol Cell Physiol* **288**, C30–C38 (2005).
34. Banes, A. J., Gilbert, J., Taylor, D., and Monbureau, O. A new vacuum-operated stress-providing instrument that applies static or variable duration cyclic tension or compression to cells in vitro. *J Cell Sci* **75**, 35–42 (1985).
35. Na, S. et al. Time-dependent changes in smooth muscle cell stiffness and focal adhesion area in response to cyclic equibiaxial stretch. *Ann Biomed Eng* **36**, 369–380 (2008).

36. Schaffer, J. L. et al. Device for the application of a dynamic biaxially uniform and isotropic strain to a flexible cell culture membrane. *J Orthop Res* **12**, 709–719 (1994).
37. Hung, C. T. and Williams, J. L. A method for inducing equi-biaxial and uniform strains in elastomeric membranes used as cell substrates. *J Biomech* **27**, 227–232 (1994).
38. Sotoudeh, M., Jalali, S., Usami, S., Shyy, J. Y., and Chien, S. A strain device imposing dynamic and uniform equi-biaxial strain to cultured cells. *Ann Biomed Eng* **26**, 181–189 (1998).
39. Kaunas, R., Nguyen, P., Usami, S., and Chien, S. Cooperative effects of Rho and mechanical stretch on stress fiber organization. *Proc Natl Acad Sci USA* **102**, 15895–15900 (2005).
40. Huang, L., Mathieu, P. S., and Helmke, B. P. A stretching device for high-resolution live-cell imaging. *Ann Biomed Eng* **38**, 1728–1740 (2010).
41. Buck, R. C. Reorientation response of cells to repeated stretch and recoil of the substratum. *Exp Cell Res* **127**, 470–474 (1980).
42. Wang, J. H., Goldschmidt-Clermont, P., Wille, J., and Yin, F. C. Specificity of endothelial cell reorientation in response to cyclic mechanical stretching. *J Biomech* **34**, 1563–1572 (2001).
43. Naruse, K., Yamada, T., and Sokabe, M. Involvement of SA channels in orienting response of cultured endothelial cells to cyclic stretch. *Am J Physiol* **274**, H1532–H1538 (1998).
44. Morioka, M. et al. Microtubule dynamics regulate cyclic stretch-induced cell alignment in human airway smooth muscle cells. *PLoS One* **6**, e26384 (2011).
45. Faust, U. et al. Cyclic stress at mHz frequencies aligns fibroblasts in direction of zero strain. *PLoS One* **6**, e28963 (2011).
46. Hsu, H. J., Lee, C. F., and Kaunas, R. A dynamic stochastic model of frequency-dependent stress fiber alignment induced by cyclic stretch. *PLoS One* **4**, e4853 (2009).
47. Jungbauer, S., Gao, H., Spatz, J. P., and Kemkemer, R. Two characteristic regimes in frequency-dependent dynamic reorientation of fibroblasts on cyclically stretched substrates. *Biophys J* **95**, 3470–3478 (2008).
48. Tondon, A., Hsu, H. J., and Kaunas, R. Dependence of cyclic stretch-induced stress fiber reorientation on stretch waveform. *J Biomech* **45**, 728–735 (2012).
49. Katsumi, A. et al. Effects of cell tension on the small GTPase Rac. *J Cell Biol* **158**, 153–164 (2002).
50. Hsu, H. J., Lee, C. F., Locke, A., Vanderzyl, S. Q., and Kaunas, R. Stretch-induced stress fiber remodeling and the activations of JNK and ERK depend on mechanical strain rate, but not FAK. *PLoS One* **5**, e12470 (2010).
51. Chen, B., Kemkemer, R., Deibler, M., Spatz, J., and Gao, H. Cyclic stretch induces cell reorientation on substrates by destabilizing catch bonds in focal adhesions. *PLoS One* **7**, e48346 (2012).
52. Goldyn, A. M., Rioja, B. A., Spatz, J. P., Ballestrem, C., and Kemkemer, R. Force-induced cell polarisation is linked to RhoA-driven microtubule-independent focal-adhesion sliding. *J Cell Sci* **122**, 3644–3651 (2009).
53. Cukierman, E., Pankov, R., Stevens, D. R., and Yamada, K. M. Taking cell-matrix adhesions to the third dimension. *Science* **294**, 1708–1712 (2001).
54. Baker, B. M. and Chen, C. S. Deconstructing the third dimension: How 3D culture microenvironments alter cellular cues. *J Cell Sci* **125**, 3015–3024 (2012).
55. Nieponice, A., Maul, T. M., Cumer, J. M., Soletti, L., and Vorp, D. A. Mechanical stimulation induces morphological and phenotypic changes in bone marrow-derived progenitor cells within a three-dimensional fibrin matrix. *J Biomed Mater Res A* **81**, 523–530 (2007).
56. Nekouzadeh, A., Pryse, K. M., Elson, E. L., and Genin, G. M. Stretch-activated force shedding, force recovery, and cytoskeletal remodeling in contractile fibroblasts. *J Biomech* **41**, 2964–2971 (2008).
57. Hu, J. J., Humphrey, J. D., and Yeh, A. T. Characterization of engineered tissue development under biaxial stretch using nonlinear optical microscopy. *Tissue Eng Part A* **15**, 1553–1564 (2009).
58. Henshaw, D. R., Attia, E., Bhargava, M., and Hannafin, J. A. Canine ACL fibroblast integrin expression and cell alignment in response to cyclic tensile strain in three-dimensional collagen gels. *J Orthop Res* **24**, 481–490 (2006).

59. Brown, R. A., Prajapati, R., McGrouther, D. A., Yannas, I. V., and Eastwood, M. Tensional homeostasis in dermal fibroblasts: Mechanical responses to mechanical loading in three-dimensional substrates. *J Cell Physiol* **175**, 323–332 (1998).

60. Zahalak, G. I., Wagenseil, J. E., Wakatsuki, T., and Elson, E. L. A cell-based constitutive relation for bio-artificial tissues. *Biophys J* **79**, 2369–2381 (2000).

61. Wang, N. et al. Cell prestress. I. Stiffness and prestress are closely associated in adherent contractile cells. *Am J Physiol Cell Physiol* **282**, C606–C616 (2002).

62. Reinhart-King, C. A. Endothelial cell adhesion and migration. *Methods Enzymol* **443**, 45–64 (2008).

63. Reinhart-King, C. A., Dembo, M., and Hammer, D. A. Cell-cell mechanical communication through compliant substrates. *Biophys J* **95**, 6044–6051 (2008).

64. Engler, A. J., Sen, S., Sweeney, H. L., and Discher, D. E. Matrix elasticity directs stem cell lineage specification. *Cell* **126**, 677–689 (2006).

65. Trepat, X. et al. Universal physical responses to stretch in the living cell. *Nature* **447**, 592–595 (2007).

66. Mason, T. G. and Weitz, D. A. Linear viscoelasticity of colloidal hard sphere suspensions near the glass transition. *Phys Rev Lett* **75**, 2770–2773 (1995).

67. Sollich, P., Lequeux, F., Hebraud, P., and Cates, M. E. Rheology of soft glassy materials. *Phys Rev Lett* **78**, 2020–2023 (1997).

68. Krishnan, R. et al. Fluidization, resolidification, and reorientation of the endothelial cell in response to slow tidal stretches. *Am J Physiol Cell Physiol* **303**, C368–C375 (2012).

69. Chen, C. et al. Fluidization and resolidification of the human bladder smooth muscle cell in response to transient stretch. *PLoS One* **5**, e12035 (2010).

70. Throm Quinlan, A. M., Sierad, L. N., Capulli, A. K., Firstenberg, L. E., and Billiar, K. L. Combining dynamic stretch and tunable stiffness to probe cell mechanobiology in vitro. *PLoS One* **6**, e23272 (2011).

71. Tondon, A. and Kaunas, R. The direction of stretch-induced cell and stress fiber orientation depends on collagen matrix stress. *PLoS One* **9**, e89592 (2014).

72. Gavara, N., Roca-Cusachs, P., Sunyer, R., Farre, R., and Navajas, D. Mapping cell-matrix stresses during stretch reveals inelastic reorganization of the cytoskeleton. *Biophys J* **95**, 464–471 (2008).

73. De, R., Zemel, A., and Safran, S. A. Dynamics of cell orientation. *Nat Phys* **3**, 655–659 (2007).

74. De, R., Zemel, A., and Safran, S. A. Do cells sense stress or strain? Measurement of cellular orientation can provide a clue. *Biophys J* **94**, L29–L31 (2008).

75. Krishnan, R. et al. Reinforcement versus fluidization in cytoskeletal mechanoresponsiveness. *PLoS One* **4**, e5486 (2009).

76. Pirentis, A. P., Peruski, E., Iordan, A. L., and Stamenovic, D. A model for stress fiber realignment caused by cytoskeletal fluidization during cyclic stretching. *Cell Mol Bioeng* **4**, 67–80 (2011).

77. Kaunas, R., Hsu, H. J., and Deguchi, S. Sarcomeric model of stretch-induced stress fiber reorganization. *Cell Health Cytoskelet* **3**, 13–22 (2011).

78. Hill, A. V. The heat of shortening and the dynamic constants of muscle. *Proc R Soc Lond B Biol Sci* **126**, 136–195 (1938).

79. Howard, J. Molecular motors: Structural adaptations to cellular functions. *Nature* **389**, 561–567 (1997).

80. Bell, G. I. Models for the specific adhesion of cells to cells. *Science* **200**, 618–627 (1978).

81. Wei, Z., Deshpande, V. S., McMeeking, R. M., and Evans, A. G. Analysis and interpretation of stress fiber organization in cells subject to cyclic stretch. *J Biomech Eng* **130**, 031009 (2008).

82. Kaunas, R., Usami, S., and Chien, S. Regulation of stretch-induced JNK activation by stress fiber orientation. *Cell Signal* **18**, 1924–1931 (2006).

83. Obbink-Huizer, C. et al. Computational model predicts cell orientation in response to a range of mechanical stimuli. *Biomech Model Mechanobiol* **13**, 227–236 (2013).

84. Vernerey, F. J. and Farsad, M. A constrained mixture approach to mechano-sensing and force generation in contractile cells. *J Mech Behav Biomed Mater* **4**, 1683–1699 (2011).

85. Saez, A., Ghibaudo, M., Buguin, A., Silberzan, P., and Ladoux, B. Rigidity-driven growth and migration of epithelial cells on microstructured anisotropic substrates. *Proc Natl Acad Sci USA* **104**, 8281–8286 (2007).

86. Zhong, Y., Kong, D., Dai, L., and Ji, B. Frequency-Dependent Focal Adhesion Instability and Cell Reorientation Under Cyclic Substrate Stretching. *Cell Mol Bioeng* **4**, 442–456 (2011).

87. Kong, D., Ji, B., and Dai, L. Stabilizing to disruptive transition of focal adhesion response to mechanical forces. *J Biomech* **43**, 2524–2529 (2010).

88. Kaunas, R. and Hsu, H. J. A kinematic model of stretch-induced stress fiber turnover and reorientation. *J Theor Biol* **257**, 320–330 (2009).

89. Chen, B. and Gao, H. Motor force homeostasis in skeletal muscle contraction. *Biophys J* **101**, 396–403 (2011).

90. Chancellor, T. J., Lee, J., Thodeti, C. K., and Lele, T. Actomyosin tension exerted on the nucleus through nesprin-1 connections influences endothelial cell adhesion, migration, and cyclic strain-induced reorientation. *Biophys J* **99**, 115–123 (2010).

91. Simmons, C. S., Ribeiro, A. J., and Pruitt, B. L. Formation of composite polyacrylamide and silicone substrates for independent control of stiffness and strain. *Lab Chip* **13**, 646–649 (2013).

92. Lee, S. L. et al. Physically-induced cytoskeleton remodeling of cells in three-dimensional culture. *PLoS One* **7**, e45512 (2012).

93. Foolen, J., Deshpande, V. S., Kanter, F. M. W., and Baaijens, F. P. T. The influence of matrix integrity on stress-fiber remodeling in 3D. *Biomaterials* **33**, 7508–7518 (2012).

7

Mechanics of Cell-Seeded ECM Scaffolds

Guy M. Genin and Elliot L. Elson

CONTENTS

7.1 Introduction

Among their many other roles, cells in solid tissues function to stabilize or actuate tissues mechanically. In organs such as the heart, kidney, and lungs, cells such as smooth muscle cells and fibroblasts function in part to develop and define shape and structure. Cells in these tissues also serve dynamic functions such as muscle contraction and wound healing. Understanding many of the important functions of these tissues therefore requires an understanding of the mechanical properties and functioning of the cells that comprise them.

Much early work on cell mechanics derived from assays performed on cells that were cultured on 2D substrata. However, a defining feature of each of the cellular functions listed earlier is that they occur while the cell is situated in a 3D environment. For many cell functions and indeed cell types, a 2D substratum provides an unrealistic environment, and increasing evidence indicates that cells in 2D cultures differ substantially in structure and function from cells in a natural tissue (e.g., Abbott 2003; Lee et al. 2012), with substantial differences evident in signal transduction, migration, growth, development, and, as is the focus of this chapter, mechanical interaction with extracellular matrix (ECM) (Cukierman et al. 2002; Nelson and Bissell 2005; Paszek et al. 2005; Vidi et al. 2013). Since the mechanical environment of cells dictates many features of development, the absence of out-of-plane stiffness and the rigidity of the substratum may be important factors in these differences (Stegemann et al. 2005; Asnes et al. 2006). Cells exhibit much greater stiffness when stressed through their natural structural networks in a 3D tissue construct than when probed in a 2D culture (Marquez et al. 2005a,b). Even the subcellular structures that are believed to define the mechanics of cells and the interactions of cells with their mechanical environment appear to differ between 2D and 3D culture. One prominent example is the debate about the role, morphology, and even existence of stress fibers and focal adhesions in 2D and 3D (e.g., Cukierman et al. 2001; Grinnell 2003; Fraley et al. 2010).

Testing of mechanical responses of living cells in natural tissues would then seem to be the logical course of action. However, the heterogeneity and structure of these tissues makes testing of specific cellular responses difficult. Additionally, in adult tissues, cellular mechanical responses are often small in magnitude compared to the passive responses of the ECM proteins they synthesize and maintain. Tissue constructs with specifiable cell and ECM constituents provide model systems in which responses of cells can be better resolved, can be better defined, and are more flexible, less expensive, and simpler than natural tissues. Such tissue constructs have therefore proven to be useful models for studying cellular mechanics and function (Wakatsuki et al. 2000; Zahalak et al. 2000; Legant et al. 2009; Lee et al. 2012; Lee and Vandenburgh 2013). Tissue constructs provide important advantages such as the ability to construct a tissue model containing a single cell type or a defined mixture of cell types with specified cell density and ECM composition. Hence, they could be designed to behave as models for disease conditions in which normal cell balance is lost.

Although these tissue constructs overcome basic challenges associated with the dimensionality of the cellular environment, they introduce a host of new challenges. A central challenge is that cellular behaviors must be deduced from ensemble behaviors of the entire tissue construct. Overcoming this challenge requires models of cells, ECM, and their interactions, along with tightly integrated experiments that enable identification of model parameters. These models and their application to some key questions in cellular biomechanics are the focus of this chapter.

The chapter begins with a brief perspective on founding, early work in the mechanics of living cells conducted on 2D substrata, and then progresses to describe a number of experimental systems that have been developed to probe cell mechanics in 3D. The chapter then describes mathematical models that have been developed to deduce mechanical responses of cells from some of these experimental systems. The chapter concludes with three examples of cellular behaviors deduced from tissue constructs, drawn from our own work. The first is an estimate of the mechanical traction that a fibroblast cells exert on a remodeled ECM scaffold. An important question is whether this varies with straining of a cell in 3D, and careful analysis enables the active cellular contraction to be separated from passive effects such as viscoelastic relaxation. The second is a characterization of the way that contractile fibroblast cells appear to modulate both their effective stiffness and that of the ECM in which they are embedded. This appears to be regulated by the volume fraction of cells within a tissue construct; contractile fibroblast cells and their ECM have matched effective relaxed moduli of approximately 10–20 kPa when the volume fraction of cells within the tissue construct is that associated with the percolation threshold. Additionally, cells appear to proliferate or die off to reach this volume fraction. The third observation is one of mechanical tractions resolved to the level of individual cells. All cells studied apply high tractions to the ECM over only a fraction of their bodies, with a magnitude of stressing that varies with cell type and morphology.

7.2 A Very Brief Survey of Cell Mechanics in 2D

The foundations of our understanding of cellular mechanics in 3D culture systems are embedded in both early and ongoing efforts to study cells cultured upon 2D substrata.

We briefly describe here some of the early, founding efforts and, for details, refer the reader to an early review of experimental methods and theoretical approaches that have been used to understand these founding efforts in terms of cell structures (Elson 1988).

One of the most important early protocols for measuring cell mechanics is the Cole experiment involving compression of living cells between parallel plates (Cole 1932). The response to stretching or compression of a single cell held between two plates continues to provide important data about cellular viscoelasticity over timescale ranges pertinent to elastic and contractile responses (Thoumine et al. 1995). A similar approach was used to quantify for the first time the putative contributions of collagen, titin, microtubules, and intermediate filaments to the passive tension of individual cardiac muscle cells (Granzier and Irving 1995). A strength of this approach is that unlike the 3D tissue construct approaches that are the focus of this chapter, the results of this protocol do not need to be interpreted through complicated models to identify mechanical contributions of cells. However, a weakness is that the cells are loaded in an unnatural environment.

The use of micropipettes, which dates to the 1950s (Mitchison and Swann 1954), is a step more complicated to interpret. Micropipette aspiration has been extensively used to study circulating cells such as erythrocytes and leukocytes (e.g., Evans and Skalak 1980; Skalak et al. 1984; Elgsaeter et al. 1986; Evans and Yeung 1989). These studies explained how erythrocytes with diameters of 7–8 µm could squeeze through capillaries with much smaller diameters and demonstrated that resistance of an erythrocyte to shear and bending was weak compared to resistance to membrane area expansion. A strength of these approaches is that the aspiration force transduced in the experiment is one transmitted directly to the cell. However, interpreting aspiration forces in terms of cytoskeletal and membrane responses remains an ongoing challenge (Chen et al. 2009; Shao 2009).

An additional class of techniques to study cells cultured upon 2D substrata involves poking cells using an atomic force microscope (AFM). The very first AFM-like instrument of which we are aware was in fact invented specifically for this purpose (Petersen et al. 1982; Zahalak et al. 1990). This device was also used to measure the area expansion modulus of erythrocytes (Daily et al. 1984) and to study the biophysical roles of subcellular mechanical proteins such as the role of myosin II in the capping process in lymphocytes and *Dictyostelium* amoebae (Pasternak and Elson 1985; Pasternak et al. 1989). The device was also used to establish the contribution of cell stiffening to the retention of neutrophils during acute inflammatory processes (Worthen et al. 1989; Downey et al. 1990, 1991). AFM continues to be used to study the mechanical properties of adherent cells in a variety of contexts (e.g., Radmacher 2002). Like micropipette aspiration, the force transduced in such experiments arises directly from contact with a cell. However, again like micropipette aspiration, the mode of loading of a cell by AFM is highly unnatural and must be interpreted through intricate mechanical models (Daily et al. 1984; Weafer et al. 2013).

The past 15 years have witnessed an explosion of techniques for studying cells on 2D substrata, including deformable substrata, micropost arrays, and traction microscopy systems involving tracking of fluorescent beads embedded within or upon a substrate. These all work off of the basic principle that mechanical forces exerted by cells can be deduced from deformations of a cell's mechanical environment. A notable early application of this to nonmuscle cells is an experiment that showed the ability of fibroblasts and other types of cells to exert contractile force through active wrinkling of a thin, soft silicone rubber layer (Harris et al. 1980). Much work has emerged from this study, some of which is summarized in Chapter 10 of this book.

Allowing cells to attach to an array of flexible microposts enables the determination not only of the dependence of cell shape on the spatial arrangement of the posts but also of the force exerted by the cells measured in terms of the bending of the posts (Tan et al. 2003). Cells display a distribution of SFs that is determined by the pattern of the posts and exert forces on the posts that are greater at the cell periphery than in the interior (McGarry et al. 2009). Much progress has been made recently on extending these technologies to incorporate additional loading and topographical schemes (e.g., Chen et al. 2010; le Digabel et al. 2010; Sniadecki 2010; Han et al. 2012; Lam et al. 2012a,b; Mann et al. 2012; Polacheck et al. 2013). Studies of cells on flexible substrata and associated models have elucidated the mechanisms by which endothelial cells align in response to pulsatile stretch, and uncovered that stress fibers gradually align in a direction of minimal substrate deformation (Kaunas et al. 2005, 2006; Hsu et al. 2009, 2010; Kaunas and Hsu 2009). This work and the body of research it has founded are the focus of Chapter 6. Studies of cells on substrates containing fluorescent beads have uncovered a range of dynamic responses of single cells and cell sheets (Dembo et al. 1996; Dembo and Wang 1999; Butler et al. 2002; Bursac et al. 2005; Trepat et al. 2009; Chen et al. 2010), including several that mimic results observed in 3D (Nekouzadeh et al. 2008). A strength of this entire family of approaches is that the models required to infer cell forces can be quite straightforward and accurate: such models rely on the often linear elastic mechanics of the substratum, which can be known to great accuracy. As a consequence, very intricate predictive models of cell behavior in 2D have been fit to the results of tests performed on flexible and structured 2D substrata (Deshpande et al. 2006, 2007, 2008; Pathak et al. 2008; Wei et al. 2008; McGarry et al. 2009; Ronan et al. 2012). However, a weakness is that for many types of cells, a 2D substratum is a poor mimic of a solid tissue.

These measurements on individual cells in 2D have provided valuable information about the mechanical properties of both circulating cells and isolated tissue cells. However, the link between cellular mechanics and environment cannot be fully captured on a 2D substratum. Therefore, systems to study cellular mechanics in a 3D scaffold that mimics a natural environment are essential. The next section describes several approaches.

7.3 Probing Cell Mechanics Using Cell-Seeded ECM Scaffolds in Three Dimensions

Three major classes of approaches exist for probing cell mechanics in 3D tissue constructs, all involving cell-seeded ECM scaffolds: (1) monitoring of cell-seeded collagen microspheres, (2) mechanical testing of ECM scaffold tensile specimens, and (3) imaging of living cells within such scaffolds. An advantage of all of these systems over 2D assays of cell mechanics is that they present a 3D structure that cells can manipulate and remodel into a more natural and realistic microenvironment. A challenge with all of these systems is that the models needed to quantify the mechanical responses and properties of cells are much more complicated than those needed for 2D assays. The ways that cells manipulate their local environment complicate these models and measurements further.

Collagen microspheres have been one of the most important tools for tracking the progression of ECM remodeling by cells and comprise an important tool for studying development, ECM remodeling, and cell traction (Barocas et al. 1995; Moon and Drubin 1995; Barocas and Tranquillo 1997a,b; Knapp et al. 1999). A feature and challenge of all approaches to estimate cell traction in 3D environments is the need to fit a series of models. In the case of the collagen microsphere assay, simplified viscoelastic representations of the ECM are fit using straightforward creep tests, and the tractions exerted by cells on the ECM are estimated from the changes in morphology of cells and microspheres over time (Barocas et al. 1995). The estimates are sensitive to the rate of proliferation and spreading of cells over time, but models for these factors can be fit using independent measurements. The tool has persisted for 20 years as the most important laboratory model of development, and techniques for producing highly consistent microspheres continue to improve, in part due to advances in microfluidics (Hong et al. 2012).

The other two methods for determining cell mechanics in 3D focus on active cellular forces and cellular mechanics deduced from observations over timescales that are short compared to those associated with tissue remodeling (see, e.g., Moon and Tranquillo 1993; Tranquillo and Murray 1993, 1992; Barocas et al. 1995; Barocas and Tranquillo 1997a,b; Tranquillo 1999) and cellular remodeling (Zemel et al. 2006; Zemel and Safran 2007). These methods rely largely on tightly integrated mechanical and biochemical study. In these approaches, tissue constructs containing cells and defined ECM are monitored, while the tissue construct is perturbed mechanically through some combination of stretching of the tissue construct (what Provenzano terms "outside–in," cf. Chapter 14 of this book) and active stressing of the construct by the cells within it (what Provenzano terms "inside–out"). Inside–out and outside–in mechanobiological signaling of the same magnitude can have very different effects (cf. Chapter 14), and similarly from a mechanical perspective, they must be modeled quite differently. The same tissue construct is then monitored while a promoter or inhibitor of cellular contraction is applied or while the cells themselves are lysed using a detergent.

Early examples of this type of approach involved measurements of stress at the level of an entire 3D tissue construct, and many important early observations of cell biophysics arose from these ensemble measurements. Delvoye et al. directly measured the contractile force developed in rectangular fibroblast-containing collagen tissue constructs by attaching one end of a tissue construct to a strain gauge–based force transducer and the other to a movable micrometer that could be used to stretch the construct (Delvoye et al. 1991). This system enabled the first observation that the ensemble-averaged contractile force of a population of 3D increases with the number of cells and with the application of an agonist (fetal calf serum) and decreases with application of an actin cytoskeleton inhibitor (cytochalasin B). Kolodney and Wysolmerski applied a similar approach to provide greater detail on the role of cytoskeletal organization (Kolodney and Wysolmerski 1992). Kolodney and Elson applied this approach to demonstrate the role of regulatory light chain phosphorylation of myosin II in contraction of a tissue construct containing fibroblasts (Kolodney and Elson 1993). Nekouzdeh et al. (2008) applied this approach combined with a time course of tissue construct fixation and staining to show the effect of mechanical loading rate on the nature of the ensemble-averaged active force recovery following loadings sufficient to depolymerize the actin cytoskeletons of cells within the tissue construct.

Extending beyond ensemble averages requires application of additional mathematical and experimental techniques, and these are the focus of this chapter. These techniques and some key applications of them are the focus of the remainder of this chapter.

7.3.1 ECM Scaffold–Based Tensile Specimens

Interpreting the ensemble mechanics of a tissue construct to identify the mechanics of the individual cells that comprise it requires a tightly integrated set of models and experiments.

The ideal specimens would be designed to distribute mechanical loads evenly over the cross sections of the specimens, but achieving this in practice is a challenge. An important hurdle is ensuring that the Williams-type singularity (Williams 1952) and other artifacts associated with gripping do not hopelessly confuse the interpretation of data. One approach that has worked well for structural engineering materials is the use of dog-bone-like specimens, in which the material a short distance from the position of gripping is reduced in cross-sectional area relative to that at the position of gripping itself (e.g., Roeder et al. 2002). The idea here is to minimize the strain of the specimen associated with stress concentrations at the point of gripping. The approach has not been used as extensively for testing of tissue constructs, perhaps in part because of difficulties associated with remodeling of the ECM by cells in the vicinity of the tapered region: nonuniformity associated with such remodeling can propagate the entire length of a specimen as cells react to their local mechanical environments (for a review of such effects on the cells themselves, see Elson and Genin, 2013).

An additional technique for reducing artifacts associated with clamping tensile specimens is the use of ring-shaped tissue constructs that can loop over bars attached to stepper motors and force transducers (Figure 7.1). Although artifacts arise due to compression of these rings at the loading bars, techniques such as optical strain measurement allow most of these to be overcome, and the system has seen widespread use. Ring-shaped specimens are formed in annular molds (Wakatsuki et al. 2000) from a suspension of cells in a solution of collagen kept in liquid form at low temperature. The solution is poured into the annular space (typically 2–3 mm wide) between the inner surface of outer cylindrical mold and a central mandrel. When the mold is placed in a 37°C incubator, the collagen gels and the cells remodel and compress the collagen to form a tissue construct (typically reducing its volume to ~10% of its original value within 24 h). Removal of the construct from the

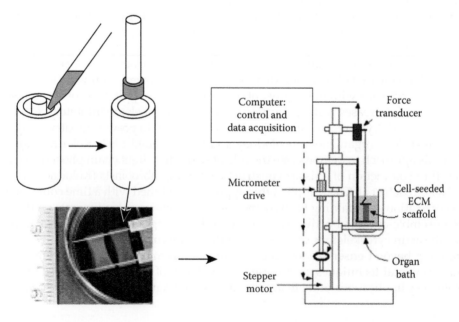

FIGURE 7.1
Ring-shaped cell-seeded ECM scaffolds can be formed from cell solutions incubated in annular molds. After a few days of culture at 37°C, the ring-shaped specimens can be removed and placed on spacers for further culture or directly onto a tensile testing apparatus.

mandrel yields a ring-shaped specimen that can be suspended between an isometric force transducer and a linear microstepping motor. The force transducer then measures inside–out force generated by contraction of muscle or nonmuscle cells and also outside–in forces developed when the tissue is stretched to yield the tissue stiffness at various magnitudes and rates of strain.

The most straightforward test used to characterize these specimens is a viscoelastic relaxation test. Here, one end of the specimen is displaced rapidly, while the other is held fixed. A force transducer, usually at the fixed end, then monitors the time course of the isometric force needed to sustain the specimen isometrically at its new length. After adequate time for the specimen to reach a steady-state isometric force, the test is repeated with additional increments of strain to identify the strain dependence of the response. At a prescribed time, the specimen is unloaded entirely and allowed to relax back to its original shape, and then a drug or detergent is applied to depolymerize or deactivate the cells or to decellularize the ECM. The test is then repeated to assess the mechanics of the ECM.

These constructs have been used to study the distinct contributions of cells and ECM to the viscoelastic properties of cell-seeded ECM scaffolds. Typical force relaxation data for a cell-seeded or decellularized ECM scaffold show an initial rise and then a logarithmic decay with respect to time over an hour or more of isometric relaxation (Figure 7.2a), as would be expected for a material that expresses a continuous spectrum of relaxation times (Marquez et al. 2006). Although error bars are large due to specimen-to-specimen variation, the peak force and relaxation rate scale both scale approximately linearly with linearized strain ε (Marquez et al. 2006):

$$F\left(t,\varepsilon,f_c\right) \approx a_0\left(f_c\right) + a_1\left(f_c\right)\varepsilon + b_1\left(f_c\right)\varepsilon \ln\left(-t/t_0\right), \qquad (7.1)$$

where

a_0, a_1, and b_1 are functions of the final cell volume fraction f_c only
t is time
t_0 is an arbitrary normalization constant

Note that this expression is not valid at $t=0$ but that in practice, this does not cause difficulty because the first useful data point in a relaxation test is recorded in a finite time after the completion of stretching.

Several interesting observations can be made from very simple analysis of the data in Figure 7.2a. The first is that the contribution of active cellular contraction, represented by a_0, is largely independent of strain over the range tested, although it increases exponentially with increasing final cell concentration. As discussed in the following, a particularly useful cell concentration to study is the percolation threshold associated with the formation of a continuous network of connected cells. A tissue construct with normalized 3D cell concentration $C_{3D} = Nl_0^3 = 11.8$ is just above the percolation threshold; here, N is the number of cells per unit volume, and l_0 is the nominal length of a (spindle-like) cell. For this, a value of $a_0 = 2.2$ mN was extracted, corresponding to an active stress of approximately $\sigma_0 = 410$ Pa in the construct, independent of mechanical strain.

The second aspect of these data that is important is that the effective modulus of a tissue construct estimated immediately after stretching increases with cell concentration. Although this is not surprising, an aspect of this that could not be predicted is that both cells and ECM increase their contributions to the effective modulus of the tissue construct with increasing cell concentration (Figure 7.2b through d). This observation is made by dividing the isometric force at a prescribed (relatively short) time following stretch of a tissue construct by the

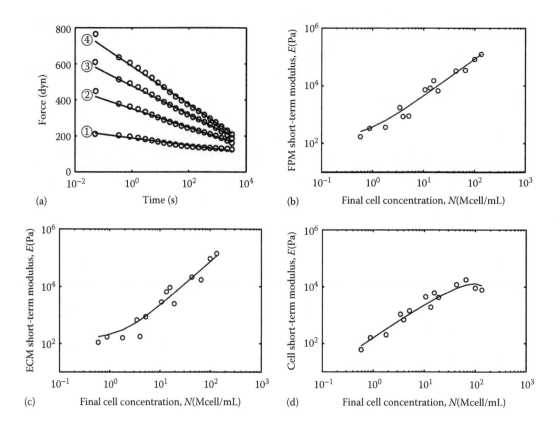

FIGURE 7.2

Results of a series of experiments in which fibroblast cell-seeded ECM scaffolds (FPMs) were stretched a pre-scribed amount and then held isometrically while force was monitored. Specimens were preconditioned with a stretch of 10% over 150 s, returned to their reference length at this same rate, then allowed to relax for 3600 s. The specimens were then stretched rapidly (over 10–15 ms) to prescribed nominal strains. Isometric force was monitored over a 3600 s relaxation period. FPMs were stretched to nominal strains of 2%, 8%, 14%, and 20%; the specimens were not unloaded between stretches. Tests were performed after 3 days of tissue culture. (a) Isometric force decayed logarithmically and increased with strain level. (b through d) The contributions of cells and ECM to overall FPM modulus increased with increasing cell concentration. (Graphs reprinted with permission from Marquez, J.P. et al., *Ann. Biomed. Eng.*, 34(9), 1475, 2006.)

initial cross-sectional area of the tissue construct and by the strain level. ECM estimates were obtained by repeating the procedure following treatment with an inhibitor, and cell contributions were estimated from the difference between these two tests.

Extending beyond this ensemble interpretation to estimate the responses of individual cells requires modeling of the structure, organization, and composition of the tissue construct. Techniques appropriate for this build from three classes of models developed through study of the mechanics of composite materials and paper. The first class is statistical approaches like the Zahalak model (Zahalak et al. 2000) and its extensions (Marquez et al. 2005a,b; Marquez et al. 2010). These relate the active and passive mechanical responses of cells and ECM to the overall mechanic responses of tissue construct through consideration of the statistics of cell morphology and distribution. Related methods trace back to Flory and have enjoyed broad usage for incorporating collagen fiber geometry into macroscale constitutive models of soft tissues (Lanir 1983; Horowitz et al. 1988; Lanir et al. 1988;

Hurschler et al. 1997; Wren and Carter 1998; Sacks 2003). The second class involves unit cell or periodic microfield models based upon numerical simulation of repeating microstructures (Bao et al. 1991; Bahei-El-Din 1996). These have been used widely to study the mechanical environment of chondrocytes (Wu et al. 1999; Wu and Herzog 2000). The third class is homogenization methods, based largely upon the Eshelby solution (Eshelby 1957, 1959), including self-consistent (mean field) approaches (Budiansky 1965; Hill 1965), in which elastic fields within each constituent are replaced by average values. These approaches and their extensions (Mori and Tanaka 1973; Chou 1992; Budiansky and Cui 1995; Chen and Cheng 1996; Tucker and Liang 1999; Milton 2002; Genin and Birman 2009) relate cell shape, cell and ECM mechanical properties, and, to some degree, the statistics of the cell distribution to the overall mechanics of a fibroblast-populated matrix (FPM). The methods described in the following involve a suite of models that incorporate all three approaches.

7.3.1.1 Statistical Approaches

Statistical approaches such as the Zahalak model (Zahalak et al. 2000) and its extensions describe the active and passive mechanics of a tissue construct containing of a single type of cell. Zahalak estimated the ensemble response expected from defined spatial and orientation distributions of cells and developed a framework for fitting this to results of mechanical tests performed on tissue constructs in combination with treatment using biochemical inhibitors. Zahalak idealized cells as thin contractile rods, appropriate for spindle-shaped cells in tissue constructs (Wakatsuki et al. 2000), and performed statistical averaging of the passive and active contributions of the cells and ECM to overall tissue construct mechanics. These contributions were taken to be in parallel, so that the stress at any material point is the sum of matrix (m) and cell (c) contributions:

$$\sigma_{ij} = \left(1 - f_c\right)\sigma_{ij}^{(m)} + f_c\sigma_{ij}^{(c)}, \tag{7.2}$$

where $\sigma_{ij}^{(m)}$ and $\sigma_{ij}^{(c)}$, the components of the engineering stress tensors, are weighted by cell and ECM volume fractions f_c and $f_m = 1 - f_c$. Zahalak also assumed that deformations are sufficiently characterized by the infinitesimal strain tensor: $\varepsilon_{ij} = 1/2(u_{i,j} + u_{j,i})$, where u_i is the macroscopic continuum displacement of the tissue and commas indicate partial differentiation. This is adequate for the small strains (<10%) that are of most interest.

Zahalak modeled myofibroblast cells with the Hill model (Hill 1938) for skeletal muscle including three elements (cf. Figure 7.3): a parallel elastic component accounting for static, passive mechanical properties, a series elastic element to account for increased dynamic stiffness, and a damped contractile element that embodies the active force generating and quasi-viscous force–velocity response of the cell. Zahalak adopted a linearized version of the Hill model, which yields the following equation governing the mechanical behavior of a cell:

$$\frac{dF}{dt} + \frac{F}{\tau_c} + \frac{F_0}{\tau_c} + l_0\left(k_s + k_p\right)\frac{1}{l_0}\frac{dl}{dt} + \frac{l_0 k_p}{\tau_c}\left(\frac{l}{l_0} - 1\right), \tag{7.3}$$

where
  F is the total force a cell exerts on the ECM
  l and l_0 are the current and initial cell lengths, respectively
  k_s and k_p are the series and parallel elastic stiffness values of the cell, respectively
  $\tau_c = \eta_c/k_s$ is the cell time constant, in which η_c is the cell's effective viscosity

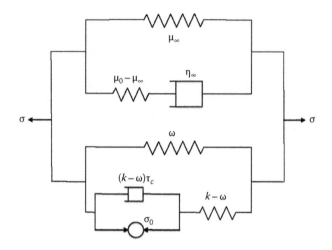

FIGURE 7.3
In a 1D idealization of the Zahalak model, cells (Hill element, lower grouping) are treated as acting in parallel with ECM (upper grouping, here represented as a linear viscoelastic solid).

The cell-generated contractile force, F_0, can rise in response to activators such as fetal bovine serum or thrombin and fall in response to inhibitors such as Y27632.

To model a tissue of these cells, Zahalak considered an idealized tissue with N cells per unit volume, each directed along a unit vector $\mathbf{n}$ according to a (measurable) orientation distribution $p(\mathbf{n})$, which is generally provided that $\int_\Omega p(\mathbf{n})d\Omega = 1$, where Ω is the surface of a unit sphere, and provided that it possesses all of the symmetries needed for a spherical distribution of fibers:

$$\sigma_{ij}^{(c)} = N l_0 \int_\Omega F(\Omega) n_i n_j p(\Omega)\, d\Omega. \tag{7.4}$$

After much manipulation, an integral form of the constitutive law can be written as

$$\sigma_{ij}(t) = \left(1 - f_c\right)\sigma_{ij}^{(m)}(t) + f_c \int_{-\infty}^{t} \exp\left(\frac{t - t'}{\tau_c}\right)\left\{\frac{3}{\tau_c}\sigma_0 A_{ij} + \left(\kappa\frac{\partial}{\partial t'} + \frac{\omega}{\tau_c}\right)\left(B_{ijpq}\varepsilon_{pq}\right)\right\} dt', \tag{7.5}$$

where $\sigma_{ij}^{(m)}(t)$ is the mechanical response of the ECM at time t; $\kappa = N l_0^2 \left(k_s + k_p\right)$, $\omega = N l_0^2 k_p$, and $\sigma_0 = (N l_0 F_0 / 3)$ are tissue-level mechanical parameters associated with cell-level mechanical responses; and the two "cell anisotropy tensors" $A_{ij} = \int_\Omega n_i n_j p(\Omega)d\Omega$ and $B_{ijpq} = \int_\Omega n_i n_j n_p n_q p(\Omega)\, d\Omega$ translate the anisotropy of the microscopic cell distribution into anisotropic mechanical behavior of the macroscopic tissue. The parameter ω represents the macroscopic manifestation of cellular parallel elasticity ("slow" elasticity), and κ the combination of parallel and series elasticity ("fast" elasticity) (Figure 7.3). Applying this model with the Fung's quasi-linear viscoelastic (QLV) model to represent $\sigma_{ij}^{(m)}(t)$, and assuming that cellular contractile force was independent of cell orientation, Zahalak et al. (2000) estimated the active component of cell stress as $\sigma_0 \approx 420$ Pa, which is on the same order as more recent estimates using traction microscopy and also as the very simple analysis presented at the start of this section.

Marquez et al. extended this type of model to situations in which cells and ECM interact and do not stretch in parallel by correcting average strain fields experienced by cells and ECM. The strain factor, S, defined as the relationship between the macroscopic strain applied to a tissue (resolved in a cell's axial direction) and the axial strain experienced by the cell is approximately constant for all cell orientations and is surprisingly insensitive to cell anisotropy (Marquez et al. 2010). In terms of this factor, Equation 7.5 can be rewritten:

$$\sigma_{ij}(t) = \left(1 - f_c\right)\sigma_{ij}^{(m)}(t) + f_c \int_{-\infty}^{t} \exp\left(\frac{t - t'}{\tau_c}\right)\left[\frac{3}{\tau_c}\sigma_0 A_{ij} + \left(\kappa\frac{\partial}{\partial t'} + \frac{\omega}{\tau_c}\right)B_{ijpq}S(t')\varepsilon_{pq}\right]dt'. \quad (7.6)$$

Note that the ECM response $\sigma_{ij}^{(m)}(t)$ must also be adjusted for the fact that cells and ECM do not in general stretch the same amount. For example, if the ECM is represented by the Fung's QLV model, the response to a strain history $\varepsilon_{ij}(t)$ would have to be adapted as follows:

$$\sigma_{ij}^{(m)}(t) = -p\delta_{ij} + \int_{-\infty}^{t} 2G(t - t')M(t')\left(\frac{\partial\sigma_{ij}^{(e)}\left(\varepsilon_{ij}(t')\right)}{dt}\right)dt', \quad (7.7)$$

where the factor $M(t) = \varepsilon_{ij}^{(m)}(t)\varepsilon_{ij}^{-1}(t) = 1 + \left(1 - S(t)\right)f_c / \left(1 - f_c\right)$ provides the ratio between the average strain field $\varepsilon_{ij}^{(m)}(t)$ in the ECM and the average strain field $\varepsilon_{ij}(t)$ in the tissue construct as a whole, δ_{ij} is the Kronecker's delta, p is a hydrostatic pressure, $G(t)$ is the Fung's reduced relaxation function, and $\sigma_{ij}^{(e)}$ is a function that defines the fast-loading constitutive response of the tissue.

From Equation 7.6, calculation of the instantaneous and relaxed effective moduli of cells and ECM is straightforward: the problem reduces to linear elasticity, with the caveat that the moduli must be understood as tangent moduli at a particular strain level (Prager 1969). For this case, and limiting to the case of isotropic cells and ECM, the strain factor for a thin tissue construct (thickness on the order of cell length) can be written in closed form using a self-consistent-type approach (Budiansky 1965; Hill 1965; Marquez et al. 2005a):

$$S \approx \frac{1}{1 + 2\left(d/l_0\right)\left(E_c/E_m\right)} \equiv \frac{1}{1 + KY} \quad (7.8)$$

where
   E_c is the elastic modulus of the cells
   E_m is the corresponding average tissue modulus
   d is the (elliptical) cell width
   l_0 is the (elliptical) cell length
   $Y = dE_c/l_0E_m$
   K is a constant that equals 2 for long, slender elliptical cells

The ratio E_c/E_m can be expected to differ for the instantaneous and fully relaxed responses of the cell and ECM, and S would then vary as a function of time. Applying this to tissue constructs shows that cells in tissue constructs appear either to become stiffer or to generate locally stiff regions of highly remodeled collagen when seeded sparsely in an ECM scaffold, as will be discussed in the following.

7.3.1.2 Unit Cell Approaches

Determining K requires an additional set of simulations involving study of idealized, periodic microstructures of a tissue construct. Such simulations show that $K \approx 2.2$ for "rectangular" planar cells; since this is very close to the value of 2 for very thin elliptical cells, we can conclude that the solution is not highly sensitive to the details of cell shape. This also appears to be the case in 3D (Marquez et al. 2005a). Such analyses show that as cells encroach upon their neighbors, several types of collective action are observed in tissue constructs. The first is an active effect as cells make contact and link through adherens and gap junctions, leading to structural and electrical connectivity. The second is a passive effect due to mechanical interactions between cells (cf. Bug et al. 1985a). As is often the case, the most interesting effects occur as cells approach the volume fraction associated with percolation, a threshold that can be predicted from certain idealizations of cell shape (Bug et al. 1985b). Simulations show a sharp rise in the strain factor near the critical cell concentration that depends strongly on the cell geometry but only weakly on the cell mechanical properties. At low cell concentrations, cells behave as if fully isolated. In the case of very stiff cells, the axial strain experienced by the cells is small compared to that of the nearby ECM. At the percolation threshold, the strain factor rises close to a value of 1, and the cells and ECM deform nearly in registry.

At these higher cell concentrations, two changes occur to the cellular environment. First, the local ECM modulus changes due to passive effects of neighboring cells. A self-consistent estimate of this stiffening to account for this, for the case of cells stiffer than ECM, is $E_m^{\text{eff}} = E_m + E_m^{\text{eff}} = E_m + \frac{3}{8} S \left(E_c - E_m \right) C d / l_0$, where, for a thin tissue construct, the dimensionless cell concentration is $C = A_c N l_0^2 / d = f_c l_0 / d$, in which A_c is the nominal cross-sectional area of a cell. Second, the effective length of cells changes because of physical connections. A reasonable model for this accounts for the number of nearby cells that a cell can expect to intersect, which rises as C/π (Kallmes and Corte 1960; Marquez et al. 2005a): $l_{\text{eff}} = l_0 (1 + C/2\pi)$. Expressions exist for cells in thicker tissue constructs as well and depend on orientation distributions (Marquez et al. 2005a). Defining an effective modulus ratio $Y^* = t E_c / l_{\text{eff}} E_m^{\text{eff}}$ and using this in place of Y in the scaling law (Equation 7.7), a very good estimate of the strain factor is obtained (Figure 7.4, $K = 2.2$); the model predicts the results of finite element analyses on representative periodic random cell distributions.

The rapid rise in S in Figure 7.4 near $C = 4$ (note that in a thicker tissue construct with a uniform 3D fiber orientation distribution, this corresponds to $N l_0^3 \approx 11$) corresponds to percolation and occurs just above the value of C at which cells can be expected to intersect with a single neighbor. The Zahalak model assumption of $S = 1$ is reasonable for very high cell concentrations; however, it is a poor approximation for lower cell concentrations, especially in tissues with a large stiffness mismatch between cells and matrix (high Y^*).

With these limits established, estimating the cell modulus is possible from the definition of Y^* if the matrix properties and cell dimensions are characterized. The recipe is as follows. A relaxation test is performed on a tissue construct, then repeated following an adequate relaxation interval and deactivation of the contractile apparatus (and hence a setting of $\sigma_0 = 0$) using an inhibitor such as Y27632. This second test enables estimation of the quantity ω (cf. Figure 7.3). Thereafter, the relaxation test is repeated following an adequate interval for relaxation and application of a detergent such as deoxycholate. This last test enables characterization of the viscoelastic response of the ECM. With this information, plus structural information needed to generate a lookup chart such as that of Figure 7.4 and knowledge of the cell volume fraction and average dimensions, one can use the following relationship to estimate the quantity SY^* (Zahalak et al. 2000):

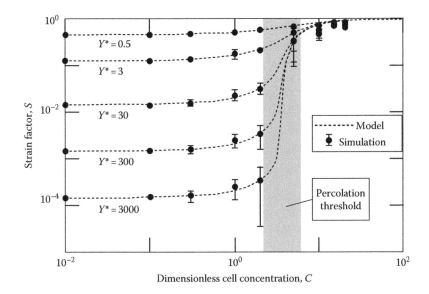

FIGURE 7.4
The strain factor indicates how much cells strain as a function of the average strain tensor at a material point within a tissue construct. At the percolation threshold (here $C \approx 4$), cells and ECM begin to stretch in registry ($S \approx 1$).

$$SY^* = \left(\omega N^{2/3}\right)\big/\left(E_m l_0^2\right). \qquad (7.9)$$

The relationship between Y^* and SY^* can then be found numerically from Equation 7.8 a function of dimensionless cell concentration C, and given the modulus of the matrix and a description of the cell morphology and arrangement, the effective cell modulus E_c can be calculated from the definition of Y^*.

7.3.1.3 Homogenization Approaches

Information is also available from tissue construct tensile test data through application of classical homogenization methods. The experiments, as earlier, involve relaxation tests performed before and after treatment with deoxycholate to lyse cells. From the relaxation responses, two mechanics challenges must then be addressed. The first is estimating the effective mechanical response of the ECM from the measured relaxation response of the deoxycholate-treated construct. This involves subtracting out the effect of fluid-filled voids from the measured response. The second is to apply this information to estimate cell moduli from relaxation data acquired from the untreated tissue construct. In the following, we will summarize approaches taken to overcome these challenges for the instantaneous and fully relaxed responses of the tissue construct, which are the simplest to address because the framework of linear elasticity can be applied rigorously.

For the first challenge, the starting point is the models of Pabst (Pabst and Gregorov·2004) for porous ceramics, in which the modulus $E_p(f_c)$ of a porous linear elastic solid containing a volume fraction f_c of void ceramic is approximated as

$$E_p\left(f_c\right)\big/E_m = \left(1-f_c\right)\left(1-f_c/\phi_o\right), \qquad (7.10)$$

where $\phi_o = 0.872$ can be interpreted as the pore volume fraction beyond which the ECM is no longer continuous. This was shown through Monte Carlo simulations to be a good fit for idealized tissue constructs with isotropic, nearly incompressible constituents (Poisson's ratio $\nu = 0.49$) containing spindle-shaped cells with a nominally uniform spherical orientation distribution (Thomopoulos et al. 2011). For the second, Marquez et al. found that a generalization of the model of Pabst et al. was accurate for all tissue constructs in which cells are more compliant than ECM:

$$\frac{E_t(f_c)}{E_m} = 1 + \left(1 - \frac{E_c}{E_m}\right)\left(\frac{1 + \alpha E_c/E_m}{1 + \beta E_c/E_m}\right)\left(\frac{E_p(f_c)}{E_m} - 1\right), \tag{7.11}$$

where $E_p(f_c)/E_m$ is given by Equation 7.10 and $\alpha = 11.57$ and $\beta = 15.00$ are dimensionless scaling parameters obtained by least squares fitting. For cases of cells stiffer than the ECM, homogenization bounds are very broad, and no adequate approximation exists. However, a curve midway between the three point bounds for overlapping spherical cells (Milton 2002) provides an adequate approximation (Thomopoulos et al. 2011), even for cells much stiffer than the ECM ($R^2 = 0.85$ for $E_c/E_m = 1000$).

Within this modeling framework, questions can be asked about the ways that interactions between cells and ECM affect cell and ECM mechanics. Of particular interest is how mechanics of cells and ECM interact during the process of tissue remodeling. Remodeling has long been known to be dependent upon culture conditions, with cells reducing the volume of a tissue construct during remodeling in proportion to their number density. What does this mean for cell and ECM mechanics? The passive, incremental tangent moduli of the ECM increase monotonically with increasing cell concentration (Figure 7.5), at a rate

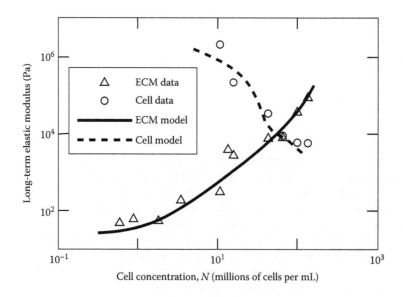

FIGURE 7.5

In fibroblast cell-seeded collagen ECMs, the average modulus of the ECM increases with cell concentration, and the effective modulus of cells decreases. The crossover point is near the percolation threshold at which cells start to form a contiguous network. Cells in these scaffolds proliferate or die off to enable the remodeled cell-seeded scaffold to reach the percolation threshold. (Graph redrawn from data in Marquez, J.P. et al., *Phil. Trans. A Math. Phys. Eng. Sci.*, 368(1912), 635, 2010.)

well beyond what could be accounted for simply through compression of the collagen network, suggesting that other factors such as cross-linking of collagen and collagen production by cells might play a role (Marquez et al. 2010). Note that the modulus in Figure 7.5 was calculated from the second two terms in Equation 7.1, an approximation that is most valid when a_0 is small compared to these terms.

The effective modulus of cells, however, decreases with increasing cell concentration. At the lowest cell concentrations considered, the model of a cell as a contractile rod in a homogeneous ECM is likely invalid, as we and others have observed pockets of enhanced remodeling around cells in sparsely populated tissue constructs. Indeed, for the very lowest concentrations in Figure 7.5, a contractile rod that would mimic the cellular contribution to tissue construct mechanics would require infinite stiffness. As the cell concentration increases, the reliability of this model increases because the ECM becomes more spatially homogenous through the collective action of cells. At the highest cell concentrations considered, the model shows cells are more compliant than the ECM.

The crossover point at which cell and ECM moduli are equal is particularly interesting. At this point, the cells match the modulus of the ECM to their own modulus. The cell concentration at which this occurs corresponds roughly to the percolation threshold, suggesting that this matching of modulus is a result of collective action by cells. The magnitude of modulus at which this occurs is also of interest: the long-term modulus (tangent modulus 3600 s after stretch) of the tissue construct is just over 10 kPa, which is the substrate modulus observed by Engler et al. (2004) to steer mesenchymal stem cells to differentiate into myofibroblastic cells. The crossover modulus shortly after stretch is approximately four times greater (Marquez et al. 2010).

Cells in constructs that initially contain lower concentrations of cells divide and multiply quickly compared to those near the percolation threshold, while cells in those with initially higher concentrations decrease in number. These data together hint that contractile fibroblasts in tissue constructs might modify themselves and their ECM to approach the state in which they exist in a network that is just at the percolation threshold and in which cell and ECM modulus match. This is not unexpected in the context of wound healing, in which formation of a contiguous myofibroblast network is desirable. Such a picture of cell proliferation relating to mechanical environment is consistent with the work of Zhu et al. (2001), who observed differences in the fate of cells within tissue constructs as a function of mechanical constraints during incubation. The target modulus of approximately 10 kPa is particularly interesting because it coincides with the range of substrate moduli known to drive stem cells toward a fibroblastic lineage. However, we note that this is not likely to be a feature of mechanics alone and that interrelated features such as nutrient transport, steric effects of ECM, and the concentration of ECM are also factors. Models for flow through fibrous networks are increasingly mature (e.g., Stylianopoulos et al. 2008), and these offer hope for dissecting these effects through coupled modeling and experimentation.

7.3.2 Single-Cell Imaging Methods

A series of technologies have emerged recently for the purpose of optically monitoring the actions of live cells in a cell-seeded ECM. Live cell imaging following stretch has been performed by Lee et al. with the goal of understanding how stress fiber kinetics are affected by a sustained mechanical stretch (Lee et al. 2010). The experiments were accomplished by developing an apparatus to perform tensile tests on ring-shaped tissue constructs while simultaneously imaging fluorescently labeled cytoskeletal proteins over the objective of a confocal fluorescence microscope. The Kaunas group has recently developed an innovative technology to enable stress fiber kinetics models such as those described in Chapter 6 to be tested on cyclically stretched cells in 3D culture (Tondon and Kaunas 2014). The fundamental challenge is that collagen ECM scaffolds typically present important viscoelastic time constants in the 0.1–10 s range that is of the greatest interest for stress fiber kinetics (Pryse et al. 2003; Nekouzadeh et al. 2007; Nekouzadeh and Genin 2013). A tissue construct can therefore not relax adequately following loadings in the 0.1–10 Hz range for a cyclic test to be performed. The strategy used by the Kaunas group was to coat a flexible elastic substratum with a cell-seeded collagen layer. By choosing a substratum that was stiff relative to the cell-seeded collagen layer and ensuring a well-bonded interface, the collagen layer could be forced to reverse direction at appropriate frequencies.

Technologies are just now becoming available for estimating mechanical tractions around individual cells within a cell-seeded ECM. The approach that is the focus of this section is one in which cells are cultured within a well-defined hydrogel scaffold and cell tractions are inferred from the motion of fluorescent particles embedded within the hydrogel. A central challenge is to design the hydrogel scaffold so that cells can remodel their environment and adopt realistic morphologies. Although this chemistry is not the focus of this section, the details of it are fascinating, and the reader is referred to the work of Legant et al. (2010) for a discussion. Note that bead tracking had to account for the effect of the cell-lysing agent on the strain in the scaffold: very minor changes in molarity can have tremendous effect on volumetric strain. The strategy was to include dilatation and rigid body motion of the scaffold in the optimization algorithm for tracking beads from the reference to the deformed configuration. Thereafter, bead displacements were interpolated onto a cubic with characteristic length on the order of the mean bead spacing, and the deformation gradient **F** was estimated from the displacements of these grid points.

An experiment of this character involves allowing cells to remodel their local environment over some prescribed interval and then taking a stack of confocal images of fluorescent beads in the vicinity of a cell of interest. These images define the "deformed" configuration of the beads. The cell is then imaged based upon a fluorescent probe in its cytoplasm or on its membrane. Thereafter, the cells are washed away using a detergent such as deoxycholate, and the imaging is repeated to define the "reference" (undeformed) configuration of the beads. A host of methods exist for identifying the positions of beads from the confocal stacks, and approaches such as the radius-limited feature vector approach (Legant et al. 2010) are effective at matching beads from the reference to the deformed image.

With this displacement field and a discretization of the cell's outer boundary, the obvious strategy is to estimate strain fields, extrapolate them to the surface of the cell, estimate stresses, and then estimate tractions from these stresses. However, this "forward" problem is fraught with difficulty, in part because of the sensitivity of the stresses estimated in a hydrogel to small errors in tracking: in a nearly incompressible ECM, small errors in displacement can translate to enormous errors in estimated pressure.

Legant et al. (2010) therefore developed a regularized inverse method, based upon readily available tools for this (cf. Schwarz et al. 2002; Hansen 2007). A Green's function was generated numerically that relates cell surface tractions from each of a prescribed number of triangular patches, defined to approximate the cell surface, to displacements within the gel. These enabled estimation of how a particular set of surface tractions would displace the population of beads in the neighborhood of the cell of interest, resulting in a linear set of equations whose influence factors could be optimized to best fit the displacement field. However, especially given that hydrogels are relatively stiff in hydrostatic compression, the problem is not well posed: a great number of traction fields can yield similar levels of fitting error. Indeed, this is an issue in many inverse problems (Schwarz et al. 2002).

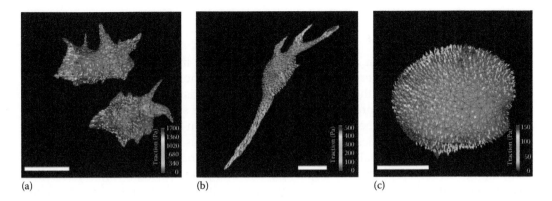

FIGURE 7.6
Cellular tractions measured within a 3D ECM scaffold differ by cell type. (a) NIH 3T3 fibroblasts after 3 days of culture show peak stresses that are much higher than the ensemble average of 400 Pa usually observed for fibroblasts in tissue constructs. However, these peak stresses are localized, and averages over a surface are closer to 400 Pa. (b) Human mesenchymal stem cells invaded into the surrounding hydrogel with slender and occasionally branched protrusions. Inward traction levels were much lower and were localized predominantly at near the tips of these protrusions. (c) Lewis lung carcinoma cells did not invade this ECM but instead proliferated to form a multicellular spheroid. All images were taken after 3 days of culture. All scale bars = 20 μm. (Reprinted with permission from Legant, W.R. et al., *Nat. Methods.*, 7(12), 969, 2010.) Traction magnitudes are indicated by the lengths of arrows, scaled to the maximum in the legend of each panel (color available online).

The solution is regularization: the solution chosen is that which minimizes the total magnitude of a vector **T** that incorporates all of the components of the tractions applied to all of the triangular cell surface elements that represent the cell:

$$\min\left\{\left|\mathbf{u}_{\text{estimated}}(\mathbf{T}) - \mathbf{u}_{\text{measured}}\right|^2 + \lambda^2 \left|\mathbf{T}\right|^2\right\}, \tag{7.12}$$

where $\mathbf{u}_{\text{estimated}}(\mathbf{T})$ is a vector containing estimates of all the components of the displacements of all the beads, $\mathbf{u}_{\text{measured}}$ is a vector containing all the measured components of the displacements of all the beads, and λ is a regularization parameter.

With this method, and modest computational power, Legant et al. (2010) obtained for the first time realistic estimates of tractions exerted by several different types of cells embedded within a 3D ECM scaffold (Figure 7.6). Results show that cellular tractions measured within a 3D ECM scaffold differ by cell type. NIH 3T3 fibroblasts after 3 days of culture show peak stresses that are much higher than the ensemble average of 400 Pa usually observed for fibroblasts in tissue constructs. However, these peak stresses are localized, and averages over a surface are closer to 400 Pa (Figure 7.6a). Human mesenchymal stem cells (Figure 7.6b) invaded the surrounding hydrogel with slender and occasionally branched protrusions. Inward traction levels were much lower and were localized predominantly at near the tips of these protrusions. Lewis lung carcinoma cells did not invade this ECM, but instead proliferated to form a multicellular spheroid (Figure 7.6c).

This method is highly effective, but computationally tedious, and requires much expertise on the part of the scientist in the form of designing the ECM scaffold, discretizing the cell geometry, generating the Green's functions, and interpreting the mechanical data. This is an area with much promise for the application of simplified numerical algorithms, and that will benefit much as computational speed increases.

7.4 Concluding Remarks

Mathematical tools exist that allow the mechanics of cells in cell-seeded ECM scaffolds to be deduced, but only in very specially designed experiments. We conclude with a few words about ways in which these tools are at present unsatisfactory and about directions we consider promising. At present, a specific limitation of such experiments is that the cells must be perturbed biochemically if enough information is to be provided to fitting of models of cell mechanics. These perturbations are enormous and often involve lysing the cells irreversibly. A major challenge for future work is estimating cell mechanics without eliminating the cells. Although collagen spheroid assays are a promising tool for such measurements in a highly uniform environment, they hold little promise for unraveling the complicated interplay between cell and ECM in development of more complicated tissues. As an example, a stem cell–seeded ECM model of flexor tendon development shows that stem cell fate, organization, and proliferation interact with ECM and mechanical loading in complicated ways that vary with the nature of mechanical loading (Thomopoulos et al. 2011). Beginning to address these mechanisms requires real-time estimates of how cells and ECM share mechanical loads. However, promising technologies involving fluorescent ECM probes are on the horizon (e.g., Legant et al. 2012).

A major limitation of the models presented in this chapter is that they assume some degree of homogeneity in the collagen ECM, whereas cells likely alter their local environment differently that they do the collagen that is more distal. Techniques to measure and estimate the effects of this "functional grading" are at present lacking, with the ability to model multiple-phase composites quite low.

Three-dimensional cell-seeded ECM scaffolds continue to serve as the simplest way to study mechanics of a cell in a realistic 3D environment and have provided much fundamental insight into cell behavior expected within tissues. With the emergence of future observational and mathematical techniques, we expect this contribution to continue to grow.

References

Abbott, A. 2003. Cell culture: Biology's new dimension. *Nature* 424(6951):870–872.

Asnes, C. F., J. P. Marquez, E. L. Elson, and T. Wakatsuki. 2006. Reconstitution of the Frank-Starling mechanism in engineered heart tissues. *Biophysical Journal* 91(5):1800–1810.

Bahei-El-Din, Y. A. 1996. Finite element analysis of viscoplastic composite materials and structures. *Mechanics of Advanced Materials and Structures* 3(1):1–28.

Bao, G., J. W. Hutchinson, and R. M. McMeeking. 1991. Particle reinforcement of ductile matrices against plastic flow and creep. *Acta Metallurgica et Materialia* 39(8):1871–1882.

Barocas, V. H., A. G. Moon, and R. T. Tranquillo. 1995. The fibroblast-populated collagen microsphere assay of cell traction force—Part 2: Measurement of the cell traction parameter. *Journal of Biomechanical Engineering* 117(2):161–170.

Barocas, V. H. and R. T. Tranquillo. 1997a. An anisotropic biphasic theory of tissue-equivalent mechanics: The interplay among cell traction, fibrillar network deformation, fibril alignment, and cell contact guidance. *Journal of Biomechanical Engineering* 119(2):137–145.

Barocas, V. H. and R. T. Tranquillo. 1997b. A finite element solution for the anisotropic biphasic theory of tissue-equivalent mechanics: The effect of contact guidance on isometric cell traction measurement. *Journal of Biomechanical Engineering* 119(3):261–268.

Budiansky, B. 1965. On the elastic moduli of some heterogeneous materials. *Journal of the Mechanics and Physics of Solids* 13:223–227.

Budiansky, B. and Y. L. Cui. 1995. Toughening of ceramics by short aligned fibers. *Mechanics of Materials* 21:139–146.

Bug, A. L. R., S. A. Safran, G. S. Grest, and I. Webman. 1985a. Do interactions raise or lower a percolation threshold? *Physical Review Letters* 55(18):1896–1899.

Bug, A. L. R., S. A. Safran, and I. Webman. 1985b. Continuum percolation of rods. *Physical Review Letters* 54(13):1412–1415.

Bursac, P., G. Lenormand, B. Fabry, M. Oliver, D. A. Weitz, V. Viasnoff, J. P. Butler, and J. J. Fredberg. 2005. Cytoskeletal remodelling and slow dynamics in the living cell. *Nature Materials* 4(7):557–561.

Butler, J. P., I. M. Tolic-Norrelykke, B. Fabry, and J. J. Fredberg. 2002. Traction fields, moments, and strain energy that cells exert on their surroundings. *American Journal of Physiology* 282(3):C595–C605.

Chen, C., R. Krishnan, E. Zhou, A. Ramachandran, D. Tambe, K. Rajendran, R. M. Adam, L. Deng, and J. J. Fredberg. 2010. Fluidization and resolidification of the human bladder smooth muscle cell in response to transient stretch. *PLoS One* 5(8):e12035.

Chen, C. H. and C. H. Cheng. 1996. Effective elastic moduli of misoriented short-fiber composites. *International Journal of Solids and Structures* 33:2519–2539.

Chen, Y., B. Liu, G. Xu, and J.-Y. Shao. 2009. Validation, in-depth analysis, and modification of the micropipette aspiration technique. *Cellular and Molecular Bioengineering* 2(3):351–365.

Chou, T. W. 1992. *Microstructural Design of Fiber Composites*. Cambridge, U.K.: Cambridge University Press.

Cole, K. S. 1932. Surface forces of the arbacia egg. *Journal of Cellular and Comparative Physiology* 1(1):1–9.

Cukierman, E., R. Pankov, D. R. Stevens, and K. M. Yamada. 2001. Taking cell-matrix adhesions to the third dimension. *Science* 294(5547):1708–1712.

Cukierman, E., R. Pankov, and K. M. Yamada. 2002. Cell interactions with three-dimensional matrices. *Current Opinion in Cell Biology* 14(5):633–640.

Daily, B., E. L. Elson, and G. I. Zahalak. 1984. Cell poking. Determination of the elastic area compressibility modulus of the erythrocyte membrane. *Biophysical Journal* 45(4):671–682.

Delvoye, P., P. Wilquet, J.-L. Leveque, B. V. Nusgens, and C. M. Lapiere. 1991. Measurement of mechanical forces generated by skin fibroblasts embedded in a three-dimensional collagen gel. *Journal of Investigative Dermatology* 97:898–902.

Dembo, M., T. Oliver, A. Ishihara, and K. Jacobson. 1996. Imaging the traction stresses exerted by locomoting cells with the elastic substratum method. *Biophysical Journal* 70(4):2008–2022.

Dembo, M. and Y. L. Wang. 1999. Stresses at the cell-to-substrate interface during locomotion of fibroblasts. *Biophysical Journal* 76(4):2307–2316.

Deshpande, V. S., R. M. McMeeking, and A. G. Evans. 2006. A bio-chemo-mechanical model for cell contractility. *Proceedings of the National Academy of Sciences of the United States of America* 103(38):14015–14020.

Deshpande, V. S., R. M. McMeeking, and A. G. Evans. 2007. A model for the contractility of the cytoskeleton including the effects of stress-fibre formation and dissociation. *Proceedings of the Royal Society A: Mathematical, Physical and Engineering Science* 463(2079):787–815.

Deshpande, V. S., M. Mrksich, R. M. McMeeking, and A. G. Evans. 2008. A bio-mechanical model for coupling cell contractility with focal adhesion formation. *Journal of the Mechanics and Physics of Solids* 56(4):1484–1510.

Downey, G. P., D. E. Doherty, B. D. Schwab, E. L. Elson, P. M. Henson, and G. S. Worthen. 1990. Retention of leukocytes in capillaries: Role of cell size and deformability. *Journal of Applied Physiology* 69(5):1767–1778.

Downey, G. P., E. L. Elson, B. D. Schwab, S. C. Erzurum, S. K. Young, and G. S. Worthen. 1991. Biophysical properties and microfilament assembly in neutrophils: Modulation by cyclic amp. *Journal of Cell Biology* 114(6):1179–1190.

Elgsaeter, A., B. T. Stokke, A. Mikkelsen, and D. Branton. 1986. The molecular basis of erythrocyte shape. *Science* 234(4781):1217–1223.

Elson, E. and G. Genin. 2013. The role of mechanics in actin stress fiber kinetics. *Experimental Cell Research* 319(16):2490–2500.

Elson, E. L. 1988. Cellular mechanics as an indicator of cytoskeletal structure and function. *Annual Review of Biophysics and Biophysical Chemistry* 17:397–430.

Engler, A., L. Bacakova, C. Newman, A. Hategan, M. Griffin, and D. Discher. 2004. Substrate compliance versus ligand density in cell on gel responses. *Biophysical Journal* 86(1 Pt 1):617–628.

Eshelby, J. D. 1957. The determination of the elastic field outside an ellipsoidal inclusion and related problems. *Proceedings of the Royal Society of London A* 241:376–396.

Eshelby, J. D. 1959. The elastic field outside an ellipsoidal inclusion. *Proceedings of the Royal Society of London, Series A, Mathematical and Physical Sciences* 252:561–569.

Evans, E. and A. Yeung. 1989. Apparent viscosity and cortical tension of blood granulocytes determined by micropipet aspiration. *Biophysical Journal* 56(1):151–160.

Evans, E. A. and R. Skalak. 1980. *Mechanics and Thermodynamics of Biomembranes*. Boca Raton, FL: CRC Press.

Fraley, S. I., Y. Feng, R. Krishnamurthy, D.-H. Kim, A. Celedon, G. D. Longmore, and D. Wirtz. 2010. A distinctive role for focal adhesion proteins in three-dimensional cell motility. *Nature Cell Biology* 12(6):598–604.

Genin, G. M. and V. Birman. 2009. Micromechanics and structural response of functionally graded, particulate-matrix, fiber-reinforced composites. *International Journal of Solids and Structures* 46(10):2136–2150.

Granzier, H. L. and T. C. Irving. 1995. Passive tension in cardiac muscle: Contribution of collagen, titin, microtubules, and intermediate filaments. *Biophysical Journal* 68(3):1027–1044.

Grinnell, F. 2003. Fibroblast biology in three-dimensional collagen matrices. *Trends in Cell Biology* 13(5):264–269.

Han, S. J., K. S. Bielawski, L. H. Ting, M. L. Rodriguez, and N. J. Sniadecki. 2012. Decoupling substrate stiffness, spread area, and micropost density: A close spatial relationship between traction forces and focal adhesions. *Biophysical Journal* 103(4):640–648.

Hansen, P. C. 2007. Regularization tools version 4.0 for matlab 7.3. *Numerical Algorithms* 46(2):189–194.

Harris, A. K., P. Wild, and D. Stopak. 1980. Silicone rubber substrata: A new wrinkle in the study of cell locomotion. *Science* 208(4440):177–179.

Hill, A. 1938. The heat of shortening and the dynamic constants of muscle. *Proceedings of the Royal Society of London. Series B, Biological Sciences* 126(843):136–195.

Hill, R. 1965. A self-consistent mechanics of composite materials. *Journal of the Mechanics and Physics of Solids* 13:213–222.

Hong, S., H.-J. Hsu, R. Kaunas, and J. Kameoka. 2012. Collagen microsphere production on a chip. *Lab on a Chip* 12(18):3277–3280.

Horowitz, A., Y. Lanir, F. C. Yin, M. Perl, I. Sheinman, and R. K. Strumpf. 1988. Structural three-dimensional constitutive law for the passive myocardium. *Journal of Biomechanical Engineering* 110(3):200–207.

Hsu, H. J., C. F. Lee, and R. Kaunas. 2009. A dynamic stochastic model of frequency-dependent stress fiber alignment induced by cyclic stretch. *PLoS One* 4(3):e4853.

Hsu, H. J., C. F. Lee, A. Locke, S. Q. Vanderzyl, and R. Kaunas. 2010. Stretch-induced stress fiber remodeling and the activations of jnk and erk depend on mechanical strain rate, but not fak. *PLoS One* 5(8):e12470.

Hurschler, C., B. Loitz-Ramage, and R. Vanderby, Jr. 1997. A structurally based stress-stretch relationship for tendon and ligament. *Journal of Biomechanical Engineering* 119(4):392–399.

Kallmes, O. and H. Corte. 1960. The structure of paper: The statistical geometry of an ideal two-dimensional fiber network. *Tappi* 43:737–752.

Kaunas, R. and H.-J. Hsu. 2009. A kinematic model of stretch-induced stress fiber turnover and reorientation. *Journal of Theoretical Biology* 257(2):320–330.

Kaunas, R., P. Nguyen, S. Usami, and S. Chien. 2005. Cooperative effects of rho and mechanical stretch on stress fiber organization. *Proceedings of the National Academy of Sciences of the United States of America* 102(44):15895–15900.

Kaunas, R., S. Usami, and S. Chien. 2006. Regulation of stretch-induced jnk activation by stress fiber orientation. *Cellular Signalling* 18(11):1924–1931.

Knapp, D. M., E. F. Helou, and R. T. Tranquillo. 1999. A fibrin or collagen gel assay for tissue cell chemotaxis: Assessment of fibroblast chemotaxis to grgdsp. *Experimental Cell Research* 247(2):543–553.

Kolodney, M. S. and E. L. Elson. 1993. Correlation of myosin light chain phosphorylation with isometric contraction of fibroblasts. *Journal of Biological Chemistry* 268(32):23850–23855.

Kolodney, M. S. and R. B. Wysolmerski. 1992. Isometric contraction by fibroblasts and endothelial cells in tissue culture: A quantitative study. *The Journal of Cell Biology* 117:73–82.

Lam, R. H., Y. Sun, W. Chen, and J. Fu. 2012a. Elastomeric microposts integrated into microfluidics for flow-mediated endothelial mechanotransduction analysis. *Lab on a Chip* 12(10):1865–1873.

Lam, R. H., S. Weng, W. Lu, and J. Fu. 2012b. Live-cell subcellular measurement of cell stiffness using a microengineered stretchable micropost array membrane. *Integrative Biology: Quantitative Biosciences from Nano to Macro* 4(10):1289–1298.

Lanir, Y. 1983. Constitutive equations for fibrous connective tissues. *Journal of Biomechanics* 16(1):1–12.

Lanir, Y., E. L. Salant, and A. Foux. 1988. Physico-chemical and microstructural changes in collagen fiber bundles following stretch in-vitro. *Biorheology* 25(4):591–603.

le Digabel, J., M. Ghibaudo, L. A. Trichet, A. Richert, and B. Ladoux. 2010. Microfabricated substrates as a tool to study cell mechanotransduction. *Medical & Biological Engineering & Computing* 48(10):965–976.

Lee, C. F., C. Haase, S. Deguchi, and R. Kaunas. 2010. Cyclic stretch-induced stress fiber dynamics— Dependence on strain rate, rho-kinase and mlck. *Biochemical and Biophysical Research Communications* 401(3):344–349.

Lee, P. H. and H. Vandenburgh. 2013. Skeletal muscle atrophy in bioengineered skeletal muscle: A new model system. *Tissue Engineering* 19:2147–2155.

Lee, S. L., A. Nekouzadeh, B. Butler, K. M. Pryse, W. B. McConnaughey, A. C. Nathan, W. R. Legant et al. 2012. Physically-induced cytoskeleton remodeling of cells in three-dimensional culture. *PLoS One* 7(12):e45512.

Legant, W. R., C. S. Chen, and V. Vogel. 2012. Force-induced fibronectin assembly and matrix remodeling in a 3D microtissue model of tissue morphogenesis. *Integrative Biology* 4(10):1164–1174.

Legant, W. R., J. S. Miller, B. L. Blakely, D. M. Cohen, G. M. Genin, and C. S. Chen. 2010. Measurement of mechanical tractions exerted by cells in three-dimensional matrices. *Nature Methods* 7(12):969–971.

Legant, W. R., A. Pathak, M. T. Yang, V. S. Deshpande, R. M. McMeeking, and C. S. Chen. 2009. Microfabricated tissue gauges to measure and manipulate forces from 3D microtissues. *Proceedings of the National Academy of Sciences of the United States of America* 106(25):10097–10102.

Mann, J. M., R. H. Lam, S. Weng, Y. Sun, and J. Fu. 2012. A silicone-based stretchable micropost array membrane for monitoring live-cell subcellular cytoskeletal response. *Lab on a Chip* 12(4):731–740.

Marquez, J. P., E. L. Elson, and G. M. Genin. 2010. Whole cell mechanics of contractile fibroblasts: Relations between effective cellular and extracellular matrix moduli. *Philosophical Transactions. Series A, Mathematical, Physical, and Engineering Sciences* 368(1912):635–654.

Marquez, J. P., G. M. Genin, K. M. Pryse, and E. L. Elson. 2006. Cellular and matrix contributions to tissue construct stiffness increase with cellular concentration. *Annals of Biomedical Engineering* 34(9):1475–1482.

Marquez, J. P., G. M. Genin, G. I. Zahalak, and E. L. Elson. 2005a. The relationship between cell and tissue strain in three-dimensional bio-artificial tissues. *Biophysical Journal* 88(2):778–789.

Marquez, J. P., G. M. Genin, G. I. Zahalak, and E. L. Elson. 2005b. Thin bio-artificial tissues in plane stress: The relationship between cell and tissue strain, and an improved constitutive model. *Biophysical Journal* 88:765–777.

McGarry, J. P., J. Fu, M. T. Yang, C. S. Chen, R. M. McMeeking, A. G. Evans, and V. S. Deshpande. 2009. Simulation of the contractile response of cells on an array of micro-posts. *Philosophical Transactions. Series A, Mathematical, Physical, and Engineering Sciences* 367(1902):3477–3497.

Milton, G. 2002. *The Theory of Composites.* Cambridge, U.K.: Cambridge University Press.

Mitchison, J. M. and M. M. Swann. 1954. The mechanical properties of the cell surface. *Journal of Experimental Biology* 31(3):443–460.

Moon, A. and D. G. Drubin. 1995. The ADF/Cofilin proteins: Stimulus-responsive modulators of actin dynamics. *Molecular Biology of the Cell* 6(11):1423–1431.

Moon, A. G. and R. T. Tranquillo. 1993. The fibroblast-populated collagen microsphere assay of cell traction force: Part 1. Continuum model. *AIChE Journal* 39:163–177.

Mori, T. and K. Tanaka. 1973. Average stress in matrix and average elastic energy of materials with misfitting inclusions. *Acta Metallurgica* 21:571–574.

Nekouzadeh, A. and G. M. Genin. 2013. Adaptive quasi-linear viscoelastic modeling. In *Computational Modeling in Tissue Engineering*, Liesbet Geris, ed. Installment in the series, *Studies in Mechanobiology, Tissue Engineering and Biomaterials*, Amit Gefen, ed. Berlin, Germany: Springer.

Nekouzadeh, A., K. M. Pryse, E. L. Elson, and G. M. Genin. 2007. A simplified approach to quasi-linear viscoelastic modeling. *Journal of Biomechanics* 40(14):3070–3078.

Nekouzadeh, A., K. M. Pryse, E. L. Elson, and G. M. Genin. 2008. Stretch-activated force shedding, force recovery, and cytoskeletal remodeling in contractile fibroblasts. *Journal of Biomechanics* 41(14):2964–2971.

Nelson, C. M. and M. J. Bissell. 2005. Modeling dynamic reciprocity: Engineering three-dimensional culture models of breast architecture, function, and neoplastic transformation. *Seminars in Cancer Biology* 15:342–352.

Pabst, W. and E. Gregorov·. 2004. Effective elastic properties of alumina-zirconia composite ceramics-part 2. Micromechanical modeling. *Ceramics Silikaty* 48(1):14–23.

Pasternak, C. and E. L. Elson. 1985. Lymphocyte mechanical response triggered by cross-linking surface receptors. *Journal of Cell Biology* 100(3):860–872.

Pasternak, C., J. A. Spudich, and E. L. Elson. 1989. Capping of surface receptors and concomitant cortical tension are generated by conventional myosin. *Nature* 341(6242):549–551.

Paszek, M. J., N. Zahir, K. R. Johnson, J. N. Lakins, G. I. Rozenberg, A. Gefen, C. A. Reinhart-King, S. S. Margulies, M. Dembo, and D. Boettiger. 2005. Tensional homeostasis and the malignant phenotype. *Cancer Cell* 8(3):241–254.

Pathak, A., V. S. Deshpande, R. M. McMeeking, and A. G. Evans. 2008. The simulation of stress fibre and focal adhesion development in cells on patterned substrates. *Journal of the Royal Society Interface* 5(22):507–524.

Petersen, N. O., W. B. McConnaughey, and E. L. Elson. 1982. Dependence of locally measured cellular deformability on position on the cell, temperature, and cytochalasin b. *Proceedings of the National Academy of Sciences of the United States of America* 79(17):5327–5331.

Polacheck, W. J., R. Li, S. G. Uzel, and R. D. Kamm. 2013. Microfluidic platforms for mechanobiology. *Lab on a Chip* 13(12):2252–2267.

Prager, W. 1969. On the formulation of constitutive equations for living soft tissues. *Applied Mathematics* 27:128–132.

Pryse, K. M., A. Nekouzadeh, G. M. Genin, E. L. Elson, and G. I. Zahalak. 2003. Incremental mechanics of collagen gels: New experiments and a new viscoelastic model. *Annals of Biomedical Engineering* 31(10):1287–1296.

Radmacher, M. 2002. Measuring the elastic properties of living cells by the atomic force microscope. *Methods in Cell Biology* 68:67–90.

Roeder, B. A., K. Kokini, J. E. Sturgis, J. P. Robinson, and S. L. Voytik-Harbin. 2002. Tensile mechanical properties of three-dimensional type I collagen extracellular matrices with varied microstructure. *Journal of Biomechanical Engineering* 124(2):214–222.

Ronan, W., V. S. Deshpande, R. M. McMeeking, and J. Patrick McGarry. 2012. Numerical investigation of the active role of the actin cytoskeleton in the compression resistance of cells. *Journal of the Mechanical Behavior of Biomedical Materials* 14:143–157.

Sacks, M. S. 2003. Incorporation of experimentally-derived fiber orientation into a structural constitutive model for planar collagenous tissues. *Journal of Biomechanical Engineering* 125(2):280–287.

Schwarz, U., N. Balaban, D. Riveline, A. Bershadsky, B. Geiger, and S. Safran. 2002. Calculation of forces at focal adhesions from elastic substrate data: The effect of localized force and the need for regularization. *Biophysical Journal* 83(3):1380–1394.

Shao, J.-Y. 2009. Biomechanics of leukocyte and endothelial cell surface. *Current Topics in Membranes* 64:25–45.

Skalak, R., S. Chien, and G. W. Schmid-Schonbein. 1984. Viscoelastic deformation of white cells: Theory and analysis. *Kroc Foundation Series* 16:3–18.

Sniadecki, N. J. 2010. Minireview: A tiny touch: Activation of cell signaling pathways with magnetic nanoparticles. *Endocrinology* 151(2):451–457.

Stegemann, J. P., H. Hong, and R. M. Nerem. 2005. Mechanical, biochemical, and extracellular matrix effects on vascular smooth muscle cell phenotype. *Journal of Applied Physiology* 98(6):2321–2327.

Stylianopoulos, T., A. Yeckel, J. J. Derby, X. J. Luo, M. S. Shephard, E. A. Sander, and V. H. Barocas. 2008. Permeability calculations in three-dimensional isotropic and oriented fiber networks. *Physics of Fluids* 20(12):123601.

Tan, J. L., J. Tien, D. M. Pirone, D. S. Gray, K. Bhadriraju, and C. S. Chen. 2003. Cells lying on a bed of microneedles: An approach to isolate mechanical force. *Proceedings of the National Academy of Sciences of the United States of America* 100(4):1484–1489.

Thomopoulos, S., R. Das, V. Birman, L. Smith, K. Ku, E. L. Elson, K. M. Pryse, J. P. Marquez, and G. M. Genin. 2011. Fibrocartilage tissue engineering: The role of the stress environment on cell morphology and matrix expression. *Tissue Engineering. Part A* 17(7–8):1039–1053.

Thoumine, O., R. M. Nerem, and P. R. Girard. 1995. Oscillatory shear stress and hydrostatic pressure modulate cell-matrix attachment proteins in cultured endothelial cells. In vitro *Cellular & Developmental Biology. Animal* 31(1):45–54.

Tondon, A. and R. Kaunas. 2014. The direction of stretch-induced cell and stress fiber orientation depends on collagen matrix stress. *PLoS One* 9(2):e89592.

Tranquillo, R. T. 1999. Self-organization of tissue-equivalents: The nature and role of contact guidance. *Biochemical Society Symposium* 65:27–42.

Tranquillo, R. T. and J. D. Murray. 1992. Continuum model of fibroblast-driven wound contraction: Inflammation-mediation. *Journal of Theoretical Biology* 158(2):135–172.

Tranquillo, R. T. and J. D. Murray. 1993. Mechanistic model of wound contraction. *Journal of Surgical Research* 55(2):233–247.

Trepat, X., M. R. Wasserman, T. E. Angelini, E. Millet, D. A. Weitz, J. P. Butler, and J. J. Fredberg. 2009. Physical forces during collective cell migration. *Nature Physics* 5(6):426–430.

Tucker, C. L. and E. Liang. 1999. Stiffness prediction for unidirectional short-fiber composites: Review and evaluation. *Composites Science and Technology* 59:655–671.

Vidi, P.-A., M. J. Bissell, and S. A. Lelievre. 2013. Three-dimensional culture of human breast epithelial cells: The how and the why. *Methods in Molecular Biology* 945:193–219.

Wakatsuki, T., M. S. Kolodney, G. I. Zahalak, and E. L. Elson. 2000. Cell mechanics studied by a reconstituted model tissue. *Biophysical Journal* 79(5):2353–2368.

Weafer, P., W. Ronan, S. Jarvis, and J. McGarry. 2013. Experimental and computational investigation of the role of stress fiber contractility in the resistance of osteoblasts to compression. *Bulletin of Mathematical Biology* 75(8):1284–1303.

Wei, Z., V. S. Deshpande, R. M. McMeeking, and A. G. Evans. 2008. Analysis and interpretation of stress fiber organization in cells subject to cyclic stretch. *Journal of Biomechanical Engineering* 130(3):031009.

Williams, M. 1952. Stress singularities resulting from various boundary conditions in angular corners of plates in extension. *Journal of Applied Mechanics* 19(4):526–528.

Worthen, G. S., B. d. Schwab, E. L. Elson, and G. P. Downey. 1989. Mechanics of stimulated neutrophils: Cell stiffening induces retention in capillaries. *Science* 245(4914):183–186.

Wren, T. A. and D. R. Carter. 1998. A microstructural model for the tensile constitutive and failure behavior of soft skeletal connective tissues. *Journal of Biomechanical Engineering* 120(1):55–61.

Wu, J. Z. and W. Herzog. 2000. Finite element simulation of location- and time-dependent mechanical behavior of chondrocytes in unconfined compression tests. *Annals of Biomedical Engineering* 28(3):318–330.

Wu, J. Z., W. Herzog, and M. Epstein. 1999. Modelling of location- and time-dependent deformation of chondrocytes during cartilage loading. *Journal of Biomechanics* 32(6):563–572.

Zahalak, G. I., W. B. McConnaughey, and E. L. Elson. 1990. Determination of cellular mechanical properties by cell poking, with an application to leukocytes. *Journal of Biomechanical Engineering* 112(3):283–294.

Zahalak, G. I., J. E. Wagenseil, T. Wakatsuki, and E. L. Elson. 2000. A cell-based constitutive relation for bio-artificial tissues. *Biophysical Journal* 79(5):2369–2381.

Zemel, A., I. B. Bischofs, and S. A. Safran. 2006. Active elasticity of gels with contractile cells. *Physical Review Letters* 97(12):128103.

Zemel, A. and S. A. Safran. 2007. Active self-polarization of contractile cells in asymmetrically shaped domains. *Physical Review. E, Statistical, Nonlinear, and Soft Matter Physics* 76(2 Pt 1):021905.

Zhu, Y. K., T. Umino, X. D. Liu, H. J. Wang, D. J. Romberger, J. R. Spurzem, and S. I. Rennard. 2001. Contraction of fibroblast-containing collagen gels: Initial collagen concentration regulates the degree of contraction and cell survival. *In Vitro Cellular & Developmental Biology—Animal* 37(1):10–16.

8

Cell Motility in 3D Matrices

Yasha Sharma and Muhammad H. Zaman

CONTENTS

Abbreviations

2D   Two Dimensional
3D   Three Dimensional
ECM   Extracellular Matrix
EMT   Epithelial to Mesenchymal Transition
GAGs   Glycosaminoglycans
MMPs   Matrix Metalloproteases
RGD   Arginylglycylaspartic acid
ROCK   Rho-associated protein kinsase

8.1 Introduction

The advent of 3D constructs and confocal imaging in the past two decades has allowed researchers in fields from cell biology to biological engineering to probe cells with an additional and previously unexplored advantage. The third dimension allows for characterization of cells with a real and functional surface area, volume, and steric hindrance, thus providing cell surface protein interactions in every direction. Indeed, cells in 3D are not flat and spread like their 2D counterparts. Instead, they are spherical or ellipsoid bodies of a smaller diameter, sometimes with oblong protrusions in various directions leading to an arborized appearance.[1] Three-dimensional constructs comprising of different materials with different physical and chemical properties have allowed for the characterization of cells in a vast array of conditions.[2] From these studies, cells emerge as increasingly complex systems, influenced by and responding to a host of available mechanisms and parameters. Some fundamental concepts of cell motility, however, remain uniform and can broadly categorize these hosts of behaviors.

In order to move inside a 3D system, a cell must undergo adhesion and invasion. Cell adhesion to the extracellular matrix (ECM) is accomplished when surface proteins on the cell attach to proteins in the ECM. Invasion refers to cells moving to occupy a part of their environment that they did not previously occupy. Invasion is usually accomplished by one of two processes. Cells may break down ECM components in order to create a path in a process called *proteolysis*. Alternatively, cells may squeeze through available pores or holes in the ECM and find an already available path.

Conventionally, three characteristic steps have been used to describe cell motility:

1. Polarization and attachment to the front
2. Contraction of cell body forward
3. Detachment from the rear

Polarization refers to cells forming temporary differences in states across the cell body, generally to form a front and rear along the direction of motion. Cells in 3D do not have a dorsal–ventral polarity as observed in 2D and instead form an active *leading front* or edge surface. Adhesion proteins in the cell membrane move toward this leading front, and many active membrane protrusions reach out into the ECM. In the second step, adhesion proteins in membrane protrusions at the leading front attach to ECM components.

These protrusions may also assist cells in breaking down local ECM proteins to carve a path and are rich in cytoskeletal proteins that align along the direction of protrusions. The cytoskeleton must then exert force within the cell to contract the cell body forward in the third step. For the final step, adhesions in the rear are detached and either deposited in the matrix or relocated along the cell membrane.[3]

Cells in 3D are highly dynamic. The three characteristic steps mentioned earlier can occur concurrently; a cell may polarize and adhere in one direction and polarize and adhere in a different direction immediately after. In order to have a single successful step of cell motility, successive polarization, adhesion, contraction, and detachment of a cell must occur along the same direction.

Cell motility is also crucial to cancer metastasis. Cancer cells acquire enhanced properties similar to those of stem cells and developmental cells, and the initial steps of cell motility allow cancer cells to escape from a primary tumor site and invade the surrounding environment. These cells may also then migrate to secondary sites on the host body and create secondary tumors. Understanding cell motility in 3D has long been considered a crucial aspect of understanding cancer dissemination and metastasis.

8.1.1 Extracellular Matrix

In vivo, the ECM and environment of cells is dynamic, diverse, rich, and complex and varies among different tissues in its composition, structure, and mechanical properties.[4] It comprises of proteins and polysaccharides that cells both respond to and modify, and therefore, the ECM is constantly being *remodeled* by cells. Transmembrane proteins known as integrins bind to ECM components to couple the cell cytoskeleton with the ECM. These cell–matrix adhesions are the fundamental drivers of ECM remodeling and cell fate, for they allow communication between the cell and ECM using the flow of signaling molecules in both directions.[5,6] The ECM provides structural integrity, assists tissue organization, determines cell fate either directly or indirectly, and provides a substrate for motility. ECM proteins also regulate, distribute, and activate soluble growth factors as well as present them to cells.[7,8]

The primary protein in the ECM is collagen, which also makes up more than 30% of all proteins in the human body. Collagen is structurally very important for it provides support, controls stiffness, and resists tension in the ECM. It is also able to bind to adhesion factors such as fibronectin and growth factors that are soluble in the interstitial fluid.[8] It can form a highly porous ECM structure by attaching to other collagen fibrils by covalent bonds or cross-links.[9] The other major structural protein in the ECM is elastin, especially in more elastic connective tissues. Nonstructural adhesive glycoproteins in the ECM are fibronectin, laminin, and vitronectin among others. Integrins bind to these proteins to form cell–matrix adhesions and start signal transduction into the cell from the ECM. The ECM also contains many glycosaminoglycans (GAGs) such as heparin sulfate, chondroitin sulfate, keratin sulfate, and hyaluronic acid, most of which can assemble into proteoglycans. GAGs impart functional as well as structural properties to tissues.[10] Combinations of these ECM components result in a variety of functional, structural, and mechanical properties of ECM throughout different tissues in the body.

8.1.2 3D Cell Culture

Three-dimensional models of ECM allow us to attempt to bridge the gap between in vitro and in vivo cell studies. Three-dimensional models refer to 3D gel-like structures in which cells can be suspended, cultured, and observed. They pose promising for therapeutic purposes as well as from a tissue engineering perspective, allowing medical professionals to envision repairing and administering treatment for cell–ECM composites.[11] The types of 3D culture systems can be classified as natural or synthetic.

8.1.2.1 Models with Natural Matrices

Natural 3D culture systems are usually harvested from cells or animals. When harvesting from animals, natural 3D models are derived by decellularizing tissues. Cell-derived matrices are created by cells in culture systems and later decellularized.[12,13] Matrices or gels can be made from reconstituted collagen, fibrin, fibronectin, elastin, and fibrinogen, among others.[14] The advantage these gels offer over synthetic matrices is that they contain biochemical cues for the cells. Collagen gels can be tuned for pore size, stiffness, and ligand density by varying collagen concentration and pH.[15] Ligand density refers to the number of available sites that the cells can cleave via proteolysis in a given volume of collagen. Denser and larger fibers of collagen result in stiffer matrices. Architecture control in collagen gels is achieved by aligning gels that are stretched. Tissue transglutaminases are used to increase cross-linking, thus making collagen stiffer without increasing ligand density. However, collagen gels are very heterogeneous on the micron scale; thus, a cell may experience extremely different properties across its length or in different parts of a collagen gel.

8.1.2.2 Models with Synthetic Matrices

Controlled or synthetic gels provide the option of molecular control and homogeneity in the matrix structure. However, these gels are created at the expense of in vivo like biochemistry. Synthetic gels are often polymers such as poly(ethylene glycol), poly(vinyl alcohol), and polyacrylates such as poly(2-hydroxyethyl methacrylate). These are typically soft and elastic, thermodynamically compatible with water, and thus useful in many biomedical applications.[16] Various properties such as network mesh size, fiber size, extent of cross-linking, gradients, and known sequences for adhesion and degradation have been controlled in different synthetic matrices. Cells are viable in many such matrices but may not always be migrating; however, the presence of degradable arginylglycylaspartic acid (RGD) elements allows cells in many such matrices to cleave RGD and migrate through the hydrogel.[17–21]

8.1.3 Dimensionality

As we are beginning to see, cells in vivo occupy a diverse environment in which they are surrounded by multiple ECM proteins, other cells, and cell-secreted factors from different cell types. Remarkable and numerous differences exist between these systems and traditional monolayer culture and have been described by some as a *double-edged sword*.[22] Two-dimensional monolayer systems allow for reductionist methods and simple and categorical classification of different cell types and properties. However, it is becoming increasingly apparent that these monolayer systems do not clearly represent their in vivo counterparts.

Dimensionality accounts for differences between monolayer, 3D in vitro, and in vivo systems of cells. Since cells sense, respond to, and modify their environment using a number of mechanisms, the effect of dimensionality cannot be understood as a single parameter

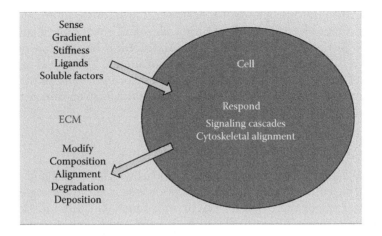

FIGURE 8.1
Cells have a highly dynamic relationship with the ECM. At a given point of time, a cell is able to sense, respond, and modify the ECM concurrently. Other cells may also be interacting with nearby ECM similarly, so cells can also interact with each other through the ECM.

(Figure 8.1). For example, while most cells move slower in 3D collagen gels than they do on 2D glass or collagen-coated dishes,[15] neutrophils that are almost still on glass dishes in 2D cultures display motility in 3D collagen gels.[23] Within 3D too, certain cells migrate proteolytically while others migrate by finding a path through squeeze through.[24] These observations, along with others, have led to the understanding that dimensionality must be explored while taking every factor of experimental conditions into account.

8.2 Initial Steps: Polarization and Adhesion

The polarization of cells (Figure 8.2) is a decisive first step in all migration modes, regardless of dimension. Little is known about the exact initiator of polarization in 3D cell migration; however, we can confidently assess that due to a combination of various cell

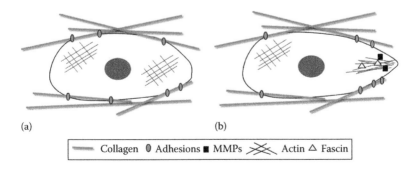

(a)   (b)

Collagen 〇 Adhesions ■ MMPs ⧓ Actin △ Fascin

FIGURE 8.2
Cell polarization. (a) An unpolarized cell represented in 3D. (b) A polarized cell. It has a leading front with organized and bundled actin, more adhesions, and more MMPs.

and ECM properties, two codependent processes occur in order to initiate cell motility—actin polymerization to form membrane protrusions and clustering of adhesive proteins at the leading edge. Cells are extremely dynamic structures on the scale of seconds; their membrane is constantly transient, probing the environment all around as far as possible given the state of the cell. If the membrane, while exploring its environment, comes into contact with a cue that is able to differentiate the cell along a given direction, the polarization process can be initiated.

8.2.1 Actin Polymerization and Membrane Protrusions

Cells form membrane extensions in 3D called protrusive pseudopodia or finger-like filopodia at the leading edge. High actin polymerization is seen at the leading edge and generates a propulsive force in these bodies.[25] Actin monomers (G-actin) are globular proteins that spontaneously assemble to form filaments (F-actin) and have been extensively studied in the cytoskeletal context.[26] These filaments have a barbed end, the site of addition of G-actin monomers and growth, as well as a pointed end, the site of removal of G-actin monomers. Many F-actin filaments are bundled together by cross-linking proteins such as fascin, and the barbed end of collective fibers is oriented toward the membrane. Thus, continual actin polymerization at the barbed end results in the membrane protruding outward to form filopodia.[27] A number of proteins in the cell and ECM can initiate this polarization process directly or by upregulating other proteins, some examples being the actin-related protein (ARP) complex, cofilin, Rho GTPases, Rac1, and Cdc42.[28–31] Filopodia probe the environment for extracellular cues, and the active polymerization initiates clustering of integrins to the leading edge of the cell. Some integrins and other cell surface receptors are transferred to filopodia tips by the cytoskeletal protein myosin-X.[32]

8.2.2 Integrins and Focal Adhesions

Cells in 3D form adhesions in all directions, and more adhesions are recruited to the leading edge as the cell polarizes.[31,33] Integrins are transmembrane heterodimeric proteins containing two distinct subunits called alpha (α) and beta (β) subunits. So far, 18 α and 8 β subunits have been characterized for mammals. Both subunits can enter the plasma membrane, and both have small cytoplasmic portions as well.[34] Combinations of these subunits allow cells to bind to a number of proteins: collagen, laminin, fibronectin, vitronectin, fibrillin, fibrinogen, and more. When integrins transport to the leading edge after formation of filopodia and pseudopodia, their β units bind to ECM proteins. This results in integrin activation, and anchorage proteins in the cytoplasmic portions of integrins bind to the actin cytoskeleton.[5] Actin-stabilizing proteins vinculin, α-actinin, and paxillin assist in this binding and recruit signaling proteins such as focal adhesion kinase (FAK) and Src.[6] This starts a signaling cascade that results in recruiting additional integrins into the region, yielding a cluster of integrins or a focal complex.

Over 50 distinct molecules are associated with adhesions, which are relevant to cell morphology, proliferation, and proteolysis, among other processes.[35,36] They are also crucial in force generation for the second step of cell motility. Much is known about focal complexes in 2D cell motility. They have distinct molecular structure and generally contain paxillin, FAK, and the integrins $\alpha5\beta1$ and $\alpha5\beta3$.[4] In most 3D systems, focal adhesion proteins do not form complexes but instead diffuse throughout the cell membrane. However, despite not forming complexes, focal adhesion proteins drive cell motility in 3D by affecting protrusion activity and membrane deformation.[37] Two-dimensional-like adhesions were only observed in more rigid 3D matrices, or in cells that expressed active GTPase RhoA, consistent with observations that focal adhesion complexes form on stiff 2D substrates.[38–40]

8.3 Intermediate Steps: Force Generation and Contractility

The cellular skeleton or cytoskeleton is responsible for generating force inside of cells. Probing intracellular components in 3D systems presents a technical challenge, and there is still much to learn about the exact mechanism of force generation and contractility. The most important cytoskeletal proteins in the context of cell motility are actin, myosin, and microtubules. Disruption of any of these has been shown to decrease intracellular forces.[41] Live imaging has shown that the actin protrusive force generated in filopodia oscillates with a retraction of actin in the rear to generate motion.[42] In some cases, actomyosin inhibitors showed that forward protrusive force by actin polymerization could alone allow cells to move, but these forces are of much smaller magnitude than produced by the actomyosin system.[43] Actomyosin is formed when myosin motors are translocated along actin filaments using ATP, an extremely well-characterized system that elucidates muscular contraction for us. The Rho/Rho-associated protein kinase (ROCK) pathway is crucial to this process.

8.3.1 Rho/ROCK Pathway

The ROCK is a downstream effector of a small GTPase, RhoA. In 3D, this pathway is essential to cell motility; it causes collagen fibers in the ECM to align by remodeling in the direction of motion and induces phosphorylation of the myosin light chain to generate force.[44] This process of aligning the matrix fibers along the direction of motion is called contact guidance. Reducing ROCK results in cells shaped dendritically with reduced motility compared to standard spindle-shaped cells. A dendritic shape implies that cells have plenty of protrusions in various directions with no preferred direction of movement. Active ROCK results in more polarity and matrix contraction.[1,45,46]

8.3.2 Growth Factors

We have yet to form a complete picture of how growth factors and other small molecules assist force generation in 3D systems; however, some studies of 3D force generation have been conducted using 3D traction force microscopy.[47] A biphasic response of force generated was seen after adding increasing amounts of the growth factor TGF-β to cells in collagen gels. This implies that increasing TGF-β initially increased traction force generated in cells, and, after peaking as a bell-shaped curve, decreased traction formation. Two forms of TGF-β, TGF-β1 and TGF-β3, together increased the biphasic response of force generated, suggesting a cooperative mechanism of action.[48]

8.3.3 Adhesions

In both 2D and 3D systems, cell motility is biphasic with regard to adhesions. Adhesions allow cells to exert a frictional force opposite to the direction of motion and move forward. When there are no adhesions, cells are not able to push against any substrate, and do not move forward. With increasing adhesions, cells increase their frictional force. However, this frictional force does not have the same linear relationship with motility as in classical mechanics. Too many adhesions imply that there are too many cell–matrix connections that the cells may not be able to completely detach in order to move forward. Thus, increasing adhesions results in a bimodal motility behavior, with an initial increase in motility leading to a peak and then a decrease in motility.[49]

8.4 Terminal Steps: Detachment in the Rear

For 3D motility, the exact mechanism of detachment and retraction in the rear of the cell is unknown. Some research indicates that actomyosin contractility actively contracts and releases the rear.[50,51] Over 80% of the cell integrins are left behind in the ECM during this process, while the rest will be used and recycled by the cell for a next motility step.[52] The ligands may be weakened simply by weakened kinetic affinity and a low binding state.[3] Integrins can also be detached or broken down by proteases in the cell, and this debris can leave migration tracks that provide cues and guides for other cells so they follow the same path.[53] This is an example of the kind of natural biochemical signal that could be available to cells when using naturally derived matrices as opposed to synthetic and well-controlled constructs. Rear retraction and detachment is considered to be the rate-limiting step in cell motility, implying that the speed of cell motion is primarily determined by how effective the cell is in detaching from the matrix, and not how protrusive it is, or how well it generates force.

8.5 Cell Migration in 3D

When a cell undergoing motility becomes directed and moves toward a given direction with an average speed and velocity, the process is called cell migration. There are two primary modes of cell migration in 3D (Figure 8.3). The first is mesenchymal migration, in which cells invade into the matrix by creating a path in the matrix around them and is dependent on proteolysis and actin polymerization. The second is amoeboid

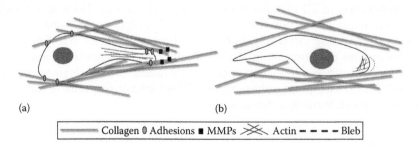

(a)   (b)

Collagen ⊙ Adhesions ▪ MMPs ⨳ Actin − − − − Bleb

FIGURE 8.3
Migration phenotypes. (a) Mesenchymal migration employs adhesions to stick to the matrix, MMPs to break and align ECM, and cytoskeletal propulsive forces. (b) Amoeboid migration starts from blebs that become detached from the cortical cytoskeleton and involve the cell squeezing through holes in the matrix.

TABLE 8.1

Possible Migration Modes for Given Matrix and Cell Properties

	Mesenchymal	Amoeboid
Matrix		
Nondegradable	✘	✓
Stiff	✓	✓
Small pores	✓	✘
Cells		
Deformable	✓	✘
Large volume	✓	✘
Nuclei		
Deformable	✓	✘
Large volume	✓	✘

migration in which cells squeeze through any path available to them using membrane protrusions called blebs (Table 8.1).

8.5.1 Mesenchymal Migration

Mesenchymal migration can be said to have four characteristic steps. In addition to the three steps mentioned earlier, cells break down the matrix around them after forming protrusions and remodel it along their path. The proteolysis and matrix remodeling is followed by force generation and rear retraction. As we are beginning to see however, these processes are not completely independent of each other, and all evidence to date indicates that they are correlated with each other and even positively regulating each other. For instance, the Rho/ROCK pathway directly controls actin contractility, but seems to have a strong impact on matrix remodeling. Cells were observed to change the orientation of collagen fibers and align them along their direction of motion. However, when the same cells were ROCK inhibited, they did not display such behavior. Interestingly, when the same cells were seeded in pre-aligned collagen, they did not reorient the fibers and did not display much motility.[54]

8.5.1.1 Proteolysis and Extracellular Matrix Remodeling

Proteolysis is perhaps the most important step in 3D cell migration. Since there are so many steric hindrances in 3D, the ability of cells to break the matrix down around them and create a new path is critical for migration. It allows cells to create a path in extremely dense matrices, with pore sizes through which they would not be able to migrate otherwise. In fact, when proteases were inhibited in synthetic gels with a pore size of 0.025 μm, cells were not able to migrate at all.[18] Many cancer researchers are extensively focusing on protease activity in cancer cells to probe the critical contributors to cancer invasion and metastasis.[55,56] Chief proteases responsible for proteolysis are matrix metalloproteases (MMPs).

8.5.1.1.1 Matrix Metalloproteases

MMPs are zinc-containing enzymes that can be either membrane bound or soluble. Their inhibition has been shown to decrease the speed and directional persistence of cells in 3D systems.[18,57,58] Directional persistence refers to the amount of time, on average, for which cells move in the same direction. Increased MMPs are commonly associated with tumor cells. Increased MMP activity can be induced in cells by adding growth factors or tumor necrosis factors that increase the number of moving cells as well as speed and directional persistence in 3D.[59]

8.5.1.1.2 *Podosomes and Invadopodia*

Proteolysis during 3D migration is focused and directed. This can occur when MMPs are recruited to invasive cell and membrane protrusions called podosomes and invadopodia. Podosomes and invadopodia in 2D are rounded structures containing F-actin, myosin, microtubules, and proteolytic MMPs.[60–62] These structures allow matrix degradation in the leading edge of the cell, effectively widening the migration path of the cell by remodeling the ECM. In some cases, these may even facilitate collective cell migration by creating a path through which more cells can pass to migrate collectively.

8.5.1.2 Amoeboid Migration

The other primary mode of cell migration and motility is amoeboid migration. It is characterized by rounder cells, with fewer protrusions, less structure in the cytoskeleton, no proteolysis, and no co-clustering of adhesion proteins.[63] It can commonly be observed in cancer cells, leukocytes, and embryonic cells. Cells form blebs, which are protruding portions of the cell membrane that are disconnected from the cortical cytoskeleton because of pressure gradients. These are distinct from filopodia, which protrude outward because they are connected to a rapidly expanding cortical cytoskeleton. In this case, however, the cortical cytoskeleton may rupture, or a high pressure may cause part of the cell membrane to protrude outward. Once blebs are formed, actomyosin contractility allows cells to move forward and migrate.[24,64,65] Some argue that amoeboid motility is not physiologically relevant because it has only been observed in collagen matrices that are not cross-linked, a scenario we are unlikely to find in vivo. When cells were inhibited for MMPs and placed in collagen containing cross-linkers, cells reduced speed but did not migrate using an amoeboid mobility.[66,67] However, leukocytes use amoeboid motility with higher actin contractility instead of bleb formation.[68] Similarly, dendritic cells can migrate with all integrin expression knocked out, indicating that their migration is occurring independent of adhesion and therefore likely by an amoeboid mechanism. However, this phenomenon is only observed in 3D matrices, suggesting perhaps that cells in 3D and in vivo have versatility and plasticity in migration modes.[69]

8.5.1.3 Determinants of Migration Mode

Amoeboid migration is often observed in cells that have undergone an epithelial–mesenchymal transition (EMT). HT-1080 fibrosarcoma cells and MDA-MB-231 carcinoma cells generally adopt a mesenchymal migration mechanism and have been shown to have proteolytic tracks, integrin clustering, and integrin colocalization with MMPs. However, when MMPs are inhibited, these cells are able to switch migration strategies.[24] Similarly, MMP-inhibited cells in collagen or fibrin gels with large pore sizes of 0.5 and 7.4 μm, respectively, were unaffected in migration.[18,70]

Cellular properties that determine the mode of migration are the volume and deformability of the cytoplasm and the volume and deformability of the nucleus. If either the cytoplasm or the nucleus is not deformable to the appropriate degree, the cell will be unable to squeeze through pores in the matrix. ECM properties that determine the mode of migration are porosity, alignment, and stiffness or elasticity of the matrix. If the ECM does not have large enough pores, amoeboid migration is impossible. Stiffer gels may force cells to change the mode of migration as well. Material properties are also crucial to migration mode; if a synthetic and nondegradable material is used, proteolysis is impossible.[71]

8.6 Cell Motility in Cancer

We are now beginning to see that cells display plasticity in migration modes and have a variety of equipment available at their disposal when they need to move. This is especially important in the context of cancer metastasis. Cancer metastasis refers to the ability of cancer cells to escape from their native environment, travel to a secondary site, and create a new secondary tumor. Cell motility and plasticity is at the crux of cancer metastasis.

EMT occurs when epithelial cells become disconnected from neighboring epithelial cells due to downregulation of the cell–cell junction protein E-cadherin and acquire more invasive properties. Numerous studies of cell motility aim to identify aspects of cell motility that are heightened in cancer cells. For instance, we know specifically that MT1-MMP is critical to matrix degradation and invasion by ovarian cancer cells.[72] We also know that the kinase PDK1 indirectly promotes ROCK to regulate cancer motility.[73] The proteins 14-3-3zeta and ErbB2 have been shown to induce EMT in cells and form invasive ductal carcinomas.[74] Such studies have resulted in a vast amount of biological and clinical data, some very specific to particular types of cancer, and some shared across different cancer species. A common thread running through these findings is that enhanced cell motility is an aspect of cancer metastasis. Computational models provide an avenue to integrate clinical findings with biological data.

8.7 Computational Modeling of Cell Motility

The scales and complexities involved in cell motility provide a challenge to computational modeling, especially in 3D systems. Recently though, new data continues to increase and expand our approaches to mathematical models. Computational models are able to replicate behavior from real experiments and have predictive power for trends and dependence on various parameters. A few categories of computational models for 3D motility are described below.

8.7.1 Force-Based Models

Force-based models utilize the characteristic steps of mesenchymal migration to calculate migration speeds. They account for traction forces, matrix density, cell–matrix adhesivity, and drag force from the matrix and have been further extrapolated to include proteolysis and small cell clusters. These models are thus able to calculate cell migration speed dependence on stiffness, adhesion, ligand characteristics among other variables.[75,76]

The first force-based 3D migration model uses a force balance equation, as shown in Equation 8.1, to calculate instantaneous speed of the cell for each time step.[76] The traction force ($F_{traction}$) comprises of at least two components: one in the direction of the movement and the other opposing the movement, represented as F_{trac-f} and F_{trac-r} forward and rearward, respectively, and calculated using Equations 8.2 through 8.5. In these equations, the traction force in either direction is dependent upon the force per receptor ligand complex (F_{R-L}) that has a linear dependence (with a constant c_1 representing area) on Young's modulus (E_{mod}) of surrounding microenvironment up to a certain maximum value, after which the traction force is a constant maximum (c_2). The traction force also scales with a time-dependent adhesivity term $\beta(t)$ that can account for differences in the front and rear of the cell. The protrusive force (F_{prot}) is a function of actin polymerization, and drag force (F_{drag}) is dependent on medium viscosity. The direction of protrusion is randomly chosen after every time step, equivalent to 600 s that is the time required for cells to form stable protrusions that result in cell motility. Smaller protrusions and edge dynamics generally occur at timescales of a few seconds and can be excluded from consideration:

$$F_{total} = F_{prot} + F_{drag} + F_{traction} = 0 \tag{8.1}$$

$$F_{trac-f} = F_{R-L} * \beta_f(t) \tag{8.2}$$

$$F_{trac-b} = F_{R-L} * \beta_r(t) \tag{8.3}$$

$$F_{R-L} = c_1 * E_{mod}(E_{mod} < 1\,\mathrm{MPa}) \tag{8.4}$$

$$F_{R-L} = c_2(E_{mod} \geq 1\,\mathrm{MPa}) \tag{8.5}$$

The adhesivity factor $\beta(t)$ is a dimensionless parameter and is calculated according to Equations 8.6 and 8.7. $\beta_f(t)$ is a multiple of n that is the number of receptors in the front or rear, $[L]$ that is the concentration of ligands, and k the binding constant in the front or rear (subscripts f and r, respectively). Depending on the cell type and microenvironment being considered, these values can be varied to result in either equal or varied adhesivity in the front and rear. The ligand concentration is assumed to be constant for the first model. The drag force (F_{drag}) in Equation 8.8 can now be used to calculate a speed (v) of the cell assuming the cell is a sphere with known radius (r) in a viscoelastic medium with a calculated effective viscosity (η):

$$\beta_f(t) = k_f * n_f * \left[L_f\right] \tag{8.6}$$

$$\beta_r(t) = k_r * n_r * \left[L_r\right] \tag{8.7}$$

$$F_{drag} = 6\pi * r * \eta v \tag{8.8}$$

This model was able to successfully replicate the bimodal dependence of cell speed on ligand density, adhesivity, and detachment force (traction force at the rear). This model also considers asymmetry of cells and shows that the higher degree of asymmetry, the lower cell speed is likely to be. However, this initial model is simplistic in assuming uniform ligand density in the environment of the cell, and successful improvements on the model involve the inclusion of proteolysis to modify ligand density.[77] MMPs in proteolysis degrade the matrix around them, and under the assumption that ligand concentration is linearly related to MMP concentrations, two possible profiles of ligand concentration in relation to MMP concentration ($[M]$) were investigated: linear and log linear as shown in Equations 8.9 and 8.10, respectively. In both of these cases, only the ligand concentration in the front of the cell was investigated since MMPs are recruited to the front or leading edge of the cell in the first step of migration. Since cell adhesion receptors or integrins also regulate MMP expression, the number of receptors available to the cell is a function of MMP concentration ($[M]$) and total receptors available (r) as shown in Equation 8.11. Of these, 95% of receptors are assigned to the front of the cell while the remaining are assigned to the rear according to Equations 8.12 and 8.13. With these modifications, the model predicts a bimodal relationship of cell speed with MMP concentration as well as ligand concentration:

$$\left[L_f \right] = \left[L_{f1} \right] - [M] \tag{8.9}$$

$$\left[L_f \right] = \left[L_{f1} \right] * (1 - [M]) \tag{8.10}$$

$$n \propto r * [M] \tag{8.11}$$

$$n_f = 0.95 * n \tag{8.12}$$

$$n_r = n - n_f \tag{8.13}$$

This model was also successfully modified to track a single-cell cluster speed in 3D.[78] Cell clusters comprising of five cells were investigated in seven different geometries with clusters (Figure 8.4), six of which are more planar geometries and one of which has a 3D arrangement of cells. Protrusion forces and traction forces are calculated for all cells in the cluster and added to obtain the force balance in Equation 8.1.

In order to account for collective motion, a leader cell and a trailing cell are assigned to the cluster in accordance with theoretical and experimental observations. A single cell was assigned as the trailing cell for all geometries. Any cell, based on the randomized direction of the protrusion force, can be assigned as the leading cell for each time step. In any given time step, there is a small probability of cells that are not either leading or trailing to exchange positions; this accounts for cellular translocation within the cluster. All non-trailing cells that do not switch positions contribute traction forces to the cluster. The leader cell contributes the major protrusive force, and all non-trailing cells contribute 25% of the protrusive force of the leader cell.

Since cells are represented with spherical geometry and various cluster structures are studied, the drag force is modified to account for this change as shown in Equation 8.14. In this equation, ρ is the medium density, v is the speed of the cluster, C_D is a dimensionless drag coefficient, and A is the projected area of cluster in the direction of motion:

$$F_{drag} = \frac{1}{2}\rho v^2 * C_D * A \tag{8.14}$$

The incorporation of different cluster geometries into the simulation resulted in noisy data for average cluster speed over 100 runs of the simulation. While bimodal trend of data

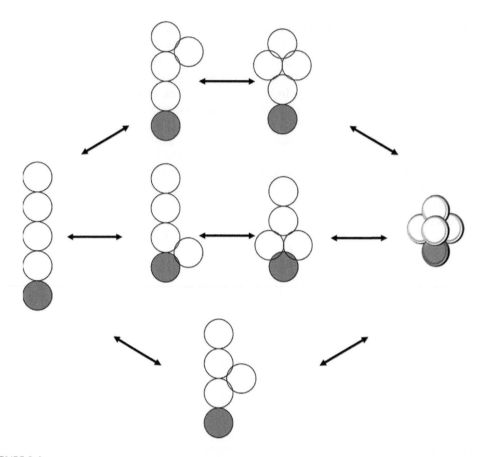

FIGURE 8.4
The cluster geometries explored in the model. Gray circle represents the trailing edge; arrows connect geometries that can interchange within a single time point.[78]

previously observed was preserved in some cases of studying the tetrahedral geometry alone with no changes, incorporating transitions in geometry gives more complicated and realistic data for cluster behavior.

Force-based models have also been modified in clusters to account for cell–cell adhesion and account for cell–pair interactions in clusters as big as 50 cells.[79] Recently, a stress-based model in 3D has also been adapted to account for actin flow, adhesion, and morphology.[80] This model accounts for protrusions into the ECM as flat lamellipodia and has a diffusion gradient across the membrane. Actin is treated as a viscous fluid, and contractile stresses are generated for both actin and myosin protrusions. Focal adhesions are created at random across the cell interior in proportional concentrations to local actin density. This model was effective at predicting morphology for stationary and moving cells and biphasic dependence on adhesion among other factors.

8.7.2 Stochastic Models of Persistent Random Walks

Stochastic random walk models treat each cell as having a persistent random walk[81] and are effective at describing and predicting the behavior of a population. They describe cells within a cubic volume element and find the path taken by each cell according to Equation

8.15 where S_n is the root mean square speed, $D^2(t)$ is the mean square displacement of the cell as a function of time, a is the dimensionality of the system, and P is the persistence time, which is defined as the time span in which cells do not change direction. These models have been successfully used to predict neutrophil motility in 3D matrices.[82] Other models utilizing this method have incorporated anisotropy and stimulus gradients.[83] These models have been successfully used to predict neutrophil motility in 3D matrices and have been adapted to incorporate anisotropy and stimulus gradients:[81,83]

$$\left\langle D^2(t) \right\rangle = a(S_n)^2 P \left[t - P + Pe^{-(t/p)} \right] \tag{8.15}$$

8.7.3 Monte Carlo Models

Monte Carlo methods are based on lattice site occupation of cells and, when applied to model cell motility and migration, are useful for adding ECM factors and accounting for other parameters qualitatively. Some lattice-based models assign space to lattice sites that represent the cell and the ECM, where the cell can be proliferative, quiescent, or necrotic depending on nutrient levels and local conditions, and the ECM can be in fibrous or nonfibrous states.[84] Cells can change between the three conditions with necrosis being a terminal state. Proliferating cells result in more cells replacing ECM sites. A differential adhesion hypothesis that states that cells move until they reach an arrangement with lowest adhesion energy is applied. Cell–cell and cell–ECM adhesion sites are given different energies based on experimental data, and the lattice evolves until the lowest energy state is obtained. Although this is a 2D simulation, it captures cell contractility in cases where cell mass goes through ECM channels smaller than the cell body and thus can be considered a 2D slice of a 3D experiment. Such models have also been expanded to 3D lattices with varying pore size that allow cells to choose between forms of mesenchymal or amoeboid migration, effectively capturing features of early tumor metastasis for a single cell.[85]

8.7.4 Reaction Diffusion Models

Reaction diffusion models treat a cluster of cells as spheroids and account for effects of nutrient and oxygen availability to drive cell movement and cell–cell and cell–matrix adhesions to drive stability. Such models have been successfully used to model cancer invasion by obtaining model parameters from tumor growth and invasion in cell culture plates.[86]

8.8 New Horizons

8.8.1 Multiscale Computational Modeling

As we have explored, various events occur at various timescales and length scales to provide motility to cells. Molecular events such as cytoskeletal reorganization and recruitment of invasive proteins to the leading edge occur within microseconds or milliseconds at the scale of angstroms and nanometers. Cells in 3D show movement of hours and days within microns of movement. In the body however, resulting invasive and metastatic events may occur over months and years, and cells may have traversed large distances of centimeters and decimeters. Multiscale computational models provide an avenue to connect clinical and experimental data over these two scales.[87,88] Algorithms that integrate effects over various scales and relations between them will not only provide essential answers to clinical and biological questions; they will also address the problem of clinical data management across multiple studies and results.[89]

8.8.2 Collective Cell Motility

The dynamic relationship of cells with the ECM allows individual cells to interact with each other via the ECM. Cells may also interact with each other by mechanical coupling via protein complexes called cadherins. Collective cell motility is observed when cells move together while affecting one another, making contact either temporarily or permanently.[90] In three dimensions, collective migration has been observed in the form of vascular sprouting for angiogenesis, branching for morphogenesis in mammary glands, free groups of border cells, detached clusters, and multicellular 3D invasion strands in cancer.[91]

Collective decision making and sensing of environmental signals represent adaptive behaviors believed to offer advantage to tumors.[92] Collective motion in tumors has many behaviors similar to morphogenesis and development. This has been observed to occur by two different mechanisms. On one hand, protruding sheets and strands can maintain contact with the initial tumor and then generate local invasion. On the other hand, a cluster of cells may detach from the original tumor and extend across gaps in tissue along the path of least resistance.[93]

Physical observations have led to the belief that collective migration is the principal method of invasion in epithelial cancers. This is especially the case for epithelial cancer cells that do not undergo EMT, which used to be considered the only form of tumor dissemination and invasion.[94–96] Cells that retain their epithelial characteristics are more inclined to retain cell–cell contacts and migrate in a cluster or collective, as has been observed in squamous carcinoma cells (SCCs). When only a few of these cells acquire invasive properties, they become leader cells and can allow an otherwise noninvasive cluster to become invasive, as has been shown in some in vitro studies.[97–99] In fact, it is believed that the larger a group or cluster, the fewer leader cells are required to guide it. Thus, EMT is no longer considered a requirement for a cancer to be invasive.[90,100,101]

Future studies on collective cell motility in 3D will address cell collectives in 3D environments in vitro and attempt to display collective motility. Collective motility is observed when the system behaves as *larger than the sum of its parts* and has properties unlikely to exist within single cells.[102] It has been observed in many biological systems such as bacteria, locusts, fish, and birds and will be studied in cells as well. Leadership has been characterized statistically for these systems by finding a unit that is responsible for leading one or many other units of the system along its general path and direction.[103] Collective 3D studies will adapt techniques used to study collective motility in other systems for cells to identify leader and malignant cells in collective cell cancer systems. Such research will bridge the gap between current studies of collective migration at the cellular and molecular level with those at the systems level that study interactions between cells, density-dependent phases, and the emergence of collective properties. It will identify the essential mechanisms to collective migration and invasion in epithelial-derived cancer, thus providing crucial information to cancer treatment and therapeutic research in addition to progressing basic understanding of cell motility.

8.9 Conclusion

Research on cells in 3D is progressing faster than ever before with newer biomaterials and imaging modalities being explored and heightened interest in understanding fundamental questions and their relevance to disease. While there is much to be learned, we can summarize what we know so far into a few key observations:

1. Cells in 3D are highly dynamic systems.
2. Three-dimensional cell motility does share some salient features with 2D behavior; however, cells in 3D have plasticity: they can use multiple pathways and adopt different modes of migration.
3. The cytoskeleton and the ECM are coupled mechanically to facilitate motility.
4. The same molecules can initiate cytoskeletal as well as extracellular organization and alignment. The Rho/ROCK pathway in particular initiates both actin propulsion and collagen alignment.
5. Various promoters of high motility display biphasic trends.

As more pieces of the puzzle that is cell motility in 3D start to fit in, multiscale modeling approaches improve, and collective cell motility is understood, we expect we will get closer and closer to understanding cell motility in vivo and in diseased systems.

Acknowledgment

The authors would like to acknowledge the support of NIH grant 1U01CA177799-01.

References

1. Roy, P., Petroll, W. M., Cavanagh, H. D., Chuong, C. J., and Jester, J. V. An in vitro force measurement assay to study the early mechanical interaction between corneal fibroblasts and collagen matrix. *Exp. Cell Res.* **232**, 106–117 (1997).
2. Wang, W. et al. 3D spheroid culture system on micropatterned substrates for improved differentiation efficiency of multipotent mesenchymal stem cells. *Biomaterials* **30**, 2705–2715 (2009).
3. Friedl, P. and Bröcker, E. B. The biology of cell locomotion within three-dimensional extracellular matrix. *Cell. Mol. Life Sci.* **57**, 41–64 (2000).
4. Pedersen, J. A. and Swartz, M. A. Mechanobiology in the third dimension. *Ann. Biomed. Eng.* **33**, 1469–1490 (2005).
5. Baker, E. L. and Zaman, M. H. The biomechanical integrin. *J. Biomech.* **43**, 38–44 (2010).
6. Berrier, A. L. and Yamada, K. M. Cell–matrix adhesion. *J. Cell. Physiol.* **213**, 565–573 (2007).
7. Griffith, L. G. and Swartz, M. A. Capturing complex 3D tissue physiology in vitro. *Nat. Rev. Mol. Cell Biol.* **7**, 211–224 (2006).
8. Hynes, R. O. The extracellular matrix: Not just pretty fibrils. *Science* **326**, 1216–1219 (2009).
9. Eyre, D. R. and Wu, J.-J. Collagen cross-links. *Top. Curr. Chem.* **247**, 207–229 (2005).

10. Nia, H. T., Han, L., Li, Y., Ortiz, C., and Grodzinsky, A. Poroelasticity of cartilage at the nanoscale. *Biophys. J.* **101**, 2304–2313 (2011).
11. Schmeichel, K. L. and Bissell, M. J. Modeling tissue-specific signaling and organ function in three dimensions. *J. Cell Sci.* **116**, 2377–2388 (2003).
12. King, S. J. and Parsons, M. Imaging cells within 3D cell-derived matrix. *Methods Mol. Biol.* **769**, 53–64 (2011).
13. Ott, H. C. et al. Perfusion-decellularized matrix: Using nature's platform to engineer a bioartificial heart. *Nat. Med.* **14**, 213–221 (2008).
14. Hooper, S., Marshall, J. F., and Sahai, E. Tumor cell migration in three dimensions. *Methods Enzymol.* **406**, 625–643 (2006).
15. Harjanto, D., Maffei, J. S., and Zaman, M. H. Quantitative analysis of the effect of cancer invasiveness and collagen concentration on 3D matrix remodeling. *PLoS One* **6**, e24891 (2011).
16. Slaughter, B. V., Khurshid, S. S., Fisher, O. Z., Khademhosseini, A., and Peppas, N. A. Hydrogels in regenerative medicine. *Adv. Mater.* **21**, 3307–3329 (2009).
17. Schwartz, M. P. et al. A synthetic strategy for mimicking the extracellular matrix provides new insight about tumor cell migration. *Integr. Biol. (Camb).* **2**, 32–40 (2010).
18. Raeber, G. P., Lutolf, M. P., and Hubbell, J. A. Molecularly engineered PEG hydrogels: A novel model system for proteolytically mediated cell migration. *Biophys. J.* **89**, 1374–1388 (2005).
19. Guarnieri, D. et al. Covalently immobilized RGD gradient on PEG hydrogel scaffold influences cell migration parameters. *Acta Biomater.* **6**, 2532–2539 (2010).
20. Kenawy, E. R. et al. Electrospinning of poly(ethylene-co-vinyl alcohol) fibers. *Biomaterials* **24**, 907–913 (2003).
21. Luo, Y. and Shoichet, M. S. A photolabile hydrogel for guided three-dimensional cell growth and migration. *Nat. Mater.* **3**, 249–253 (2004).
22. Baker, B. M. and Chen, C. S. Deconstructing the third dimension: How 3D culture microenvironments alter cellular cues. *J. Cell Sci.* **125**, 3015–3024 (2012).
23. Brown, A. F. Neutrophil granulocytes: Adhesion and locomotion on collagen substrata and in collagen matrices. *J. Cell Sci.* **58**, 455–467 (1982).
24. Wolf, K. et al. Compensation mechanism in tumor cell migration: Mesenchymal-amoeboid transition after blocking of pericellular proteolysis. *J. Cell Biol.* **160**, 267–277 (2003).
25. Mogilner, A. and Oster, G. Cell motility driven by actin polymerization. *Biophys. J.* **71**, 3030–3045 (1996).
26. Kim, T., Hwang, W., and Kamm, R. D. Computational analysis of a cross-linked actin-like network. *Exp. Mech.* **49**, 91–104 (2007).
27. Vignjevic, D. et al. Role of fascin in filopodial protrusion. *J. Cell Biol.* **174**, 863–875 (2006).
28. Yamaguchi, H., Wyckoff, J., and Condeelis, J. Cell migration in tumors. *Curr. Opin. Cell Biol.* **17**, 559–564 (2005).
29. Okeyo, K. O., Adachi, T., and Hojo, M. Mechanical regulation of actin network dynamics in migrating cells. *J. Biomech. Sci. Eng.* **5**, 186–207 (2010).
30. Buccione, R., Orth, J. D., and McNiven, M. A. Foot and mouth: Podosomes, invadopodia and circular dorsal ruffles. *Nat. Rev. Mol. Cell Biol.* **5**, 647–657 (2004).
31. Chhabra, E. S. and Higgs, H. N. The many faces of actin: Matching assembly factors with cellular structures. *Nat. Cell Biol.* **9**, 1110–1121 (2007).
32. Arjonen, A., Kaukonen, R., and Ivaska, J. Filopodia and adhesion in cancer cell motility. *Cell Adh. Migr.* **5**, 421–430 (2011).
33. Lundquist, E. A. The finer points of filopodia. *PLoS Biol.* **7**, e1000142 (2009).
34. Humphries, J. D., Byron, A., and Humphries, M. J. Integrin ligands at a glance. *J. Cell Sci.* **119**, 3901–3903 (2006).
35. Zamir, E. and Geiger, B. Molecular complexity and dynamics of cell-matrix adhesions. *J. Cell Sci.* **114**, 3583–3590 (2001).
36. Lock, J. G., Wehrle-Haller, B., and Strömblad, S. Cell-matrix adhesion complexes: Master control machinery of cell migration. *Semin. Cancer Biol.* **18**, 65–76 (2008).

37. Fraley, S. I. et al. A distinctive role for focal adhesion proteins in three-dimensional cell motility. *Nat. Cell Biol.* **12**, 598–604 (2010).
38. Paszek, M. J. et al. Tensional homeostasis and the malignant phenotype. *Cancer Cell* **8**, 241–254 (2005).
39. Peyton, S. R., Kim, P. D., Ghajar, C. M., Seliktar, D., and Putnam, A. J. The effects of matrix stiffness and RhoA on the phenotypic plasticity of smooth muscle cells in a 3-D biosynthetic hydrogel system. *Biomaterials* **29**, 2597–2607 (2008).
40. Pelham, R. J. and Wang, Y.-L. Cell locomotion and focal adhesions are regulated by substrate flexibility. *Proc. Natl. Acad. Sci. USA* **94**, 13661–13665 (1997).
41. Kraning-Rush, C. M., Carey, S. P., Califano, J. P., Smith, B. N., and Reinhart-King, C. A. The role of the cytoskeleton in cellular force generation in 2D and 3D environments. *Phys. Biol.* **8**, 015009 (2011).
42. Starke, J., Maaser, K., Wehrle-Haller, B., and Friedl, P. Mechanotransduction of mesenchymal melanoma cell invasion into 3D collagen lattices: Filopod-mediated extension-relaxation cycles and force anisotropy. *Exp. Cell Res.* **319**, 2424–2433 (2013).
43. Fournier, M. F., Sauser, R., Ambrosi, D., Meister, J. J., and Verkhovsky, A. B. Force transmission in migrating cells. *J. Cell Biol.* **188**, 287–297 (2010).
44. Provenzano, P. P., Inman, D. R., Eliceiri, K. W., and Keely, P. J. Matrix density-induced mechano-regulation of breast cell phenotype, signaling and gene expression through a FAK-ERK linkage. *Oncogene* **28**, 4326–4343 (2009).
45. Gunzer, M. et al. Migration of dendritic cells within 3-D collagen lattices is dependent on tissue origin, state of maturation, and matrix structure and is maintained by proinflammatory cytokines tissue origins within 3-D collagen lattices through. *J. Leukoc. Biol.* **67**, 622–629 (2000).
46. Kim, A., Lakshman, N., and Petroll, W. M. Quantitative assessment of local collagen matrix remodeling in 3-D culture: The role of Rho kinase. *Exp. Cell Res.* **312**, 3683–3692 (2006).
47. Franck, C., Maskarinec, S. A, Tirrell, D. A., and Ravichandran, G. Three-dimensional traction force microscopy: A new tool for quantifying cell-matrix interactions. *PLoS One* **6**, e17833 (2011).
48. Brown, R. A. et al. Enhanced fibroblast contraction of 3D collagen lattices and integrin expression by TGF-beta1 and -beta3: Mechanoregulatory growth factors? *Exp. Cell Res.* **274**, 310–322 (2002).
49. Zaman, M. H. et al. Migration of tumor cells in 3D matrices is governed by matrix stiffness along with cell-matrix adhesion and proteolysis. *Proc. Natl. Acad. Sci. USA* **103**, 10889–10894 (2006).
50. Clow, P. A. and McNally, J. G. In vivo observations of myosin II dynamics support a role in rear retraction. *Mol. Biol. Cell* **10**, 1309–1323 (1999).
51. Kirfel, G., Rigort, A., Borm, B., and Herzog, V. Cell migration: Mechanisms of rear detachment and the formation of migration tracks. *Eur. J. Cell Biol.* **83**, 717–724 (2004).
52. Palecek, S. P., Schmidt, C. E., Lauffenburger, D. A., and Horwitz, A. F. Integrin dynamics on the tail region of migrating fibroblasts. *J. Cell Sci.* **109 (Pt 5)**, 941–952 (1996).
53. Palecek, S. P., Huttenlocher, A., Horwitz, A. F., and Lauffenburger, D. A. Physical and biochemical regulation of integrin release during rear detachment of migrating cells. *J. Cell Sci.* **111 (Pt 7)**, 929–940 (1998).
54. Provenzano, P. P., Inman, D. R., Eliceiri, K. W., Trier, S. M., and Keely, P. J. Contact guidance mediated three-dimensional cell migration is regulated by Rho/ROCK-dependent matrix reorganization. *Biophys. J.* **95**, 5374–5384 (2008).
55. Pietras, K. and Ostman, A. Hallmarks of cancer: Interactions with the tumor stroma. *Exp. Cell Res.* **316**, 1324–1331 (2010).
56. Friedl, P. and Wolf, K. Tube travel: The role of proteases in individual and collective cancer cell invasion. *Cancer Res.* **68**, 7247–7249 (2008).
57. Kim, H. et al. Epidermal growth factor-induced enhancement of glioblastoma cell migration in 3D arises from an intrinsic increase in speed but an extrinsic matrix- and proteolysis-dependent increase in persistence. *Mol. Biol. Cell* **19**, 4249–4259 (2008).

58. Fisher, K. E. et al. Tumor cell invasion of collagen matrices requires coordinate lipid agonist-induced G-protein and membrane-type matrix metalloproteinase-1-dependent signaling. *Mol. Cancer* **5**, 69 (2006).

59. Wolf, K. et al. Multi-step pericellular proteolysis controls the transition from individual to collective cancer cell invasion. *Nat. Cell Biol.* **9**, 893–904 (2007).

60. Rottiers, P. et al. TGFbeta-induced endothelial podosomes mediate basement membrane collagen degradation in arterial vessels. *J. Cell Sci.* **122**, 4311–4318 (2009).

61. Artym, V. V, Zhang, Y., Seillier-Moiseiwitsch, F., Yamada, K. M., and Mueller, S. C. Dynamic interactions of cortactin and membrane type 1 matrix metalloproteinase at invadopodia: Defining the stages of invadopodia formation and function. *Cancer Res.* **66**, 3034–3043 (2006).

62. Deryugina, E. I. and Quigley, J. P. Matrix metalloproteinases and tumor metastasis. *Cancer Metastasis Rev.* **25**, 9–34 (2006).

63. Guck, J., Lautenschläger, F., Paschke, S., and Beil, M. Critical review: Cellular mechanobiology and amoeboid migration. *Integr. Biol. (Camb).* **2**, 575–583 (2010).

64. Charras, G. and Paluch, E. Blebs lead the way: How to migrate without lamellipodia. *Nat. Rev. Mol. Cell Biol.* **9**, 730–736 (2008).

65. Narumiya, S. and Watanabe, N. Migration without a clutch. *Nat. Cell Biol.* **11**, 1394–1396 (2009).

66. Sabeh, F., Shimizu-Hirota, R., and Weiss, S. J. Protease-dependent versus -independent cancer cell invasion programs: Three-dimensional amoeboid movement revisited. *J. Cell Biol.* **185**, 11–19 (2009).

67. Bloom, R. J., George, J. P., Celedon, A., Sun, S. X., and Wirtz, D. Mapping local matrix remodeling induced by a migrating tumor cell using three-dimensional multiple-particle tracking. *Biophys. J.* **95**, 4077–4088 (2008).

68. Renkawitz, J. et al. Adaptive force transmission in amoeboid cell migration. *Nat. Cell Biol.* **11**, 1438–1443 (2009).

69. Lämmermann, T. et al. Rapid leukocyte migration by integrin-independent flowing and squeezing. *Nature* **453**, 51–55 (2008).

70. Raeber, G. P., Lutolf, M. P., and Hubbell, J. A. Mechanisms of 3-D migration and matrix remodeling of fibroblasts within artificial ECMs. *Acta Biomater.* **3**, 615–629 (2007).

71. Wolf, K. and Friedl, P. Extracellular matrix determinants of proteolytic and non-proteolytic cell migration. *Trends Cell Biol.* **21**, 736–744 (2011).

72. Sodek, K. L., Ringuette, M. J., and Brown, T. J. MT1-MMP is the critical determinant of matrix degradation and invasion by ovarian cancer cells. *Br. J. Cancer* **97**, 358–367 (2007).

73. Pinner, S. and Sahai, E. PDK1 regulates cancer cell motility by antagonising inhibition of ROCK1 by RhoE. *Nat. Cell Biol.* **10**, 127–137 (2008).

74. Lu, J. et al. 14-3-3zeta Cooperates with ErbB2 to promote ductal carcinoma in situ progression to invasive breast cancer by inducing epithelial-mesenchymal transition. *Cancer Cell* **16**, 195–207 (2009).

75. Zaman, M. H., Kamm, R. D., Matsudaira, P., and Lauffenburger, D. A. Computational model for cell migration in three-dimensional matrices. *Biophys. J.* **89**, 1389–1397 (2005).

76. Zaman, M. H. Multiscale modeling of tumor cell migration. *Phys. Biol.* **851**, 117–122 (2006).

77. Harjanto, D. and Zaman, M. H. Computational study of proteolysis-driven single cell migration in a three-dimensional matrix. *Ann. Biomed. Eng.* **38**, 1815–1825 (2010).

78. Vargas, D. A. and Zaman, M. H. Computational model for migration of a cell cluster in three-dimensional matrices. *Ann. Biomed. Eng.* **39**, 2068–2079 (2011).

79. Frascoli, F., Hughes, B. D., Zaman, M. H., and Landman, K. A. A computational model for collective cellular motion in three dimensions: General framework and case study for cell pair dynamics. *PLoS One* **8**, e59249 (2013).

80. Shao, D., Levine, H., and Rappel, W.-J. Coupling actin flow, adhesion, and morphology in a computational cell motility model. *Proc. Natl. Acad. Sci. USA* **109**, 6851–6856 (2012).

81. Dickinson, R. B. and Tranquillo, R. T. Optimal estimation of cell movement indices from the statistical analysis of cell tracking data. *AIChE J.* **39**, 1995–2010 (1993).

82. Parkhurst, M. R. and Saltzman, W. M. Quantification of human neutrophil motility in three-dimensional collagen gels. Effect of collagen concentration. *Biophys. J.* **61**, 306–315 (1992).

83. Dickinson, R. B. A generalized transport model for biased cell migration in an anisotropic environment. *J. Math. Biol.* **40**, 97–135 (2000).

84. Rubenstein, B. M. and Kaufman, L. J. The role of extracellular matrix in glioma invasion: A cellular Potts model approach. *Biophys. J.* **95**, 5661–5680 (2008).

85. Zaman, M. H. A multiscale probabilistic framework to model early steps in tumor metastasis. *Mol. Cell. Biomech.* **4**, 133–141 (2007).

86. Frieboes, H. B. et al. An integrated computational/experimental model of tumor invasion. *Cancer Res.* **66**, 1597–1604 (2006).

87. Stolarska, M. A., Kim, Y., and Othmer, H. G. Multi-scale models of cell and tissue dynamics. *Philos. Trans. A. Math. Phys. Eng. Sci.* **367**, 3525–3553 (2009).

88. Cai, A. Q., Landman, K. A., and Hughes, B. D. Multi-scale modeling of a wound-healing cell migration assay. *J. Theor. Biol.* **245**, 576–594 (2007).

89. Zaman, M. H. The role of engineering approaches in analysing cancer invasion and metastasis. *Nat. Rev. Cancer* **13**, 596–603 (2013).

90. Rørth, P. Collective cell migration. *Annu. Rev. Cell Dev. Biol.* **25**, 407–429 (2009).

91. Friedl, P. and Gilmour, D. Collective cell migration in morphogenesis, regeneration and cancer. *Nat. Rev. Mol. Cell Biol.* **10**, 445–457 (2009).

92. Deisboeck, T. S. and Couzin, I. D. Collective behavior in cancer cell populations. *Bioessays* **31**, 190–197 (2009).

93. Mierke, C. T. Physical break-down of the classical view on cancer cell invasion and metastasis. *Eur. J. Cell Biol.* **92**, 89–104 (2013).

94. Thiery, J. P. and Lim, C. T. Tumor dissemination: An EMT affair. *Cancer Cell* **23**, 272–273 (2013).

95. Thiery, J. P. Epithelial-mesenchymal transitions in tumour progression. *Nat. Rev. Cancer* **2**, 442–454 (2002).

96. Thiery, J. P. Metastasis: Alone or together? *Curr. Biol.* **19**, R1121–R1123 (2009).

97. Gaggioli, C. Collective invasion of carcinoma cells: When the fibroblasts take the lead. *Cell Adh. Migr.* **2**, 45–47 (2008).

98. Gaggioli, C. et al. Fibroblast-led collective invasion of carcinoma cells with differing roles for RhoGTPases in leading and following cells. *Nat. Cell Biol.* **9**, 1392–1400 (2007).

99. Carey, S. P., Starchenko, A., McGregor, A. L., and Reinhart-King, C. A. Leading malignant cells initiate collective epithelial cell invasion in a three-dimensional heterotypic tumor spheroid model. *Clin. Exp. Metastasis* **30**, 615–630 (2013).

100. Kabla, A. J. Collective cell migration: Leadership, invasion and segregation. *J. R. Soc. Interface* **9**, 3268–3278 (2012).

101. Khalil, A. A. and Friedl, P. Determinants of leader cells in collective cell migration. *Integr. Biol. (Camb).* **2**, 568–574 (2010).

102. Vicsek, T. and Zafeiris, A. Collective motion. *Phys. Rep.* **517**, 71–140 (2012).

103. Nagy, M., Akos, Z., Biro, D., and Vicsek, T. Hierarchical group dynamics in pigeon flocks. *Nature* **464**, 890–893 (2010).

9

Collective Cell Migration

Nir S. Gov

CONTENTS

9.1 Introduction

What do we mean by "collective cell migration"? The scope of this phenomenon includes different forms of motion of groups of cells moving together. In this review, we focus only on eukaryotic cells that belong to a multicellular organism, thereby not treating the collective motion of bacteria. Most cells of a multicellular adult organism are rather stationary, but situations of collective migration of cells arise during the normal embryonic development process [1–3] and the physiological responses during wound healing or immune response. It also plays an important role during pathologies such as cancer metastasis [4]. During these processes, cells have to be both motile and have a certain degree of adhesion to one another [5], so that the collectivity of the migration is maintained.

While single-cell motility has been extensively studied, much less is understood about the mechanisms of collective cell migration [6]. This field has seen a huge increase in attention over the past decade [7]. On the experimental side, there is an increasing use of patterning techniques for the fabrication of well-controlled geometries [8–16]. These are complemented by techniques to probe the motion and the traction forces during the migration [17–20]. These experiments have enabled to probe collective cell migration in vitro in great detail. Collective cell migration in vivo is of course much more complex, and ultimately the knowledge gained from in vitro experiments will benefit the understanding of such systems as well.

The theoretical modeling of this phenomenon has explored several approaches, most notably the use of particle-based simulations that include some form of orientational ordering interactions [21–27], similar to the descriptions used for flocking and swarming phenomena [28].

The ability of cells to move collectively, even within a dense tissue, naturally raises the question of how is collective motion of cells different from the motility of solitary cells, as far as traction force production and dynamics. In addition, a major open puzzle in this field relates to the way cells communicate with each other to produce organized collective migration and the determinants that control the length scale of coherent and correlated cellular migration. These open questions are the focus of much current research.

There is at present relatively little knowledge about the molecular regulators of collective cell migration. This field presents a challenge both for biologists and for physicists, which are collaborating in order to extend and deepen our understanding. There have been several recent reviews of this field, which cover in detail the in vivo, biological, and experimental aspects. This chapter will be therefore mostly concerned with the physical aspects of collective cell migration, therefore focusing on in vitro experiments and *theoretical* models. It aims to place this scientific inquiry within the larger context of biological physics and active matter [29], where biological phenomena require and inspire new physical concepts and ideas.

9.2 From Single-Cell Motility to Collective Motion: What Are the Relevant Forces

The mechanisms that drive the motility of a single eukaryotic cell are varied (see recent reviews [6,30]). While the exact details that control cell motility are still actively explored and not fully understood, it seems that there are two main categories: (1) lamellipodia-driven motility which utilizes the formation of a single lamellipodia at the cell leading edge (this structure provides traction forces through the friction of treadmilling actin over adhesion complexes; this mechanism is quite universal, and to be effective, it involves the polarization of the cell such that there is only one dominant lamellipodia) and (2) amoeboid-like motility which utilizes the contractile force of myosin at the cell trailing edge to "pump" the cell cytoplasm into an expanding bleb at the cell front. Different cells may switch between these two classes of motility mechanisms, depending on the metabolic conditions or due to the geometry of the space in which they move. For example, weakly adhering cells that produce ineffective lamellipodia and move very slowly on a flat surface become fast-moving cells when confined in a narrow channel, where they move by utilizing strong contractility at their trailing edge [31–33].

How do cells move around when they are in contact with neighboring cells, for example, within a confluent monolayer [5,34]? It is observed that individual cells move more slowly when in contact with neighboring cells. The origins for this slowing down can be varied: When cells "bump" into each other, they physically inhibit their ability to extend large lamellipodia, called "contact inhibition of locomotion" [35], similar to the inhibition cells exhibit when an external force is applied artificially [36]. Some "cryptic" lamellipodia, which extend beneath the neighboring cells, are still able to form and provide traction forces [37]. We therefore expect the traction forces produced by cells within a dense collective to be lower than for individual cells. Nevertheless, large traction forces, comparable to those produced by isolated cells, are still observed beneath confluent monolayers [38–40].

Another source for lower cell velocity could arise from the larger restoring forces exerted on each cell by its neighbors. Even with the traction forces unchanged, a particle moving in a potential landscape with deeper barriers will end up moving more slowly. This is illustrated by simulations of hard-core repulsive particles, where a lower diffusion coefficient is observed as the density increases [41].

Finally, the physical linkages that form between neighboring cells, such as cadherin-based cell–cell junctions, contribute to the friction forces acting on cells that slide past each other. The transient formation, stretching, and unbinding of these bonds induce an effective friction force [42]. Summarizing these effects, we can write the following equation of motion for a cell within a cell layer:

$$\vec{v}_{cell} = \frac{\vec{F}_{tract} + \sum_{nn} \vec{F}_{pot}}{\lambda_{surf} + \lambda_{cell-cell}} \tag{9.1}$$

where
  F_{tract} is the active traction force produced by the cell
  F_{pot} is the force due to the effective potential of the nearest-neighbor cells
  λ_{surf} and $\lambda_{cell-cell}$ are the friction coefficients due to the cell–surface contact and the cell–cell contacts, respectively

Note that the effective cell–cell and cell–substrate friction is itself velocity dependent, which then makes Equation 9.1 an implicit equation for the cell velocity v.

All of these effects may of course occur simultaneously, and current research is aimed at deciphering how they are combined and in which situations one is dominant. For example, it has been observed that the mean velocity of cells within a confluent monolayer decreases as the density of the cells increases [13,40,43–45]. The increased density could give rise to all the three effects mentioned earlier, by increasing the inhibition of cryptic lamellipodia and by inducing stronger confining potentials and intercellular friction. The decreased lamellipodia formation due to increased density was estimated using a simple model in Ref. [13]: The traction force that each cell produces can be regarded as linearly proportional to the contact area of these lamellipodia around the cell rim, that is, $F_{tract} \approx l\sqrt{A}$, where l is the protrusion extension of the lamellipodia beneath the neighboring cell and A is the area of the cell contact with the substrate (the areal density of cells is $\rho = 1/A$). If the cellular volume is constant, $V_{cell} = Ah$, then we can replace the cell layer thickness h by $h = V_{cell}/A$. The friction force on the cell is $F_{fric} \sim \lambda A v_{cell}$, where λ is the friction per unit area of cell–substrate contact and v is the cell velocity. Equating these two forces, we get $v_{cell} \propto l/\sqrt{A} \sim l\sqrt{\rho}$. The extension of the cryptic lamellipodia l decreases with increased cell density, due to the larger restoring force exerted by the dense layer. Consider that a subcellular lamellipodia is driven by a constant force per unit length along the cell rim f_0. The elasticity of the cell layer resisting the deformation upward that the cryptic lamellipodia induces increases linearly with the layer thickness h. The lamellipodia thickness δ, which we consider to be linearly related to its lateral extent l, is therefore given by $\delta \sim f_0/(h\,E_b)$, where E_b is the effective elastic bulk modulus of the cell layer. If the cell volume is constant, then we have $l \sim 1/\rho$. Using this result in Equation 9.1, we get that $v_{cell} \propto \rho^{-1/2}$, while the observed relation is with exponent ~ -0.7 [13]. This calculation should be treated as a mere estimate, since it is difficult to disentangle the different processes that contribute to the cell velocity.

Beyond the slowing down of cells moving within a tissue, there are also more subtle interactions observed among cells [5,34,46]. Most prominent is the tendency for cells to move in a coordinated fashion. Such behavior, which has been compared to the flocking and swarming motion of larger organisms [28,29], is revealed by velocity–velocity correlations that extend over a range of tens of cells [10,40,43–45,47–49]. Cells move collectively in a coordinated manner, following each other, presumably due to mechanochemical signaling that converts the forces that they exert on each other into reorganization of the internal cytoskeleton machinery involved in the cell motility. One possible link between the applied forces and the internal chemistry of the cytoskeleton is through the cadherin-based cell–cell junctions [50–52]. Another study has attributed the coordination of motion between cells to communication mediated by the stress fields that cells induce in the underlying substrate [48], while a further study has revealed a possible tendency of cells to minimize the shear stresses that form between them. This later form of coordination was termed "plithotaxis" [39]. The details of the intercellular interactions that lead to coordinated motion are not clear at present. For example, when a cellular layer is expanding (see next section), the cells in the bulk feel a pull exerted on them by the edge cells and respond by directing their motility toward the free edge [10,44,47]. However, as recently demonstrated in one system, cells in small isolated groups may respond to a pulling force by trying to pull in the opposite direction [53], that is, becoming anticorrelated with their neighbors. The nature of these cell–cell interactions and how they are integrated to control the collective cell migration is the major open puzzle in this field.

The description we have given earlier for the forces determining cell motion within a confluent monolayer is modified for the cells at the free edge, as we describe in the next section.

9.3 Cells on the Edge: Collective Motion in Confined Geometries and in Expanding Systems

9.3.1 Confined Geometries

In order to study collective cell motion, it is useful to have a well-controlled geometry for the confluent monolayer. A few such geometries are shown in Figure 9.1. In Figure 9.1a, we see isolated "islands" of cells grown on circular adhesive patches, and this is an example that illustrates the difference in behavior between the cells at the monolayer edge and those in its interior [13,14]. Cells at the edge of these cellular islands seem to be elongated, and overall more dynamic than the bulk cells, exerting larger traction forces on the substrate [18].

Furthermore, it is found that for small enough radii, the cells tend to circulate coherently within the circular island [8,9,13]. This circulation resembles solid-body rotation, such that the cellular velocity is tangential, and its value increases linearly with the radius. This form of coherent motion is lost when the islands have a radius larger than ~100–200 μm (for Madin-Darby canine kidney (MDCK) cells). For cells with weaker cell–cell adhesion, such as when adherens junctions are downregulated and in cancerous cell types, this correlation length is significantly smaller. This experiment is a larger version of the "yin-yang" experiment of two corotating cells performed before [8,9] and is a 2D version of the 3D rotation observed in vivo [1,54,55]. These observations highlight the fact that cells are

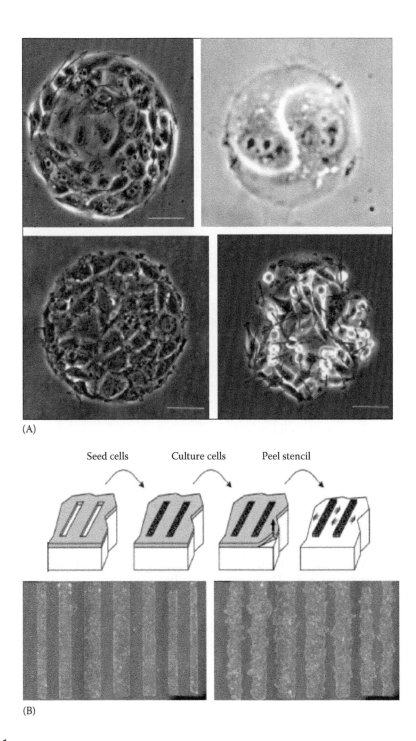

(A)

(B)

FIGURE 9.1
Different in vitro experiments using patterning techniques for the fabrication of well-controlled geometries: (A) circular islands of cells (top left and bottom [13], top right [9]). Top left panel indicates highly correlated, solid body–like rotation, while the bottom panels show uncorrelated motion of malignant cancerous cells. (From Doxzen, K. et al., *Integr. Biol.*, 5, 1026, 2013.) Expanding layers of cells, confined using (B) a physical barrier. (From Poujade, M. et al., *Proc. Natl. Acad. Sci. USA*, 104, 15987, 2007.) *(Continued)*

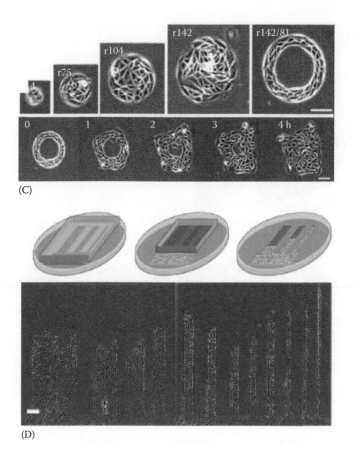

FIGURE 9.1 (*Continued*)
Different in vitro experiments using patterning techniques for the fabrication of well-controlled geometries: (C) a chemical barrier. (From Rolli, C.G. et al., *Biomaterials*, 33, 2409, 2012.) (D) Cells expanding into a series of strips of different widths. (From Vedula, S.R.K. et al., *Proc. Natl. Acad. Sci. USA*, 109, 12974, 2012.)

able to move such that they minimize the shear between them, which means that they attempt to move as a solid body. The tendency of cells to move such that the shear between them is minimized was also observed for cells flowing within an expanding cellular layer and termed "plithotaxis" [39].

Since the velocity at the island rim is roughly constant [13], it suggests that the rotation is driven by a breaking of the symmetry of the edge cells, but this remains to be confirmed by future experiments.

Let us note the relevance of such confined artificial geometries with those encountered in vivo: The isolated islands of cells shown to rotate in vitro in two dimensions [8,9,13] are similar in their behavior to the spontaneously rotating follicular epithelium during Drosophila oogenesis [1]. This comparison brings up a further issue, which is the applicability of 2D experiments in vitro to the 3D environment in vivo. A recent study [56] demonstrates that even within the body groups of cells, here, cancer cells are observed to migrate collectively along internal 2D surfaces. While these surfaces may be highly curved, such as in blood vessels and along connective tissue, cells seem to employ the same locomotion mechanisms observed in in vitro experiments.

9.3.2 Expanding Geometries

An even more striking difference between the bulk and edge cells appears in expanding cellular cultures. In Figure 9.1b and c, we see a geometry in which cells are initially confined but then released and allowed to expand freely on a uniform surface. Such experiments have been conducted where the confinement was a physical barrier [10–12,38,49] (Figure 9.1b) or chemical [14,57] (Figure 9.1c). In this geometry, the cells at the edge are able to spread onto the free surface, thereby changing their morphology: they become thinner, with larger area, and often develop large lamellipodia that extend toward the free surface. The motility of cells at the edge, especially those cells that develop large lamellipodia, is strongly biased toward the free surface, that is, in the direction of the outward pointing normal along the monolayer edge [58,59]. Depending on the conditions, it is often observed that sparse cells along the edge form highly prominent lamellipodia, developing large traction forces and forming "fingers" or cellular columns that extend onto the free surface ahead of the average location of the culture's edge [10,60] (Figures 9.1b and 9.2). Such cells have been termed "leader cells" and they indeed lead a cohort of cells behind them, over a coherence length that is cell-type specific (Figure 9.2). This length seems to depend strongly on the cell–cell adhesiveness, and for MDCK, it is of the order ~100–200 μm [44,47]. The mechanisms leading to this instability of the cellular layer's edge are not entirely clear. It is observed that this phenomenon can be eliminated by treatment of the cells with a myosin II inhibitor (such as blebbistatin), where all the edge cells develop large lamellipodia and the edge remains smooth [58]. Recently, it was observed that the initiation of leader cells may be influenced by the curvature of the monolayer edge [14], such that convex regions are more likely to initiate leaders compared to concave regions.

Along the sides of the "fingers" of cells, it is observed that cells form a continuous actin–myosin belt [10] (Figure 9.2b) that spans many cells at the monolayer edge. Specifically, this belt is under tension, thereby adding a line tension that tends to straighten the shape of the edge. This belt is observed only where the edge has a concave or straight shape and is absent where the edge is convex. There seems to be some exclusivity between the ability of edge cells to form protrusive lamellipodia and a contractile actin–myosin belt [10,12,23].

Leader cells have been found to undergo a form of epithelial-to-mesenchymal transition (EMT), whereby they express a mesenchymal phenotype [10] (Figure 9.2b). In experiments where the expanding monolayer meets another expanding monolayer, in a "wound-healing"-type experiment, the leader cells lose their unique properties as soon as they are not at the free edge (the "wound" closes). They return to normal epithelial behavior, demonstrating that there is strong feedback between the location of the cell at the free edge and its entire metabolism and genetic expression.

Behind the expanding free edge of the monolayer, cells within the bulk tend to follow the edge cells and move toward the free edge. This collective flow of cells may be gradual or initiated in the form of a propagating wave of acceleration [61–63]. The exact origin of these waves is still being actively investigated, but it seems that they form when there are strong cadherin-based cell–cell junctions and when the cells at the free edge are highly motile and exert a strong pull. Note that the motion of the bulk cells is not due to being simply pulled by the edge cells forward but rather that they respond to the edge motion and orient their internal motility apparatus to coincide with the outward expansion. Understanding the mechanisms leading to the coordinated motion of the cells is one of the biggest open puzzles in this field at present [11].

Another recent geometry, which is somewhat in between a fully confined and the freely expanding geometries described earlier, is shown in Figure 9.1d [11]. In this experiment,

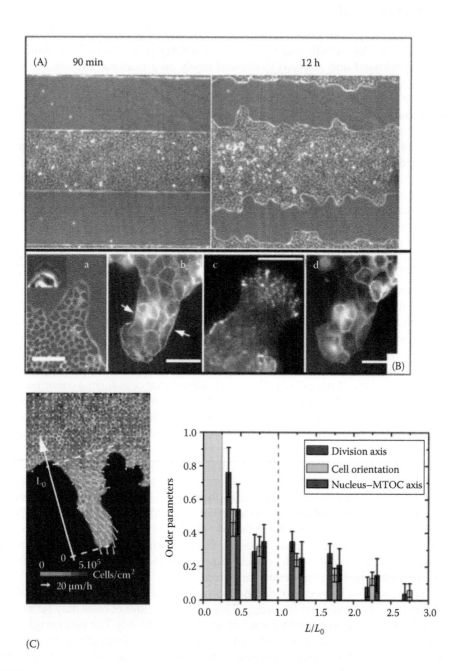

FIGURE 9.2
Instability of an expanding cellular layer: (A) a large-scale view of the instability in a strip geometry. (From Poujade, M. et al., *Proc. Natl. Acad. Sci. USA*, 104, 15987, 2007.) "Fingers" of cells protrude from the colony edge, following (B) "leader cells," which have a mesenchymal-like morphology [10]: (a) large lamellipodia, (c) enhanced substrate adhesion, and (d) somewhat reduced cell–cell contacts. (C) Long-range velocity correlations among cells following the "leader cell" [44], as demonstrated by following different markers of cellular polarity.

a confined monolayer was allowed to expand onto a series of narrow adhesive strips. It was found that in the narrowest strips, the cells at the leading tip maintain a leader-like behavior that allows for a faster overall progression of the monolayer. The critical thickness mentioned earlier, which the behavior becomes independent of the strip confinement, was found to be of the order ~100–200 µm. This value is similar to the velocity–velocity correlation length found for these cells in open geometries [10,23,44].

9.3.3 Converging Geometries: Hole Closures

Another behavior is observed for edge cells around small concave "holes" (Figure 9.3). Such geometries have been explored recently [12,14]. It is found that for large holes, larger than ~20–30 µm in diameter, the edge cells develop lamellipodia as discussed earlier.

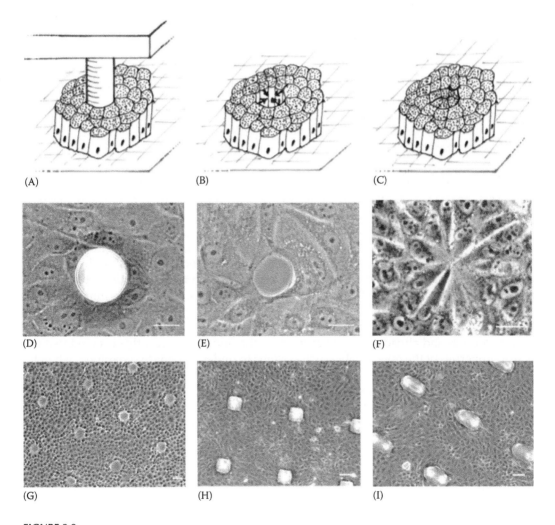

(A) (B) (C)

(D) (E) (F)

(G) (H) (I)

FIGURE 9.3
Hole-closing geometry [12]: (A–C) using a pillar as a physical barrier to create the initial hole in the cellular layer. (D–F) Following the removal of the pillar, the hole closes, sometimes leaving a rosette arrangement of the cells. (G–I) Initial conditions of a field of holes of various shapes.

However, for smaller holes, the lamellipodia formation is suppressed and cells tend to form a continuous actin–myosin cable that spans through the cells along the hole perimeter. This cable acts as a "purse string" and was previously observed in the context of wound healing [64,65] and embryogenesis [66]. The tension generated within this cable exerts an inward-directed force that pulls together the edge cells to close the hole.

9.4 Models of Collective Cell Migration

We now discuss the theoretical models proposed to describe the variety of observations of collective cell motion described earlier.

9.4.1 Cells as Self-Propelled Particles and "Glass"

Before treating the complexities of cellular motion that are particular to cells, we start by considering cells as another example of self-propelled particles [28,29]. The study of such systems has expanded greatly in recent years, with motivation arising from the study of flocking and swarming behaviors of animals such as fish and birds. Nonliving systems have also been realized [67,68] where chemical reactions drive the propulsion of colloidal particles near the boundary of the fluid. Self-propelled particles of the simplest kind have the following properties:

- They are able to produce an active motile force, of fixed magnitude, F_{tract}.
- This force acts along the same direction for an average duration τ_p, which is called the "persistence time."

The actual realization of such properties may be through a continuous diffusion of the orientation of the active force, with rotational diffusion coefficient, $D_r \approx \tau_p^{-1}$, or by choosing a random new direction after each pulse of average duration τ_p. The activity is therefore in the form of a "colored" noise that, unlike thermal noise, has a given force amplitude and temporal correlations.

At the simplest level, such active particles may have purely repulsive, hard-core (or soft-core) potentials and already exhibit surprising behavior: it was recently observed that the active properties of the noise enable such systems to undergo a phase separation, whereby the particles form aggregates and macroscopic clusters [41,69,70]. Note that due to the purely repulsive interactions, such clustering is not expected in thermal equilibrium and appears only due to the active noise. These model systems are being intensively explored to reveal their full range of behaviors and will be useful references for more complex models of collective cellular migration.

In a dense system of such particles [71,72], large dynamical heterogeneities have been found in simulations. These resemble the behavior in out-of-equilibrium amorphous systems such as glasses. The analogy between the properties of solid-like cellular layers and glasses has been proposed recently [38,40,43]. While some features are indeed similar, it is too early to judge how far this analogy can be useful.

As an example, we note the observed exponential distribution of cellular traction forces observed in a moving cell layer [38]. It follows previous observations of an exponential velocity distribution within cellular layers [26,49,73,74]. This distribution resembles the force distribution in nonbiological nonequilibrium systems, such as sheared granular materials, while also possibly arising from the exponential distribution of traction forces produced on the cellular and subcellular levels [34]. Using a completely different approach, such a distribution was also found recently using a driven heterogeneous elastic model [75]. The relation between the properties of a cellular layer, such as the traction force distribution, and amorphous materials remains an exciting direction for future research.

9.4.2 Mechanisms for Collective Migration

Models for collective cell migration are broadly divided between particle-based and cellular Potts models, which we briefly describe in the following.

In particle-based models [21,23,24,76,77], the active particles described earlier have, in addition to the repulsive core, also an attractive potential that describes the ability of cells to form cell–cell junctions. Note that while such junctions are quite robust, they are also highly dynamic and allow cells to move past each other without getting dissociated. The biological complexity of these junctions is not fully captured by an attractive potential, but this is the simplest way to incorporate cell–cell adhesion.

The second ingredient implemented in such models is the ability of cells to interact with each other and thereby influence the traction forces that they produce. A general scheme for describing collective behavior of self-propelled particles has been proposed by Vicsek et al. [78], which was since analyzed in detail. The essence of this approach is to add to Equation 9.1 a term that phenomenologically describes the tendency of cells to align their traction forces with those of their neighbors. It can be written as

$$\vec{v}_{cell} = \frac{\vec{F}_{tract} + \sum_{nn} \vec{F}_{pot} + \beta \sum_{nn} \left(\vec{v}_{nn} - \vec{v}_{cell} \right)}{\lambda_{surf} + \lambda_{cell-cell} - \beta N_{nn}}$$

(9.2)

where β is the coupling parameter describing the strength of the orientational interactions, and the friction coefficient has been renormalized by this parameter and the number of nearest neighbors N_{nn} so that the mean cell velocity is not affected by these interactions.

One way to physically interpret this term, in the context of cell–cell interactions, is to treat the relative velocity between neighboring cells as applying a strain on the cell–cell junctions that connect them. Assuming that the cell–cell junctions extend over a width a, the strain rate applied to these molecular complexes is simply $\dot{\varepsilon} = |\vec{v}_{nn} - \vec{v}_{cell}|/a$. As a microscopic picture for these cellular connections, we consider each linker as a spring of stiffness k, making association/dissociation dynamics with k_{on} independent of the strain and k_{off} reduced when it is stretched so that the linker breaks up more easily under strain. The simplest form is

$$k_{off} = k_{off,0} e^{-\left[\Delta E - ka^2 \left(\dot{\varepsilon}/k_{off} \right)^2 \right]/k_B T}$$

(9.3)

where
  ΔE is the energy gap for dissociation
  $k_{off,0}$ is the unstrained dissociation rate, and we took that the average time that the spring is attached and therefore strained is given by k_{off}^{-1}

Equation 9.3 needs to be solved self-consistently for k_{off}, from which the average occupation of the cell–cell linkers as a function of the relative cellular velocities can be found.

The relative occupation of such linkers along the sides of the cell, in response to the velocities of its neighbors, can serve as the chemical signaling that is polarizing the cytoskeleton of the cell along the direction of motion of its neighbors. It remains as an open challenge to relate such microscopic mechanosensitive mechanisms to the simplified Vicsek-type terms used in models of collective cell motility. In other words, whether the Vicsek term in Equation 9.2 can be derived from a microscopic consideration (such as Equation 9.3).

From Equation 9.3, it is trivial to obtain a linear relation between a small change in the relative velocity between the neighboring cells and a resulting change in the dissociation rate. What is possibly more interesting is that the self-consistent solution for k_{off} is

$$k_{off} = \frac{\left|\vec{v}_{nn} - \vec{v}_{cell}\right|\sqrt{2k}}{\sqrt{W\left(\dfrac{2k\left|\vec{v}_{nn} - \vec{v}_{cell}\right|^2 e^{2\Delta E}}{k_{off,0}^2}\right)}} \tag{9.4}$$

where W is the product log or Lambert W function. This expression shows that over a large range of relative velocities the occupation of the linker proteins is linear in the relative velocity (the denominator depends only logarithmically on the relative velocity). If the motility apparatus of the cell is recruited (or inhibited) in response to this occupation along the cell periphery, we can suggest a chemomechanical mechanism that gives rise to the simple linear form of the Vicsek term in Equation 9.2.

Works based on cellular Potts models essentially use the same Vicsek-type interactions but now implemented in a semicontinuum version [22,79–81], where the cell polarity is made to follow the motion of the cell induced by its neighbors. In such models, since the cell shape is described as a polygonal object, it is often found that there is a correlation between the polarization of the cell motion and its elongated shape, as is observed to some extent in experiments [38,47].

Implementing the orientational ordering interaction in a continuum model, several works [82–84] have attempted to describe the collective motion of a cell layer in terms of "active-matter" models [29]. In these models, the individual cells are not described, and the monolayer is treated as an active liquid crystal that has an orientational "director field," which responds to the active stress terms that describe the cellular motility. The advantage of such models over cell-based models is that the computation becomes much simpler and can use powerful techniques from continuum modeling. The disadvantage is that the discrete nature of the cells is of course lost.

Another approach to describe the emergence of coherent cellular migration, in cell-based models, was proposed in Refs. [24,78]. In these works, the cell either chooses always the path of least resistance [78] or abandons more readily active forces along directions of higher resistance [24]. The resulting behavior seems to resemble very much that which arises in Vicsek-type models, but it remains to be shown if indeed all such models can be effectively mapped into each other.

A recent direction attempts to model the collective migration by utilizing detailed models for single-cell motility, which is based on a description of the individual cellular protrusions that determine the resultant traction force of the cell [25,26]. In these models, the individual subcellular protrusions experience contact inhibition of locomotion [35], when attempting to protrude into a neighboring cell. While this form of physical interaction surely exists, it may oversimplify the real cellular interactions: for example, cryptic lamellipodia may polarize cells to move in synchrony, while cells pulling on each other may respond by trying to pull in the opposite direction, that is, becoming anticorrelated with their neighbors [53]. Models with subcellular-scale details point toward the future direction, whereby the modeling of collective cell motility will be directly connected to the modeling of single-cell motility, avoiding the need for "black-box" phenomenological models.

9.4.3 Expanding Geometries: Leader-Cell Instability

As we described earlier, the cells at the edge of an expanding cell layer behave in a distinct manner from the bulk cells. Several models have attempted to describe the observed behavior of these cells, namely, the formation of "leader cells" and the consequent migration of strands of cells that follow these leaders (Figures 9.1b and 9.2).

One approach to this problem was based on a continuum description of the edge of the cellular layer while not explicitly describing the motion of the bulk cells at all [85]. The monolayer edge is treated as a continuous 1D "membrane" that is pushed by the motile forces of the cells while experiencing both friction force and restoring forces due to the tendency of cells to resist bending and stretching. At the same time, a feedback between the ability of the edge cell to produce effective traction forces and its shape was postulated. This feedback assumed that the traction force pointing toward the outward normal increases linearly with the convex (negative) curvature of the edge, up to some maximal value

$$
F_n = \begin{cases} F_0, & H > 0 \\ F_0 + \alpha |H|, & -H_{max} < H < 0 \\ F_{max} = F_0 + \alpha |H_{max}|, & -H_{max} > H \end{cases} \tag{9.5}
$$

where

F_0 is the background expansion force for flat and concave curvature

$\alpha = (F_{max} - F_0)/|H_{max}|$ is the relation between the traction force and the force (units of membrane tension, force/length)

H_{max} is the maximal curvature that a cell can experience

This form (Equation 9.5) of the expansion force leads to an instability, which is reminiscent of the Mullins–Sekerka instability found for dendritic crystal growth [86]. Similar to the case of crystal growth, it was found that this model predicts the phenomenon of "tip splitting," whereby the growing fingers split at their tip into two fingers growing sideways. While this is sometimes observed in the experiments [85] (Figure 9.4), it is not as prevalent as this simple model would suggest. However, this model predicts that under strong anisotropy, the tip splitting can be replaced by the phenomenon of side branching. Indeed, in narrow strips [11], where the single cell at the tip is "protected" by the geometry, it is observed that the tendency for instability is transformed from tip splitting to periodic widening of the finger, similar to side branching in anisotropic crystal growth.

Note that it is implicitly assumed in this model that the cells of the bulk follow the motion of the edge cells, so that perfect correlation is assumed, at least over a sizable region near the edge cells. According to this model, the typical length scale of the instability, that is, finger separation, can be estimated using linear stability analysis $l = \sqrt{\kappa/(\alpha - \sigma)}$, where κ and σ are, respectively, the effective bending modulus and membrane tension of the cells at the colony's edge.

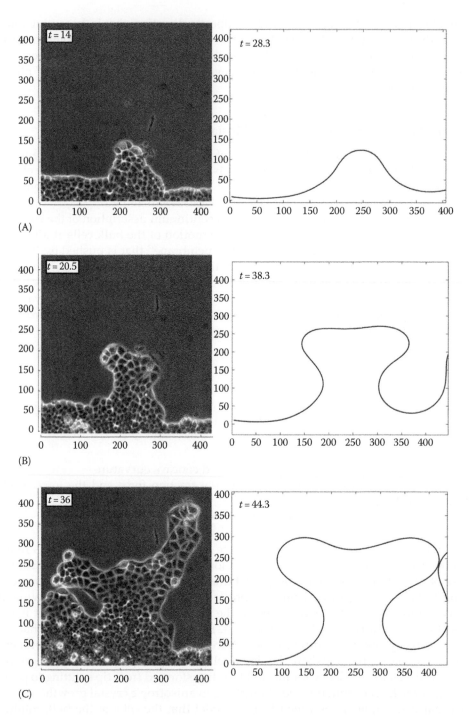

FIGURE 9.4
Tip splitting of a moving "finger" of cells [84]: the left panels show the dynamical instability at the edge of an expanding cellular layer, in the form of a "finger" of cells. The right panels show a simulation of the evolution of the contour of the edge of the monolayer. While the model predicts that all such "fingers" should become unstable and undergo tip splitting as the contour evolves ((a) to (c)), the example on the left is in fact rare, with most real "fingers" having a fairly stable tip.

This model also predicts that if all cells become highly motile, irrespective of their shape, that is, taking $F_0 \rightarrow F_{max}$, $\alpha \rightarrow 0$, then the instability vanishes. This may be the situation when the cells are treated with blebbistatin [11,58]. The feedback between cell motility and geometry may also explain the observed stabilization of the leader cell in narrow strips, where the geometry maintains the high curvature of the edge cell [11]. Finally, recent experiments [14,87] suggest that there is indeed a feedback between the curvature of the edge cells and their propensity to become leader cells. It remains to be explored further how central is this mechanism in the formation of the observed instability of moving cellular fronts.

Other approaches to explain this observed instability are based on descriptions that couple the cell motility with chemical signaling. Such diffusing signals that are emitted by the cells, either only at the edge or throughout the colony, and guide the cell motility by chemotaxis can lead to unstable fronts [80,84,88]. For example, in Ref. [80], the cell motility depends on the local chemical gradient, and this gradient becomes sharper for cells at the tips of small undulations, which therefore become unstable. The exact details of the chemomechanical coupling are different between these models and require validation in future experimental testing.

Finally, edge roughening of an expanding cellular layer was also observed to a certain extent using particle-based simulations of models that include cell–cell adhesion [89] and orientational ordering [24].

9.4.4 Converging Geometries: Hole Closures

One simple physical description of the hole-closure process [12,14] (Figure 9.3) is as follows: We assume that the edge moves uniformly due to the average combined effects of the tension in the actin–myosin cable and due to lamellipodia protrusion. We also assume that the movement of the monolayer behind the edge exerts a constant effective friction-like force, which means that there are no large elastic stresses in the cell layer. This assumption effectively means that the cells move collectively with a high degree of correlation behind the edge. Using these assumptions, we can write the following equation of motion for the location of the edge radius R:

$$\dot{R}(t) = -\frac{1}{\lambda}\left[F_{lam} + \frac{\gamma}{R(t)}\right] \Rightarrow \frac{F_{lam}}{\lambda}\frac{\partial t(R)}{\partial R} = -\left(1 + \frac{R_c}{R}\right)^{-1} \tag{9.6}$$

where
F_{lam} is the force due to lamellipodia-driven traction of the edge cells
λ is the effective friction experienced by the edge cells
γ is the tension in the actin cable
$R_c = \gamma / F_{lam}$

Solving Equation 9.6, we find that

$$t(R) = \frac{\lambda}{F_{lam}}\left[R_0 - R + R_c\log\left(\frac{R + R_c}{R_0 + R_c}\right)\right] \Rightarrow t_c = v_0\left[R_0 + R_c\log\left(\frac{R_c}{R_0 + R_c}\right)\right] \tag{9.7}$$

where
R_0 is the initial radius of the hole
$v_0 = \lambda / F_{lam}$ is the velocity of the cells at the edge in the limit of very large holes

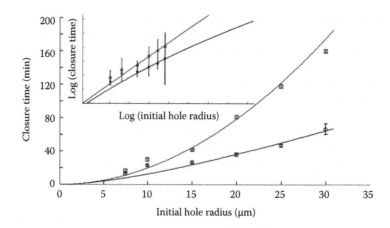

FIGURE 9.5

Hole-closure dynamics: using the simple model for the motion of the hole edge, we calculate the dependence of the closure time on the initial hole radius (Equation 9.6, solid lines) and compare to the experimental observations. (From Anon, E. et al., *Proc. Natl. Acad. Sci. USA*, 109, 10881, 2012.) The normal MDCK cells are indicated by black circles and the Rac-inhibited cells by the red squares. The agreement is very good, using a single set of parameters and assuming that the Rac inhibition effectively eliminates the protrusive force of lamellipodia. The inset shows that for normal cells at $R_0 \gg R_c$, the relation is linear (slope 1), while in the Rac-inhibited case, the relation is quadratic (slope 2).

The closure time t_c is given in Equation 9.7, for $R = 0$. This expression gives that for large holes $(R_0 \gg R_c)$, the closure time increases linearly with the hole radius, as observed [12], while for small holes, the closure time should increase quadratically with R_0. When the cells are treated with an inhibitor of Rac, which suppresses lamellipodia formation $(F_{lam} \to 0)$, we find that the closure time increases quadratically with R_0, $t_c \to (\lambda/2\gamma)R_0^2$, as is shown in Figure 9.5. For normal cells, the velocity is measured [12], $v_0 \approx 0.25$ μm/min, and we find a good fit using $R_c \approx 10$ μm. It is nice that these same values can be used to fit the Rac-inhibited cells, where now $\lambda/\gamma \approx 0.4$(min/μm²), showing that this description is consistent with the observations.

We emphasize again that this calculation assumes implicitly that the cells behind the moving edge, inside the bulk, move coherently with the edge cells, so that no elastic restoring forces form within the cellular layer. This assumption can surely break down for solid-like layers.

9.5 Conclusion

Collective cell migration is a major open puzzle for biology and biological physics. It presents the physicist with an elegant system of self-propelled, active matter, which undergoes dynamical transitions from fluidlike flow to an amorphous, glass-like solid. The intercellular interactions that give rise to the collective behavior couple chemical and mechanical signaling, directly between cells and through the surrounding medium and underlying solid substrate. This coupling makes this system difficult to unravel both at the biological level and at the physical and theoretical level. The problem is compounded by the complexity of single-cell motility, which is better understood but also not fully solved.

Collective cell migration is an example where a rather universal problem in biology, which appears in numerous biological contexts and in many cell types, also inspires research into new classes of out-of-equilibrium physical systems. Current advances in micromanipulation of cells and high-resolution measurements will provide the information needed to build a deeper theoretical understanding of this phenomenon.

References

1. Rørth, P. Fellow travellers: Emergent properties of collective cell migration. *EMBO Reports* 13(11) (2012): 984–991.
2. Rørth, P. Collective cell migration. *Annual Review of Cell and Developmental Biology* 25 (2009): 407–429.
3. Montell, D.J., Yoon, W.H., and Starz-Gaiano, M. Group choreography: Mechanisms orchestrating the collective movement of border cells. *Nature Reviews Molecular Cell Biology* 13(10) (2012): 631–645.
4. Friedl, P. et al. Classifying collective cancer cell invasion. *Nature Cell Biology* 14(8) (2012): 777–783.
5. Theveneau, E. and Mayor, R. Cadherins in collective cell migration of mesenchymal cells. *Current Opinion in Cell Biology* 24(5) (2012): 677–684.
6. Trepat, X., Chen, Z., and Jacobson, K. Cell migration. *Comprehensive Physiology* (2012).
7. Travis, J. Mysteries of the cell: Cell biology's open cases. *Science* 334(6059) (2011): 1051.
8. Brangwynne, C. et al. (2000) Symmetry breaking in cultured mammalian cells. *In Vitro Cellular & Developmental Biology - Animal* 36: 563–565.
9. Huang, S. et al. (2005) Symmetry-breaking in mammalian cell cohort migration during tissue pattern formation: Role of random-walk persistence. *Cell Motility and the Cytoskeleton* 61: 201–213.
10. Poujade, M. et al. Collective migration of an epithelial monolayer in response to a model wound. *Proceedings of the National Academy of Sciences of the United States of America* 104(41) (2007): 15987–15993.
11. Vedula, S.R.K. et al. Emerging modes of collective cell migration induced by geometrical constraints. *Proceedings of the National Academy of Sciences of the United States of America* 109(32) (2012): 12974–12979.
12. Anon, E. et al. Cell crawling mediates collective cell migration to close undamaged epithelial gaps. *Proceedings of the National Academy of Sciences of the United States of America* 109(27) (2012): 10881–10886.
13. Doxzen, K. et al. Guidance of collective cell migration by substrate geometry. *Integrative Biology* 5(8) (2013): 1026–1035.
14. Rolli, C.G. et al. Switchable adhesive substrates: Revealing geometry dependence in collective cell behavior. *Biomaterials* 33(8) (2012): 2409–2418.
15. Marel, A.K. et al. Arraying cell cultures using PEG-DMA micromolding in standard culture dishes. *Macromolecular Bioscience* 13(5) (2013): 595–602.
16. Riahi, R. et al. Advances in wound-healing assays for probing collective cell migration. *Journal of Laboratory Automation* 17(1) (2012): 59–65.
17. Du Roure, O. et al. Force mapping in epithelial cell migration. *Proceedings of the National Academy of Sciences of the United States of America* 102(7) (2005): 2390–2395.
18. Saez, A. et al. Traction forces exerted by epithelial cell sheets. *Journal of Physics: Condensed Matter* 22(19) (2010): 194119.
19. Munevar, S., Wang, Y., and Dembo, M. Traction force microscopy of migrating normal and H-ras transformed 3T3 fibroblasts. *Biophysical Journal* 80(4) (2001): 1744–1757.

20. Tambe, D.T. et al. Monolayer stress microscopy: Limitations, artifacts, and accuracy of recovered intercellular stresses. *PloS One* 8(2) (2013): e55172.
21. Szabo, B. et al. Phase transition in the collective migration of tissue cells: Experiment and model. *Physical Review E* 74(6) (2006): 061908.
22. Szabó, A. et al. Collective cell motion in endothelial monolayers. *Physical Biology* 7(4) (2010): 046007.
23. Sepúlveda, N. et al. Collective cell motion in an epithelial sheet can be quantitatively described by a stochastic interacting particle model. *PLoS Computational Biology* 9(3) (2013): e1002944.
24. Basan, M. et al. Alignment of cellular motility forces with tissue flow as a mechanism for efficient wound healing. *Proceedings of the National Academy of Sciences of the United States of America* 110(7) (2013): 2452–2459.
25. Coburn, L. et al. Tactile interactions lead to coherent motion and enhanced chemotaxis of migrating cells. *Physical Biology* 10(4) (2013): 046002.
26. Vedel, S. et al. Migration of cells in a social context. *Proceedings of the National Academy of Sciences of the United States of America* 110(1) (2013): 129–134.
27. Rey, R. and García-Aznar, J.M. A phenomenological approach to modelling collective cell movement in 2D. *Biomechanics and Modeling in Mechanobiology* 12(6) (2013): 1089–1100.
28. Vicsek, T. and Zafeiris, A. Collective motion. *Physics Reports* 517(3) (2012): 71–140.
29. Marchetti, M.C. et al. Hydrodynamics of soft active matter. *Reviews of Modern Physics* 85(3) (2013): 1143.
30. Danuser, G., Allard, J., and Mogilner, A. Mathematical modeling of eukaryotic cell migration: Insights beyond experiments. *Annual Review of Cell and Developmental Biology* 29(1) (2013): 501–528.
31. Lämmermann, T. and Sixt, M. Mechanical modes of 'amoeboid' cell migration. *Current Opinion in Cell Biology* 21(5) (2009): 636–644.
32. Faure-André, G. et al. Regulation of dendritic cell migration by CD74, the MHC class II-associated invariant chain. *Science* 322(5908) (2008): 1705–1710.
33. Hawkins, R.J. et al. Pushing off the walls: A mechanism of cell motility in confinement. *Physical Review Letters* 102(5) (2009): 058103.
34. Gov, N.S. Traction forces during collective cell motion. *HFSP Journal* 3(4) (2009): 223–227.
35. Mayor, R. and Carmona-Fontaine, C. Keeping in touch with contact inhibition of locomotion. *Trends in Cell Biology* 20(6) (2010): 319–328.
36. Zaidel-Bar, R., Kam, Z., and Geiger, B. Polarized downregulation of the paxillinp130CAS-Rac1 pathway induced by shear flow. *Journal of Cell Science* 118(Pt 17) (2005): 3997–4007.
37. Farooqui, R. and Fenteany, G. Multiple rows of cells behind an epithelial wound edge extend cryptic lamellipodia to collectively drive cell-sheet movement. *Journal of Cell Science* 118(1) (2005): 51–63.
38. Trepat, X. et al. Physical forces during collective cell migration. *Nature Physics* 5(6) (2009): 426–430.
39. Trepat, X. and Fredberg, J.J. Plithotaxis and emergent dynamics in collective cellular migration. *Trends in Cell Biology* 21(11) (2011): 638–646.
40. Tambe, D.T. et al. Collective cell guidance by cooperative intercellular forces. *Nature Materials* 10(6) (2011): 469–475.
41. Fily, Y. and Marchetti, M.C. Athermal phase separation of self-propelled particles with no alignment. *Physical Review Letters* 108(23) (2012): 235702.
42. F. Gerbal, V. et al. On the 'listeria' propulsion mechanism. *Pramana*, 53(1) (1999): 155–170.
43. Angelini, T.E. et al. Glass-like dynamics of collective cell migration. *Proceedings of the National Academy of Sciences of the United States of America* 108(12) (2011): 4714–4719.
44. Petitjean, L. et al. Velocity fields in a collectively migrating epithelium. *Biophysical Journal* 98(9) (2010): 1790–1800.
45. Puliafito, A. et al. Collective and single cell behavior in epithelial contact inhibition. *Proceedings of the National Academy of Sciences of the United States of America* 109(3) (2012): 739–744.
46. Kim, J.H., Dooling, L.L., and Asthagiri, A.R. Intercellular mechanotransduction during multicellular morphodynamics. *Journal of the Royal Society Interface* 7(Suppl 3) (2010): S341–S350.

47. Reffay, M. et al. Orientation and polarity in collectively migrating cell structures: Statics and dynamics. *Biophysical Journal* 100(11) (2011): 2566–2575.
48. Angelini, T.E. et al. Cell migration driven by cooperative substrate deformation patterns. *Physical Review Letters* 104(16) (2010): 168104.
49. Nnetu, K.D. et al. Directed persistent motion maintains sheet integrity during multi-cellular spreading and migration. *Soft Matter* 8(26) (2012): 6913–6921.
50. Drees, F. et al. α-Catenin is a molecular switch that binds E-cadherin-β-catenin and regulates actin-filament assembly. *Cell* 123(5) (2005): 903–915.
51. Ng, M.R. et al. Substrate stiffness regulates cadherin-dependent collective migration through myosin-II contractility. *The Journal of Cell Biology* 199(3) (2012): 545–563.
52. Basan, M. et al. A reaction-diffusion model of the cadherin-catenin system: A possible mechanism for contact inhibition and implications for tumorigenesis. *Biophysics Journal* 98(12) (2010): 2770–2779.
53. Weber, G.F., Bjerke, M.A., and DeSimone, D.W. A mechanoresponsive cadherin-keratin complex directs polarized protrusive behavior and collective cell migration. *Developmental Cell* 22(1) (2012): 104–115.
54. Haigo, S.L. and Bilder, D. Global tissue revolutions in a morphogenetic movement controlling elongation. *Science* 331 (2011): 1071–1074.
55. Tanner, K. et al. Coherent angular motion in the establishment of multicellular architecture of glandular tissues. *Proceedings of the National Academy of Sciences of the United States of America* 109 (2012): 1973–1978.
56. Alexander, S. et al. Preclinical intravital microscopy of the tumour-stroma interface: Invasion, metastasis, and therapy response. *Current Opinion in Cell Biology* 25(5) (2013): 659–671.
57. Marel, A.K., Alberola, A.P., and Rädler, J.O. Proliferation and collective migration of small cell groups released from circular patches. *Biophysical Reviews and Letters* 7(1,2) (2012): 15–28.
58. Lim, J.I. et al. Protrusion and actin assembly are coupled to the organization of lamellar contractile structures. *Experimental Cell Research* 316(13) (2010): 2027–2041.
59. Kim, J.H. et al. Propulsion and navigation within the advancing monolayer sheet. *Nature Materials* 12(9) (2013): 856–863.
60. Gov, N.S. Collective cell migration patterns: Follow the leader. *Proceedings of the National Academy of Sciences of the United States of America* 104(41) (2007): 15970–15971.
61. Serra-Picamal, X. et al. Mechanical waves during tissue expansion. *Nature Physics* 8(8) (2012): 628–634.
62. Zaritsky, A. et al. Emergence of HGF/SF-induced coordinated cellular motility. *PloS One* 7(9) (2012): e44671.
63. Weiger, M.C. et al. Real-time motion analysis reveals cell directionality as an indicator of breast cancer progression. *PloS One* 8(3) (2013): e58759.
64. Fenteany, G., Janmey, P.A., and Stossel, T.P. Signaling pathways and cell mechanics involved in wound closure by epithelial cell sheets. *Current Biology* 10(14) (2000): 831–838.
65. Tamada, M. et al. Two distinct modes of myosin assembly and dynamics during epithelial wound closure. *The Journal of Cell Biology* 176(1) (2007): 27–33.
66. Martin, P. and Parkhurst, S.M. Parallels between tissue repair and embryo morphogenesis. *Development* 131(13) (2004): 3021–3034.
67. Palacci, J. et al. Living crystals of light-activated colloidal surfers. *Science* 339(6122) (2013): 936–940.
68. Buttinoni, I. et al. Dynamical clustering and phase separation in suspensions of self-propelled colloidal particles. *Physical Review Letters* 110(23) (2013): 238301.
69. Redner, G.S., Hagan, M.F., and Baskaran, A. Structure and dynamics of a phase-separating active colloidal fluid. *Physical Review Letters* 110(5) (2013): 055701.
70. Bialké, J., Löwen, H., and Speck, T. Microscopic theory for the phase separation of self-propelled repulsive disks. *Europhysics Letters* 103(3) (2013): 30008.
71. Berthier, L. Nonequilibrium glassy dynamics of self-propelled hard disks. *Preprint arXiv*1307(0704) (2013).

72. Fily, Y., Henkes, S., and Marchetti, M.C. Freezing and phase separation of self-propelled disks. *Preprint arXiv* 1309(3714) (2013).
73. Czirók, A. et al. Exponential distribution of locomotion activity in cell cultures. *Physical Review Letters* 81(14) (1998): 3038.
74. Selmeczi, D. et al. Cell motility as persistent random motion: Theories from experiments. *Biophysical Journal* 88(2) (2005): 912–931.
75. Bameta, T. et al. Broad-tailed force distributions and velocity ordering in a heterogeneous membrane model for collective cell migration. *Europhysics Letters* 99(1) (2012): 18004.
76. Drasdo, D., Hoehme, S., and Block, M. On the role of physics in the growth and pattern formation of multi-cellular systems: What can we learn from individual-cell based models? *Journal of Statistical Physics* 128(1–2) (2007): 287–345.
77. Bindschadler, M. and McGrath, J.L. Sheet migration by wounded monolayers as an emergent property of single-cell dynamics. *Journal of Cell Science* 120(5) (2007): 876–884.
78. Vicsek, T. et al. Novel type of phase transition in a system of self-driven particles. *Physical Review Letters* 75(6) (1995): 1226.
79. Kabla, A.J. Collective cell migration: Leadership, invasion and segregation. *Journal of the Royal Society Interface* 9(77) (2012): 3268–3278.
80. Ouaknin, G.Y. and Bar-Yoseph, P.Z. Stochastic collective movement of cells and fingering morphology: No maverick cells. *Biophysical Journal* 97(7) (2009): 1811–1821.
81. Salm, M. and Pismen, L.M. Chemical and mechanical signaling in epithelial spreading. *Physical Biology* 9(2) (2012): 026009.
82. Lee, P. and Wolgemuth, C.W. Crawling cells can close wounds without purse strings or signaling. *PLoS Computational Biology* 7(3) (2011): e1002007.
83. Lee, P. and Wolgemuth, C. Advent of complex flows in epithelial tissues. *Physical Review E* 83(6) (2011): 061920.
84. Köpf, M.H. and Pismen, L.M. A continuum model of epithelial spreading. *Soft Matter* 9 (2013): 3727–3734.
85. Mark, S. et al. Physical model of the dynamic instability in an expanding cell culture. *Biophysical Journal* 98(3) (2010): 361–370.
86. Langer, J.S. Instabilities and pattern formation in crystal growth. *Reviews of Modern Physics* 52(1) (1980): 1.
87. Tse, J.M. et al. Mechanical compression drives cancer cells toward invasive phenotype. *Proceedings of the National Academy of Sciences of the United States of America* 109(3) (2012): 911–916.
88. Amar, M.B. Chemotaxis migration and morphogenesis of living colonies. *The European Physical Journal E* 36(6) (2013): 1–13.
89. Khain, E. et al. Migration of adhesive glioma cells: Front propagation and fingering. *Physical Review E* 86(1) (2012): 011904.

10

Connective Tissue Development

Albert Harris

CONTENTS

10.1 Introduction

Connective tissues hold our bodies together. They include tendons, ligaments, fascia, organ capsules, the tunica media of blood vessels, the dermis (leathery inner layer of the skin), cartilage, bone, and both skeletal and smooth muscles. The first two figures both show attachments between cartilages and skeletal muscles, Figure 10.1 in the head of a salamander and Figure 10.2 in the leg of a mouse.

Connective tissues are the load-bearing and force-exerting networks of the body. Despite their diversity, all connective tissues consist of interwoven networks of collagen and cells, and for some of them, cartilage and bone. Without connective tissues, anatomy would have little strength and would collapse. Indeed, there are genetic syndromes in which certain connective tissues collapse for lack of physical support (Remington et al. 2009).

The formation of connective tissues depends on weaving collagen fibers together in certain spatial patterns. Good examples of this are formation of bones and cartilages. Figure 10.3 is a histological section through an area of bone formation in the leg of a rabbit. Cartilage is at the upper right and appears yellow. Bone is forming at the lower left, spreading toward the right. Orientations of collagen fibers are made visible by polarized light: Blue indicates where collagen fibers are oriented parallel to the long axis of the bone, running from lower left toward the upper right; yellow color is caused by collagen fibers that are perpendicular to the long axis. It is remarkable how abruptly the collagen fibers become realigned by 90°, just in front of where cartilage begins to be destroyed and replaced by bone.

The causal mechanisms of geometrically diverse arrangements of collagen remain a subject of conjecture, debate, and experiment. Although there are about 30 different molecular species of collagen, each with characteristic locations and properties (e.g., most of the collagen inside cartilage is type II collagen), nevertheless, about a third of all the protein in our bodies is type I collagen. A special paradox is that this one protein somehow gets arranged

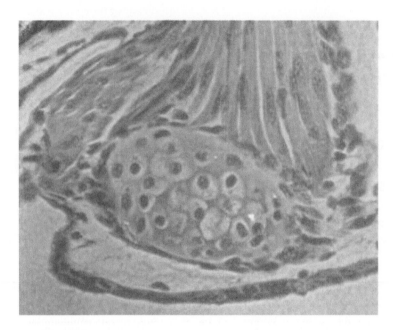

FIGURE 10.1
Histological section of an attachment between a muscle (above) and a cartilage (center) in the head of a larval salamander. Stained with hematoxylin and eosin.

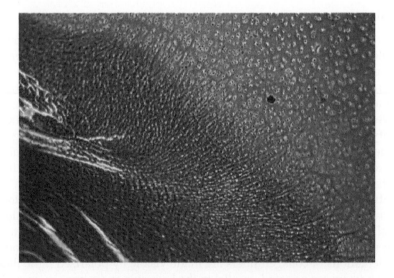

FIGURE 10.2
Histological section of an attachment between cartilage (upper right) and a tendon (center and left), in the leg of a mouse. Stained with hematoxylin and eosin and photographed by phase-contrast microscopy.

in many different patterns—some places densely packed, other places a loose mesh, in tendons aligned, and in organ capsules as sheets. An even more tantalizing puzzle is the near-perpendicular alignment of adjacent layers of collagen that form during ossification of long bones (as seen in Figure 10.3), in the cornea of the eye, the sheath of the notochord, and the dermis of amphibians, among other locations. Functionally, being perpendicular

FIGURE 10.3
Epiphysis of a rabbit bone, photographed by polarized light with quarter-wave filters, so that colors of birefringent materials will vary as a function of direction. Cartilage in the upper right area is gradually being destroyed and replaced by bone, advancing from the lower left, into the center. The yellow color results from collagen fibers in the cartilage being oriented transverse to the advancing bone. The dark blue streaks in the center show that collagen has become reoriented by 90°, so as to become parallel to the direction of the advancing bone. Stacks of enlarged chondrocytes are visible in the pale areas between the blue streaks.

can be an optimal arrangement; but that does not explain its causation. None of the theories to be discussed in the following has a good explanation for how cells produce perpendicular layers, nor an experimental method capable of replicating them.

The locations where cells secrete collagen are a large part of the story. In addition, mesenchymal cells are effective movers of collagen that has already been secreted (Stopak and Harris 1982). Secretion in place has been extensively documented by Birk and Trelstad (1986) and Birk et al. (1989, 1990), primarily by the use of transmission electron microscopy and taking advantage of the uniquely regular banded appearance of type I collagen fibers, as seen by the electron microscope. Birk and Trelstad proposed that the locations and directions of collagen fibers are created in the way that a spider builds a web, with fibroblasts playing the role of the spider and with collagen patterns being equivalent to the web.

Based on this hypothesis, then the formation of tendons and ligaments ought to be accompanied by intensive back and forth locomotion of collagen-secreting cells. Such movement has not yet been observed or reported. Likewise, alternating perpendicular migration of cells is to be expected at locations like the notochord sheath and perhaps the cornea. This, again, has yet to be observed. The great limitation of electron microscopy, is its blindness to movement. Nor can it see physical forces. Fluorescence microscopy of living tissues, combined with photobleaching of selected cells, will probably be the best way to determine with certainty whether cell migration during secretion is or is not the main cause of connective tissue structure. Parallel locomotion is to be expected at places where tendons and ligaments are being formed; perpendicular migrations should occur around the developing notochord and cornea.

Another category of explanation is self-assembly of molecules (causally equivalent to crystallization). Gross (1956) discovered that type I collagen can be caused to self-assemble in several dramatically different geometries. When he solubilized type I collagen from tendons, and then reprecipitated aliquots of this dissolved collagen, the presence of glyco-proteins, or of ATP, or the pH, caused assembly of collagen into fibers having dramatically different microstructures and spacing distances (i.e., between the bands seen in electron microscopy). Gross' discovery greatly increased the plausibility that normal differences between, for example, the collagen of tendons and of the cornea could be caused by dif-ferences in the local chemical environment at the time of their deposition. Although the diverse self-assembly patterns he produced don't correspond to anatomical differences observed in the body, he definitely proved that type I collagen has the ability to self-assemble in so many alternative geometries. Surely, these diverse patterns of self-assembly must have functions.

10.2 Development of Connective Tissues in Organ Culture

A third category of pattern-generating mechanisms is the physical rearrangement of colla-gen by cellular forces. This idea originated from Harrison's (1907, 1910, 1912, 1914) original invention of tissue culture. Harrison always used clotted lymph (i.e., fibrin) as the substra-tum for his cells. His primary motivation was to prove beyond doubt that nerve axons form as actively motile strands of cytoplasm, which crawl through tissues to their synapses.

Until then, many had believed that axons were assembled from prefabricated cylindrical rods, stuck together end to end.

Ross Harrison, and later Honor Fell and her collaborators, cultured many kinds of differ-entiated cells, in addition to nerves, and soon noticed that some cells cause dramatic radial reorientations of fibrin (Fell 1925; Fell and Canti 1934; Fell and Robison 1929; Strangeways and Fell 1926). They also discovered that embryonic cartilages, dissected out and cultured in nutrient media, expand autonomously to approximately their normal shapes (though often distorted after the fashion of Salvador Dali). This set of discoveries was the triumph of early research on organ culture. These classic papers are still worth reading, surely con-tain the seeds of new discoveries, and are now available on the Internet.

Starting with research in Harrison's laboratory, Weiss (1928, 1929) made these fibrin orientations the focus of his attention for many years. Among other things, Weiss pro-posed that equivalent patterns should be produced inside developing embryos and could serve the function of guiding nerve fibers. He coined the term "nervenbahn" for these hypothetical nerve roads, on analogy to German autobahns. Figure 10.4 shows an unpub-lished study by David Stopak in 1982 in which nerve axons are crawling through a colla-gen gel and the fibers are being oriented between two clumps of mesenchymal cells. The paths of the nerve fibers are strongly oriented relative to the collagen, although to a lesser degree than one sees in the actual anatomy of an animal. Notice that Weiss, and everyone else at that time, tacitly assumed that collagen fibers in tissues would reorient in the same geometric patterns as they observed in fibrin clots. This turned out to be true. In fact, gels made of reprecipitated collagen align even more dramatically, and over much longer dis-tances, than fibrin gels. Figure 10.5 shows an example of long-range pattern formation in a collagen gel being contracted by fibroblast cells.

Unfortunately, Weiss guessed wrong in respect to the physical forces that cause realign-ment of fibrous protein substrata. He surmised that fibrin clots get pulled by their own shrinkage, rather than by cell contractility. Cell displacements were long regarded

FIGURE 10.4

"Nervenbahn." Low-magnification, dark-field microscopy view of living nerve axons crawling through a collagen gel, from an autonomic ganglion at the lower right corner toward the center and left. On the left side of this field, collagen has been strongly aligned (from lower left to upper center) by two clumps of strongly contractile mesenchymal cells (of which only the lower clump is in the field). The point is that the nerve axons are only slightly aligned. Live, unstained tissue.

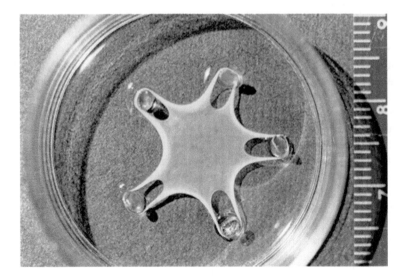

FIGURE 10.5

Contraction of a collagen gel by traction forces exerted by fibroblasts suspended in the collagen before it gelled. Five cylindrical pegs of polystyrene were attached to the bottom of the petri dish and are holding back the contraction of the gel. This is an ordinary 35 mm polystyrene petri dish, and the small divisions on the scale bar on the right are millimeters. The point is that cell traction can align collagen at distances as large as several centimeters and as large as or larger than the dimensions of embryonic cartilages and muscles. Live, unstained tissue culture.

as caused largely by this hypothetical shrinkage of extracellular matrix. The only evidence was that nerve cells produced much less distortion of fibrin clots than fibroblasts. (Subsequently, nerves were found to be less strongly contractile.) In addition, Weiss claimed that adhesion of cells to fibers produced a pulling force on the cell, like capillarity, rather than by anything like actomyosin contractility. Weiss and his theories were dominant for decades, to the extent of having veto power over grant funding. That situation is illustrated by an exchange of interesting letters to the editor of *Science* between Weiss and Katzberg (1952). They agreed that traction could not possibly move tissue cells.

10.3 Flexible Substrata as a Method for Mapping Propulsive and Contractile Forces

Those who deserve most credit for the eventual overthrow of these gel shrinkage dogmas were Michael Abercrombie (Abercrombie et al. 1954, 1970) and Trinkaus (1984), both of whom regarded embryonic cell migrations as a special form of amoeboid locomotion. In addition, the publications of Norman Wessells and his students were decisive in converting people to the concept of actomyosin-based cell locomotion (Wessells et al. 1971). In particular, Abercrombie proposed that distortion of fibrin substrata by crawling cells results from the contractile forces by which these cells crawl, in other words, traction. This led Graham Dunn and me to study rearward force exertion by individual cells (Harris and Dunn 1972). We mapped rearward transport of micrometer-sized particles across the surfaces of crawling fibroblasts. This retrograde transport had been discovered by Ingram (1969) and by Abercrombie et al. (1970).

Graham Dunn's and my contribution was to find that this particle transport occurs on both surfaces of crawling fibroblasts—in other words, not just on the "upper" side of the cell, away from the substratum, but also in the narrow spaces between cell and substratum, where propulsive forces are exerted. This was followed by time lapse movies of individual differentiated cells crawling across the upper surfaces of fibrin gels (Harris 1973). Particles of carbon black had been mixed into the fibrin before it clotted. These carbon particles were found to be temporarily displaced each time a cell crawled over them, and the direction of these displacements was always rearward relative to the axis of each cell's locomotion. This began the method of using distortions of flexible gels, rubber sheets, and viscous gels to measure traction force of individual cells (Harris et al. 1980). Figure 10.6 shows traction exerted by living mesenchymal cells cultured on a thin sheet of silicone rubber, producing a combination of stretching (at the top) and compression (in the middle and below). Several square centimeters of rubber sheet are being compressed into a tiny volume of a few cubic millimeters. In skilled hands, the silicone rubber technique has a spatial resolving power as small as individual sarcomeres of cardiac muscle cells, which is only 2 μm (Danowski et al. 1992). An excellent computer-aided version of the flexible substratum method is now marketed under the name "traction force microscopy" (Munevar et al. 2001), and there have been studies of the transmission of traction tangentially through plasma membranes (Dembo and Harris 1981). One of the most exciting and unexpected discoveries has been that changing the flexibility of culture substrata can control cell differentiation (Discher et al. 2005; Engler et al. 2004, 2006).

The effects of cell traction on collagen gels were first studied by Tom Elsdale and Jonathan Bard (Elsdale and Bard 1972), Eugene Bell (Bell et al. 1979), and others, who made

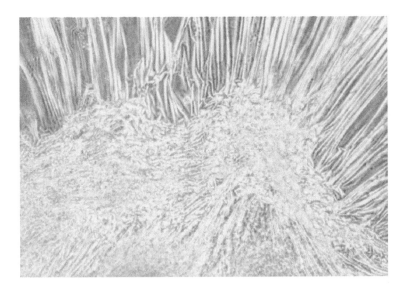

FIGURE 10.6
A thin sheet of silicone rubber (cross-linked polymethyl silicone fluid) is being pulled toward the bottom of the field by cell traction. Many tension wrinkles run vertically across the upper part of the field, and even more compression wrinkles fill the center and lower part of the field. Several square centimeters of rubber are in the process of being compressed. Live, unstained tissue culture, photographed by polarized light.

time lapse movies and photographs of realignments of collagen gels. Many of the geometric patterns observed using collagen closely resemble those that Weiss had discovered in plasma clots. Indeed, when you use collagen gels, alignment is much longer and more dramatic as compared with the effects that are produced from plating clumps of fibroblasts onto a blood plasma clot.

Subsequently, David Stopak, Patricia Wild, and I studied effects of crawling cells on meshworks of reprecipitated collagen (Harris et al. 1981, 1984). We noticed similarities between the distortion effects of cell traction on collagen gels, as compared with their previous observations of cells wrinkling very thin layers of silicone rubber (Harris et al. 1980). In particular, fibroblasts exert much stronger traction than epithelial cells and enormously much stronger traction than macrophages and other kinds of white blood cells. This was a paradox, because white blood cells crawl ten to a hundred times faster than the locomotion of fibroblasts. The latter exerted much more force, while only managing to crawl at less than a tenth the speed of leukocytes. This seemed to be a terrible waste of energy. A second paradox was why these stronger forces didn't interfere with the unknown mechanisms that control the positioning of collagen in the body, for example, its alignment over long distances to form tendons. Only later did it occur to us that cell traction might itself be that unknown mechanism.

10.4 Is Fibroblast Traction Primarily a Method for Rearranging Collagen?

A solution to both paradoxes was suggested to us by the similarity of traction-aligned collagen gels to normal tendons and ligaments. I was teaching an anatomy course at the time. We proposed a somewhat radical explanation, that the main function of fibroblast traction is displacement and alignment of collagen so as to form tendons, organ capsules, and other anatomical structures (Harris et al. 1984). The idea is that fibroblast locomotion is a side effect of this mechanism of moving collagen. Testable predictions of this theory have been confirmed in several studies and contradicted by another, as will be described.

One experiment consisted of explanting chicken embryo cartilages into a collagen gel, near pieces of embryonic back muscles (Stopak and Harris 1982). The results were photographed over a 6-day period, during which time mesenchymal cells at various points near the ends of the cartilages exerted concentrated traction on the collagen. One effect was creation of ligament-like strands of aligned collagen connecting between the ends of the cartilages. A more dramatic effect was alignment and concentration of collagen fibers to form tendon-like strands that connected the cartilages to the explants of muscle cells. These tendon-like strands slowly shortened and pulled the masses of muscle several millimeters through the collagen gel. By 6 days of culture, functional tendons had been formed.

A series of related experiments was then carried out in collaboration with Norman Wessells (Stopak et al. 1985). These tests of the theory consisted of solubilizing type I collagen from the tail tendons of rats, covalently bonding fluorescein directly to the collagen molecules, and then injecting this labeled collagen into the limb buds of embryonic chickens. The embryos were allowed to continue development for a couple of days; then fixed, sectioned, and stained by ordinary histological methods; and photographed through a fluorescence microscope. Notice that this was the reverse of our previous experiments, in that we were putting collagen gels into living embryos, instead of putting pieces of living tissues into collagen gels.

The key question was whether this injected collagen would become rearranged to form parts of normal anatomical structures. To the extent that either local self-assembly or secretion in place is the true cause of collagen arrangement, then there would be no reason to expect injected collagen to be rearranged (and no evident way to explain it). On the other hand, if cells physically rearrange collagen, then the fluorescent injected collagen should be subject to the same forces as normal collagen.

The results were better than we had hoped. The fluorescent collagen became rearranged to be part of whichever connective tissues were forming at the sites of injection. The tight wrapping of this collagen around developing arteries was especially dramatic. That particular phenomenon is deserving of further research, for example, in relation both to atherosclerotic blockages of arteries and to arterial stenoses.

Another study that seems to us to support our hypothesis, but whose authors interpreted very differently, was performed by Grim and Wachtler (1991). It was a surgical transplantation, done on early chicken embryos. Nonmuscle tissue from the periphery of embryos was relocated into developing limb buds, where it became rearranged and attached to the skeleton, as if it were a specific muscle. Although the grafted tissue did not become contractile, it acquired the exact shape and attachments of a particular muscle that glories in the name "extensor medius longus." Grim and Wachtler interpreted this as a result of "positional information," in the sense of control of cell behavior by chemical signals, and didn't regard mechanical forces as being necessary or relevant to the movement and reshaping of cells. I think equivalent experiments should be tried in which pieces of flexible plastic or rubber are grafted into limb buds. This would be a logical extension of the gel and rubber substratum approach, with the flexible materials being put into embryos. To the extent that nonliving plastics also become remolded into the semblance of particular normal muscles, that would either cast doubt on positional information as the explanation, or would indicate that positional information has powers not previously imagined.

Some data that seriously challenge our (i.e., Stopak and Harris) theory have been published by Eckes et al. (1998, 2000). They studied force exertion by cells of mice lacking genes for the cytoskeletal protein vimentin. These mice develop normal connective tissues, including tendons, muscles, and cartilages, but cells cultured from this same strain of mice exert much weaker traction than equivalent cells from normal mice, probably less than half as much, based on their reduced ability to compress collagen gels. The only other phenotypic abnormality of these mutant mice is unusually slow wound closure. A reasonable interpretation of their studies is that Stopak and I misinterpreted myofibroblasts, in the sense of cells whose contractile strength was amplified as a wound closure mechanism, and that those cells' production of normal-looking anatomical patterns was only coincidental. They have a good point, because explanted cells really do become more strongly contractile during the first days in culture. On the other hand, Eckes et al. suggest no explanation for the rearrangement of injected collagen. They call our theory "the dogma," which is premature, but we are nevertheless grateful for the written proof that not everybody already knew it all along. Another explanation is that half the normal amount of traction force is enough to cause formation of tendons.

Another important question is whether tendons and muscles develop independently of each other, in the embryo (Christ et al. 1983; Christ and Brand-Saberi, 2002). They are definitely derived from different cell populations: muscles from somites and tendons from perichondria. In an excellent paper, Kardon (1998) tracked muscle and tendon formation in cleared whole mounts of chick hind limbs. Besides normally developing limbs, she also studied limb buds that had never been reached by muscle cells and limbs in which tendon rudiments had been surgically removed. These experiments demonstrated that in the avian hind limb, the initial morphogenetic events, formation of tendon primordia, and initial differentiation of myogenic precursors occur autonomously with respect to one another. However, later morphogenetic events, such as subdivision of muscle masses and segregation of tendon primordia into individual tendons, do require to various degrees reciprocal interactions between muscle and tendon. Kardon's observations neither contradict nor support Stopak's theory, which postulated that cells interact mechanically via cellular tractions, but they do contradict traditional belief in independent development of muscles relative to tendons (see reference in Kardon 1998). She also describes the surprisingly early alignment of muscle cells and the fascinating splitting of tendons and muscles from one end to the other. Whether traction can produce such splitting has not yet been studied.

This field of study greatly needs experimental methods for changing amounts and directions of tension exerted within developing embryos. Extra tendons ought to form where tensile stress is made strongest. For example, if you surgically implanted nodules of especially strong myofibroblasts, to arbitrary anatomical locations, then supernumerary tendons and ligaments should form, connecting these implants to normal origin and insertion sites on skeletal surfaces. Furthermore, we should seek evidence of cartilage differentiation being stimulated at locations where compressive stress is maximized. Thus, we need optical or other methods that can "see" relative amounts of tension and compression.

If you look at any cartilage using polarization microscopy, you see complex and beautiful geometric patterns (as in Figure 10.3). These patterns map the predominant directions of collagen fibers and thus indirectly map physical tensions. Almost any compound microscope can be used to see these patterns, by inserting two polaroid filters in combination with quarter-wave plates at appropriate locations along the light path. The geometric patterns of birefringence would be even more beautiful, as well as more informative, if the optical effects of perpendicular collagen fibers did not cancel each other.

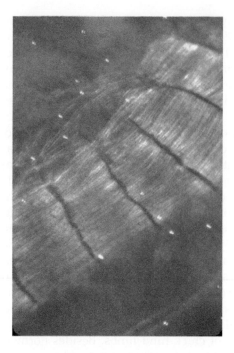

FIGURE 10.7
Chicken embryo somites. Whole mount of living tissue, viewed by polarized light microscopy. The myotome part of each of the five somites appears bright white, mostly because of alignment of cytoplasmic actomyosin.

Also, observers should be aware that fibers running parallel to the light path experience no birefringence. The colors that one sees result from more fibers running in some directions than in other directions. There is a well-developed subfield of engineering, called "photoelasticity" (Frocht 1965), which uses these principles to map stresses in models of machinery, dams, and buildings. For example, scale models made of gelatin have been used to predict stress distributions inside dams.

Photoelasticity won't be adequate for visualization of tension and stress in embryos until some (nontoxic and nonhydrolyzing) substance becomes available, which is substantially more sensitive than collagen or actomyosin, in respect to how much birefringence gets produced per amount of stress exerted. The need is to distinguish causes from effects. For example, the myotome subdivisions of somites become strongly birefringent very early in development. This is easy to see by polarized light (Figure 10.7), but is alignment of actin and collagen the result, or the cause, of tension? Both could be true, in a feedback cycle. It is not difficult to make videos of birefringence developing where tendons will form, but that is not proof that tension precedes alignment of collagen.

10.5 Difficult Physics of Cartilage

The causal chain from genes to anatomical shapes depends very much on how cartilages get their many different shapes. This cannot be determined without understanding the relevant biophysical forces. Of all connective tissues, cartilage is the most important contributor to the shape of the skeleton and the growth of the body. The ability of cartilage to resist compression, expand, and reshape itself depends on a form of osmosis that does not involve semipermeable membranes. This means of pressure generation is called electroosmosis (Glasstone 1948; Reynaud and Quinn 2006). Unfortunately, the word electroosmosis also refers to some seemingly unrelated phenomena. Each of these is caused by unequal distributions of counterions, as will be explained. To minimize confusion, these other examples of electroosmosis will be summarized at the end of this section.

Cartilage consists of a mixture of collagen and very high-molecular-weight sulfated polysaccharides, especially chondroitin sulfate. These sugar chains have very high densities of sulfate and carboxylic side chains. At any but the most acidic pH, sulfates and carboxyl groups ionize. Their ionization releases positive counterions, including sodium ions, potassium ions, hydrogen ions, and calcium ions. Ionization also creates negative charges on the sulfate and carboxylic side chains. These negative charges cannot disperse, because they are covalently bonded to sugar chains. Ultimately, their dispersal is mechanically constrained by collagen fibers, some running through the interior of cartilages, and others wrapped around their surfaces. Locations and orientations of fibers can be seen by polarized light.

The positive counterions cannot diffuse away, out of cartilage, because they are held back by the negative charges of the ionized sulfate and carboxylic groups—that is, by Coulomb attraction. Pressure is produced by water molecules diffusing to locations of highly concentrated counterions. This produces just as much osmotic pressure as would result if the same concentration of counterions were held in place by being surrounded by a semipermeable membrane. The key fact is that osmotic pressure is created by any situation in which an elevated concentration of diffusible ions or molecules is prevented from dispersing, but into which water is free to diffuse.

For these reasons, the interior of every cartilage contains a strongly hypertonic solution of positive counterions. The effect on cell volumes of ordinary hypertonic salt solutions is familiar to every biologist, but there is less awareness that cells enclosed in cartilage experience the same osmotic effect. In effect, water is constantly being sucked out of them, just as if they were in hypertonic saline. More research is needed on the internal osmotic pressure of chondrocytes. In particular, this is relevant to the large increases in volume of chondrocytes seen near regions of ossification. These expansions are 10-fold or greater, are easily observed, and there are pictures of them in every anatomy textbook.

It is unknown to what extent swelling of chondrocytes results from increased cytoplasmic pressure, as opposed to decreased matrix pressure. The hypertonicity of cartilage is also relevant to its penetration by capillaries. Inhibitory proteins are not the only obstacle to vascularization of cartilage. Osmotic pressure also has to be overcome.

Unfortunately, nearly all scientists interested in cartilage attribute its swelling pressure to the Gibbs–Donnan effect, also known as the Donnan equilibrium. For example, see Myers and Mow (1983), Lai et al. (1991), and Mow et al. (2005). This belief is misguided. The titles of Frederick Donnan's two classic papers on this subject were *The theory of membrane equilibria* (1924) and before that the original German version: *Theorie der Membrangleichgewichte und Membranpotentiale...* (1911). The titles of these papers accurately reflect their contents, which are specifically concerned with phenomena that depend on semipermeable membranes. Except for the plasma membranes surrounding each individual chondrocyte, the cartilage is not itself surrounded by a membrane. Furthermore, Donnan's interest was ratios of ion concentrations, not mechanical pressures. The swelling pressure of cartilage is not caused by a Donnan equilibrium. This misunderstanding delays research progress. For example, strengths of osmotic pressures cannot possibly vary as a function of direction (any more than a chemical concentration can have a direction). Concentrations often change as you move in a direction, but cannot themselves possess a direction; therefore, electroosmotic pressures can't push more strongly in one direction than any other. This is important in relation to the directional elongation of the skeleton. Textbooks generally attribute such directionality to the pressure. However, to the extent that the pressure is electroosmotic, directional differences can only be produced by different resistances to the pressure (e.g., from the collagenous perichondrium) being stronger in one direction than another.

In their excellent paper titled *Growth of Cartilage*, Hinchliffe and Johnson (1983) wrote that there are only four mechanisms by which a cartilage can expand in volume. They listed these four mechanisms as

1. Cell division
2. Matrix secretion
3. Increase in cell volume
4. Recruitment of cells from the surroundings

They did not consider the following fifth possibility:

5. Weakening or disconnection of the collagen fibers that resist matrix swelling

This fifth method is analogous to increasing the bulging of a balloon by weakening the rubber and illustrates the importance of understanding the physics of cartilage. If you use polarization microscopy to look at a section through an elongating bone (Figure 10.3), the patterns of birefringence indicate localized and directional weakening of collagen. These patterns of changing birefringence show that the swelling pressure within cartilages varies gradually from place to place, in complex patterns (Figure 10.8, lower right photograph).

Differential swelling is how cartilage performs its embryological function of shaping the skeleton (Thorogood 1983). Just as described by Wolpert (1982), cutting through the collagen sheath of an early embryonic cartilage results in masses of rounded chondrocytes flowing out, like toothpaste out of a tube (Figure 10.8, upper left). Incidentally, this is a good way to isolate very pure populations of chondrocytes.

The femur starts as a humble blob of tissue; it then differentiates into cartilage, which in turn expands directionally by combinations of actual growth (cell growth and secretion of more matrix) plus considerable electroosmotic inflation (Archer et al. 1983; Rooney et al. 1984). Weakening of collagen fibers, both those surrounding a cartilage and those within, results in expansion and elongation of cartilages. It is not yet known how this differential weakening is produced in normal development. In principle, it could be a reduction in either Young's modulus (which is the proportionality constant that measures how much force is needed to produce a given percentage of elastic stretching) or it could be a lessening of the yield stress (how much force is needed to cause plastic, as opposed to elastic, distortion). Detachment of collagen molecules from each other or from cells is another way to weaken resistance to swelling forces.

More difficult to explain is the formation of thin, flat sheets of cartilage in the side walls of the eyeball of birds (and all other vertebrates, except mammals). Figure 10.9 shows the complex pattern of birefringence in a histological section of a rabbit ear. The numerous blue lines are bundles of collagen that connect the two surfaces of the ear. This same pattern of collagen alignment should also be looked for in all flat cartilages, including those of bird eyeballs. Weiss and Amprino (1940; also Weiss and Moscona 1958) discovered that cells of these cartilages will reform flat sheets, even after having been dissociated into random collections of individual cells. Other phenomena not yet well understood include

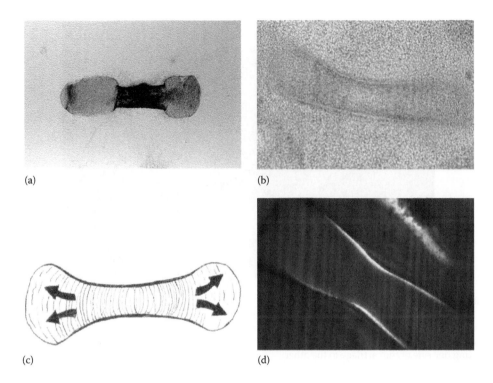

FIGURE 10.8
(a) The shape of long bones (such as the femur, ulna, and phalanges) is created by electroosmotic swelling of cartilage. Collagen fibers are wrapped circumferentially around cartilage, the pressure of which squeezes chondrocytes and cartilage so that they balloon outward at both ends. A chicken embryo cartilage had both ends cut off, allowing thousands of chondrocytes to be pushed out the ends. (c) Drawing of the spatial distribution of forces in an elongating cartilage. (b) Formation of a leg cartilage of a chicken embryo. Whole mount of living tissue. Phase contrast microscopy. This is typical of the initial condensations of cartilages. (d) Leg "bone" (at this stage entirely made of cartilage) seen by polarized light microscopy. Areas in which collagen fibers are predominantly longitudinal appear blue, while yellowish red indicates that collagen fibers are predominantly circumferential. The colors in this photograph indicate that there are many longitudinal fibers, but more circumferential fibers, perhaps 55% circumferential and 45% longitudinal. Living tissue.

the initial condensation of cartilages, the control of their directional elongation, side-to-side fusion of certain long cartilages, and the rearrangement and reshaping of cartilages in regenerating legs of salamanders.

As was mentioned previously, the name "electroosmosis" also applies to phenomena that seem unrelated to the swelling pressure of cartilage. For example, direct current electric voltages can be used to pull water directionally through masonry and soil. The reason this process is also called electroosmosis is because it results from positive counterions being pulled by a voltage gradient and osmosis pulling water along with the counterions. Conversely, if you push water through sand or soil, you create a small voltage gradient. It's those positive counterions again: they get created by release from particles of sand or clay, and they get carried along by a flow of water. Therefore, the name electroosmosis gets applied to those phenomena. If you experiment with galvanotaxis in tissue culture (e.g., Harris et al. 1990), you watch steady sideways flows of water and debris, pulled along with positive counterions, next to the negatively charged glass or plastic substrata. That, too, is electroosmosis. Furthermore, electroosmosis is used as a very refined separation method

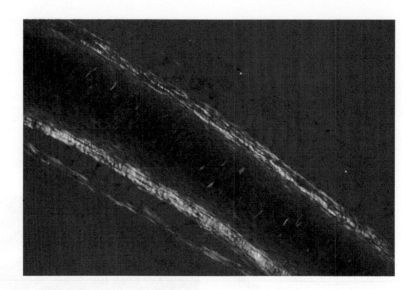

FIGURE 10.9
Cross section of a rabbit ear, photographed by polarized light microscopy. Mammal ears, including human ears, are supported by elastic cartilage. Tissues appear yellow where collagen and elastin fibers are oriented predominantly along the length of the ear. The blue areas are locations where collagen and elastin fibers are connecting one side of the cartilage to the other.

by chemists, complementary to electrophoresis (Jorgenson and Lukacs, 1981). These are all manifestations of a single phenomenon.

Localized variations in speeds of electroosmotic flow past cell surfaces could be developed into methods for measuring densities of negatively ionized groups on cell surfaces. That could be more informative than the method of "cell electrophoresis," briefly in vogue during the late 1960s. Likewise, squeezing cartilage produces substantial but brief voltages (see Reynaud and Quinn 2006) that dwarf those produced by the piezoelectric properties of bone. Voltages are caused by positive counterions getting carried along with the water, not only when water is squeezed out but also when it flows back in. Herniated disks and damaged articular cartilages cause so much human suffering! Their physical mechanism deserves more attention. Every biophysics and embryology textbook ought to have a chapter on electroosmosis, but none do.

10.6 Regeneration of Skeleton and Muscles: Rearrangement or Redifferentiation?

More is known about regeneration of connective tissues than about their original formation. This is especially true of salamanders, because they can regenerate their legs (among surprisingly many other organs that they can regenerate). During regeneration of salamander legs, both cartilage cells and muscle cells appear to lose their differentiated state, in the sense that they become indistinguishable from each other, even by transmission electron microscopy. On this basis, one might have predicted that regenerated cartilages would be partly made of descendants of muscle cells and newly formed muscles made partly of former chondrocytes. However, the evidence from nuclear labeling (Steen 1968, 1970) is that neither of these conversions occurs, especially not to the extent that would be needed to contribute to the shaping of regenerating cartilages and regeneration of muscles. For example, the radius and ulna, the carpals, and the phalanges are mostly made of cells that had been part of the humerus (although many of the new chondrocytes had been dermal fibroblasts). To an even greater extent, all the nuclei of the regenerated front foot are descendants of nuclei that had been part of shoulder and upper arm muscles. Repeated and meticulous nuclear labeling studies have consistently shown that dedifferentiated cartilage cells redifferentiate almost entirely into cartilage cells (see Stocum and Cameron 2011).

Among other implications, this means that the chondrocytes of new regenerated carpals used to be chondrocytes of the humerus. It also means that muscles of the regenerated foot are syncytia whose nuclei had previously composed muscles of the upper limb. This is equivalent to making hand muscles from pieces of shoulder muscle. More to the point, it means that shoulder muscle cells have been rearranging themselves to become hand muscles, despite their very different skeletal attachment sites (origins and insertions). Furthermore, and even more contrary to what has been assumed, these nuclear labeling results seem (to me, at least) to mean that leg regeneration can only be understood in terms of mechanisms of cell rearrangement. Unless cells were switching differentiated cell type, which the nuclear labeling experiments proved they don't, then the new cartilages and muscles are formed by cell rearrangement. That means limb regeneration is a special case of cell sorting.

Stocum and Cameron (2011) recently published a wonderfully inclusive review of over 300 publications about salamander limb regeneration. These include a dozen nuclear labeling studies equivalent to Trygve Steen's, which reached the same conclusions about nuclei remaining loyal to their original cell type. Amidst so much effort, few papers asked what physical forces serve to pull or push redifferentiating cartilage and muscle cells into functional locations and attachments. The closest anyone has come to this was the important but ambiguous discovery by Nardi and Stocum (1983) that regenerating limb buds will engulf each other and sort out based on their proximal versus distal origin.

Among the theories of cell sorting, they considered only Malcolm Steinberg's differential adhesion hypothesis (Steinberg 1963) as a possible cause of sorting behavior. A considerable literature (cited in Stocum and Cameron's review) has developed concerning adhesion gradients in regenerating limb buds. But it is now realized that differences in contractility can also affect sorting (Brodland 2002, 2004; Green 2008), besides having the advantage of being able to shape cartilages and muscles (as this review has stressed). Either way, it remains a tantalizing challenge to account for the shaping of new phalanges, carpals, radius, and ulna, made out of redifferentiating chondrocytes (about half descended from the former humerus, with the other half derived from former dermal cells). I am puzzled that few researchers in this area seem to regard skeletal and muscle regeneration as questions of rearrangement of cells that were already committed to which cell type they are going to resume being. But unless cells switch cell type, rearrangement is the only way for them to form new muscles and cartilages.

10.7 Conclusion

Physical forces play major roles in the shaping and development of connective tissues, including especially cartilage and the musculature. Muscle cells crawl actively all through the body, starting from their sites of initial differentiation in the somites. In contrast, cartilages condense from local mesenchymal tissues. Nevertheless, the eventual shapes of cartilages result largely from differential electroosmotic swelling. These movements and shape changes, which readily continue in organ culture, include crawling locomotion of cells (including myoblasts), condensation and swelling of cartilages, and also active rearrangement of collagen gels to form strands that not only resemble ligaments and tendons but which also spontaneously interconnect cartilages to muscles. Collagen also gets actively rearranged if isolated from one animal and injected into developing embryos, even of other species. Mechanical forces, especially cell traction and electroosmosis, are recurring causes of anatomical shapes and connective tissue development.

References

Abercrombie, M., M. H. Flint, and D. W. James. 1954. Wound contraction in relation to collagen formation in scorbutic guinea-pigs. *J. Embryol. Exp. Morph.* 4: 167–174.

Abercrombie, M., J. E. Heaysman, and S. M. Pegrum. 1970. The locomotion of fibroblasts in culture. 3. Movements of particles on the dorsal surface of the leading lamella. *Exp. Cell Res.* 62: 389–398.

Archer, C. W., A. Hornbruch, and L. Wolpert. 1983. Growth and morphogenesis of the fibula in the chick embryo. *J. Embryol. Exp. Morph.* 75: 101–116.

Bell, E., B. Ivarsson, and C. Merrill. 1979. Production of a tissue-like structure by contraction of collagen lattices by human fibroblasts of different proliferative potential in vitro. *Proc. Natl. Acad. Sci. USA* 76: 1274–1278.

Birk, D. E. and R. L. Trelstad. 1986. Extracellular compartments in tendon morphogenesis: Collagen fibril, bundle, and macroaggregate formation. *J. Cell Biol.* 103: 231–240.

Birk, D. E., E. K. Zycband, D. A. Winkelmann, and R L. Trelstad. 1990. Collagen fibrillogenesis *in situ*. Discontinuous segmental assembly in extracellular compartments. *Ann. N. Y. Acad. Sci.* 580: 176–194.

Birk, E. E., E. I. Zycband, D. A. Winkelmann, and R. L. Trelstad. 1989. Collagen fibrillogenesis *in situ*: Fibril segments are intermediates in matrix assembly. *Proc. Natl. Acad. Sci. USA* 86: 4549–4553.

Brodland, G. W. 2002. The Differential Interfacial Tension Hypothesis (DITH): A comprehensive theory for the self-rearrangement of embryonic cells and tissues. *J. Biomech. Eng.* 124: 188–197.

Brodland, G. W. 2004. Computational modeling of cell sorting, tissue engulfment and related phenomena: A review. *Appl. Mech. Rev.* 57: 47–76.

Christ, B. and B. Brand-Saberi. 2002. Limb muscle development. *Int. J. Dev. Biol.* 46: 905–914.

Christ, B., M. Jacob, and H. J. Jacob. 1983. On the origin and development of the ventrolateral abdominal muscles in the avian embryo. An experimental and ultrastructural study. *Anat. Embryol. (Berl).* 166: 87–101.

Danowski, B. A., K. Imanaka-Yoshida, J. M. Sanger, and J. W. Sanger. 1992. Costameres are sites of force transmission to the substratum in adult rat cardiomyocytes. *J. Cell Biol.* 118: 1411–1420.

Dembo, M. and A. K. Harris. 1981. Motion of particles adhering to the leading lamella of crawling cells. *J. Cell Biol.* 91: 528–536.

Discher, D. E., P. Janmey, and Y.- L. Wang. 2005. Tissue cells feel and respond to the stiffness of their substrate. *Science* 310: 1139–1143.

Donnan, F. G. 1911. Theorie der Membrangleichgewichte und Membranpotentiale bei Vorhandensein von nicht dialysierenden Elektrolyten. Ein Beitrag zur physikalisch-chemischen Physiologie. [The theory of membrane equilibrium and membrane potential in the presence of a non-dialyzable electrolyte. A contribution to physical-chemical physiology]. *Z. Elektrochemie angewandte physikalische Chemie* 17: 572–581.

Donnan, F. G. 1924. The theory of membrane equilibria. *Chem. Rev.* 1: 73–90.

Eckes, B., E. Colucci-Guyon, H. Smola, S. Nodder, C. Babinet, T. Krieg, and P. Martin. 2000. Impaired wound healing in embryonic and adult mice lacking vimentin. *J. Cell Sci.* 113: 2455–2462.

Eckes, B., D. Dogic, E. Colucci-Guyon, N. Wang, A. Maniotis, D. Ingber, A. Merckling et al. 1998. Impaired mechanical stability migration and contraction capacity in vimentin-deficient fibroblasts. *J. Cell Sci.* 111: 1895–1907.

Elsdale, T. and J. Bard. 1972. Collagen substrata for studies on cell behavior. *J. Cell Biol.* 54: 626–637.

Engler, A. J., M. A. Griffin, S. Sen, S. Bönnemann, H. L. Sweeney, and D. E. Discher. 2004. Myotubes differentiate optimally on substrates with tissue-like stiffness: Pathological implications for soft or stiff microenvironments. *J. Cell Biol.* 166: 877–887.

Engler, A. J., S. Sen, H. L. Sweeney, and D. E. Discher. 2006. Matrix elasticity directs stem cell lineage specification. *Cell* 126: 677–689.

Fell, H. B. 1925. The histogenesis of cartilage and bone in the long bones of the embryonic fowl. *J. Morphol. Physiol.* 40: 417–458.

Fell, H. B. and R. G. Canti. 1934. Experiments on the development in vitro of the avian knee joint. *Proc. R. Soc. London, Ser. B* 116: 316–351.

Fell, H. B. and R. Robison. 1929. Thee growth, development and phosphatase activity of embryonic avian femora and limb-buds cultivated in vitro. *Biochem. J.* 23: 767–784.

Frocht, M. M. 1965. *Photoelasticity.* London, U.K.: J. Wiley & Sons.

Glasstone, S. 1948. *Textbook of Physical Chemistry.* New York: Van Nostrand.

Green, J. B. A. 2008. Sophistications of cell sorting. *Nat. Cell Biol.* 10: 375–377.

Grim, M. and F. Wachtler. 1991. Muscle morphogenesis in the absence of myogenic cells. *Anat. Embryol. (Berl).* 183: 67–70.

Gross, J. 1956. The behavior of collagen units as a model of morphogenesis. *J. Biophys. Biochem. Cytol.* 2: 261–274.

Harris, A. K. 1973. Behavior of cultured cells on substrata of variable adhesiveness. *Exp. Cell Res.* 77: 285–297.

Harris, A. K. and G. A. Dunn. 1972. Centripetal transport of attached particles on both surfaces of moving fibroblasts. *Exp. Cell Res.* 73: 519–523.

Harris, A. K., N. K. Pryer, and D. Paydarfar. 1990. Effects of electric fields on fibroblast contractility and cytoskeleton. *J. Exp. Zool.* 253: 163–176.

Harris, A. K., D. Stopak, and P. Warner. 1984. Generation of spatially periodic patterns by a mechanical instability: A mechanical alternative to the Turing model. *J. Embryol. Exp. Morph.* 80: 1–20.

Harris, A. K., D. Stopak, and P. Wild. 1981. Fibroblast traction as a mechanism for collagen morphogenesis. *Nature* 290: 249–251.

Harris, A. K., P. Wild, and D. Stopak. 1980. Silicone rubber substrata: A new wrinkle in the study of cell locomotion. *Science* 208: 177–179.

Harrison, R. 1907. Observations on the living developing nerve fiber. *Anat. Rec.* 1: 116–118.

Harrison, R. 1910. The outgrowth of the nerve fiber as a mode of protoplasmic movement. *J. Exp. Zool.* 9: 787–846.

Harrison, R. 1912. The cultivation of tissues in extraneous media as a method of morphogenetic study. *Anat. Rec.* 6: 181–193.

Harrison, R. 1914. The reaction of embryonic cells to solid structures. *J. Exp. Zool.* 17: 521–544.

Henderson, J. H. and D. R. Carter. 2002. Mechanical induction in limb morphogenesis: The role of growth-generated strains and pressures. *Bone* 31: 645–653.

Henderson, J. H., L. de la Fuente, D. Romero, C. I. Colnot, S. Huang, D. R. Carter, and J. A. Helms. 2007. Rapid growth of cartilage may generate perichondrial structures by mechanical induction. *Biomechan. Model Mechanobiol.* 6: 127–137.

Hinchliffe, J. R. and D. R. Johnson. 1983. The growth of cartilage. In *Cartilage. Development, Differentiation, and Growth*, B. K. Hall, ed., pp. 255–295. New York: Academic Press.

Ingram, V. M. 1969. A side view of moving fibroblasts. *Nature* 222: 641–644.

Jorgenson, J. W. and K. D. Lukacs. 1981. High-resolution separations based on electrophoresis and electroosmosis. *J. Chromatogr.* 218: 209–216.

Kardon, G. 1998. Muscle and tendon morphogenesis in the avian hind limb. *Development* 125: 4019–4032.

Lai, W. M., J. S. Hou, and V. C. Mow. 1991. A triphasic theory for the swelling and deformation behaviors of articular cartilage. *J. Biomech. Eng.* 113: 245–258.

Mow, V. B., W. Y. Gu, and F. H. Chen. 2005. Structure and function of articular cartilage and meniscus. In *Basic Orthopaedic Biomechanics and Mechano-Biology*, 3rd edn., pp. 181–258. Philadelphia, PA: Lippincott, Williams & Wilkins.

Munevar, S., Y.-L. Wang, and M. Dembo. 2001. Traction force microscopy of migrating normal and H-ras transformed 3T3 fibroblasts. *Biophys. J.* 80: 1744–1757.

Myers, E. R. and V. C. Mow. 1983. Biomechanics of cartilage and its response to biomechanical stimuli. In *Cartilage, Volume 1. Structure, Function, and Biochemistry*, B. K. Hall, ed., pp. 313–341. New York: Academic Press.

Nardi, J. B. and D. L. Stocum. 1983. Surface properties of regenerating limb cells: Evidence for gradation along the proximodistal axis. *Differentiation* 25: 27–31.

Oster, G. F., J. D. Murray, and A. K. Harris. 1983. Mechanical aspects of mesenchymal morphogenesis. *J. Embryol. Exp. Morph.* 78: 83–125.

Pacifici, M., E. Koyama, and M. Iwamoto. 2005. Mechanisms of synovial joint and articular cartilage formation: Recent advances, but many lingering mysteries. *Birth Defects Res. C* 75: 237–248.

Remington, J., X. Wang, Y. Hou, H. Zhou, J. Burnett, T. Muirhead, J. Uitto, D. R. Keene, D. T. Woodley, and M. Chen. 2009. Injection of recombinant human type VII collagen corrects the disease phenotype in a murine model of dystrophic epidermolysis bullosa. *Mol. Ther.* 17: 26–33.

Reynaud, B. and T. M. Quinn. 2006. Tensorial electrokinetics in articular cartilage. *Biophys. J.* 91: 2349–2355.

Rooney, P., C. Archer, and L. Wolpert. 1984. Morphogenesis of cartilaginous long bone rudiments. In *The Role of Extracellular Matrix in Development*, R. L. Trelstad, ed., pp. 305–322. New York: Alan R. Liss.

Steen, T. P. 1968. Stability of chondrocyte differentiation and contribution of muscle to cartilage during limb regeneration in the axolotl (*Siredon mexicanum*). *J. Exp. Zool.* 167: 49–77.

Steen, T. P. 1970. Origin and differentiative capacities of cells in the blastema of the regenerating salamander limb. *Am. Zool.* 10: 119–132.

Steinberg, M. S. 1963. Reconstruction of tissues by dissociated cells. Some morphogenetic tissue movements and the sorting out of embryonic cells may have a common explanation. *Science* 141: 401–408.

Stocum, D. L. and J. A. Cameron. 2011. Looking proximally and distally: 100 years of limb regeneration and beyond. *Dev. Dyn.* 240: 943–968.

Stopak, D. and A. K. Harris. 1982. Connective tissue morphogenesis by fibroblast traction I. Tissue culture observations. *Dev. Biol.* 90: 383–398.

Stopak, D., N. K. Wessells, and A. K. Harris. 1985. Morphogenetic rearrangement of injected collagen in developing limb buds. *Proc. Natl. Acad. Sci. USA* 82: 2804–2808.

Strangeways, T. S. P. and H. B. Fell. 1926. Experimental studies on the differentiation of embryonic tissues growing in vivo and in vitro. I. The development of the undifferentiated limb-bud (a) when subcutaneously grafted into the post-embryonic chick and (b) when cultivated in vitro. *Proc. R. Soc. London, Ser. B* 99: 340–364.

Thorogood, P. 1983. Morphogenesis of cartilage. In *Cartilage. Volume 1. Development, Differentiation, and Growth*, B. K. Hall, ed., 223–254. New York: Academic Press.

Trinkaus, J. P. 1984. *Cells Into Organs. The Forces That Shape The Embryo.* 2nd edn. Englewood Cliffs, NJ: Prentice-Hall.

Weiss, P. 1928. Experimentelle Organizierungen ueber des Gewebewachstums in vitro. *Biol. Zentralbl.* 48: 551–566.

Weiss, P. 1929. Erzwingung elementarer Strukturverschiedenheiten am in vitro wachsenden gewebe. Die Wirkung mechanisher Spannung auf Richtung und Intensität des Gewebwachstums und ihre Analysis. *Roux Archiv Entwicklungsmechanik der Organismen* 116: 438–554.

Weiss, P. and R. Amprino. 1940. The effect of mechanical stress on the differentiation of scleral carti-lage in vivo and in the embryo. *Growth* 4: 245–258.

Weiss, P. and A. A. Katzberg. 1952. Comments and communications: "Attraction fields" between growing tissue cultures. *Science* 115: 293–296.

Weiss, P. and A. Moscona. 1958. Type-specific morphogenesis of cartilages developed from dissoci-ated limb and scleral mesenchyme. *J. Embryol. Exp. Morphol.* 6: 238–246.

Wessells, N. K., B. S. Spooner, J. F. Ash, M. O. Bradley, M. A. Luduena, E. L. Taylor, J. T. Wrenn, and K. Yamada. 1971. Microfilaments in cellular and developmental processes. *Science* 171: 135–153.

Wolpert, L. 1982. Cartilage morphogenesis in the limb. In *Cell Behaviour*, R. Bellairs, A. S. G. Curtis, and G. Dunn, eds., pp. 359–372. Cambridge, U.K.: Cambridge University Press.

11

Cellular Forces in Morphogenesis

Larry A. Taber

CONTENTS

11.1 Introduction

During development, the embryo undergoes dramatic and relatively rapid changes in form. The shape changes that characterize morphogenesis are not simply a manifestation of regional differences in growth rates. Rather, many are caused primarily by complex 3-D tissue deformations driven by active cellular and subcellular forces.[1-6] Embryologists, biophysicists, and engineers have expended considerable effort studying these mechanisms, but one surprising aspect of this field may be how much remains unknown about the physical mechanisms of morphogenesis. This is especially true for understanding how cell-level activity is coordinated to change tissue-level morphology. The main purpose of this chapter is to provide an overview of the current state of knowledge in this field.

Researchers now know much about the behavior of individual cells, shedding light on their potential large-scale morphogenetic behavior. Cells can divide or change size, change shape by actomyosin contraction or actin polymerization, migrate, and adhere to other cells and matrix. Like the number of individual genes and molecular signals, the number of morphogenetic activities available to cells during development is relatively small, but these activities can be combined in a myriad of ways to generate animal form. In general, collective cell behavior can be considerably more complex than the behavior of an isolated cell. For example, if all cells in an unconstrained and unstressed tissue grow isotropically at equal rates via multiplication or increased cell volume, the tissue as a whole would increase in volume but remain stress free macroscopically, although individual cytoskeletal and matrix constituents always bear some stress. However, regional differences in growth rate generate residual stress that can alter the shapes of individual cells, while the global shape of the tissue remains relatively unchanged, for example, when a small tumor grows inside the brain.

In general, development involves three fundamental processes: growth, remodeling, and morphogenesis.[3] *Growth* is defined here as a change in material volume.* Positive growth can occur by cell multiplication (hyperplasia), increase in cell size (hypertrophy), creation or swelling of extracellular matrix (ECM), or by outside cells migrating into the tissue. Programmed cell death (apoptosis) is an important mechanism for negative growth, for example, to eliminate webs between fingers and toes or the common wall between two tubes that fuse into a single tube.[1] *Remodeling* entails a change in material properties, which often is associated with changes in cytoskeletal or matrix structure. Remodeling of the cytoskeleton and nucleus generally accompanies cellular differentiation.[7-11] Finally, *morphogenesis* is a change in shape. Although all of these processes are coupled to some extent, it is useful to consider them separately. This chapter focuses on morphogenesis.

The emphasis here is on fundamental morphogenetic processes. For illustration, we discuss aspects of the development of specific tissues and organs, but there is no attempt to be all inclusive. In addition, we deal primarily with physical mechanisms, neglecting the crucial effects of biochemistry, genetics, and molecular signaling.

11.2 Fundamental Building Blocks of an Embryo

11.2.1 Cells and Cell Sheets

Embryos contain two types of cells. *Mesenchymal cells* move and deform as individual cells or 3-D aggregates, whereas *epithelial cells* move and deform together as sheets. Bones and some organs, for example, kidneys and liver, are constructed primarily from mesenchyme. Epithelia play a major role in the early stages of heart, brain, and lung development.[12]

To create structures, mesenchymal cells and matrix condense and stick together to form aggregates, which are then molded into the appropriate shapes. Cells adhere to other cells and ECM by expressing cell–cell adhesion molecules (e.g., cadherins) and cell–matrix adhesion molecules (e.g., integrins), respectively.[13] Cell–cell adhesion also can be promoted by matrix degradation, which allows cells initially separated by matrix to come into contact.[1] These aggregates can take the form of either 3-D masses or epithelia. Embryonic epithelia typically consist of one or more cell layers attached to a basement membrane composed of ECM.

---

* Some authors use somewhat different terminology or definitions for these processes; for example, some define growth as a change in mass. During morphogenesis, however, mass usually is not a crucial variable. For example, replacing a cell by a heavier cell has relatively little effect on stress, whereas replacing a cell by a bigger cell can perturb stress significantly.

The creation of an epithelium from mesenchymal cells is an example of *mesenchymal–epithelial transition* (MET).[14] One example occurs after the first few cell divisions in the embryo as the cells coalesce into a small mass, and then fluid pumped into the center of the mass expands the cells into a fluid-filled epithelial shell called a *blastula*.[12] Another example is found in early heart development, when mesenchymal cells undergo MET to form a pair of bilateral epithelia that fold and fuse to create the primitive heart tube.[15]

The opposite process, termed *epithelial–mesenchymal transition* (EMT), is no less important.[14] During EMT, cells lose adhesion to their neighbors and leave the epithelium to migrate elsewhere. In fact, cells often go back and forth between mesenchyme and epithelia during development. In the embryo, EMT creates neural crest cells, which emerge from the neural ectoderm and migrate to various locations, where they differentiate into several different tissue types.[12] EMT also plays an important role in the development of heart valves as they form from mounds of ECM called endocardial cushions, which initially are encased by an epithelium and function as primitive valves in the early heart tube.[16] Endothelial cells invade the ECM, where they contract and mold the tissue into valve leaflets. Recent studies have suggested that this process is regulated in part by hemodynamic loads.[17,18] EMT also can have deleterious effects, such as in cancer metastasis.[14]

11.2.2 Mechanical Behavior of Embryonic Tissues

On the macroscopic level, embryonic tissues generally exhibit the classical characteristics of viscoelastic materials,[19–22] that is, their mechanical response to loads resembles a mix of an elastic solid and a viscous fluid. In general, the response of embryonic tissues to external loads can be considered as solid-like on short timescales but fluid-like on the relatively long timescales of most morphogenetic processes.[23,24] These properties can change considerably over the course of development.[25–27]

For the most part, mathematical models for morphogenetic processes have treated embryonic tissues as either elastic solids or viscous fluids.[28] When tissues are modeled as fluids, morphogenesis is assumed to be driven by surface tension, possibly modulated in part by active contraction (see section 11.3.1). This assumption, however, precludes the possibility that the free surface of an embryonic tissue cannot be in a state of in-plane compression, contrary to experimental results for the heart.[29] Hence, developing tissues exhibit some solid-like behavior even over long timescales.

11.3 Pattern Formation

After the blastula forms, the work of building an embryo with a specific form or *pattern* begins. Patterning involves spatial differentiation of cells into specific types, as well as the creation of the basic organization and shape of organs, bones, and other structures. Blastula formation can be considered as the beginning of this procedure. Next is the morphogenetic process of *gastrulation*, which reorganizes the blastula into a multilayered structure consisting of three primary germ layers. From outside to inside, these layers are called the *ectoderm*, *mesoderm*, and *endoderm*.[12] The mesoderm is initially composed of mesenchymal cells, while the ectoderm and endoderm are epithelia. Among other structures, the ectoderm becomes the skin, brain, and spinal cord; the mesoderm forms the heart, bones, muscles, and kidneys; and the endoderm creates the gut and lungs.

11.3.1 Cell Sorting

A remarkable property of embryonic cells was revealed by the classic experiments of Johannes Holtfretter, who mixed together dissociated germ cells from amphibian embryos (see review by Steinberg[30]). During subsequent culture, these cells sorted themselves into layers with mesoderm situated between ectoderm and endoderm, as occurs in vivo. The relevance of cell sorting to the process of gastrulation is not clear, however.[31]

Consistent with the view that embryonic tissues possess fluid-like properties on long timescales, Steinberg[32] noted that this behavior resembles that of a mixture of immiscible liquids, which separate according to the relative strengths of intermolecular adhesion that generates surface tension parallel to the liquid interface. For example, if two liquids are mixed together, the one with the greater adhesion (surface tension) becomes engulfed by the other. According to this differential adhesion hypothesis (DAH), a mixture of cells of different types organizes itself into layers of increasing cell–cell adhesion strength toward the inside of the mixture. The DAH eventually gained widespread acceptance, as it was supported by experimental measurements of intercellular adhesion strength[33] and the relative expression levels of cadherin molecules,[34] as well as by computational modeling.[35]

In a prescient critique of DAH, however, Harris[36] argued that cells are fundamentally different from passive fluids. He acknowledged that cell sorting is likely driven by a type of surface tension, but he proposed that this tension is produced by contraction of the cell cortex, rather than adhesion. In this view, cells with the strongest contractile force would pull themselves together at the center of a mixture of cells. Harris showed that cell sorting driven by active forces rather than passive forces is more realistic from a thermodynamic viewpoint.

About 25 years later, Brodland[37] developed a phenomenological cell-level computational model based on this idea. He pointed out that adhesive and contractile forces actually cause opposite effects, as increased adhesion tends to lengthen the contact area between two cells while cortical contraction decreases contact area. Hence, these two effects should produce different behavior. Brodland's model consists of a 2-D layer of polygonal cells with the net interfacial tension given by the product of the length of the boundary and the difference between cortical tension and the equivalent tangential stress due to adhesion. Increasing this net interfacial tension shortens the boundary and vice versa, with neighbor exchange occurring when a boundary shrinks to zero length. For different boundary conditions and parameter values, the model generates sorted patterns much like those observed experimentally (Figure 11.1).[37]

Recent experimental studies appear to support this alternate interpretation of DAH, which Brodland[37] calls the differential interfacial tension hypothesis (DITH). Apparently, Steinberg himself eventually supported extending DAH to include the effects of cortical contractility.[38] In fact, new evidence suggests that cortical tension dominates the effects of adhesion.[31,39,40] This does not mean that intercellular adhesion is not crucial for cell sorting, as adhesive forces are required to maintain the integrity of the cell aggregate. Moreover, recent studies have indicated that forces between adhesions and cytoskeletal elements strengthen and stabilize both cellular components.[5,41–43] Thus, adhesion strength may indirectly affect cell sorting by modulating cortical tension.

Finally, we mention recent studies of the self-organizing behavior of embryonic stem (ES) cells. In an intriguing set of experiments, Eiraku et al.[44] mixed mouse ES cells with growth factors that are essential for eye development. The cells differentiated into retinal tissue, sorted themselves, and self-assembled into a hollow sphere. Part of the surface then evaginated to form a structure resembling an optic vesicle, which then invaginated to create an

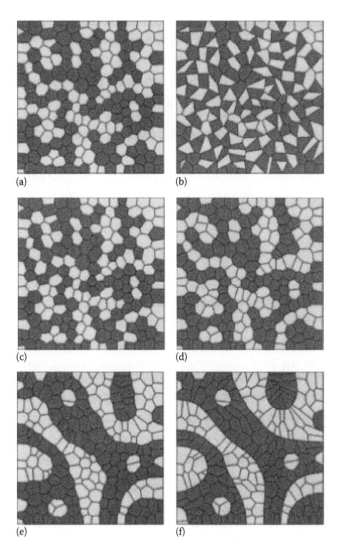

FIGURE 11.1
Computational model for 2-D cell sorting. The initial state consists of a random mixture of light and dark cells. Contraction and adhesion forces at cell boundaries drive neighbor exchanges that gradually generate a pattern. Panels (a)–(f) show results for various values of the interfacial tensions. (Reprinted from Brodland, G.W., *J. Biomech. Eng.*, 124, 188, 2002 with permission from ASME.)

optic cup (Figure 11.2). Although there may be some mechanistic differences with normal eye development, this behavior is remarkably similar to that observed in vivo. The field of tissue engineering could benefit greatly from these types of experiments.[45]

11.3.2 Cell Differentiation

The next major step in embryogenesis is the creation of more detailed pattern within the germ layers. According to the positional information hypothesis of Wolpert,[46] cells somehow sense their location in the embryo and then differentiate or perform other activities accordingly. What they may sense was unknown, but researchers speculated

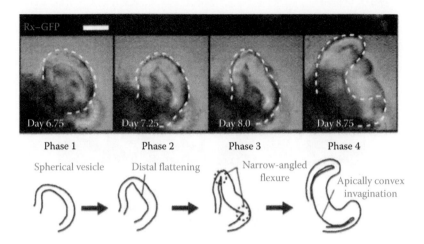

FIGURE 11.2
Cultured ES cells from mouse retinal epithelium recapitulate early eye development. *Top row*: images of cultured ES cells as they form an optic vesicle, which then invaginates to create an optic cup. *Bottom row*: four phases of the invagination process. (Reprinted by permission from Macmillan Publishers Ltd. *Nature*. Eiraku, M. et al., Self-organizing optic-cup morphogenesis in three-dimensional culture, 472, 51–56, copyright 2011.)

that cells respond to the local concentration of molecules called *morphogens*, which diffuse through the embryo and establish spatial gradients.[47–50] Some investigators also suggested that complex patterns of morphogen concentration could be created through a reaction–diffusion system of biochemicals. This idea stems from the classic paper of Turing,[51] who found that such systems can produce spatial patterns in chemical concentration that resemble patterns found in nature, for example, stripes and spots in animal coats. Although the first morphogen was not found until many years later,[52] this idea became widely accepted following the publication of additional mathematical analyses and simulations.[53,54]

Later studies indicated, however, that the actual patterning mechanism is more complicated, at least for cellular differentiation. While researchers generally agree that morphogens are important conveyors of signals during pattern formation, spatial patterns of cell differentiation do not coincide in general with morphogen concentration.[49] Many now believe that gene regulatory network activity transforms morphogen concentrations into the signals that directly regulate pattern.[49,55] This topic continues to be debated.[49,56]

Recent studies have shown that physical forces also can play an important role in differentiation. For example, investigators have shown that substrate stiffness influences the differentiation of ES cells.[57] Soft and hard substrates induce differentiation into brain and bone cells, respectively.

11.3.3 Structural Patterns

Patterning can refer to virtually anything in the embryo that acquires characteristic genetic, biochemical, mechanical, or geometric properties. Mammalian and avian hearts, for example, have two atria and two ventricles arranged in a specific spatial pattern. The architecture of the skeleton is another example. Here, we discuss branching patterns in tubular structures, for example, blood vessels, lungs, kidneys, and glands, with the focus being primarily on blood vessels. Although much has been learned about the genetic and molecular aspects of this problem,[58,59] the physical mechanisms of branching morphogenesis are incompletely understood.

Vascular networks form by two main processes.[60,61] Before the heart begins to beat, *vasculogenesis* creates a polygonal network of blood vessels. Later, in both embryos and adults, branches sprout from existing blood vessels in the process called *angiogenesis*. Tubular branching in organs and glands is generally similar to angiogenesis.

During vasculogenesis, endothelial cells migrate as mesenchyme along tracks through the ECM and aggregate along lines, which then hollow out to create tubes.[62] According to the conventional view, these cells are guided by genetic or biochemical prepatterns. Experiments have shown, however, that endothelial cells cultured in collagen gels can create tubular networks resembling those of blood vessels in vivo, even though there is no prepattern.[61,62]

To explain this finding, Manoussaki et al.[63] proposed a mechanical mechanism based on ideas published by Harris et al.,[64] who speculated that mechanical instability explains patterns of aggregated fibroblasts seen when cells are cultured in collagen gels. Manoussaki et al.[63] illustrated the plausibility of this hypothesis using a computational model based on the Murray–Oster theory for mesenchymal morphogenesis.[65,66] When the initial distribution of cell density in the model is given a small random perturbation, cells contract and pull neighboring cells toward them, regionally increasing the total contractile force and creating local accumulations of high cell density. Cells then migrate and accumulate along lines of tension between these dense regions, forming networks that strongly resemble those observed in culture.[63,67] Mathematically, the governing equations take the form of a reaction–diffusion system with cellular contraction providing the activator and matrix elasticity the inhibitor[65]; bifurcations produce Turing-like patterns. Related models have been proposed for skeletal patterning and branching in the lung.[68,69]

After the heart begins to beat, the vascular bed expands and remodels in response to hemodynamic forces.[70–73] Much of this expansion occurs through angiogenesis, which can be of enormous benefit (e.g., create alternate paths for blood around blocked arteries) or detriment (e.g., supply blood to cancerous tumors).

Angiogenesis and tubular branching in other systems can occur through several different mechanisms, including sprouting of new branches and splitting of old ones.[1] In general, the process involves cellular contraction, proliferation, migration, and rearrangement, as well as matrix degradation and contraction.[69] Recent experiments suggest that branches initiate at sites of high stress concentrations induced in part by contractility.[74] However, relatively little is yet understood about specific mechanisms.

11.4 Epithelial Morphogenesis

Epithelia play important roles in the morphogenesis of most organs and many other structures in the embryo. In the adult, epithelia are specialized to perform certain functions, such as the exchange of substances between compartments, for example, gas exchange in the lung. In the embryo, however, morphogenesis is a primary function of epithelia.

There are three main types of epithelia in the embryo (Figure 11.3). Simple epithelia, for example, the gut tube, consist of a single cell layer attached to a basement membrane (ECM). Stratified epithelia, for example, the myocardium of the heart tube, contain two or more cell layers. Finally, pseudostratified epithelia, for example, the brain tube, contain

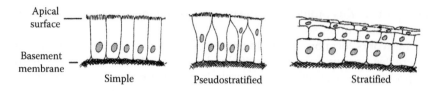

Apical surface

Basement membrane

Simple Pseudostratified Stratified

FIGURE 11.3
Types of epithelia. (Reprinted from *Mechanisms of Morphogenesis: The Creation of Biological Form*, Davies, J.A., Copyright 2005, with permission from Elsevier.)

only one layer of cells, but the nuclei are arranged at different levels, giving it the appearance of being stratified.[1]

Epithelial cells are polarized with the apical side facing open space and the basal side facing the basement membrane.[13] These cells are joined together and to the basement membrane by junctions containing specific types of adhesion and associated molecules. *Tight junctions*, located near the cell apex, function primarily as barriers to substances crossing the membrane, while *adherens junctions*, located just below tight junctions, are more important for morphogenesis. Adherens junctions join cells via cadherins and transmit forces through attachments to the actin cytoskeleton as well as to an actomyosin belt that encircles the cell apex (Figure 11.4). These attachments transmit contractile and other forces from cell to cell, enabling epithelia to undergo large-scale deformations.[75] We next consider specific mechanisms by which epithelia undergo morphogenetic changes.

11.4.1 Global Changes in Epithelial Morphology

An epithelium can change its overall size and shape using a variety of means. The mechanisms by which epithelia can spread or elongate include the following[1,58,76]:

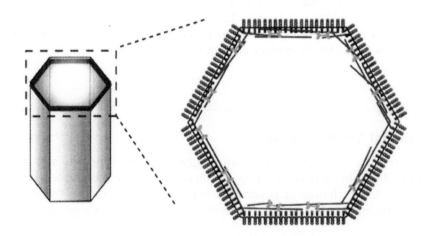

FIGURE 11.4
Schematic of epithelial cell. Actin filaments (red) and myosin II (green) form a contractile belt connected to adherens junctions (blue) around cell apex. (Reprinted from *Dev. Biol.*, 341, Martin, A.C., Pulsation and stabilization: Contractile forces that underlie morphogenesis, 114–125, Copyright 2010, with permission from Elsevier.)

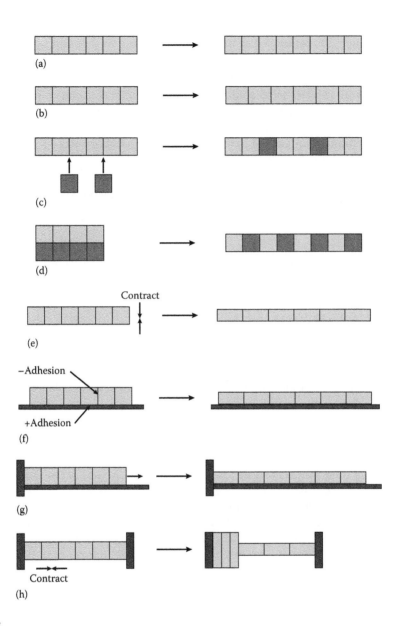

FIGURE 11.5
Mechanisms for epithelial elongation. (a) Cell division. (b) Cellular hypertrophy. (c) MET. (d) Cell intercalation. (e) Active cell-shape change by transverse contraction. (f) Decreased cell–cell adhesion or increased cell–matrix adhesion. (g) Marginal cells crawling and pulling trailing cells. (h) Constrained contraction of cells in one region.

- Cells divide or hypertrophy parallel to the membrane (Figure 11.5a and b). Examples: looping and expansion of the heart tube[77] and closure of the neural tube (NT).[78]

- Outside cells enter the membrane by MET (Figure 11.5c). Examples: blastulation[12] and formation of presumptive myocardium in bilateral heart fields.[79]

- Cells rearrange within the epithelium via intercalation (Figure 11.5d). Examples: epiboly,[80] *Drosophila* germ band extension,[81] and frog gastrulation.[82]

- Active cellular elongation, for example, by transverse contraction, would thin and spread the layer (Figure 11.5e). Examples: second phase of ascidian gastrulation[83] and evagination of *Drosophila* imaginal leg disk.[84]

- Decreased cell–cell adhesion could cause a decrease in cell–cell contact area, forcing the epithelium to spread. Increased adhesion between cells and substrate also could induce spreading (Figure 11.5f). Example: This mechanism underlies the DAH[34]; constrained spreading through changes in adhesion has been proposed as a mechanism of invagination.[4]

- Boundary cells crawl forward, stretching the sheet behind them (Figure 11.5g). Examples: later stages of epiboly in zebrafish[85]; contractile ring activity during postequatorial epiboly[86,87] or dorsal closure in *Drosophila*[88] represents a similar mechanism.

- Local in-plane contraction of a constrained epithelium would shorten and thicken the contracted region and stretch the adjacent region (Figure 11.5h). Example: proposed (but not yet confirmed) mechanism for placode formation (local thickening of an epithelium).[89]

Many of these examples are discussed later in this chapter.

It is not always clear why epithelial morphogenesis has evolved to use one particular mechanism over others. Sometimes, speed may be an issue, and multiple possibilities offer the advantage of redundancy to ensure normal development. Two of the most studied problems of this type are epiboly and convergent extension (CE), which are now discussed in some detail.

11.4.1.1 Epiboly

During *epiboly*, an outer cell sheet spreads to enclose deeper parts of the embryo. Here, we consider epiboly in zebrafish, where 3-D geometry plays a role.

The zebrafish blastoderm initially consists of an aggregate of cells (blastomeres) sitting on top of the yolk at the animal pole of the embryo (Figure 11.6a). The upper surface of this mound is an epithelium called the enveloping layer (EVL), and the underlying *deep cells* later develop into the embryonic tissues.[86,87] Cells along the margin of the blastoderm are connected to a relatively narrow band of cytoplasm called the yolk cytoplasmic layer. Near the 1000-cell stage, these marginal blastomeres merge with the cytoplasmic layer to form the yolk syncytial layer (YSL).[80]

During epiboly, the blastoderm spreads toward the vegetal pole and eventually surrounds the entire yolk. This process begins when the yolk pushes upward (*domes*) locally into the blastoderm, causing the loosely connected deep cells to thin and spread by radial intercalation.[80,86,87,90] What happens next is not entirely clear. The EVL is connected to the YSL membrane via tight junctions and follows the YSL as it moves vegetally, suggesting that the YSL pulls the EVL margin forward. The forces that drive the motion of the YSL are incompletely understood, however.

One force causing EVL spreading may come from microtubules, which extend from the margin of the YSL into the yolk cytoplasmic layer. During epiboly, these microtubules shorten, suggesting that they pull on the YSL.[86,87] However, while disrupting the microtubules or stabilizing their length using taxol stops YSL advancement, it only delays the motion of the EVL. Thus, microtubule shortening may be just one of several mechanisms that drive epiboly.

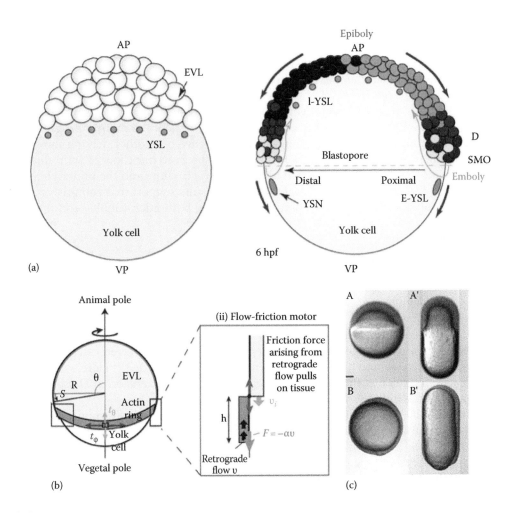

FIGURE 11.6

Zebra fish epiboly. (a) Embryo is a mound of cells that spreads over the yolk. EVL, enveloping layer; YSL, yolk syncytial layer. (Reprinted from *Curr. Opin. Gen. Develop.*, 16, Solnica-Krezel, L., Gastrulation in zebrafish—All just about adhesion?, 433–441, Copyright 2006, with permission from Elsevier.) (b) An actomyosin ring and frictional forces generated by retrograde actin flow pull the EVL forward. (c) The EVL margin continues to advance past the equator when the embryo is forced into a cylindrical shape. (Panels (b) and (c) are from Behrndt, M. et al., Forces driving epithelial spreading in zebrafish gastrulation, *Science*, 338, 257–260, 2012. Reprinted with permission from AAAS.)

Actin-associated processes appear to provide other mechanisms. For example, as the EVL margin approaches the equator of the embryo, a circumferential actomyosin ring begins to form along the EVL boundary (Figure 11.6b). Researchers have speculated that, once the leading edge of the EVL passes the equator of the spherical zebrafish embryo, this ring contracts to pull the EVL vegetally the rest of the way.[85–87] (Note that contraction above the equator would push the EVL margin in the wrong direction.) Such a supracellular contractile ring has been implicated in a number of morphogenetic processes, including dorsal closure in *Drosophila* and embryonic wound healing.[88,91] Supporting an important role for the contractile ring in epiboly are data showing that disrupting actomyosin activity or cutting the ring with a laser delays EVL progression.[85,92]

Recently, Behrndt et al.[85] studied the mechanics of zebrafish epiboly in considerable detail. Using dissection experiments to probe stress, they found that the magnitude of meridional tension in the EVL is greater than what would be expected by contraction of the circumferential ring alone. Moreover, in an interesting experiment, they changed the shape of the embryo from spherical to cylindrical by inserting it into a pipette with a smaller diameter (Figure 11.6c). Even though actomyosin ring contraction would not be able to pull the EVL downward over the cylindrical region, the margin progressed at a relatively normal rate. This result suggests the presence of another driving force, perhaps due to microtubules. However, Behrndt et al.[85] observed backflow of actin near the EVL margin and speculated that friction between this actin flow and the underlying yolk would pull the EVL forward. Taken together, these results suggest that zebrafish epiboly is driven by multiple redundant mechanisms involving both microtubules and the actin cytoskeleton.

11.4.1.2 Convergent Extension

During CE, an epithelium narrows along one axis and extends along the other axis within the plane of the membrane.[82] Possible morphogenetic mechanisms for CE include passive stretch, anisotropic growth, active cell-shape change, and cell intercalation.[1,2,82] The specific mechanism is system dependent. For example, changes in cell shape dominate CE during evagination of the imaginal leg disk* in *Drosophila*,[84] while both directed cell division and cell intercalation contribute to CE in the neural plate (NP) of birds.[93,94] Here, we focus on cell intercalation, which is the primary driver of germ band extension in *Drosophila*,[81] as well as gastrulation in the frog† and sea urchin.[82,95–97]

Cell intercalation can produce large changes in shape via relatively small but coordinated cellular motions. Consider, for example, the merging of two rows of cells into one (Figure 11.5d). Cells in one row insert themselves between those in the other row, resulting in a membrane twice as long and half as wide.

Although the mechanisms for intercalation are not completely understood, experiments suggest variations of two major themes. In the first, each cell extends a relatively stiff protrusion laterally between two neighboring cells, pushing them apart and grabbing them via cell–cell adhesions.[82,95,97] The protrusion then retracts, pulling the cell between its neighbors. The cytoskeletal activity driving this process is unknown, but actin polymerization and actomyosin contraction may extend and retract the protrusion, respectively. The second method involves neighbor exchanges similar to those discussed earlier for cell sorting.[2,6,81] In this case, contraction-driven interfacial tension would be greatest along cell boundaries oriented toward the direction of convergence. Intense accumulations of myosin have been observed along the appropriate cell boundaries during CE, but these accumulations also could be along similarly oriented protrusions. Computational models for CE based on these ideas have been proposed recently.[98,99] As always, the mechanism of choice may be species or location dependent.

---

* Imaginal disks are groups of undifferentiated cells in insect larvae that form legs, wings, and other structures during the transformation to the adult.[12] Before extension, the imaginal leg disk roughly has the shape of a circular disk consisting of concentric folds of epithelia with circumferentially elongated cells. To create a leg, these cells shorten circumferentially (possibly by contraction) and lengthen axially to extend the folds like a telescope, with the central folds creating the most distal segments.[84]

† Convergent extension in amphibian embryos mainly involves mesenchymal cells of the mesoderm, but the prospective ectoderm also undergoes CE through cell intercalation.[82]

It is important to note that an epithelium undergoing CE often faces resistance from adjacent tissues. This can produce significant compressive and shear stresses that can buckle the membrane. To avert this situation, tissues tend to stiffen during CE through an increase in material modulus or adhesion stiffness.[82,100] An increased modulus can be produced by cytoskeletal remodeling or the contraction that drives CE.

11.4.2 Local Changes in Epithelial Morphology

Localized bending is the primary mode by which epithelia create 3-D structures. As in spreading of epithelia, embryos employ several different mechanisms to bend membranes into a wide variety of shapes.[4] In general, these mechanisms consist of two types: (1) bending caused by intrinsic changes in the shape of individual cells (Figure 11.7a) and (2) bending driven by constrained expansion or shrinkage of a sheet of cells or matrix (Figure 11.7b). Both mechanisms generally produce wedge-shaped cells, with the cell surface area becoming smaller at the inner curvature and larger at the outer curvature of the bent region. However, while cell wedging actively generates type 1 bending, it is a passive result of type 2 bending.

The most common mechanism for cell wedging during type 1 bending is apical constriction. Constriction of the cell apex typically occurs by contraction of actomyosin fibers that encircle the apex (called a *purse string*).[2,101] During *Drosophila* gastrulation, however, apical constriction is driven in part by contractile fibers that span the surface of each epithelial cell.[75] This deformation pushes cytoplasm toward the base, causing it to expand if cell elongation is restricted, for example, in the bottle cells that initiate gastrulation in the frog.[102] Another possible mechanism for apical constriction is removal of portions of the plasma membrane by endocytosis.[4,103]

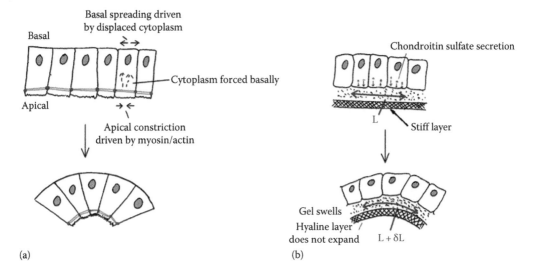

(a) (b)

FIGURE 11.7
Invagination of epithelium. (a) Apical constriction. (b) Swelling of matrix constrained by relatively stiff (bottom) layer. (Reprinted from *Mechanisms of Morphogenesis: The Creation of Biological Form*, Davies, J.A., Copyright 2005, with permission from Elsevier.)

Basal expansion alone also can cause cell wedging, although the specific mode of action is not clear. Membrane packets can be added to the basal side of the cell, but it seems unlikely that the cell membrane can push with sufficient force to expand the base without buckling. Another possibility is that the nucleus moves toward the base via interkinetic nuclear migration,[104] forcing it to expand. As discussed in the following, this latter mechanism may play a role in neurulation.

Type 2 epithelial bending often involves differences in growth rate between adjacent layers (Figure 11.7b). This mechanism is similar to that of a bimetallic strip, where two bonded metal layers expand different amounts on heating, causing the strip to bend with the layer that expands the least located at the inner curvature. In the embryo, one layer would be composed of epithelial cells, while the other can be composed of either cells or matrix, for example, the basement membrane.[4] If either layer expands or contracts with no slippage between layers, the composite structure will bend. Bending also can occur if membrane expansion is resisted by relatively stiff constraints, causing it to buckle. We now consider some specific examples.

11.4.2.1 Gastrulation

Besides organizing the blastula into three primary germ layers, the process of gastrulation includes an invagination that creates the primitive gut tube. This process has been studied extensively in sea urchin, *Drosophila*, and frog embryos. Here, because of their relatively simple geometries, we consider the first two species.

At the blastula stage, the sea urchin embryo is a fluid-filled spherical shell composed of four layers.[95] From inside to outside, these layers are basal lamina, epithelium, apical lamina, and hyaline layer (ECM). Gastrulation consists of a primary and a secondary phase.[95] During primary invagination, part of the wall flattens and thickens to form a placode called the *vegetal plate*, which then invaginates to create an inward dimple (Figure 11.8a). Next, during secondary invagination, filopodia extend and attach to the wall on the opposite side of the blastula. These protrusions guide and pull on the dimple, and cell

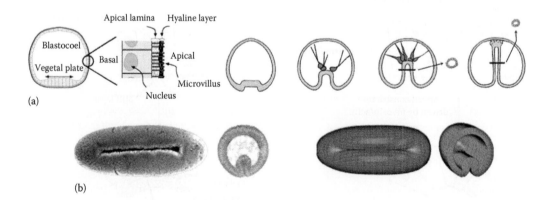

FIGURE 11.8
Gastrulation. (a) Schematic of sea urchin embryo morphology and sequence of invagination. (Reprinted from *Differentiation*, 71, Keller, R., Davidson, L.A., and Shook, D.R., How we are shaped: The biomechanics of gastrulation, 171–205, Copyright 2003, with permission from Elsevier.) (b) Experiment (left) and finite-element model (right) of ventral furrow formation in *Drosophila*. (Reprinted from *J. Mech. Behav. Biomed. Mater.*, 1, Conte, V., Munoz, J. J., and Miodownik, M., A 3D finite element model of ventral furrow invagination in the *Drosophila melanogaster* embryo, 188–198, Copyright 2008, with permission from Elsevier.)

intercalation extends the dimple (archenteron) across the blastula cavity (blastocoel) to create the gut tube.

Remarkably, despite intensive study spanning several decades, the mechanisms of primary invagination in the sea urchin remain unclear. Davidson et al. have combined experiments with computer modeling to help find answers to this problem. First, they used a finite-element model to study five possible invagination mechanisms[105]: (1) apical constriction/basal expansion, (2) buckling caused by radial contraction of protrusions within the invaginating region, (3) buckling caused by circumferential contraction of a ring of cells surrounding the invaginating region, (4) buckling caused by apicobasal contraction and radial spreading of cells in the invaginating region, and (5) regional swelling in the apical lamina constrained by a relatively stiff hyaline layer. After showing that all of these mechanisms can produce similar morphologies, they defined ranges of material moduli that would be needed to make each mechanism feasible. In a follow-up paper, they used stiffness measurements to rule out mechanisms 1 and 3,[20] and other experimental results seem to support mechanism 5.[106] These studies are not conclusive, however.

The *Drosophila* embryo is a popular model for studies of morphogenesis because of its relatively simple geometry, the availability of techniques for dynamic imaging at multiple scales, and the ability to manipulate gene activity and molecular signaling. For example, Martin et al.[75] recently used this system to study the details of how molecular and cytoskeletal activity integrate to generate tissue-level deformations. While much of this ongoing research is beyond the scope of this review, we note here the recent finding that actomyosin fibers exhibit a ratchet-like contractile behavior, as stepwise contractions occur between periods of relative quiescence when fibers presumably remodel to reset their zero-stress lengths.[2,107] This allows nonmuscle actomyosin fibers (and cells) to contact by as much as 80% or more of their initial length,[108] similar to airway smooth muscle cells.[109]

The *Drosophila* blastula is essentially a fluid-filled ellipsoidal shell composed of epithelial cells (Figure 11.8b). The first major event in gastrulation is the formation of a longitudinal invagination called the *ventral furrow*.[12] Differences in geometry alone make the mechanics of invagination different between sea urchin and fly, as axisymmetric dimpling of a sphere (sea urchin) requires considerably more force than cylindrical bending of a tube (*Drosophila*).[110] Hence, the bending strategies may differ. In fact, available evidence suggests that cell wedging driven by apical constriction plays a more prominent role in fly invagination than it does in the sea urchin.[20,75,101] In the fly embryo, cell wedging is caused by contraction of actomyosin fibers that span the apical surfaces of cells, with forces transmitted between cells via cell–cell adhesions located at the ends of these fibers.[75]

Several computational models have been proposed for ventral furrow formation driven by cell wedging. In an early model for morphogenesis, Odell et al.[111] represented the cross section of the *Drosophila* embryo as a circular ring of truss-like cells. The cell apex was assumed to be stretch activated, so that when stretched beyond a critical length, the apex contracted to a shorter length. When contraction was specified in one cell of the model, the apex shortened and stretched its neighbors, triggering a contractile wave that produced a localized invagination. More recent models include other effects, such as apicobasal elongation (placode formation), the constraints of internal fluid and outside vitelline membrane, and 3-D geometry (Figure 11.8b).[112–115] Morphogenetic forces were simulated in these models using the theory for finite volumetric growth of Rodriguez et al.[116] More in-depth discussion of these and other models for morphogenesis can be found in the recent review of Wyczalkowski et al.[28]

11.4.2.2 Neurulation

Development of the nervous system begins with the process of neurulation, whereby a region of ectoderm thickens into a placode called the *NP*, which then invaginates along a line to create the *NT*. Subsequently, the anterior portion of the NT expands to create the primitive brain tube, while the posterior part becomes the spinal cord.[12]

In the anterior part of the chick NP, cell wedging plays a role in NT formation, but it apparently is not the sole mechanism. In particular, wedging occurs only in localized regions of the NP, resulting in three longitudinal furrows called hinge points—one medial hinge point (MHP) and two bilateral dorsolateral hinge points (DLHPs) (Figure 11.9). Experiments have shown that these hinge points facilitate bending of the NP, similar to grooves in a piece of cardboard, but the surface ectoderm outside the NP actually supplies the driving forces.[78,117] These forces, which are generated by lateral extension of the surface ectoderm via cell intercalation and anisotropic cell division, push the edges of the NP toward the embryonic midline, causing it to bend into a closed tube. Incomplete closure of the NT can cause severe congenital defects, including spina bifida.[12]

The mechanism that drives cell wedging at the hinge points remains controversial. Some data indicate that apical contraction is involved at the DLHPs but not the MHP,[118] while other data suggest that contraction drives the formation of all three hinge points.[119,120] As a possible alternative mechanism, Smith and Schoenwolf[121] speculated that cell wedging in these regions is caused by basal expansion rather than apical constriction. This idea is based on the following observations: (1) neuroepithelial nuclei undergo mitosis near the cell apex and then migrate to the basal side (interkinetic nuclear migration)*; (2) because NP cells are generally narrower than their nuclei, the cells expand in regions where nuclei are located; and (3) the mitotic phase of the cell cycle is shorter in the hinge points than elsewhere, causing more nuclei to be basally located at any given time. On average, therefore, the nuclei would tend to expand the basal side of cells in the hinge points.

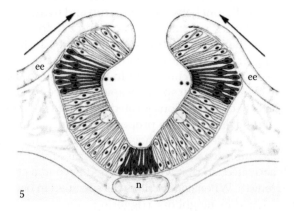

FIGURE 11.9
Neurulation in chick embryo. *Red*: MHP; *blue*: DLHPs. ee, epidermal (surface) ectoderm; n, notochord. (Reprinted with permission from Colas, J.F. and Schoenwolf, G.C., *Dev. Dyn.*, 221, 117, 2001.)

---

* The mechanism that drives interkinetic nuclear migration is the focus of ongoing studies.[104]

In models for neurulation, Clausi and Brodland[122] have shown that hinge points can be created by local instabilities that occur if the contractile strength of an apical fiber increases as it shortens. This problem warrants further study.

11.4.2.3 Other Examples of Epithelial Bending

Numerous other morphogenetic processes involve bending of epithelia. A few more examples are briefly discussed here.

First, we mention an ongoing mystery. Most invaginations are preceded by the formation of a placode (local membrane thickening). This initially may seem to be counterintuitive, as an increased thickness would increase resistance to bending. In general, this is true for bending by external loads, but not for bending caused by loads generated within the invaginating cells. Possible reasons for the appearance of placodes include the following: (1) If invagination is driven by constrained cell growth or increased cell–cell adhesion in one layer of a bilayered membrane, a placode may form as a consequence of these mechanisms.[4,123,124] (2) As cells become taller and narrower, actomyosin density may increase at the cell apex, leading to greater contractile force.[123] (3) Apical constriction with little basal expansion can create a placode, which then shortens and spreads by transverse contraction and invaginates due to being constrained by the relatively stiff apex.[83,125] We now turn to specific morphogenetic processes in the embryo.

The head fold (HF), the first major 3-D structure to develop in the vertebrate embryo, is a crescent-shape fold that forms at the anterior end of the NP (Figure 11.10a). This process initiates brain, heart, and foregut development. Recently, Varner et al.[126] used experiments on chick embryos and finite-element models to show that the HF is likely produced by a combination of three morphogenetic process: (1) CE of the NP, (2) cell wedging within the folding region, and (3) lateral shortening of ectoderm anterior to the fold.

Differential growth plays a role in large-scale bending of some epithelial structures. For example, both the heart and brain are initially tubular structures in the embryo that undergo global bending and twisting during the early stages of development (Figure 11.10b and c).[127,128] Although the mechanisms that drive these deformations remain incompletely understood, available data indicate that both the heart and brain bend in culture when isolated from the embryo.[129–131] Recent work suggests that bending of the brain tube may be caused by differential growth,[132] while bending of the heart tube is caused by a combination of differential growth and active cell-shape changes, possibly due to actin polymerization.[77,133,134] Differential growth also is thought to play a major role in folding of the cerebral cortex during later development,[135,136] as well as in buckling and folding of the gut tube and the creation of its villi.[137,138]

Finally, this review would not be complete without mentioning the remarkable transformation that occurs during development of *Volvox*, which is a type of green algae that has both plantlike and animal-like characteristics. Initially, the *Volvox* embryo is a spherical epithelial shell with flagella on the inner surface. For swimming, these flagella must be moved to the outside surface. To do this, the embryo completely turns itself inside out (Figure 11.11). This process involves dramatic changes in cell shape, cell rearrangements, and global contractile forces that have been documented in considerable detail.[139–142] Local changes in cell shape appear to be driven by microtubules rather that actomyosin fibers.[140,141] Elastic snapthrough buckling also may play a role.[140] Several hypotheses have been proposed for the mechanism of *Volvox* inversion, but to our knowledge, these ideas have not yet been tested through biomechanical modeling. This is an excellent example of a problem where a computational model can help answer fundamental questions about morphogenesis.

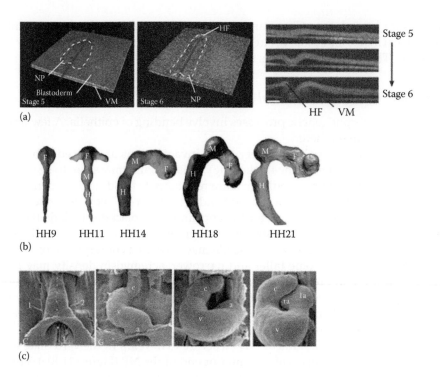

FIGURE 11.10
Organogenesis in early chick embryo. (a) HF formation. Left and center images show lengthening of NP and crescent-shaped HF. Right image shows HF formation in longitudinal section. VM, vitelline membrane. (Reprinted with permission from Varner, V.D. et al., *Development*, 137, 3801, 2010.) (b) Reconstructed brain morphology at indicated stages of development (from about 30 h to 3.5 days of incubation). Courtesy Ben Filas. (c) Looping of the heart. a, atrium; v, ventricle; c, conotruncus (outflow tract). (Reprinted with permission from Manner, J., *Anat. Rec.*, 259, 248, 2000.)

We now briefly return to a question raised in the opening paragraph: Why have the physical mechanisms of morphogenesis remained so poorly understood after so many years of study? The apparent reasons include the following: (1) the inherent complexity of morphogenesis in three dimensions can make physical intuition misleading, (2) various mechanisms often can generate similar shapes (e.g., see aforementioned discussions on epiboly and neurulation), and (3) perturbing one mechanism can trigger a backup response that may have evolved to help minimize developmental defects. While such redundancy leads to more robust development, this feature also can make it extremely difficult to identify the principal driving forces in a particular morphogenetic event.

In conclusion, much work remains to be done before we have a complete understanding of even the most basic morphogenetic processes. Much has been learned about the molecular and genetic aspects of development during the last few decades, and now, researchers have begun to investigate the links to biophysical mechanisms. These links need to be studied across spatial scales from the subcellular scale, where most morphogenetic forces are generated, to the tissue scale, where the final product of morphogenesis is most evident. It also will be important to determine the role that mechanical feedback plays in regulating morphogenesis.[124,143,144] Integrating all of these components into a complete picture of embryonic development will define the future of the field.

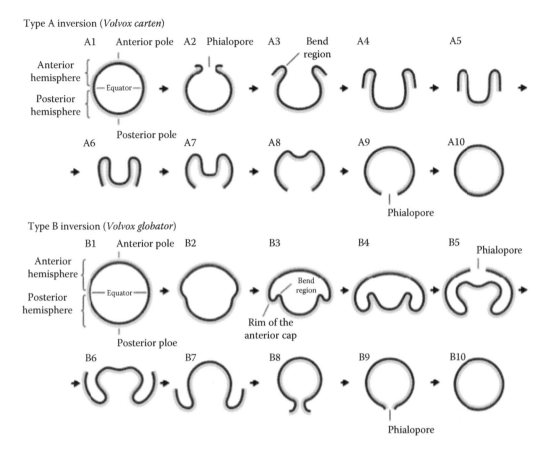

FIGURE 11.11
Two mechanisms of inversion in *Volvox*. (Reprinted with permission from Hohn, S. and Hallmann, A., *BMC Biol.*, 9, 89, 2011.)

Acknowledgment

I thank Lila Solnica-Krezel for providing comments on parts of the manuscript. This work was supported by NIH grant R01 NS070918.

References

1. Davies, J. A. *Mechanisms of Morphogenesis: The Creation of Biological Form.* (Burlington, MA: Elsevier, 2005).
2. Martin, A. C. Pulsation and stabilization: Contractile forces that underlie morphogenesis. *Dev Biol* **341**, 114–125, doi:S0012-1606(09)01290-1 [pii] 10.1016/j.ydbio.2009.10.031 (2010).
3. Taber, L. A. Biomechanics of growth, remodeling, and morphogenesis. *Appl Mech Rev* **48**, 487–545 (1995).
4. Ettensohn, C. A. Mechanisms of epithelial invagination. *Q Rev Biol* **60**, 289–307 (1985).

5. Lecuit, T., Lenne, P. F., and Munro, E. Force generation, transmission, and integration during cell and tissue morphogenesis. *Annu Rev Cell Dev Biol* **27**, 157–184, doi:10.1146/annurev-cellbio-100109-104027 (2011).

6. Lecuit, T. and Lenne, P. F. Cell surface mechanics and the control of cell shape, tissue patterns and morphogenesis. *Nat Rev Mol Cell Biol* **8**, 633–644, doi:nrm2222 [pii] 10.1038/nrm2222 (2007).

7. Hampoelz, B. and Lecuit, T. Nuclear mechanics in differentiation and development. *Curr Opin Cell Biol* **23**, 668–675, doi:S0955-0674(11)00131-1 [pii] 10.1016/j.ceb.2011.10.001 (2011).

8. Dahl, K. N., Ribeiro, A. J., and Lammerding, J. Nuclear shape, mechanics, and mechanotransduction. *Circ Res* **102**, 1307–1318, doi:102/11/1307 [pii] 10.1161/CIRCRESAHA.108.173989 (2008).

9. Pajerowski, J. D., Dahl, K. N., Zhong, F. L., Sammak, P. J., and Discher, D. E. Physical plasticity of the nucleus in stem cell differentiation. *Proc Natl Acad Sci USA* **104**, 15619–15624, doi:0702576104 [pii] 10.1073/pnas.0702576104 (2007).

10. Yu, H. et al. Mechanical behavior of human mesenchymal stem cells during adipogenic and osteogenic differentiation. *Biochem Biophys Res Commun* **393**, 150–155, doi:10.1016/j.bbrc.2010.01.107 (2010).

11. Du, A., Sanger, J. M., and Sanger, J. W. Cardiac myofibrillogenesis inside intact embryonic hearts. *Dev Biol* **318**, 236–246, doi:S0012-1606(08)00217-0 [pii] 10.1016/j.ydbio.2008.03.011 (2008).

12. Gilbert, S. F. *Developmental Biology*, 9th edn. (Sunderland, MA: Sinauer Associates, 2010).

13. Alberts, B. et al. *Molecular Biology of the Cell*. (New York: Garland Science, 2002).

14. Kalluri, R. and Weinberg, R. A. The basics of epithelial-mesenchymal transition. *J Clin Invest* **119**, 1420–1428 (2009).

15. Abu-Issa, R. and Kirby, M. L. Heart field: From mesoderm to heart tube. *Annu Rev Cell Dev Bi* **23**, 45–68 (2007).

16. Butcher, J. T. and Markwald, R. R. Valvulogenesis: The moving target. *Philos Trans R Soc Lond B Biol Sci* **362**, 1489–1503, doi:10.1098/rstb.2007.2130 (2007).

17. Tan, H. et al. Fluid flow forces and rhoA regulate fibrous development of the atrioventricular valves. *Dev Biol* **374**, 345–356, doi:10.1016/j.ydbio.2012.11.023 (2013).

18. Buskohl, P. R., Jenkins, J. T., and Butcher, J. T. Computational simulation of hemodynamic-driven growth and remodeling of embryonic atrioventricular valves. *Biomech Model Mechanobiol* **11**, 1205–1217, doi:10.1007/s10237-012-0424-5 (2012).

19. Forgacs, G., Foty, R. A., Shafrir, Y., and Steinberg, M. S. Viscoelastic properties of living embryonic tissues: A quantitative study. *Biophys J* **74**, 2227–2234 (1998).

20. Davidson, L. A., Oster, G. F., Keller, R. E., and Koehl, M. A. Measurements of mechanical properties of the blastula wall reveal which hypothesized mechanisms of primary invagination are physically plausible in the sea urchin *Strongylocentrotus purpuratus*. *Dev Biol* **209**, 221–238 (1999).

21. Yao, J. et al. Viscoelastic material properties of the myocardium and cardiac jelly in the looping chick heart. *J Biomech Eng* **134**, 024502, doi:10.1115/1.4005693 (2012).

22. Miller, C. E. and Wong, C. L. Trabeculated embryonic myocardium shows rapid stress relaxation and non-quasi-linear viscoelastic behavior. *J Biomech* **33**, 615–622 (2000).

23. Gonzalez-Rodriguez, D., Guevorkian, K., Douezan, S., and Brochard-Wyart, F. Soft matter models of developing tissues and tumors. *Science* **338**, 910–917, doi:10.1126/science.1226418 (2012).

24. Levayer, R. and Lecuit, T. Biomechanical regulation of contractility: Spatial control and dynamics. *Trends Cell Biol* **22**, 61–81, doi:S0962-8924(11)00210-8 [pii] 10.1016/j.tcb.2011.10.001 (2012).

25. Zhou, J., Kim, H. Y., and Davidson, L. A. Actomyosin stiffens the vertebrate embryo during crucial stages of elongation and neural tube closure. *Development* **136**, 677–688, doi:10.1242/dev.026211 (2009).

26. Buskohl, P. R., Gould, R. A., and Butcher, J. T. Quantification of embryonic atrioventricular valve biomechanics during morphogenesis. *J Biomech* **45**, 895–902, doi:10.1016/j.jbiomech.2011.11.032 (2012).

27. Miller, C. E., Vanni, M. A., Taber, L. A., and Keller, B. B. Passive stress-strain measurements in the stage-16 and stage-18 embryonic chick heart. *J Biomech Eng* **119**, 445–451 (1997).
28. Wyczalkowski, M. A., Chen, Z., Filas, B. A., Varner, V. D., and Taber, L. A. Computational models for mechanics of morphogenesis. *Birth Defects Res C Embryo Today* **96**, 132–152, doi:10.1002/bdrc.21013 (2012).
29. Voronov, D. A., Alford, P. W., Xu, G., and Taber, L. A. The role of mechanical forces in dextral rotation during cardiac looping in the chick embryo. *Dev Biol* **272**, 339–350 (2004).
30. Steinberg, M. S. Adhesion in development: An historical overview. *Dev Biol* **180**, 377–388, doi:10.1006/dbio.1996.0312 (1996).
31. Krieg, M. et al. Tensile forces govern germ-layer organization in zebrafish. *Nat Cell Biol* **10**, 429–436, doi:10.1038/ncb1705 (2008).
32. Steinberg, M. S. Reconstruction of tissues by dissociated cells. *Science* **141**, 401–408 (1963).
33. Foty, R. A., Pfleger, C. M., Forgacs, G., and Steinberg, M. S. Surface tensions of embryonic tissues predict their mutual envelopment behavior. *Development* **122**, 1611–1620 (1996).
34. Foty, R. A. and Steinberg, M. S. The differential adhesion hypothesis: A direct evaluation. *Dev Biol* **278**, 255–263, doi:10.1016/j.ydbio.2004.11.012 (2005).
35. Glazier, J. A. and Graner, F. Simulation of the differential adhesion driven rearrangement of biological cells. *Phys Rev E Stat Phys Plasmas Fluids Relat Interdiscip Topics* **47**, 2128–2154 (1993).
36. Harris, A. K. Is cell sorting caused by differences in the work of intercellular adhesion? A critique of the Steinberg hypothesis. *J Theor Biol* **61**, 267–285 (1976).
37. Brodland, G. W. The Differential Interfacial Tension Hypothesis (DITH): A comprehensive theory for the self-rearrangement of embryonic cells and tissues. *J Biomech Eng* **124**, 188–197 (2002).
38. Manning, M. L., Foty, R. A., Steinberg, M. S., and Schoetz, E. M. Coaction of intercellular adhesion and cortical tension specifies tissue surface tension. *Proc Natl Acad Sci USA* **107**, 12517–12522, doi:10.1073/pnas.1003743107 (2010).
39. Amack, J. D. and Manning, M. L. Knowing the boundaries: Extending the differential adhesion hypothesis in embryonic cell sorting. *Science* **338**, 212–215, doi:10.1126/science.1223953 (2012).
40. Maitre, J. L. et al. Adhesion functions in cell sorting by mechanically coupling the cortices of adhering cells. *Science* **338**, 253–256, doi:10.1126/science.1225399 (2012).
41. Hoffman, B. D., Grashoff, C., and Schwartz, M. A. Dynamic molecular processes mediate cellular mechanotransduction. *Nature* **475**, 316–323, doi:10.1038/nature10316 nature10316 [pii] (2011).
42. Nelson, C. M. and Gleghorn, J. P. Sculpting organs: Mechanical regulation of tissue development. *Annu Rev Biomed Eng* **14**, 129–154, doi:10.1146/annurev-bioeng-071811-150043 (2012).
43. le Duc, Q. et al. Vinculin potentiates E-cadherin mechanosensing and is recruited to actin-anchored sites within adherens junctions in a myosin II-dependent manner. *J Cell Biol* **189**, 1107–1115, doi:jcb.201001149 [pii] 10.1083/jcb.201001149 (2010).
44. Eiraku, M. et al. Self-organizing optic-cup morphogenesis in three-dimensional culture. *Nature* **472**, 51–56, doi:10.1038/nature09941 (2011).
45. Sasai, Y., Eiraku, M., and Suga, H. In vitro organogenesis in three dimensions: Self-organising stem cells. *Development* **139**, 4111–4121, doi:10.1242/dev.079590 (2012).
46. Wolpert, L. Positional information and the spatial pattern of cellular differentiation. *J Theor Biol* **25**, 1–47 (1969).
47. Rogers, K. W. and Schier, A. F. Morphogen gradients: From generation to interpretation. *Annu Rev Cell Dev Biol* **27**, 377–407, doi:10.1146/annurev-cellbio-092910-154148 (2011).
48. Wolpert, L. Positional information and patterning revisited. *J Theor Biol* **269**, 359–365, doi:10.1016/j.jtbi.2010.10.034 (2011).
49. Kicheva, A., Cohen, M., and Briscoe, J. Developmental pattern formation: Insights from physics and biology. *Science* **338**, 210–212, doi:10.1126/science.1225182 (2012).
50. Howard, J., Grill, S. W., and Bois, J. S. Turing's next steps: The mechanochemical basis of morphogenesis. *Nat Rev Mol Cell Biol* **12**, 392–398, doi:10.1038/nrm3120 nrm3120 [pii] (2011).
51. Turing, A. M. The chemical basis of morphogenesis. *Phil Trans R Soc London* **B237**, 37–72 (1952).

52. Frohnhofer, H. G. and Nüsslein-Volhard, C. Organization of anterior pattern in the *Drosophila* embryo by the maternal gene bicoid. *Nature* **324**, 120–125, doi:10.1038/324120a0 (1986).
53. Murray, J. D. *Mathematical Biology: Spatial Models and Biomedical Applications.* (New York: Springer-Verlag, 2002).
54. Oster, G. F. and Murray, J. D. Pattern formation models and developmental constraints. *J Exp Zool* **251**, 186–202 (1989).
55. Hironaka, K. and Morishita, Y. Encoding and decoding of positional information in morphogen-dependent patterning. *Curr opin Gen Develop* **22**, 553–561, doi:10.1016/j.gde.2012. 10.002 (2012).
56. Kondo, S. and Miura, T. Reaction-diffusion model as a framework for understanding biological pattern formation. *Science* **329**, 1616–1620, doi:329/5999/1616 [pii] 10.1126/science.1179047 (2010).
57. Engler, A. J., Sen, S., Sweeney, H. L., and Discher, D. E. Matrix elasticity directs stem cell lineage specification. *Cell* **126**, 677–689, doi:10.1016/j.cell.2006.06.044 (2006).
58. Andrew, D. J. and Ewald, A. J. Morphogenesis of epithelial tubes: Insights into tube formation, elongation, and elaboration. *Dev Biol* **341**, 34–55, doi:10.1016/j.ydbio.2009.09.024 (2010).
59. Affolter, M., Zeller, R., and Caussinus, E. Tissue remodelling through branching morphogenesis. *Nat Rev Mol Cell Biol* **10**, 831–842, doi:10.1038/nrm2797 (2009).
60. Risau, W., Feinberg, R. N., Sherer, G. K., and Auerbach, R. *The Development of the Vascular System Issues in Biomedicine*, pp. 58–68. (Basel, Switzerland: Karger, 1991).
61. Davis, G. E., Bayless, K. J., and Mavila, A. Molecular basis of endothelial cell morphogenesis in three-dimensional extracellular matrices. *Anat Rec* **268**, 252–275, doi:10.1002/ar.10159 (2002).
62. Czirok, A. and Little, C. D. Pattern formation during vasculogenesis. *Birth Defects Res C Embryo Today* **96**, 153–162, doi:10.1002/bdrc.21010 (2012).
63. Manoussaki, D., Lubkin, S. R., Vernon, R. B., and Murray, J. D. A mechanical model for the formation of vascular networks in vitro. *Acta Biotheoretica* **44**, 271–282 (1996).
64. Harris, A. K., Stopak, D., and Warner, P. Generation of spatially periodic patterns by a mechanical instability: A mechanical alternative to the Turing model. *J Embryol Exp Morph* **80**, 1–20 (1984).
65. Oster, G. F., Murray, J. D., and Harris, A. K. Mechanical aspects of mesenchymal morphogenesis. *J Embryol Exp Morph* **78**, 83–125 (1983).
66. Murray, J. D. and Oster, G. F. Cell traction models for generating pattern and form in morphogenesis. *J Math Biol* **19**, 265–279 (1984).
67. Namy, P., Ohayon, J., and Tracqui, P. Critical conditions for pattern formation and in vitro tubulogenesis driven by cellular traction fields. *J Theor Biol* **227**, 103–120 (2004).
68. Oster, G. F., Murray, J. D., and Maini, P. K. A model for chondrogenic condensations in the developing limb: The role of extracellular matrix and cell tractions. *J Embryol Exp Morph* **89**, 93–112 (1985).
69. Wan, X., Li, Z., and Lubkin, S. R. Mechanics of mesenchymal contribution to clefting force in branching morphogenesis. *Biomech Model Mechanobiol* **7**, 417–426, doi:10.1007/s10237-007-0105-y (2008).
70. Lucitti, J. L. et al. Vascular remodeling of the mouse yolk sac requires hemodynamic force. *Development* **134**, 3317–3326, doi:134/18/3317 [pii] 10.1242/dev.02883 (2007).
71. Culver, J. C. and Dickinson, M. E. The effects of hemodynamic force on embryonic development. *Microcirculation* **17**, 164–178, doi:10.1111/j.1549-8719.2010.00025.x (2010).
72. le Noble, F. et al. Control of arterial branching morphogenesis in embryogenesis: Go with the flow. *Cardiovasc Res* **65**, 619–628, doi:10.1016/j.cardiores.2004.09.018 (2005).
73. Langille, B. L. Arterial remodeling: Relation to hemodynamics. *Can J Physiol Pharmacol* **74**, 834–841 (1996).
74. Gjorevski, N. and Nelson, C. M. Endogenous patterns of mechanical stress are required for branching morphogenesis. *Integr Biol (Camb)* **2**, 424–434, doi:10.1039/c0ib00040j (2010).
75. Martin, A. C., Gelbart, M., Fernandez-Gonzalez, R., Kaschube, M., and Wieschaus, E. F. Integration of contractile forces during tissue invagination. *J Cell Biol* **188**, 735–749, doi:jcb.200910099 [pii] 10.1083/jcb.200910099 (2010).

76. Keller, R. Mechanisms of elongation in embryogenesis. *Development* **133**, 2291–2302 (2006).
77. Soufan, A. T. et al. Regionalized sequence of myocardial cell growth and proliferation characterizes early chamber formation. *Circ Res* **99**, 545–552 (2006).
78. Colas, J. F. and Schoenwolf, G. C. Towards a cellular and molecular understanding of neurulation. *Dev Dyn* **221**, 117–145, doi:10.1002/dvdy.1144 [pii] 10.1002/dvdy.1144 (2001).
79. Kirby, M. L. *Cardiac Development.* (New York: Oxford University Press, 2007).
80. Kimmel, C. B., Ballard, W. W., Kimmel, S. R., Ullmann, B., and Schilling, T. F. Stages of embryonic development of the zebrafish. *Dev Dyn* **203**, 253–310, doi:10.1002/aja.1002030302 (1995).
81. Bertet, C., Sulak, L., and Lecuit, T. Myosin-dependent junction remodelling controls planar cell intercalation and axis elongation. *Nature* **429**, 667–671, doi:10.1038/nature02590 (2004).
82. Keller, R. et al. Mechanisms of convergence and extension by cell intercalation. *Philos Trans R Soc Lond B Biol Sci* **355**, 897–922, doi:10.1098/rstb.2000.0626 (2000).
83. Sherrard, K., Robin, F., Lemaire, P., and Munro, E. Sequential activation of apical and basolateral contractility drives ascidian endoderm invagination. *Curr Biol* **20**, 1499–1510, doi:S0960-9822(10)00873-0 [pii] 10.1016/j.cub.2010.06.075 (2010).
84. Condic, M. L., Fristrom, D., and Fristrom, J. W. Apical cell shape changes during *Drosophila* imaginal leg disc elongation: A novel morphogenetic mechanism. *Development* **111**, 23–33 (1991).
85. Behrndt, M. et al. Forces driving epithelial spreading in zebrafish gastrulation. *Science* **338**, 257–260, doi:10.1126/science.1224143 (2012).
86. Solnica-Krezel, L. Gastrulation in zebrafish—All just about adhesion? *Curr Opin Gen Develop* **16**, 433–441, doi:10.1016/j.gde.2006.06.009 (2006).
87. Lepage, S. E. and Bruce, A. E. Zebrafish epiboly: Mechanics and mechanisms. *Int J Dev Biol* **54**, 1213–1228, doi:10.1387/ijdb.093028sl (2010).
88. Kiehart, D. P., Galbraith, C. G., Edwards, K. A., Rickoll, W. L., and Montague, R. A. Multiple forces contribute to cell sheet morphogenesis for dorsal closure in *Drosophila*. *J Cell Biol* **149**, 471–490 (2000).
89. Beloussov, L. V., Saveliev, S. V., Naumidi, I. I., and Novoselov, V. V. Mechanical stresses in embryonic tissues: Patterns, morphogenetic role, and involvement in regulatory feedback. *Int Rev Cytol* **150**, 1–34 (1994).
90. Warga, R. M. and Kimmel, C. B. Cell movements during epiboly and gastrulation in zebrafish. *Development* **108**, 569–580 (1990).
91. Jacinto, A., Martinez-Arias, A., and Martin, P. Mechanisms of epithelial fusion and repair. *Nat Cell Biol* **3**, E117–123, doi:10.1038/35074643 35074643 [pii] (2001).
92. Koppen, M., Fernandez, B. G., Carvalho, L., Jacinto, A., and Heisenberg, C. P. Coordinated cell-shape changes control epithelial movement in zebrafish and Drosophila. *Development* **133**, 2671–2681, doi:10.1242/dev.02439 (2006).
93. Ezin, A. M., Fraser, S. E., and Bronner-Fraser, M. Fate map and morphogenesis of presumptive neural crest and dorsal neural tube. *Dev Biol* **330**, 221–236, doi:10.1016/j.ydbio.2009.03.018 (2009).
94. Sausedo, R. A., Smith, J. L., and Schoenwolf, G. C. Role of nonrandomly oriented cell division in shaping and bending of the neural plate. *J Comp Neurol* **381**, 473–488 (1997).
95. Keller, R., Davidson, L. A., and Shook, D. R. How we are shaped: The biomechanics of gastrulation. *Differentiation* **71**, 171–205, doi:10.1046/j.1432-0436.2003.710301.x (2003).
96. Hardin, J. D. and Cheng, L. Y. The mechanisms and mechanics of archenteron elongation during sea urchin gastrulation. *Dev Biol* **115**, 490–501 (1986).
97. Keller, R., Shook, D., and Skoglund, P. The forces that shape embryos: Physical aspects of convergent extension by cell intercalation. *Phys Biol* **5**, 015007, doi:10.1088/1478-3975/5/1/015007 (2008).
98. Brodland, G. W. Do lamellipodia have the mechanical capacity to drive convergent extension? *Int J Dev Biol* **50**, 151–155, doi:10.1387/ijdb.052040gb (2006).
99. Rauzi, M., Verant, P., Lecuit, T., and Lenne, P. F. Nature and anisotropy of cortical forces orienting *Drosophila* tissue morphogenesis. *Nat Cell Biol* **10**, 1401–1410, doi:ncb1798 [pii] 10.1038/ncb1798 (2008).

100. Moore, S. W., Keller, R. E., and Koehl, M. A. The dorsal involuting marginal zone stiffens aniso-tropically during its convergent extension in the gastrula of *Xenopus laevis*. *Development* **121**, 3131–3140 (1995).

101. Sawyer, J. M. et al. Apical constriction: A cell shape change that can drive morphogenesis. *Dev Biol* **341**, 5–19, doi:10.1016/j.ydbio.2009.09.009 (2010).

102. Hardin, J. and Keller, R. The behaviour and function of bottle cells during gastrulation of *Xenopus laevis*. *Development* **103**, 211–230 (1988).

103. Lee, J. Y. and Harland, R. M. Actomyosin contractility and microtubules drive apical constriction in *Xenopus* bottle cells. *Dev Biol* **311**, 40–52, doi:10.1016/j.ydbio.2007.08.010 (2007).

104. Spear, P. C. and Erickson, C. A. Interkinetic nuclear migration: A mysterious process in search of a function. *Dev Growth Differ* **54**, 306–316, doi:10.1111/j.1440-169X.2012.01342.x (2012).

105. Davidson, L. A., Koehl, M. A., Keller, R., and Oster, G. F. How do sea urchins invaginate? Using biomechanics to distinguish between mechanisms of primary invagination. *Development* **121**, 2005–2018 (1995).

106. Lane, M. C., Koehl, M. A., Wilt, F., and Keller, R. A role for regulated secretion of apical extracel-lular matrix during epithelial invagination in the sea urchin. *Development* **117**, 1049–1060 (1993).

107. Martin, A. C., Kaschube, M., and Wieschaus, E. F. Pulsed contractions of an actin-myosin net-work drive apical constriction. *Nature* **457**, 495–499, doi:10.1038/nature07522 (2009).

108. Varner, V. D. and Taber, L. A. Not just inductive: A crucial mechanical role for the endoderm during heart tube assembly. *Development* **139**, 1680–1690, doi:10.1242/dev.073486 (2012).

109. An, S. S. and Fredberg, J. J. Biophysical basis for airway hyperresponsiveness. *Can J Physiol Pharmacol* **85**, 700–714, doi:y07-059 [pii] 10.1139/y07-059 (2007).

110. Taber, L. A. Theoretical study of Beloussov's hyper-restoration hypothesis for mechanical regu-lation of morphogenesis. *Biomech Model Mechanobiol* **7**, 427–441 (2008).

111. Odell, G. M., Oster, G., Alberch, P., and Burnside, B. The mechanical basis of morphogenesis. I. Epithelial folding and invagination. *Dev Biol* **85**, 446–462 (1981).

112. Munoz, J. J., Barrett, K., and Miodownik, M. A deformation gradient decomposition method for the analysis of the mechanics of morphogenesis. *J Biomech* **40**, 1372–1380 (2007).

113. Conte, V., Munoz, J. J., and Miodownik, M. A 3D finite element model of ventral furrow invagi-nation in the *Drosophila melanogaster* embryo. *J Mech Behav Biomed Mater* **1**, 188–198, doi:S1751-6161(07)00032-X [pii] 10.1016/j.jmbbm.2007.10.002 (2008).

114. Conte, V., Munoz, J. J., Baum, B., and Miodownik, M. Robust mechanisms of ventral furrow invagination require the combination of cellular shape changes. *Phys Biol* **6**, 016010, doi:S1478-3975(09)97865-7 [pii] 10.1088/1478-3975/6/1/016010 (2009).

115. Allena, R., Mouronval, A. S., and Aubry, D. Simulation of multiple morphogenetic movements in the *Drosophila* embryo by a single 3D finite element model. *J Mech Behav Biomed Mater* **3**, 313–323, doi:S1751-6161(10)00003-2 [pii] 10.1016/j.jmbbm.2010.01.001 (2010).

116. Rodriguez, E. K., Hoger, A., and McCulloch, A. D. Stress-dependent finite growth in soft elastic tissues. *J Biomech* **27**, 455–467 (1994).

117. Alvarez, I. S. and Schoenwolf, G. C. Expansion of surface epithelium provides the major extrin-sic force for bending of the neural plate. *J Exp Zool* **261**, 340–348, doi:10.1002/jez.1402610313 (1992).

118. Schoenwolf, G. C., Folsom, D., and Moe, A. A reexamination of the role of microfilaments in neurulation in the chick embryo. *Anat Rec* **220**, 87–102 (1988).

119. van Straaten, H. W., Sieben, I., and Hekking, J. W. Multistep role for actin in initial closure of the mesencephalic neural groove in the chick embryo. *Dev Dyn* **224**, 103–108, doi:10.1002/dvdy.10078 (2002).

120. Kinoshita, N., Sasai, N., Misaki, K., and Yonemura, S. Apical accumulation of rho in the neural plate is important for neural plate cell shape change and neural tube formation. *Mol Biol Cell* **19**, 2289–2299 (2008).

121. Smith, J. L. and Schoenwolf, G. C. Role of cell-cycle in regulating neuroepithelial cell shape dur-ing bending of the chick neural plate. *Cell Tissue Res* **252**, 491–500 (1988).

122. Clausi, D. A. and Brodland, G. W. Mechanical evaluation of theories of neurulation using computer simulations. *Development* **118**, 1013–1023 (1993).
123. Huang, J. et al. The mechanism of lens placode formation: A case of matrix-mediated morphogenesis. *Dev Biol* **355**, 32–42, doi:S0012-1606(11)00240-5 [pii] 10.1016/j.ydbio.2011.04.008 (2011).
124. Taber, L. A. Towards a unified theory for morphomechanics. *Phil Trans R Soc A Math Phys Eng Sci* **367**, 3555–3583, doi:367/1902/3555 [pii] 10.1098/rsta.2009.0100 (2009).
125. Keller, R. and Shook, D. The bending of cell sheets—From folding to rolling. *BMC Biol* **9**, 90, doi:10.1186/1741-7007-9-90 (2011).
126. Varner, V. D., Voronov, D. A., and Taber, L. A. Mechanics of head fold formation: Investigating tissue-level forces during early development. *Development* **137**, 3801–3811, doi:dev.054387 [pii] 10.1242/dev.054387 (2010).
127. Manner, J. Cardiac looping in the chick embryo: A morphological review with special reference to terminological and biomechanical aspects of the looping process. *Anat Rec* **259**, 248–262 (2000).
128. Desmond, M. E. and Jacobson, A. G. Embryonic brain enlargement requires cerebrospinal fluid pressure. *Dev Biol* **57**, 188–198 (1977).
129. Butler, J. K. An experimental analysis of cardiac loop formation in the chick. (MS thesis, Austin, TX: University of Texas, 1952).
130. Manning, A. and McLachlan, J. C. Looping of chick embryo hearts in vitro. *J Anat* **168**, 257–263 (1990).
131. Filas, B. A., Bayly, P. V., and Taber, L. A. Mechanical stress as a regulator of cytoskeletal contractility and nuclear shape in embryonic epithelia. *Ann Biomed Eng* **39**, 443–454, doi:10.1007/s10439-010-0171-7 (2011).
132. Goodrum, G. R. and Jacobson, A. G. Cephalic flexure formation in the chick embryo. *J Exp Zool* **216**, 399–408, doi:10.1002/jez.1402160308 (1981).
133. Latacha, K. S. et al. The role of actin polymerization in bending of the early heart tube. *Dev Dyn* **233**, 1272–1286 (2005).
134. Remond, M. C., Fee, J. A., Elson, E. L., and Taber, L. A. Myosin-based contraction is not necessary for cardiac c-looping in the chick embryo. *Anat Embryol (Berl)* **211**, 443–454 (2006).
135. Xu, G. et al. Axons pull on the brain, but tension does not drive cortical folding. *J Biomech Eng* **132**, 071013, doi:10.1115/1.4001683 (2010).
136. Richman, D. P., Stewart, R. M., Hutchinson, J. W., and Caviness, V. S., Jr. Mechanical model of brain convolutional development. *Science* **189**, 18–21 (1975).
137. Savin, T. et al. On the growth and form of the gut. *Nature* **476**, 57–62, doi:nature10277 [pii] 10.1038/nature10277 (2011).
138. Shyer, A. E. et al. Villification: How the gut gets its villi. *Science* **342**, 212–218, doi:10.1126/science.1238842 (2013).
139. Viamontes, G. I. and Kirk, D. L. Cell shape changes and the mechanism of inversion in *Volvox*. *J Cell Biol* **75**, 719–730 (1977).
140. Viamontes, G. I., Fochtmann, L. J., and Kirk, D. L. Morphogenesis in *Volvox*: Analysis of critical variables. *Cell* **17**, 537–550 (1979).
141. Nishii, I. and Ogihara, S. Actomyosin contraction of the posterior hemisphere is required for inversion of the *Volvox* embryo. *Development* **126**, 2117–2127 (1999).
142. Hohn, S. and Hallmann, A. There is more than one way to turn a spherical cellular monolayer inside out: Type B embryo inversion in *Volvox globator*. *BMC Biol* **9**, 89, doi:10.1186/1741-7007-9-89 (2011).
143. Pouille, P. A., Ahmadi, P., Brunet, A. C., and Farge, E. Mechanical signals trigger Myosin II redistribution and mesoderm invagination in *Drosophila* embryos. *Sci Signal* **2**, ra16, doi:scisignal.2000098 [pii] 10.1126/scisignal.2000098 (2009).
144. Beloussov, L. V. Morphogenesis as a macroscopic self-organizing process. *Biosystems* **109**, 262–279, doi:10.1016/j.biosystems.2012.05.003 (2012).

12

Mechanics of Tissue Morphogenesis

Michael J. Siedlik and Celeste M. Nelson

CONTENTS

Abbreviations

αSMA	α-Smooth muscle actin
AV	Atrioventricular
CREB	cAMP response element-binding protein
dpf	Days postfertilization
ECM	Extracellular matrix
EGFR	Epidermal growth factor receptor
ERK	Extracellular signal-regulated kinase
FGF10	Fibroblast growth factor 10
HH	Hamburger–Hamilton
hpf	Hours post-fertilization
IFP	Interstitial fluid pressure
IGF	Insulin-like growth factor
MAPK	Mitogen-activated protein kinase
miRNA	microRNA

PDGF   Platelet-derived growth factor
PIV   Particle velocimetry analysis
PI3K   Phosphoinositide 3-kinase
PKC   Protein kinase C
SRF   Serum response factor
TGF   Transforming growth factor
VEGFR   Vascular endothelial growth factor receptor

12.1 Introduction

How are we made? How does the apparently (and perhaps, deceptively) simple fertilized egg transform from a sphere to the complex geometry of a mature organism? These questions, in one form or another, have fascinated scientists and laypeople for thousands of years. Some of the earliest ideas were that organisms develop from miniature versions of their adult selves, often referred to as homunculi or animalcules, hiding within the heads of sperm. This concept, of *preformationism*, was largely abandoned in the nineteenth century when the cell theory of life became the predominant viewpoint: all living things are made of cells and so development requires that the single fertilized egg divide successively to give rise to the differentiated cell types of the mature organism.

Recent scientific pursuits have been focused on uncovering the genetic *blueprints* of morphogenesis, those genes within the fertilized egg and its subsequent progeny that direct the formation of tissues, organs, and whole organisms. This decidedly reductionist approach has been fruitful, yielding information about thousands of gene products required for embryogenesis and tissue morphogenesis. Nonetheless, the genetic blueprint paradigm has oftentimes bordered on preformationist as well, simply shifting the requirement of a map of the mature organism from one constructed of miniaturized cells and tissues to one encoded by bits of DNA. The completed sequencing of the human genome (and that of other animals) suggests that the answer to the question "How are we made?" may be more complicated than the genes themselves, since there are only ~30,000 gene products and far more morphogenetic events required to turn an egg into a person. Is it possible that all morphogenetic movements, all changes in tissue geometry, can be reduced to a series of predetermined genetically encoded routines? Is it necessary that these details be encoded precisely in the genome?

Somewhat in parallel to investigations focused on the genetics of development have been a series of studies focused on the *mechanics* of development, the forces required to fold the progeny of that single fertilized egg into the tissues and organs that make each of us complete. These studies have viewed developing tissues as physical entities, subject to the laws of matter and physics and therefore responsive to mechanical manipulations. The emerging story is that forces are essential for tissue development and that cells both exert forces and respond to them as well. In some ways, the study of the mechanics of morphogenesis is a rejection of genetic preformationism. It is an acceptance of the laws of physics that all changes in tissue geometry must result from physical forces exerted by or on the tissue that is undergoing morphogenesis.

12.2 Passive Forces in Morphogenesis

The cells and tissues of the developing embryo are exposed to both active and passive mechanical forces. Active forces result from cells and tissues pulling and pushing on each other through ATP-dependent processes. Passive forces include those that result simply from the physical environment of the embryo: pressures and flows from interstitial fluids, buckling of adjacent tissues, and forces from surface tensions. Both active and passive forces can convey morphogenetic information to the tissues upon which they act, and both have been implicated recently in studies of tissue development (Nelson and Gleghorn, 2012; Heisenberg and Bellaiche, 2013; Mammoto et al., 2013). Here, we discuss work implicating fluid forces (flow and pressure) and elastic buckling in tissue morphogenesis.

12.2.1 Fluid Flow

The mature vertebrate animal can basically be considered as a bag of tubes—vessels and ducts that partition different fluids from each other and from the outside world (Figure 12.1). Flow of these fluids through the tubular organ systems of the body is driven by convective transport down pressure gradients. In the cardiovascular system, blood flows from the high-pressure arterial system (mean of 100 mm Hg in the adult human) to the low-pressure capillary and venous system (2–5 mm Hg). This pressure gradient is driven primarily by contraction of the heart, which is an active process, but the fluid flow itself conveys mechanical information to the cells within the vessel wall (Freund et al., 2012). Flow is typically sensed by the cells lining the wall of the tube containing the fluid and is transmitted to these cells via shear stresses exerted upon them. Wall shear stress (τ)

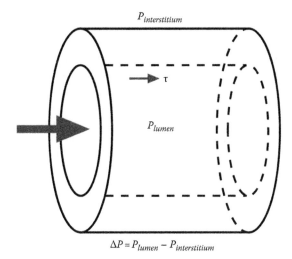

$$\Delta P = P_{lumen} - P_{interstitium}$$

FIGURE 12.1

Passive fluid forces within developing tissues. Fluid is transported within embryonic tubes during morphogenesis. The pressure within the tube (P_{lumen}) minus that outside of the tube ($P_{interstitium}$) gives the transmural pressure across the tube. Mechanical forces from flow of fluid down the tube are transmitted to the layer of cells lining the tube in part through wall shear stress, τ.

is simply the product of the fluid viscosity (μ) and the derivative of the velocity of the fluid at the wall, $\tau = \mu \left.\dfrac{\partial u}{\partial r}\right|_{wall}$.

Although it is perhaps not obvious from its mature four-chambered structure, the heart initiates its development as a linear tube (Lopez-Sanchez and Garcia-Martinez, 2011; Miquerol and Kelly, 2013). The heart is the first functioning organ in the embryo and begins contracting (as an elastic impedance pump) early in its own development and while still in a tubular form, prior to the formation of any chambers or valves (Forouhar et al., 2006). Intriguingly, the fluid flow that results from heart function is also *required* for aspects of cardiac morphogenesis in a variety of species, an excellent example of positive reciprocal feedback during embryonic development.

High-speed confocal imaging of the embryonic zebrafish heart and vasculature has enabled quantification of the hemodynamics of blood flow at both early and late stages of development (Hove et al., 2003; Anton et al., 2013; Lee et al., 2013). At 37 h postfertilization (hpf), which is prior to the development of the heart valves, particle image velocimetry (PIV) analysis revealed that the average red blood cell travels through the heart tube at a velocity of approximately 0.5 mm/s. This velocity of flow, coupled with the high viscosity of the fluid and small radius of the developing vascular system, suggests a wall shear stress greater than 1 dyn/cm². At 40–50 hpf, the heart tube loops, corresponding to a significant increase in wall shear stress, and later forms the atrium, ventricle, and atrioventricular (AV) valve (Lee et al., 2013). At 4.5 days postfertilization (dpf), PIV analysis showed that the average red blood cell travels an order of magnitude faster than at earlier stages of development, at a rate of >0.5 cm/s through the nascent valves of the heart, which would suggest wall shear stresses of >75 dyn/cm² for this region of the tube (Hove et al., 2003). PIV analysis also revealed the presence of vortical flow patterns within the heart itself.

This flow profile appears to be critical for normal morphogenesis of the zebrafish heart tube. Disrupting flow by blocking the tube with an implanted bead leads to a 10-fold reduction in the wall shear stress and prevents formation of the valves and cardiac looping (Hove et al., 2003), the morphogenetic process that builds the chambers from the single heart tube (Taber et al., 2010). Similarly, disturbed flow patterns, such as those observed during vortical flow, are critical for formation of the valves in the zebrafish heart (Vermot et al., 2009). In particular, retrograde flow induces the expression of the Krüppel-like transcription factor *klf2a* in the endothelium of the AV canal; morpholino-mediated knockdown of *klf2a* leads to defects in valvulogenesis, suggesting that this transcription factor indirectly transduces signaling from retrograde flow to the morphogenetic program (Vermot et al., 2009). This signaling pathway may be conserved: reversing flows have been reported in the developing cardiovascular system of many species (Groenendijk et al., 2008), reversing flows induce the expression of *Klf2* by vascular endothelial cells in culture (Dekker et al., 2002), and mice lacking *Klf2* exhibit heart failure and reduced cardiac output (Lee et al., 2006). Normal morphogenesis of the heart valves abolishes the reversing flow patterns and permits the establishment of unidirectional flow. (Of course, other aspects of the mechanics of zebrafish heart development are not universal: cardiac looping is impervious to disrupted blood flow in the early chicken embryo, which is more similar to the human case than is the zebrafish (Aleksandrova et al., 2012).)

In culture, cardiac endothelial cells respond to the forces of fluid flow by altering their actomyosin cytoskeleton, changing their shapes, and modifying their gene expression profile (Hahn and Schwartz, 2008; Boon and Horrevoets, 2009). Shear stress increases signaling through MAPK/ERK kinase (MEK5) (Surapisitchat et al., 2001), which is thought to enhance the stability of *Klf2* mRNA, thus leading to increased levels of Klf2 protein (Parmar et al., 2006; van Thienen et al., 2006). This signaling is mediated at least in part by microRNA (miRNA)-92a, a negative regulator of *Klf2* that is downregulated in response to flow (Bonauer et al., 2009; Wu et al., 2011). Klf2 enhances the expression of many flow-regulated genes in cultured endothelial cells, including endothelial nitric oxide synthase (eNOS) and thrombomodulin (Dekker et al., 2005; Atkins and Jain, 2007). The extent to which Klf2-mediated regulation of these genes controls mechanical regulation of cardiac morphogenesis remains unclear.

miRNAs are small noncoding RNA molecules that silence the translation of mRNAs (Chen and Rajewsky, 2007; Jackson and Standart, 2007). Several miRNAs are induced in cardiomyocytes and smooth muscle cells in response to mechanical stress (van Rooij et al., 2006; Mohamed et al., 2010). In addition to miR-92a, expression of miR-21 is induced within the endocardium of the embryonic zebrafish heart during the stages at which the valves are formed, in constricted regions of the bending tube in response to reversed blood flow (Banjo et al., 2013). This expression of miR-21 is required for valvulogenesis; miR-21 appears to act at least in part by suppressing the expression of *Sprouty-2* (*spry2*) (Banjo et al., 2013), an inhibitor of extracellular signal-regulated kinase (ERK) signaling in a variety of organs and contexts. Spry2 expression disrupts morphogenesis of the heart by altering cell proliferation.

The molecular structures within the endothelium that sense and respond to fluid flow have been under intense investigation. Several candidates have been proposed to act as mechanotransducers, including ion channels, G-protein-coupled or tyrosine kinase receptors, adhesive proteins, and the glycocalyx (Ando and Yamamoto, 2009). Shear stress has been shown to activate flow-responsive K^+ channels and Cl^- channels (Cooke et al., 1991). The outward flux of K^+ ions causes hyperpolarization of the plasma membrane and induces the inward flow of Ca^{2+} (Luckhoff and Busse, 1990), leading to Ca^{2+}-induced signaling pathways including Ca^{2+}-mediated release of eNOS (Isshiki et al., 1998). Shear stress can also directly activate vascular endothelial growth factor receptors (VEGFRs), independently of ligand binding (Shay-Salit et al., 2002; Jin et al., 2003; Lee and Koh, 2003), and induce downstream signaling through ERK and phosphoinositide 3-kinase (PI3K) (Tseng et al., 1995; Go et al., 1998). One promising candidate mechanotransducer is the primary cilium, a sensory organelle that projects from the apical surface of polarized cells, since defects in ciliogenesis within the endothelium disrupt heart development in mice (Slough et al., 2008). The primary cilia on the surfaces of endothelial cells are thought to bend in response to shear stress, and this bending increases the permeability of ion channels and thus permits the influx of Ca^{2+} (Nauli et al., 2003; Hierck et al., 2008; AbouAlaiwi et al., 2009). Flow can thus both directly and indirectly induce signaling within cells in intimate contact with fluid.

While endothelial cells lining the developing vascular system are the direct recipients of information from fluid flow, other cells not immediately adjacent to the fluid respond as well. Looping of the embryonic heart requires that the heart tube bend and twist around itself, while ballooning corresponds to the emergence of bulges within the tube that create the shape of the cardiac chambers. In the zebrafish heart tube, these morphogenetic events appear to be driven by changes in the shape of individual cardiomyocytes within the walls of the tube, with cells at the outer curvature of the nascent chambers becoming flattened and elongated (Auman et al., 2007). Importantly, blood flow is critical for these shape changes, as they are abolished when flow is disrupted. Although it remains unclear how the cardiomyocytes are sensing fluid flow, there are two obvious possibilities. First, the endothelium may be responding to flow and communicating this response chemically to the myocardium. Second, the myocardium may be sensing pressure, rather than flow, and responding to this parameter. Future work is needed to define precisely how far the forces from fluid flow can be transmitted into the vessel wall during its morphogenesis.

12.2.2 Fluid Pressure

In any system of fluid-filled tubes, pressure differentials can develop along a given tube or across its wall (Figure 12.1). The latter is known as a transmural (*across wall*) pressure and is prevalent within both mature and developing organisms. The magnitude of this pressure difference (pressure in the lumen minus pressure surrounding the tube) appears to be necessary for normal morphogenesis of several organ systems.

In the developing lung, the fetal airways are lined by one or more layers of epithelial cells that surround the lumen. The developing pulmonary epithelium secretes fluid into the lumen of the airways (Alcorn et al., 1977; Fewell et al., 1983; Harding and Hooper, 1996), and this fluid is sufficient to generate a positive distending transmural pressure of approximately 200–400 Pa (1.5–3 mm Hg) in several animal models because the fetal larynx is closed at these stages of development (Vilos and Liggins, 1982; Hooper et al., 1993; Blewett et al., 1996; Schittny et al., 2000). In fetal lambs and rabbits, chronic drainage of the airways resulting from tracheostomy leads to pulmonary hypoplasia (underdeveloped lungs with a reduced number of airways and alveoli) (Alcorn et al., 1977; Fewell et al., 1983). In contrast, tracheal ligation or laryngeal atresia results in a buildup of fluid and leads to pulmonary hyperplasia (enlarged lungs with an increased number of airways) (Alcorn et al., 1977; Wigglesworth et al., 1987; Moessinger et al., 1990; Nardo et al., 1995; Keramidaris et al., 1996; Kitano et al., 1998; Nardo et al., 1998; Kitano et al., 1999). Fetal pulmonary hypoplasia is a common finding in neonatal autopsies and often copresents with anatomic changes to the fetal chest cavity that would disrupt transmural pressure (Finegold et al., 1971; Goldstein and Reid, 1980; Liggins and Kitterman, 1981; George et al., 1987; Greenough, 2000; Jesudason, 2007).

Although the underlying molecular mechanisms remain to be uncovered, changes in transmural pressure lead to significant alterations in gene expression in the developing lung. Tracheal occlusion leads to an increase in cell proliferation and expression of insulin-like growth factor (IGF)-II in fetal lambs (Hooper et al., 1993). Conversely, draining the airways of the lung leads to a reduction in proliferation and expression of IGF-II.

At least some of the effects of pressure on morphogenesis, gene expression, and cell differentiation in the example of the developing lung appear to result from mechanical stretch. In any thick-walled viscoelastic tube, changing the relative magnitude of the pressure within the tube will change the extent to which the wall is stretched. Applying intermittent mechanical stretch was found to result in an increase in the proliferation of fetal rat lung epithelial cells and fibroblasts in culture (Liu et al., 1992), consistent with the effects of increased transmural pressure on fetal lungs in vivo (Hooper et al., 1993). Mechanical stretch led to an increase in the expression of platelet-derived growth factor (PDGF), and antisense-mediated downregulation of PDGF or treatment with anti-PDGF function-blocking antibodies blocked the effects of stretch on the proliferation of fetal rat lung cells (Liu et al., 1995a). Mechanical stretch also led to an increase in calcium influx in fetal lung cells through stretch-activated ion channels (Liu et al., 1994) and activated signaling via protein kinase C (PKC) (Liu et al., 1995b). Since PKC has been shown to regulate the expression of PDGF in endothelial cells (Hsieh et al., 1992), it is possible that stretch enhances the proliferation of the cells of the fetal lung through a similar mechanism.

At the same time as the airway epithelium is developing, the surrounding pulmonary mesenchyme is differentiating into the cell types present in the mature lung. The most prominent mesenchymal cell type to form during this period is the pulmonary smooth muscle, which differentiates in a cranial to caudal direction and forms a sheath that envelops the developing airway epithelium (Collet and Des Biens, 1974; Sparrow et al., 1999). Transmural pressure appears to be critical for development of the airway smooth muscle and also acts at least in part through stretch. Mechanical stretch induces the expression of smooth muscle–specific genes, including α-smooth muscle actin (αSMA) and smooth muscle myosin, by undifferentiated pulmonary mesenchymal cells (Yang et al., 2000). This increase in myogenesis correlates with an increase in the expression of serum response factor (SRF), as well as a decrease in the expression of the splice variant SRFΔ5 (Yang et al., 2000). SRF is a transcription factor that stimulates the expression of a wide range of smooth muscle–specific genes (Pipes et al., 2006). Conversely, SRFΔ5 blocks myogenesis in culture (Belaguli et al., 1999; Kemp and Metcalfe, 2000) and is expressed aberrantly in hypoplastic lungs from human fetuses (Yang et al., 2000), suggesting that transmural pressure regulates lung development in part by regulating alternative splicing.

Transmural pressure represents the difference in fluid pressures between a fluid-filled tube (such as the airways of the developing lung or the arterial system) and the tissue surrounding the tube. Although normally atmospheric, under some conditions, interstitial fluid can build up and generate magnitudes of interstitial fluid pressure (IFP) (Figure 12.1) greater than atmospheric (Schmid-Schonbein, 1990). Interstitial fluid accumulates between the cells within tissues due to transvascular passage of plasma fluid and is normally drained in the mature organism by the lymphatic system (Aukland and Reed, 1993). Recent studies have revealed that the development of the lymphatic vasculature, a process known as lymphangiogenesis, is regulated in part by IFP (Planas-Paz and Lammert, 2013). In the mouse, increases in IFP correlate with stretch and proliferation of lymphatic endothelial cells, and decreasing IFP reduces the activity of the major lymphatic regulator, VEGFR3 by these cells (Planas-Paz et al., 2012). Fluid pressure, whether transmural or interstitial, can thus have significant impact on tissue morphogenesis in the developing embryo.

12.2.3 Buckling

Many vertebrate organs, including the lungs, esophagus, intestine, blood vessels, exocrine glands, and kidneys, are comprised of an epithelial compartment surrounded by one or more layers of mesenchyme. This topological arrangement separates the epithelial lumen from the surrounding tissue and permits the fluid within the lumen to undergo convective transport down the tube or diffusive exchange across the wall of the tube. For organs that specialize in nutrient exchange, such as the intestine, the surface area of the epithelium is a limiting factor in the rate of transport. The luminal epithelium in these organs thus increases its surface area by adding a third dimension in the form of wrinkles or folds, in a process known as mucosal folding.

Intriguingly, these epithelial surfaces do not start out wrinkled, as most are initially smooth layers of epithelium in the form of tubes or sheets. Several mechanisms have been proposed to explain the morphogenesis of epithelial wrinkling, including localized expression of wrinkle-inducing genes in the surrounding mesenchyme (Nelson, 2013). Physically, however, epithelial tubes can be considered as elastic thick-walled cylindrical shells. When subjected to external compression, elastic shells are unstable and will buckle inward in predictable patterns that depend on the geometry of the shell (thickness and diameter) and its mechanical properties (elastic modulus and Poisson ratio) (Wang and Ertepinar, 1972; Papadaki, 2008). The compression that induces the buckling can be driven by active forces, such as something contracting around the shell, or passive forces, such as an increase in hydrostatic pressure resulting from a buildup of fluid surrounding the shell. Either way, however, the elastic shell itself undergoes a passive change in shape. Physical explanations of epithelial wrinkling thus suggest that the transformation of a smooth surface to a wrinkled one results from a mechanical instability of the inner mucosal cylinder (the epithelium) under compression from the surrounding cylinder of smooth muscle (Ben Amar and Jia, 2013).

The buckling epithelium hypothesis brings with it several predictions. First, the diameter of the epithelial ring is predicted to be larger when the mucosal epithelium is surgically separated from its surrounding mesenchyme. This is indeed the case for the porcine esophagus (Yang et al., 2007) and chicken small intestine (Shyer et al., 2013). Second, the number of folds is predicted to decrease as the stiffness of the epithelial wall relative to that of the surrounding mesenchyme increases. Put another way, to achieve any given arrangement of folds in the epithelium, the external forces applied by the surrounding smooth muscle would have to increase as the stiffness of the mucosal layer increases. This appears to be true for some disease states, including the esophagus of patients with systemic scleroderma (Villadsen et al., 2001), the bronchial airways of patients with asthma (Wiggs et al., 1997; Hogg, 2004), and the pharynx of patients with obstructive sleep apnea (Kairaitis, 2012). These conditions have been approximated computationally using continuum models of growing cylindrical surfaces (Hrousis et al., 2002; Papastavrou et al., 2013).

In the duodenal portion of the small intestine, the epithelium is surrounded by three layers of smooth muscle that differentiate from the mesenchyme during the same time period as the epithelium transforms from a smooth surface to a wrinkled sheet (Coulombre and Coulombre, 1958). In chickens and humans, the smooth epithelial sheet sequentially folds into parallel ridges, then zigzag-shaped folds, and then into pillars known as villi. The sequential genesis of these folds matches temporally the sequential differentiation of the three layers of smooth muscle that surround the epithelial wall. Formation of the intestinal villus was first proposed to result from mucosal buckling in the 1950s and was thought to depend on external compression provided by contraction of the three layers of smooth muscle (Coulombre and Coulombre, 1958). However, disrupting smooth muscle contraction by surgically removing portions of the smooth muscle wall does not disrupt epithelial wrinkling (Burgess, 1975). Nonetheless, the presence of the three layers of smooth muscle is critical for folding of the intestinal epithelium. As the surrounding mesenchyme differentiates into smooth muscle, the mesenchyme becomes stiffer; this increase in stiffness appears to be sufficient to compress the epithelium as it grows, thus causing it to buckle inward (Shyer et al., 2013). Replacing the smooth muscle with a sheath of silk of similar mechanical properties is sufficient to induce buckling of the intestinal mucosal epithelium (Shyer et al., 2013), in surprisingly the same sequence of geometric patterns as are observed in the embryo.

Similarly, the cortex (outer surface layer) of the vertebrate brain is initially smooth. In several mammals, including humans, the cortex folds during development to produce the fissures, sulci, and gyri of the mature brain (Molnar and Clowry, 2012). Cortical folding is essential for brain function, as defects are associated with severe mental disorders including autism and schizophrenia (Pavone et al., 1993; Sallet et al., 2003; Hardan et al., 2004; Nordahl et al., 2007; Cachia et al., 2008). Several hypotheses have been proposed to explain the physical mechanisms that underlie cortical folding, including buckling. In contrast to epithelial tubes, which appear to buckle along their inner surfaces, the developing brain would buckle along its outer surface (Richman et al., 1975; Raghavan et al., 1997). Such surface buckling could arise, as in the intestine, from a mechanical instability between the outer cortical layer and the deeper layers of the brain. In this case, differences in the rates of growth of the cortical layer and its *foundation* tissues would produce compressive stresses that lead to buckling. As in mucosal buckling, cortical buckling would also require that the cortical and subcortical regions of the developing brain be of different stiffness, with one model suggesting a 10-fold difference (Richman et al., 1975). However, a recent study using microindentation approaches found similar mechanical properties for both cortical and subcortical regions of the neonatal ferret brain at any given developmental stage, with a shear modulus ~40 Pa (Xu et al., 2010). A clear answer to this problem will require advanced strategies to measure in situ the mechanical properties of the different regions of the developing brain.

In lieu of purely elastic buckling, an alternate but related mechanism for cortical folding was reported recently by Taber and colleagues (Xu et al., 2010; Bayly et al., 2013). Here, the mechanical stresses between the two layers feed back to induce patterns of growth within the brain, and these patterns of differential growth in the subcortical regions are sufficient to induce folding of the cortex. The pattern of folds (wavelength and depth) depends on the relative rates of growth in the cortex, which expands as a sheet, and growth within the subcortical regions. Nonetheless, the growth rates of different regions of the cortex are presumed to be similar (Van Essen and Maunsell, 1980). Regardless of the underlying biological mechanism, tissue buckling appears to be a common mechanism to fold sheets of cells in the embryo.

12.3 Active Forces in Morphogenesis

Passive forces from fluid pressures and flows act on tissues in much the same way as they act on nonliving elastic materials, bending or buckling the material to result in a change in tissue form. Active forces, however, arise from active contractions by the cells within the tissues (Mammoto et al., 2013). These behaviors can change the shape of the tissues by changing the shapes of the individual cells relative to their neighbors, such that they lead to local or global displacements of the tissue.

12.3.1 Cell Contractions

Cells can generate forces by polymerizing and contracting their cytoskeletal machinery. Most cell-generated contractions result from the sliding of motor proteins, such as myosin, along actin filaments, which results in a local rearrangement and/or shortening of the cytoskeleton (Figure 12.2). Cytoskeletal contractions are transmitted to neighboring cells by spot welds present in the plasma membrane. The most common force-supporting intercellular junction is the adherens junction, which is comprised mainly of cadherin molecules that form homophilic interactions between neighboring cells and connect to the actin cytoskeleton via association with α- and β-catenin (Abe and Takeichi, 2008, Cavey et al., 2008). The forces that result from actomyosin contractions in one cell can thus be transmitted to neighboring cells directly via adherens junctions. When cells are connected together in tissues, actomyosin contractions can generate forces that

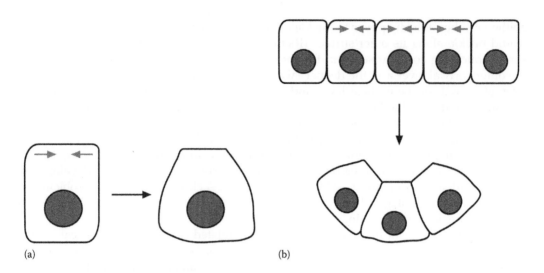

FIGURE 12.2
Active contractions induce morphogenesis. (a) Apical constriction changes the geometry of cells from cubic to trapezoid. (b) When one or more cells within a sheet undergo apical constriction, this generates sufficient force to bend the sheet.

are transmitted several hundred micrometers (several cell diameters) across the tissue (Gjorevski and Nelson, 2012). Such forces can also be transmitted indirectly through the surrounding extracellular matrix (ECM), which acts as a mesh-like substratum supporting the cells. Cells form adhesive junctions with the ECM via transmembrane complexes such as focal adhesions or hemidesmosomes. The former are comprised mainly of integrins, which also link to the actin cytoskeleton via interactions with cytoplasmic plaque proteins including vinculin and talin (Schiller and Fassler, 2013). Actomyosin contractions can thus generate forces on the underlying ECM substratum by pulling on integrins, and these forces can be transmitted to neighboring cells via deformation of the elastic ECM meshwork (Sen et al., 2009).

Coordinated and directed cellular contractions can lead to changes in tissue shape that drive early embryonic development and tissue morphogenesis (Gjorevski and Nelson, 2011). One such commonly observed contractile event is apical constriction, when the actomyosin cytoskeleton localized along the apical cortex of a cell contracts more than that along the basal or lateral surfaces (Sawyer et al., 2010; Rauzi and Lenne, 2011). Apical constriction causes the apical surface area to shrink relative to that of the basal surface and typically transforms cells from a cuboidal or rectangular geometry to a trapezoidal shape (Figure 12.2). Apical constriction drives invagination of the mesoderm during gastrulation in *Drosophila* (Leptin and Grunewald, 1990; Leptin, 1995) and ingression of the endoderm during gastrulation in *Caenorhabditis elegans* (Roh-Johnson et al., 2012). In both cases, pulsatile myosin contractions cause the apical surface to contract periodically, leading to a coordinated inward movement of the cell–cell junctions (Martin et al., 2009; Martin et al., 2010; Roh-Johnson et al., 2012). This change in cell shape causes the tissue to fold.

A similar role for apical constriction has been observed for branching morphogenesis of the airways of the avian lung (Kim et al., 2013). In the embryonic chicken, new buds form sequentially in a cranial-to-caudal direction along the dorsal surface of each primary bronchus via a process known as monopodial (or lateral) budding (Gleghorn et al., 2012).

As each new bud forms, the cross-sectional area of the primary bronchus transforms from circular shape to lemniscate (figure-eight shaped) (Kim et al., 2013). Quantitative morphometric analysis of time-lapse imaging revealed that the airway epithelium of the primary bronchus undergoes apical constriction along both the dorsal and ventral surfaces as new buds form, and both experimental and computational studies showed that this apical constriction was necessary and sufficient to induce the cylindrical tube to fold locally into a bud (Kim et al., 2013). Active mechanical forces thus cause a change in the shape of a subpopulation of cells, inducing morphogenetic folding.

Actomyosin contractions have also been found to be important for driving the early morphogenetic events in the development of the vertebrate brain. As with the heart, the early embryonic brain is initially a relatively straight tube comprised of neuroepithelium. The brain tube then bends and swells to form three primary vesicles (corresponding to the forebrain, midbrain, and hindbrain), and the hindbrain bulges into rhombomeres (Goodrum and Jacobson, 1981; Lowery and Sive, 2009). Regulated contractions of the actomyosin cytoskeleton occur on the basal surface of the neuroepithelium to form the boundary between the midbrain and hindbrain in zebrafish (Gutzman et al., 2008). In contrast, contraction of the apical surface of the neuroepithelium appears to regulate formation of the boundaries and rhombomeres in the brain tube of chicken embryos (Filas et al., 2012). How these local contractions are regulated in the development of any tissue remains unclear, but it will be interesting to determine any similarities between organs as disparate as the brain and the lung.

12.4 Long-Range Morphogenetic Signals: Mechanics, Morphogens, and Electrical Signaling

Tissue morphogenesis is a physical process that results from the changes in shapes of multicellular structures as a function of time. The regulation of this process is undoubtedly complex, but it is clear that the mechanical deformations are both a consequence and a cause of morphogenesis (Ingber, 2005; Nelson et al., 2005). Mechanical forces provide information to individual cells, but are also transmitted over long distances within tissues and whole embryos. This long-range transmission of information occurs rapidly, at the rate of elastic deformation waves (Beloussov et al., 1994), providing an information transfer–biological response system that can operate at the quick rates of embryonic development. Long-range signaling is not limited to mechanical forces, however. Morphogen gradients and electrical signaling have both been found to provide long-range signals that instruct positional cues in developing tissues. It is likely that these integrate with long-range signals from mechanical stresses in the final morphogenesis of tissues.

12.4.1 Morphogen Gradients

Morphogens are biomolecules that are produced or stored in specific locations within an embryo or tissue (Turing, 1952). Transport from their source location causes the formation of concentration gradients in space, such that some populations of cells are exposed to higher concentrations than others. In the strictest definition of the term, the local concentration of the morphogen determines the response of the cells (Wolpert, 1969; Crick, 1970), which depends on intracellular signaling networks and dynamic changes in the activity of transcription factors (Kicheva et al., 2012). Morphogen-mediated patterning has been well described for the morphogenesis of the early embryo, mammary gland, and lung.

Patterning of the early *Drosophila* embryo is driven in part by diffusion of morphogens including bicoid and dorsal (Grimm et al., 2010). In the earliest stages, this embryo is a syncytium—a single cytoplasm containing several to hundreds of nuclei, depending on the stage of development. Because of the absence of plasma membranes separating the nuclei, any molecule in principle can act as a morphogen in the early *Drosophila* embryo, including those that are normally cytoplasmic or nuclear. (For all other situations, morphogens are limited to extracellular molecules.) The well-studied morphogen, bicoid, is a transcription factor deposited as mRNA at one end of the fertilized egg by cells from its mother (Driever and Nusslein-Volhard, 1988; St Johnston et al., 1989). The protein translated from this mRNA diffuses across the length of the embryo, forming a concentration gradient; the concentration of bicoid in the nucleus of each cell along the length of the embryo defines the gene expression pattern of that cell and the resulting cell fate (Driever and Nusslein-Volhard, 1989). Investigations into the generation of the bicoid concentration gradient have revealed that this morphogen forms its long-range signaling pattern via diffusion, but that the concentration profile is modified by local degradation of the bicoid protein (Gregor et al., 2005, 2007a,b).

Branching morphogenesis of the vertebrate lung is also thought to be driven by concentration gradients of morphogens. In this case, the master regulator of epithelial branching is considered to be fibroblast growth factor 10 (FGF10). FGF10 is expressed in a focal pattern in the submesothelial mesenchyme of the lung, such that new branches in the epithelium emerge at positions adjacent to those with high expression of FGF10 (Bellusci et al., 1997; Park et al., 1998; De Moerlooze et al., 2000; Weaver et al., 2000). The extent to which diffusion might play a role in establishing the concentration profile of FGF10 is unclear. Branching occurs recursively in the lung, and focal expression of FGF10 within the mesenchyme is presumed to as well, suggesting that this protein may have a rather short half-life within the mesenchymal environment of the developing lung. Also unclear is precisely how FGF10 exerts its effects on the epithelial cells to induce branching, since some models presume a chemotactic role (Weaver et al., 2000), whereas others presume a role in proliferation (Menshykau et al., 2012) or differentiation (Volckaert et al., 2013). Computational models have shown that if the concentration of FGF10 directs morphogenesis solely by altering cell proliferation, then this could also lead to the generation of a space-filling epithelial tree (Clement et al., 2010). The critical assumption here is that patterns of proliferation are equal to patterns of morphogenesis, and which has been found to be false for most tissues and organs (Beloussov and Dorfman, 1974).

In the two examples described above, the source of the morphogen is distinct from the cells upon which it acts: bicoid is deposited maternally, whereas FGF10 is secreted by mesenchymal cells to act on adjacent epithelial cells. However, morphogens can also act on their source cells. Branching morphogenesis of the mammary epithelium is a stochastic process involving bifurcations of the terminal ends and lateral branching, depending on the species (or strain of mouse). As it develops, the epithelium synthesizes and secretes transforming growth factor β (TGFβ), which acts as an inhibitor of branching (Silberstein and Daniel, 1987; Daniel et al., 1989; Robinson et al., 1991; Daniel and Robinson, 1992; Pierce et al., 1993; Soriano et al., 1995; Bergstraesser et al., 1996; Joseph et al., 1999; Ewan et al., 2002; Crowley et al., 2005; Serra and Crowley, 2005). Since TGFβ is secreted by the epithelium itself, its diffusion forms a concentration gradient emanating along and away from the epithelium. This results in a lower concentration of TGFβ at the ends of the developing ducts and adjacent to areas of convex curvature (Silberstein et al., 1992, 1990; Ewan et al., 2002), which are precisely the regions that induce new branches (Nelson et al., 2006; Pavlovich et al., 2011; Zhu and Nelson, 2013). Diffusion of morphogens thus plays a role in instructing the morphogenetic events that build the final architectures of tissues.

12.4.2 Electrical Gradients

While chemical gradients resulting from nonuniform distributions of chemical species have a demonstrated importance in long-range morphogen signaling, electrical gradients (voltages) arising from nonuniform distributions of charge are important in biological patterning as well. A role for bioelectricity in animal physiology was first observed in the eighteenth and nineteenth centuries with the study of nerve stimulation and injury potentials, but it was not until the early twentieth century that bioelectricity was first investigated in morphogenesis (Vanable, 1991). This early work would provide evidence for qualitative correlations between measured voltages and developmental characteristics, such as the degree of growth of embryonic salamanders (Burr and Hovland, 1937) and the final shape of developing gourds (Burr and Sinnott, 1944). Bioelectric patterning has since been investigated in many morphogenetic processes (Adams and Levin, 2013), including limb generation (Altizer et al., 2001), craniofacial development (Vandenberg et al., 2011), and asymmetric left–right patterning (Levin et al., 2002; Fukumoto et al., 2005).

As in classical physics and engineering systems, electrical potential differences within biological systems result from separation of charge. Membrane proteins, predominantly ion channels and transporters, drive this charge segregation across cellular domains and membranes by generating differential ion flow (Adams, 2006). In fact, K^+ flux across the plasma membrane is thought to be a primary contributor to the membrane potential of a cell. Here, Na^+–K^+ pumps import K^+ in exchange for Na^+ to establish intracellular potassium levels that are two orders of magnitude greater than extracellular levels. In a cell without a potential difference across the membrane, this concentration difference will promote the diffusion of K^+ through the membrane via K^+ leak channels. This results in a net movement of positive charge into the extracellular space. It can thus be imagined how a resting membrane potential on the order of -60 mV arises as a by-product of balancing the transport of charged species in an electric field with transport driven by concentration differences across the membrane.

On a larger scale, transepithelial potentials can form across layered epithelia (Adams, 2006). In the skin of the adult frog, a transmembrane potential of ~100 mV is sustained due to net inward flow of Na^+ into the body through Na^+–K^+ ATPases (McCaig et al., 2005). Here, tight junctions prevent leakage so as to maintain separately charged domains. If the epithelium is damaged, this transepithelial potential is short-circuited at the wound site, creating additional electric fields along the apical and basolateral sides of the epithelium, which are thought to drive epithelial cell migration toward the wound site (Zhao, 2009). This may occur through activation of the PI3K signaling pathway at the cathodic side of the cell, where leading edge protrusions form. Other signaling pathways, such as those involving epidermal growth factor receptors (EGFRs) and mitogen-activated protein kinase (MAPK), have also been implicated in mediating a cellular response to an electric field. Furthermore, cells can also sense changes in membrane voltage through voltage-sensitive transport channels (Stock et al., 2013), which alter activity through voltage-dependent conformational changes, in addition to voltage-mediated changes in gene expression, for example, through the cAMP response element-binding protein (CREB) pathway (Deisseroth et al., 2003). In this way, cells and tissues contain the machinery necessary to create and sense voltages.

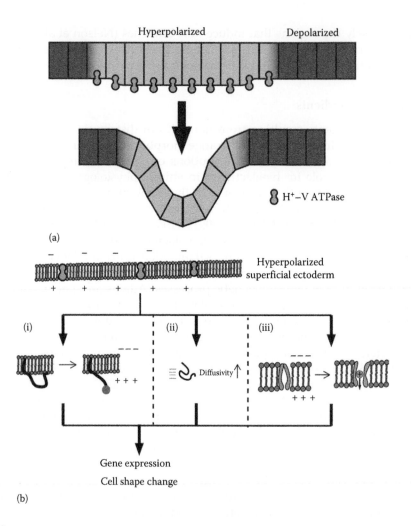

FIGURE 12.3
Membrane voltage patterns invagination and signaling during *Xenopus* neural tube closure. (a) Regions of hyperpolarization, specified by increased H⁺–V ATPase pumping activity, pattern locations that will subsequently invaginate. (b) A mechanism proposed by Vandenberg et al. (2011) suggests that the electrophysiological state of the superficial ectoderm cell could (i) (de)activate voltage-gated surface receptors, (ii) alter the diffusivity of molecular signaling components in the extracellular space, and/or (iii) de(activate) voltage-gated ion channels. As a result, this could influence gene expression and cell shape changes of that cell and, through long-range signaling, of neighboring deep ectoderm cells.

The cellular membrane potential is maintained over a long time frame (seconds to days), as opposed to an action potential operating on a millisecond timescale, and has gained attention as a key patterning mechanism in morphogenesis. In developing *Xenopus* embryos, waves of membrane hyperpolarization are observed beginning in stage 13 (Vandenberg et al., 2011). One particular wave occurs during neural tube closure, with streaks of hyperpolarized cells defining regions that will subsequently invaginate (Vandenberg et al., 2011) (Figure 12.3). The H⁺–V ATPase, which harnesses the energy of ATP hydrolysis to pump H⁺ against its electrochemical gradient, plays a central role in regulating membrane voltage and intracellular pH. Here, this transporter is required just prior to and during neural tube closure. In addition to neurulation, bioelectric patterning is also important in early asymmetric left–right patterning of vertebrate embryos. After formation of the primitive streak in Hamburger–Hamilton (HH) stage 3 chicken embryos, cells to the left of the streak become depolarized relative to those to the right of the streak (Levin et al., 2002). A voltage as large as 20 mV can be measured across the streak, although it diminishes through HH stage 4. In conjunction with this voltage, gap junctions and H^+–K^+ ATPase expression are required for correct localization of serotonin signaling and proper left–right asymmetry (Fukumoto et al., 2005). These findings suggest a possible mechanism by which differential influx of K^+, followed by K^+ loss via leakage channels in cells to the right of the streak (in chick) or ventral midline (in *Xenopus*), creates an electrical field across a cellular domain connected by gap junctions. According to this model, a low-molecular-weight determinant (perhaps serotonin) could then be transported asymmetrically via electrophoresis to initiate preferential gene expression on a particular side. The interested reader should be directed to Adams and Levin (2013) for additional examples of bioelectric patterning in morphogenesis.

12.4.3 Coupling Mechanical Signals with Morphogen Gradients and Electrical Gradients

Given the similarities in length scales upon which they act, perhaps it is not surprising that information from mechanical forces is increasingly recognized to be coupled with information from gradients in morphogens and electrical signaling. In the case of mammary epithelial branching morphogenesis, new branch sites emerge at positions that are dictated by both high mechanical stresses resulting from cellular contraction and low concentrations of TGFβ resulting from diffusion (Nelson et al., 2006; Gjorevski and Nelson, 2010). The situation is likely to be more complex than a simple Boolean logic AND gate, however, since TGFβ itself can be activated from its latent form by mechanical stress transmitted through integrins by actomyosin contraction (Annes et al., 2004; Jenkins et al., 2006). It remains unclear precisely how these two signals—mechanical and morphogen—are integrated by the tissue during morphogenesis.

There appears to be a similar coupling between mechanical forces and electrical signaling in a number of systems. During establishment of left–right asymmetry, myosin-driven leftward migration generates an asymmetric distribution of cells that secrete sonic hedgehog (SHH) and FGF8 around the node, which is induced by changes in bioelectrical activity regulated by a membrane H^+/K^+ ATPase. Mechanical loading of bone produces gradients in mechanical strains important for its morphogenesis and remodeling; these gradients in strain have been revealed to induce electrical potentials across the morphogenetic tissue (Beck et al., 2002). It will be interesting to determine the extent to which mechanical, chemical, and electrical gradients separately and synergistically regulate tissue morphogenesis.

12.5 Conclusions

One of the greatest mysteries of science is how multicellular organisms achieve their final forms. Although genes, and gene regulatory networks, clearly play a fundamental role in specifying the phenotypes of cells during morphogenesis, the products of these genes must necessarily be translated into mechanical and physical changes within and between cells to induce tissues to change their shapes. Also clear, however, is that cells and tissues respond to physical forces, which play major roles in the morphogenesis of vertebrate organs in particular. How much of this information is truly passive? To what extent can mechanically induced deformations be explained by the effects of forces acting on viscoelastic materials? How much of this information is active, requiring activation or repression of gene expression? The answers to these questions will shape our understanding of the living world and define whether tissue morphogenesis is the result of emergence or preformationism (Levin, 2012).

References

Abe, K. and Takeichi, M. 2008. EPLIN mediates linkage of the cadherin catenin complex to F-actin and stabilizes the circumferential actin belt. *Proc. Natl. Acad. Sci. USA*, 105, 13–19.

AbouAlaiwi, W. A., Takahashi, M., Mell, B. R., Jones, T. J., Ratnam, S., Kolb, R. J., and Nauli, S. M. 2009. Ciliary polycystin-2 is a mechanosensitive calcium channel involved in nitric oxide signaling cascades. *Circ. Res.*, 104, 860–869.

Adams, D. S. and Levin, M. 2006. Strategies and techniques for investigation of biophysical signals in patterning. In: Whitman, M. and Sater, A. K. (eds.) *Analysis of Growth Factor Signaling in Embryos*, CRC Press, Boca Raton, FL.

Adams, D. S. and Levin, M. 2013. Endogenous voltage gradients as mediators of cell-cell communication: Strategies for investigating bioelectrical signals during pattern formation. *Cell Tissue Res.*, 352, 95–122.

Alcorn, D., Adamson, T. M., Lambert, T. F., Maloney, J. E., Ritchie, B. C., and Robinson, P. M. 1977. Morphological effects of chronic tracheal ligation and drainage in the fetal lamb lung. *J. Anat.*, 123, 649–660.

Aleksandrova, A., Czirok, A., Szabo, A., Filla, M. B., Hossain, M. J., Whelan, P. F., Lansford, R., and Rongish, B. J. 2012. Convective tissue movements play a major role in avian endocardial morphogenesis. *Dev. Biol.*, 363, 348–361.

Altizer, A. M., Moriarty, L. J., Bell, S. M., Schreiner, C. M., Scott, W. J., and Borgens, R. B. 2001. Endogenous electric current is associated with normal development of the vertebrate limb. *Develop. Dyn.*, 221, 391–401.

Ando, J. and Yamamoto, K. 2009. Vascular mechanobiology: Endothelial cell responses to fluid shear stress. *Circ. J.*, 73, 1983–1992.

Annes, J. P., Chen, Y., Munger, J. S., and Rifkin, D. B. 2004. Integrin alphaVbeta6-mediated activation of latent TGF-beta requires the latent TGF-beta binding protein-1. *J. Cell Biol.*, 165, 723–734.

Anton, H., Harlepp, S., Ramspacher, C., Wu, D., Monduc, F., Bhat, S., Liebling, M., et al. 2013. Pulse propagation by a capacitive mechanism drives embryonic blood flow. *Development*, 140, 4426–4434.

Atkins, G. B. and Jain, M. K. 2007. Role of Kruppel-like transcription factors in endothelial biology. *Circ. Res.*, 100, 1686–1695.

Aukland, K. and Reed, R. K. 1993. Interstitial-lymphatic mechanisms in the control of extracellular fluid volume. *Physiol. Rev.*, 73, 1–78.

Auman, H. J., Coleman, H., Riley, H. E., Olale, F., Tsai, H. J., and Yelon, D. 2007. Functional modulation of cardiac form through regionally confined cell shape changes. *PLoS Biol.*, 5, e53.

Banjo, T., Grajcarek, J., Yoshino, D., Osada, H., Miyasaka, K. Y., Kida, Y. S., Ueki, Y., et al. 2013. Haemodynamically dependent valvulogenesis of zebrafish heart is mediated by flow-dependent expression of miR-21. *Nat. Commun.*, 4, 1978.

Bayly, P. V., Okamoto, R. J., Xu, G., Shi, Y., and Taber, L. A. 2013. A cortical folding model incorporating stress-dependent growth explains gyral wavelengths and stress patterns in the developing brain. *Phys. Biol.*, 10, 016005.

Beck, B. R., Qin, Y. X., Mcleod, K. J., and Otter, M. W. 2002. On the relationship between streaming potential and strain in an in vivo bone preparation. *Calcif. Tissue Int.*, 71, 335–343.

Belaguli, N. S., Zhou, W., Trinh, T. H., Majesky, M. W., and Schwartz, R. J. 1999. Dominant negative murine serum response factor: Alternative splicing within the activation domain inhibits transactivation of serum response factor binding targets. *Mol. Cell. Biol.*, 19, 4582–4591.

Bellusci, S., Grindley, J., Emoto, H., Itoh, N., and Hogan, B. L. 1997. Fibroblast growth factor 10 (FGF10) and branching morphogenesis in the embryonic mouse lung. *Development*, 124, 4867–4878.

Beloussov, L. V. and Dorfman, J. G. 1974. On the mechanics of growth and morphogenesis in hydroid polyps. *Am. Zool.*, 14, 719–734.

Beloussov, L. V., Saveliev, S. V., Naumidi, II and Novoselov, V. V. 1994. Mechanical stresses in embryonic tissues: Patterns, morphogenetic role, and involvement in regulatory feedback. *Int. Rev. Cytol.*, 150, 1–34.

Ben Amar, M. and Jia, F. 2013. Anisotropic growth shapes intestinal tissues during embryogenesis. *Proc. Natl. Acad. Sci. USA*, 110, 10525–10530.

Bergstraesser, L., Sherer, S., Panos, R., and Weitzman, S. 1996. Stimulation and inhibition of human mammary epithelial cell duct morphogenesis in vitro. *Proc. Assoc. Am. Phys.*, 108, 140–154.

Blewett, C. J., Zgleszewski, S. E., Chinoy, M. R., Krummel, T. M., and Cilley, R. E. 1996. Bronchial ligation enhances murine fetal lung development in whole-organ culture. *J. Pediatr. Surg.*, 31, 869–877.

Bonauer, A., Carmona, G., Iwasaki, M., Mione, M., Koyanagi, M., Fischer, A., Burchfield, J. et al. 2009. MicroRNA-92a controls angiogenesis and functional recovery of ischemic tissues in mice. *Science*, 324, 1710–1713.

Boon, R. A. and Horrevoets, A. J. 2009. Key transcriptional regulators of the vasoprotective effects of shear stress. *Hamostaseologie*, 29, 39–40, 41–43.

Burgess, D. R. 1975. Morphogenesis of intestinal villi. II. Mechanism of formation of previllous ridges. *J. Embryol. Exp. Morphol.*, 34, 723–740.

Burr, H. S. and Hovland, C. I. 1937. Bio-electric correlates of development in amblystoma. *Yale J. Biol. Med.*, 9, 540–549.

Burr, H. S. and Sinnott, E. W. 1944. Electrical correlates of form in cucurbit fruits. *Am. J. Bot.*, 31, 249–253.

Cachia, A., Paillere-Martinot, M. L., Galinowski, A., Januel, D., De Beaurepaire, R., Bellivier, F., Artiges, E. et al. 2008. Cortical folding abnormalities in schizophrenia patients with resistant auditory hallucinations. *Neuroimage*, 39, 927–935.

Cavey, M., Rauzi, M., Lenne, P. F., and Lecuit, T. 2008. A two-tiered mechanism for stabilization and immobilization of E-cadherin. *Nature*, 453, 751–756.

Chen, K. and Rajewsky, N. 2007. The evolution of gene regulation by transcription factors and microRNAs. *Nat. Rev. Genet.*, 8, 93–103.

Clement, R., Blanc, P., Mauroy, B., Sapin, V., and Douady, S. 2010. Shape self-regulation in early lung morphogenesis. *PLoS One*, 7, e36925.

Collet, A. J. and Des Biens, G. 1974. Fine structure of myogenesis and elastogenesis in the developing rat lung. *Anat. Rec.*, 179, 343–359.

Cooke, J. P., Rossitch, E., Jr., Andon, N. A., Loscalzo, J., and Dzau, V. J. 1991. Flow activates an endothelial potassium channel to release an endogenous nitrovasodilator. *J. Clin. Invest.*, 88, 1663–1671.

Coulombre, A. J. and Coulombre, J. L. 1958. Intestinal development. I. Morphogenesis of the villi and musculature. *J. Embryol. Exp. Morphol.*, 6, 403–411.

Crick, F. 1970. Diffusion in embryogenesis. *Nature*, 225, 420–422.

Crowley, M. R., Bowtell, D., and Serra, R. 2005. TGF-beta, c-Cbl, and PDGFR-alpha the in mammary stroma. *Dev. Biol.*, 279, 58–72.

Daniel, C. W. and Robinson, S. D. 1992. Regulation of mammary growth and function by TGF-beta. *Mol. Reprod. Dev.*, 32, 145–151.

Daniel, C. W., Silberstein, G. B., Van Horn, K., Strickland, P., and Robinson, S. 1989. TGF-beta 1-induced inhibition of mouse mammary ductal growth: Developmental specificity and characterization. *Dev. Biol.*, 135, 20–30.

De Moerlooze, L., Spencer-Dene, B., Revest, J. M., Hajihosseini, M., Rosewell, I., and Dickson, C. 2000. An important role for the IIIb isoform of fibroblast growth factor receptor 2 (FGFR2) in mesenchymal-epithelial signalling during mouse organogenesis. *Development*, 127, 483–492.

Deisseroth, K., Mermelstein, P. G., Xia, H., and Tsien, R. W. 2003. Signaling from synapse to nucleus: The logic behind the mechanisms. *Curr. Opin. Neurobiol.*, 13, 354–365.

Dekker, R. J., Van Soest, S., Fontijn, R. D., Salamanca, S., De Groot, P. G., Vanbavel, E., Pannekoek, H., and Horrevoets, A. J. 2002. Prolonged fluid shear stress induces a distinct set of endothelial cell genes, most specifically lung Kruppel-like factor (KLF2). *Blood*, 100, 1689–1698.

Dekker, R. J., Van Thienen, J. V., Rohlena, J., De Jager, S. C., Elderkamp, Y. W., Seppen, J., De Vries, C. J., et al. 2005. Endothelial KLF2 links local arterial shear stress levels to the expression of vascular tone-regulating genes. *Am. J. Pathol.*, 167, 609–618.

Driever, W. and Nusslein-Volhard, C. 1988. A gradient of bicoid protein in Drosophila embryos. *Cell*, 54, 83–93.

Driever, W. and Nusslein-Volhard, C. 1989. The bicoid protein is a positive regulator of hunchback transcription in the early Drosophila embryo. *Nature*, 337, 138–143.

Ewan, K. B., Shyamala, G., Ravani, S. A., Tang, Y., Akhurst, R., Wakefield, L., and Barcellos-Hoff, M. H. 2002. Latent transforming growth factor-beta activation in mammary gland: Regulation by ovarian hormones affects ductal and alveolar proliferation. *Am. J. Pathol.*, 160, 2081–2093.

Fewell, J. E., Hislop, A. A., Kitterman, J. A., and Johnson, P. 1983. Effect of tracheostomy on lung development in fetal lambs. *J. Appl. Physiol.*, 55, 1103–1108.

Filas, B. A., Oltean, A., Majidi, S., Bayly, P. V., Beebe, D. C., and Taber, L. A. 2012. Regional differences in actomyosin contraction shape the primary vesicles in the embryonic chicken brain. *Phys. Biol.*, 9, 066007.

Finegold, M. J., Katzew, H., Genieser, N. B., and Becker, M. H. 1971. Lung structure in thoracic dystrophy. *Am. J. Dis. Child*, 122, 153–159.

Forouhar, A. S., Liebling, M., Hickerson, A., Nasiraei-Moghaddam, A., Tsai, H. J., Hove, J. R., Fraser, S. E., Dickinson, M. E., and Gharib, M. 2006. The embryonic vertebrate heart tube is a dynamic suction pump. *Science*, 312, 751–753.

Freund, J. B., Goetz, J. G., Hill, K. L., and Vermot, J. 2012. Fluid flows and forces in development: Functions, features and biophysical principles. *Development*, 139, 1229–1245.

Fukumoto, T., Kema, I. P., and Levin, M. 2005. Serotonin signaling is a very early step in patterning of the left-right axis in chick and frog embryos. *Curr. Biol.*, 15, 794–803.

George, D. K., Cooney, T. P., Chiu, B. K., and Thurlbeck, W. M. 1987. Hypoplasia and immaturity of the terminal lung unit (acinus) in congenital diaphragmatic hernia. *Am. Rev. Respir. Dis.*, 136, 947–950.

Gjorevski, N. and Nelson, C. M. 2010. Endogenous patterns of mechanical stress are required for branching morphogenesis. *Integr. Biol.*, 2, 424–434.

Gjorevski, N. and Nelson, C. M. 2011. Integrated morphodynamic signalling of the mammary gland. *Nat. Rev. Mol. Cell Biol.*, 12, 581–593.

Gjorevski, N. and Nelson, C. M. 2012. Mapping of mechanical strains and stresses around quiescent engineered three-dimensional epithelial tissues. *Biophys. J.*, 103, 152–162.

Gleghorn, J. P., Kwak, J., Pavlovich, A. L., and Nelson, C. M. 2012. Inhibitory morphogens and monopodial branching of the embryonic chicken lung. *Dev. Dyn.*, 241, 852–862.

Go, Y. M., Park, H., Maland, M. C., Darley-Usmar, V. M., Stoyanov, B., Wetzker, R., and Jo, H. 1998. Phosphatidylinositol 3-kinase gamma mediates shear stress-dependent activation of JNK in endothelial cells. *Am. J. Physiol.*, 275, H1898–H1904.

Goldstein, J. D. and Reid, L. M. 1980. Pulmonary hypoplasia resulting from phrenic nerve agenesis and diaphragmatic amyoplasia. *J. Pediatr.*, 97, 282–287.

Goodrum, G. R. and Jacobson, A. G. 1981. Cephalic flexure formation in the chick embryo. *J. Exp. Zool.*, 216, 399–408.

Greenough, A. 2000. Factors adversely affecting lung growth. *Paediatr. Respir. Rev.*, 1, 314–320.

Gregor, T., Bialek, W., de Ruyter van Steveninck, R. R., Tank, D. W., and Wieschaus, E. F. 2005. Diffusion and scaling during early embryonic pattern formation. *Proc. Natl. Acad. Sci. USA*, 102, 18403–18407.

Gregor, T., Tank, D. W., Wieschaus, E. F., and Bialek, W. 2007a. Probing the limits to positional information. *Cell*, 130, 153–164.

Gregor, T., Wieschaus, E. F., Mcgregor, A. P., Bialek, W., and Tank, D. W. 2007b. Stability and nuclear dynamics of the bicoid morphogen gradient. *Cell*, 130, 141–152.

Grimm, O., Coppey, M., and Wieschaus, E. 2010. Modelling the Bicoid gradient. *Development*, 137, 2253–2264.

Groenendijk, B. C., Stekelenburg-De Vos, S., Vennemann, P., Wladimiroff, J. W., Nieuwstadt, F. T., Lindken, R., Westerweel, J., Hierck, B. P., Ursem, N. T., and Poelmann, R. E. 2008. The endothelin-1 pathway and the development of cardiovascular defects in the haemodynamically challenged chicken embryo. *J. Vasc. Res.*, 45, 54–68.

Gutzman, J. H., Graeden, E. G., Lowery, L. A., Holley, H. S., and Sive, H. 2008. Formation of the zebrafish midbrain-hindbrain boundary constriction requires laminin-dependent basal constriction. *Mech. Dev.*, 125, 974–983.

Hahn, C. and Schwartz, M. A. 2008. The role of cellular adaptation to mechanical forces in atherosclerosis. *Arterioscler. Thromb. Vasc. Biol.*, 28, 2101–2107.

Hardan, A. Y., Jou, R. J., Keshavan, M. S., Varma, R., and Minshew, N. J. 2004. Increased frontal cortical folding in autism: A preliminary MRI study. *Psychiatry Res.*, 131, 263–268.

Harding, R. and Hooper, S. B. 1996. Regulation of lung expansion and lung growth before birth. *J. Appl. Physiol.*, 81, 209–224.

Heisenberg, C. P. and Bellaiche, Y. 2013. Forces in tissue morphogenesis and patterning. *Cell*, 153, 948–962.

Hierck, B. P., Van der Heiden, K., Alkemade, F. E., Van de Pas, S., Van Thienen, J. V., Groenendijk, B. C., Bax, W. H., et al. 2008. Primary cilia sensitize endothelial cells for fluid shear stress. *Dev. Dyn.*, 237, 725–735.

Hogg, J. C. 2004. Pathophysiology of airflow limitation in chronic obstructive pulmonary disease. *Lancet*, 364, 709–721.

Hooper, S. B., Han, V. K., and Harding, R. 1993. Changes in lung expansion alter pulmonary DNA synthesis and IGF-II gene expression in fetal sheep. *Am. J. Physiol.*, 265, L403–L409.

Hove, J. R., Koster, R. W., Forouhar, A. S., Acevedo-Bolton, G., Fraser, S. E., and Gharib, M. 2003. Intracardiac fluid forces are an essential epigenetic factor for embryonic cardiogenesis. *Nature*, 421, 172–177.

Hrousis, C. A., Wiggs, B. J., Drazen, J. M., Parks, D. M., and Kamm, R. D. 2002. Mucosal folding in biologic vessels. *J. Biomech. Eng.*, 124, 334–341.

Hsieh, H. J., Li, N. Q., and Frangos, J. A. 1992. Shear-induced platelet-derived growth factor gene expression in human endothelial cells is mediated by protein kinase C. *J. Cell Physiol.*, 150, 552–558.

Ingber, D. E. 2005. Mechanical control of tissue growth: Function follows form. *Proc. Natl. Acad. Sci. USA*, 102, 11571–11572.

Isshiki, M., Ando, J., Korenaga, R., Kogo, H., Fujimoto, T., Fujita, T., and Kamiya, A. 1998. Endothelial $Ca2+$ waves preferentially originate at specific loci in caveolin-rich cell edges. *Proc. Natl. Acad. Sci. USA*, 95, 5009–5014.

Jackson, R. J. and Standart, N. 2007. How do microRNAs regulate gene expression? *Sci. STKE*, 2007, re1.

Jenkins, R. G., Su, X., Su, G., Scotton, C. J., Camerer, E., Laurent, G. J., Davis, G. E., Chambers, R. C., Matthay, M. A., and Sheppard, D. 2006. Ligation of protease-activated receptor 1 enhances alpha(v)beta6 integrin-dependent TGF-beta activation and promotes acute lung injury. *J. Clin. Invest.*, 116, 1606–1614.

Jesudason, E. C. 2007. Exploiting mechanical stimuli to rescue growth of the hypoplastic lung. *Pediatr. Surg. Int.*, 23, 827–836.

Jin, Z. G., Ueba, H., Tanimoto, T., Lungu, A. O., Frame, M. D., and Berk, B. C. 2003. Ligand-independent activation of vascular endothelial growth factor receptor 2 by fluid shear stress regulates activation of endothelial nitric oxide synthase. *Circ. Res.*, 93, 354–363.

Joseph, H., Gorska, A. E., Sohn, P., Moses, H. L., and Serra, R. 1999. Overexpression of a kinase-deficient transforming growth factor-beta type II receptor in mouse mammary stroma results in increased epithelial branching. *Mol. Biol. Cell*, 10, 1221–1234.

Kairaitis, K. 2012. Pharyngeal wall fold influences on the collapsibility of the pharynx. *Med. Hypotheses*, 79, 372–376.

Kemp, P. R. and Metcalfe, J. C. 2000. Four isoforms of serum response factor that increase or inhibit smooth-muscle-specific promoter activity. *Biochem. J.*, 345 Pt 3, 445–451.

Keramidaris, E., Hooper, S. B., and Harding, R. 1996. Effect of gestational age on the increase in fetal lung growth following tracheal obstruction. *Exp. Lung Res.*, 22, 283–298.

Kicheva, A., Cohen, M., and Briscoe, J. 2012. Developmental pattern formation: Insights from physics and biology. *Science*, 338, 210–212.

Kim, H. Y., Varner, V. D., and Nelson, C. M. 2013. Apical constriction initiates new bud formation during monopodial branching of the embryonic chicken lung. *Development*, 140, 3146–3155.

Kitano, Y., Davies, P., Von Allmen, D., Adzick, N. S., and Flake, A. W. 1999. Fetal tracheal occlusion in the rat model of nitrofen-induced congenital diaphragmatic hernia. *J. Appl. Physiol.*, 87, 769–775.

Kitano, Y., Yang, E. Y., Von Allmen, D., Quinn, T. M., Adzick, N. S., and Flake, A. W. 1998. Tracheal occlusion in the fetal rat: A new experimental model for the study of accelerated lung growth. *J. Pediatr. Surg.*, 33, 1741–1744.

Lee, H. J. and Koh, G. Y. 2003. Shear stress activates Tie2 receptor tyrosine kinase in human endothelial cells. *Biochem. Biophys. Res. Commun.*, 304, 399–404.

Lee, J., Moghadam, M. E., Kung, E., Cao, H., Beebe, T., Miller, Y., Roman, B. L., et al. 2013. Moving domain computational fluid dynamics to interface with an embryonic model of cardiac morphogenesis. *PLoS One*, 8, e72924.

Lee, J. S., Yu, Q., Shin, J. T., Sebzda, E., Bertozzi, C., Chen, M., Mericko, P. et al. 2006. Klf2 is an essential regulator of vascular hemodynamic forces in vivo. *Dev. Cell*, 11, 845–857.

Leptin, M. 1995. Drosophila gastrulation: From pattern formation to morphogenesis. *Annu. Rev. Cell Dev. Biol.*, 11, 189–212.

Leptin, M. and Grunewald, B. 1990. Cell shape changes during gastrulation in Drosophila. *Development*, 110, 73–84.

Levin, M. 2012. Morphogenetic fields in embryogenesis, regeneration, and cancer: Non-local control of complex patterning. *Biosystems*, 109, 243–261.

Levin, M., Thorlin, T., Robinson, K. R., Nogi, T., and Mercola, M. 2002. Asymmetries in H^+/K^+-ATPase and cell membrane potentials comprise a very early step in left-right patterning. *Cell*, 111, 77–89.

Liggins, G. C. and Kitterman, J. A. 1981. Development of the fetal lung. *Ciba Found. Symp.*, 86, 308–330.

Liu, M., Liu, J., Buch, S., Tanswell, A. K., and Post, M. 1995a. Antisense oligonucleotides for PDGF-B and its receptor inhibit mechanical strain-induced fetal lung cell growth. *Am. J. Physiol.*, 269, L178–L184.

Liu, M., Skinner, S. J., Xu, J., Han, R. N., Tanswell, A. K., and Post, M. 1992. Stimulation of fetal rat lung cell proliferation in vitro by mechanical stretch. *Am. J. Physiol.*, 263, L376–L383.

Liu, M., Xu, J., Liu, J., Kraw, M. E., Tanswell, A. K., and Post, M. 1995b. Mechanical strain-enhanced fetal lung cell proliferation is mediated by phospholipase C and D and protein kinase C. *Am. J. Physiol.*, 268, L729–L738.

Liu, M., Xu, J., Tanswell, A. K., and Post, M. 1994. Inhibition of mechanical strain-induced fetal rat lung cell proliferation by gadolinium, a stretch-activated channel blocker. *J. Cell Physiol.*, 161, 501–507.

Lopez-Sanchez, C. and Garcia-Martinez, V. 2011. Molecular determinants of cardiac specification. *Cardiovasc. Res.*, 91, 185–195.

Lowery, L. A. and Sive, H. 2009. Totally tubular: The mystery behind function and origin of the brain ventricular system. *BioEssays*, 31, 446–458.

Luckhoff, A. and Busse, R. 1990. Activators of potassium channels enhance calcium influx into endothelial cells as a consequence of potassium currents. *Naunyn. Schmiedebergs Arch. Pharmacol.*, 342, 94–99.

Mammoto, T., Mammoto, A., and Ingber, D. E. 2013. Mechanobiology and developmental control. *Annu. Rev. Cell Dev. Biol.*, 29, 27–61.

Martin, A. C., Gelbart, M., Fernandez-Gonzalez, R., Kaschube, M., and Wieschaus, E. F. 2010. Integration of contractile forces during tissue invagination. *J. Cell Biol.*, 188, 735–749.

Martin, A. C., Kaschube, M., and Wieschaus, E. F. 2009. Pulsed contractions of an actin-myosin network drive apical constriction. *Nature*, 457, 495–499.

McCaig, C. D., Rajnicek, A. M., Song, B., and Zhao, M. 2005. Controlling cell behavior electrically: Current views and future potential. *Physiol. Rev.*, 85, 943–978.

Menshykau, D., Kraemer, C., and Iber, D. 2012. Branch mode selection during early lung development. *PLoS Comput. Biol.*, 8, e1002377.

Miquerol, L. and Kelly, R. G. 2013. Organogenesis of the vertebrate heart. *Wiley Interdiscip. Rev. Dev. Biol.*, 2, 17–29.

Moessinger, A. C., Harding, R., Adamson, T. M., Singh, M., and Kiu, G. T. 1990. Role of lung fluid volume in growth and maturation of the fetal sheep lung. *J. Clin. Invest.*, 86, 1270–1277.

Mohamed, J. S., Lopez, M. A., and Boriek, A. M. 2010. Mechanical stretch up-regulates microRNA-26a and induces human airway smooth muscle hypertrophy by suppressing glycogen synthase kinase-3beta. *J. Biol. Chem.*, 285, 29336–29347.

Molnar, Z. and Clowry, G. 2012. Cerebral cortical development in rodents and primates. *Prog. Brain Res.*, 195, 45–70.

Nardo, L., Hooper, S. B., and Harding, R. 1995. Lung hypoplasia can be reversed by short-term obstruction of the trachea in fetal sheep. *Pediatr. Res.*, 38, 690–696.

Nardo, L., Hooper, S. B., and Harding, R. 1998. Stimulation of lung growth by tracheal obstruction in fetal sheep: Relation to luminal pressure and lung liquid volume. *Pediatr. Res.*, 43, 184–190.

Nauli, S. M., Alenghat, F. J., Luo, Y., Williams, E., Vassilev, P., Li, X., Elia, A. E. et al. 2003. Polycystins 1 and 2 mediate mechanosensation in the primary cilium of kidney cells. *Nat. Genet.*, 33, 129–137.

Nelson, C. M. 2013. Forces in epithelial origami. *Dev. Cell*, 26, 554–556.

Nelson, C. M. and Gleghorn, J. P. 2012. Sculpting organs: Mechanical regulation of tissue development. *Annu. Rev. Biomed. Eng.*, 14, 129–154.

Nelson, C. M., Jean, R. P., Tan, J. L., Liu, W. F., Sniadecki, N. J., Spector, A. A., and Chen, C. S. 2005. Emergent patterns of growth controlled by multicellular form and mechanics. *Proc. Natl. Acad. Sci. USA*, 102, 11594–11599.

Nelson, C. M., Vanduijn, M. M., Inman, J. L., Fletcher, D. A., and Bissell, M. J. 2006. Tissue geometry determines sites of mammary branching morphogenesis in organotypic cultures. *Science*, 314, 298–300.

Nordahl, C. W., Dierker, D., Mostafavi, I., Schumann, C. M., Rivera, S. M., Amaral, D. G., and Van Essen, D. C. 2007. Cortical folding abnormalities in autism revealed by surface-based morphometry. *J. Neurosci.*, 27, 11725–11735.

Papadaki, G. 2008. Buckling of thick cylindrical shells under external pressure: A new analytical expression for the critical load and comparison with elasticity solutions. *Int. J. Solids Struct.*, 45, 5308–5321.

Papastavrou, A., Steinmann, P., and Kuhl, E. 2013. On the mechanics of continua with boundary energies and growing surfaces. *J. Mech. Phys. Solids*, 61, 1446–1463.

Park, W. Y., Miranda, B., Lebeche, D., Hashimoto, G., and Cardoso, W. V. 1998. FGF-10 is a chemotactic factor for distal epithelial buds during lung development. *Dev. Biol.*, 201, 125–134.

Parmar, K. M., Larman, H. B., Dai, G., Zhang, Y., Wang, E. T., Moorthy, S. N., Kratz, J. R., et al. 2006. Integration of flow-dependent endothelial phenotypes by Kruppel-like factor 2. *J. Clin. Invest.*, 116, 49–58.

Pavlovich, A. L., Boghaert, E., and Nelson, C. M. 2011. Mammary branch initiation and extension are inhibited by separate pathways downstream of TGFbeta in culture. *Exp. Cell Res.*, 317, 1872–1884.

Pavone, L., Rizzo, R., and Dobyns, W. B. 1993. Clinical manifestations and evaluation of isolated lissencephaly. *Childs Nerv. Syst.*, 9, 387–390.

Pierce, D. F., Jr., Johnson, M. D., Matsui, Y., Robinson, S. D., Gold, L. I., Purchio, A. F., Daniel, C. W., Hogan, B. L., and Moses, H. L. 1993. Inhibition of mammary duct development but not alveolar outgrowth during pregnancy in transgenic mice expressing active TGF-beta 1. *Genes Dev.*, 7, 2308–2317.

Pipes, G. C., Creemers, E. E., and Olson, E. N. 2006. The myocardin family of transcriptional coactivators: Versatile regulators of cell growth, migration, and myogenesis. *Genes Dev.*, 20, 1545–1556.

Planas-Paz, L. and Lammert, E. 2013. Mechanical forces in lymphatic vascular development and disease. *Cell. Mol. Life Sci.*, 70, 4341–4354.

Planas-Paz, L., Strilic, B., Goedecke, A., Breier, G., Fassler, R., and Lammert, E. 2012. Mechanoinduction of lymph vessel expansion. *EMBO J.*, 31, 788–804.

Raghavan, R., Lawton, W., Ranjan, S. R., and Viswanathan, R. R. 1997. A continuum mechanics-based model for cortical growth. *J. Theor. Biol.*, 187, 285–296.

Rauzi, M. and Lenne, P. F. 2011. Cortical forces in cell shape changes and tissue morphogenesis. *Curr. Top. Dev. Biol.*, 95, 93–144.

Richman, D. P., Stewart, R. M., Hutchinson, J. W., and Caviness, V. S. 1975. Mechanical model of brain convolutional development. *Science*, 189, 18–21.

Robinson, S. D., Silberstein, G. B., Roberts, A. B., Flanders, K. C., and Daniel, C. W. 1991. Regulated expression and growth inhibitory effects of transforming growth factor-beta isoforms in mouse mammary gland development. *Development*, 113, 867–878.

Roh-Johnson, M., Shemer, G., Higgins, C. D., Mcclellan, J. H., Werts, A. D., Tulu, U. S., Gao, L., Betzig, E., Kiehart, D. P., and Goldstein, B. 2012. Triggering a cell shape change by exploiting preexisting actomyosin contractions. *Science*, 335, 1232–1235.

Sallet, P. C., Elkis, H., Alves, T. M., Oliveira, J. R., Sassi, E., Campi de Castro, C., Busatto, G. F., and Gattaz, W. F. 2003. Reduced cortical folding in schizophrenia: An MRI morphometric study. *Am. J. Psychiatry*, 160, 1606–1613.

Sawyer, J. M., Harrell, J. R., Shemer, G., Sullivan-Brown, J., Roh-Johnson, M., and Goldstein, B. 2010. Apical constriction: A cell shape change that can drive morphogenesis. *Dev. Biol.*, 341, 5–19.

Schiller, H. B. and Fassler, R. 2013. Mechanosensitivity and compositional dynamics of cell-matrix adhesions. *EMBO Rep.*, 14, 509–519.

Schittny, J. C., Miserocchi, G., and Sparrow, M. P. 2000. Spontaneous peristaltic airway contractions propel lung liquid through the bronchial tree of intact and fetal lung explants. *Am. J. Respir. Cell Mol. Biol.*, 23, 11–18.

Schmid-Schonbein, G. W. 1990. Microlymphatics and lymph flow. *Physiol. Rev.*, 70, 987–1028.

Sen, S., Engler, A. J., and Discher, D. E. 2009. Matrix strains induced by cells: Computing how far cells can feel. *Cell Mol. Bioeng.*, 2, 39–48.

Serra, R. and Crowley, M. R. 2005. Mouse models of transforming growth factor beta impact in breast development and cancer. *Endocr. Relat. Cancer*, 12, 749–760.

Shay-Salit, A., Shushy, M., Wolfovitz, E., Yahav, H., Breviario, F., Dejana, E., and Resnick, N. 2002. VEGF receptor 2 and the adherens junction as a mechanical transducer in vascular endothelial cells. *Proc. Natl. Acad. Sci. USA*, 99, 9462–9467.

Shyer, A. E., Tallinen, T., Nerurkar, N. L., Wei, Z., Gil, E. S., Kaplan, D. L., Tabin, C. J., and Mahadevan, L. 2013. Villification: How the gut gets its villi. *Science*, 342(6155), 203–204.

Silberstein, G. B. and Daniel, C. W. 1987. Reversible inhibition of mammary gland growth by transforming growth factor-beta. *Science*, 237, 291–293.

Silberstein, G. B., Flanders, K. C., Roberts, A. B., and Daniel, C. W. 1992. Regulation of mammary morphogenesis: Evidence for extracellular matrix-mediated inhibition of ductal budding by transforming growth factor-beta 1. *Dev. Biol.*, 152, 354–362.

Silberstein, G. B., Strickland, P., Coleman, S., and Daniel, C. W. 1990. Epithelium-dependent extracellular matrix synthesis in transforming growth factor-beta 1-growth-inhibited mouse mammary gland. *J. Cell Biol.*, 110, 2209–2219.

Slough, J., Cooney, L., and Brueckner, M. 2008. Monocilia in the embryonic mouse heart suggest a direct role for cilia in cardiac morphogenesis. *Dev. Dyn.*, 237, 2304–2314.

Soriano, J. V., Pepper, M. S., Nakamura, T., Orci, L., and Montesano, R. 1995. Hepatocyte growth factor stimulates extensive development of branching duct-like structures by cloned mammary gland epithelial cells. *J. Cell Sci.*, 108 (Pt 2), 413–430.

Sparrow, M. P., Weichselbaum, M., and McCray, P. B. 1999. Development of the innervation and airway smooth muscle in human fetal lung. *Am. J. Respir. Cell Mol. Biol.*, 20, 550–560.

St Johnston, D., Driever, W., Berleth, T., Richstein, S., and Nusslein-Volhard, C. 1989. Multiple steps in the localization of bicoid RNA to the anterior pole of the *Drosophila* oocyte. *Development*, 107 Suppl, 13–19.

Stock, C., Ludwig, F. T., Hanley, P. J., and Schwab, A. 2013. Roles of ion transport in control of cell motility. *Compr. Physiol.*, 3, 59–119.

Surapisitchat, J., Hoefen, R. J., Pi, X., Yoshizumi, M., Yan, C., and Berk, B. C. 2001. Fluid shear stress inhibits TNF-alpha activation of JNK but not ERK1/2 or p38 in human umbilical vein endothelial cells: Inhibitory crosstalk among MAPK family members. *Proc. Natl. Acad. Sci. USA*, 98, 6476–6481.

Taber, L. A., Voronov, D. A., and Ramasubramanian, A. 2010. The role of mechanical forces in the torsional component of cardiac looping. *Ann. N. Y. Acad. Sci.*, 1188, 103–110.

Tseng, H., Peterson, T. E., and Berk, B. C. 1995. Fluid shear stress stimulates mitogen-activated protein kinase in endothelial cells. *Circ. Res.*, 77, 869–878.

Turing, A. 1952. The chemical basis of morphogenesis. *Philos. Trans. R. Soc. Lond. (B)*, 237, 37–72.

Van Essen, D. C. and Maunsell, J. H. 1980. Two-dimensional maps of the cerebral cortex. *J. Comp. Neurol.*, 191, 255–281.

van Rooij, E., Sutherland, L. B., Liu, N., Williams, A. H., Mcanally, J., Gerard, R. D., Richardson, J. A., and Olson, E. N. 2006. A signature pattern of stress-responsive microRNAs that can evoke cardiac hypertrophy and heart failure. *Proc. Natl. Acad. Sci. USA*, 103, 18255–18260.

van Thienen, J. V., Fledderus, J. O., Dekker, R. J., Rohlena, J., van Ijzendoorn, G. A., Kootstra, N. A., Pannekoek, H., and Horrevoets, A. J. 2006. Shear stress sustains atheroprotective endothelial KLF2 expression more potently than statins through mRNA stabilization. *Cardiovasc. Res.*, 72, 231–240.

Vanable, J. W. 1991. A history of bioelectricity in development and regeneration. In: Dinsmore, C. E. (ed.) *A History of Regeneration Research: Milestones in the Evolution of a Science*, Cambridge, U.K.: Cambridge University Press.

Vandenberg, L. N., Morrie, R. D., and Adams, D. S. 2011. V-ATPase-dependent ectodermal voltage and pH regionalization are required for craniofacial morphogenesis. *Dev. Dyn.*, 240, 1889–1904.

Vermot, J., Forouhar, A. S., Liebling, M., Wu, D., Plummer, D., Gharib, M., and Fraser, S. E. 2009. Reversing blood flows act through klf2a to ensure normal valvulogenesis in the developing heart. *PLoS Biol.*, 7, e1000246.

Villadsen, G. E., Storkholm, J., Zachariae, H., Hendel, L., Bendtsen, F., and Gregersen, H. 2001. Oesophageal pressure-cross-sectional area distributions and secondary peristalsis in relation to subclassification of systemic sclerosis. *Neurogastroenterol. Motil.*, 13, 199–210.

Vilos, G. A. and Liggins, G. C. 1982. Intrathoracic pressures in fetal sheep. *J. Dev. Physiol.*, 4, 247–256.

Volckaert, T., Campbell, A., Dill, E., Li, C., Minoo, P., and De Langhe, S. 2013. Localized Fgf10 expression is not required for lung branching morphogenesis but prevents differentiation of epithelial progenitors. *Development*, 140, 3731–3742.

Wang, A. S. D. and Ertepinar, A. 1972. Stability and vibrations of elastic thick-walled cylindrical and spherical shells subjected to pressure. *Int. J. Non-linear Mech.*, 7, 539–555.

Weaver, M., Dunn, N. R., and Hogan, B. L. 2000. Bmp4 and Fgf10 play opposing roles during lung bud morphogenesis. *Development*, 127, 2695–2704.

Wigglesworth, J. S., Desai, R., and Hislop, A. A. 1987. Fetal lung growth in congenital laryngeal atresia. *Pediatr. Pathol.*, 7, 515–525.

Wiggs, B. R., Hrousis, C. A., Drazen, J. M., and Kamm, R. D. 1997. On the mechanism of mucosal folding in normal and asthmatic airways. *J. Appl. Physiol. (1985)*, 83, 1814–1821.

Wolpert, L. 1969. Positional information and the spatial pattern of cellular differentiation. *J. Theor. Biol.*, 25, 1–47.

Wu, W., Xiao, H., Laguna-Fernandez, A., Villarreal, G., Jr., Wang, K. C., Geary, G. G., Zhang, Y. et al. 2011. Flow-dependent regulation of kruppel-like factor 2 is mediated by microRNA-92a. *Circulation*, 124, 633–641.

Xu, G., Knutsen, A. K., Dikranian, K., Kroenke, C. D., Bayly, P. V., and Taber, L. A. 2010. Axons pull on the brain, but tension does not drive cortical folding. *J. Biomech. Eng.*, 132, 071013.

Yang, W., Fung, T. C., Chian, K. S., and Chong, C. K. 2007. Instability of the two-layered thick-walled esophageal model under the external pressure and circular outer boundary condition. *J. Biomech.*, 40, 481–490.

Yang, Y., Beqaj, S., Kemp, P., Ariel, I., and Schuger, L. 2000. Stretch-induced alternative splicing of serum response factor promotes bronchial myogenesis and is defective in lung hypoplasia. *J. Clin. Invest.*, 106, 1321–1330.

Zhao, M. 2009. Electrical fields in wound healing—An overriding signal that directs cell migration. *Semin. Cell Develop. Biol.*, 20, 674–682.

Zhu, W. and Nelson, C. M. 2013. PI3K regulates branch initiation and extension of cultured mammary epithelia, via Akt and Rac1 respectively. *Dev. Biol.*, 379, 235–245.

13

Continuum Physics of Tumor Growth

Kristen L. Mills, Shiva Rudraraju, Ralf Kemkemer, and Krishna Garikipati

CONTENTS

This chapter reviews the mechanical aspects of tumor growth with the aim of providing a point of departure for new directions, as well as further studies along some of the established lines of work. Experimental studies are reviewed first, in order to set the stage for the rest of the chapter. This is followed by a description of modeling approaches, with a focus on the continuum aspects. The chapter ends with an outline of recent studies in our group on continuum mechanical models of growth that also account for some of the reaction–transport phenomena.

13.1 Background

The process of tumorigenesis is typically thought of as one governed by biochemistry. The roots of this view are not hard to trace: the description of cell biology in the past few decades has resided in molecular chemistry, and cancer itself is understood as a malfunction of the chemical pathways that control programmed cell death [31]. However, an alternate view has been gaining ground in at least some quarters of the cancer cell biology community: that physical forces also play a role [24,32]. If viewed as an extension of cell biophysics, this latter view is not very surprising and may even seem inevitable: the area of cell mechanics is somewhat well established, and physical forces are understood to play roles in many cell functions including adhesion, migration, mitosis, phagocytosis, and chromatin separation [16,25]. Additionally, the fact of cell contractility renders the question somewhat moot: cells generate forces, themselves, so they are likely to also modify their functions under forces. Cancer cells also would be expected to have a similar response (provided the pathology of disease does not shut off force sensitivity itself). This is largely true, and there are already studies on the regulation of chemical activity in cancer cell lines when subjected to forces [3,28].

A separate line of the experimental cancer biology literature has concerned itself with the response of the tumor mass itself to physical forces. These studies were perhaps a natural outgrowth of earlier ones that ignored physical forces. From studies that consider the alteration of tumor growth rates, through the effect on metastatic potential, to the shape of growing tumors, the effect of physical forces and of tumor deformation has received attention over nearly the past two decades. All of this work has been in vitro, to the best of our knowledge. The question of how important these effects are in determining the course of the disease in vivo, however, is a much harder question because of the difficulty of carrying out controlled experiments.

The experiments on mechanical effects on tumor growth have motivated a host of mathematical and computational studies, which we review later. Many of these have positioned themselves as extensions of mathematical models of the biochemistry of tumor growth. Both continuum partial differential and discrete models have been employed [5,6,18,23,26], with mechanical effects incorporated more prominently in the former. The key idea here has been that tumor growth needs cell production and that the newly created biomass occupies greater volume, which, if not accommodated, leads to forces of constraint. Much sophistication is possible from this standpoint, including a feedback to the biochemistry that drives cell growth and division [27]; passive and active constitutive models of mechanics, reviewed in Ambrosi et al. [1]; and even recent attempts to pose broader questions on the thermodynamics [23].

Here, we review what is known of some strands of the experimental literature, touch upon the main modeling approaches, and place some of our own work in context.

13.2 Experiments

Experiments on tumor growth have traditionally been carried out using agglomerates formed of cancer cells. As they grow, they attain shapes that traditionally have been described as spheroidal, leading to the term *tumor spheroids*.

13.2.1 Tumor Growth in the Absence of Mechanical Influences

13.2.1.1 Biochemical Control

A considerable body of work now exists on controlling the nutrient, oxygen, and pH environments of growing tumor spheroids [7–10]. The nutrient used has typically been glucose. These studies have focused on varying the glucose concentration of the culture medium; the oxygen content of the reaction chamber's atmosphere; and the pH of the medium, independently; and in combination with each other. These studies have revealed that, as may be expected, as glucose concentration increases/decreases, so does the rate of tumor cell division and therefore tumor growth rate. The same holds with oxygen concentration. The cell division rate also decreases with the concentration of H^+ ions, that is, in acidic environments. In the work of Casciari et al. [2], these dependencies have been used to extract an empirical fit for the cell number doubling time as a function of the glucose, oxygen, and H^+ ion concentrations. (The notion of a cell number doubling time itself emerges from the assumption of exponential growth, which is applicable if all viable cells also divide.) The experiments suggest that the uptake of glucose and oxygen follows Michaelis–Menten kinetics, starting at some initial rate and asymptoting to a maximum as glucose and oxygen concentrations increase. In contrast, the rate of glucose (oxygen) consumption varies inversely with oxygen (glucose) concentration, suggesting that deprivation of one essential substance forces the cells to become less efficient by increasing their consumption of the other substance to maintain their energy supply. In Section 13.4, we describe the functional form of these dependencies in greater detail, in the context of recent modeling studies in our group.

13.2.1.2 Tumor Metabolism

These experiments, when run long enough, ultimately demonstrate the development of a necrotic core, leaving only a thin rim of viable cells at the boundary of the spheroid. Necrosis develops as tumor growth is diffusion limited in spheroids. As the tumor grows, glucose and oxygen diffusion is insufficient to maintain the viability of cells in the spheroid core. These cells die and form the nectotic core. However, rather than leading to tumor death due to diffusion limitation, the resulting acidic environment may be related to the release of angiogenetic factors. In vivo tumors then develop vascularization and a new lease on viability [15,31]. The death of large numbers of cancer cells during this diffusion-limited growth phase leaves behind the surviving fitter cells, preselected for robustness to acidic environments, and with fewer competitors for glucose and oxygen. This effect is properly modeled by introducing distinct cell populations with different division rates that have sensitivity to the environmental conditions, such as H^+ ion concentration. The *fitter* cells can then develop a competitive advantage for resources (glucose and oxygen). This aspect has been incorporated via a cancer stem cell population in recent work [27]; details are in Section 13.4.2.

The metabolism driving tumor growth follows the glycolytic pathway, producing only 2 molecules of ATP for each molecule of glucose, rather than the oxidative phosphorylation pathway, which produces between 30 and 36 molecules of ATP [31]. Notably, however, the glycolytic pathway proceeds faster than the oxidative phosphorylation pathway. This phenomenon, known as the Warburg paradox [29], is somewhat understood as evidence of a switch to a less efficient, but faster, metabolic pathway that the cells adopt in order to fuel the rapidly growing tumor. It is considered a signature of tumor metabolism. Our group has made connections with this and other aspects of tumor metabolism in a quantitative context, which we review in Section 13.4.

13.2.2 Mechanical Influence on Tumor Growth

The force response of individual cells is now a mainstream topic in the biophysical literature [16,25,28,32], and as alluded to in Section 13.1, it is natural to find evidence of it in tumor growth. However, the regimes of response of cell adhesion, migration, and mitosis demonstrate a complexity that depends on the environment: whether a cell is attached to other cells by adherens junctions or to the extracellular matrix (ECM) by integrins or focal adhesions, and the transmission of force and deformation through the mass of cells. Cancer cells, by virtue of being transformed, are expected to demonstrate deviations from the response of untransformed cells. Partly for this reason, the literature has remained focused on the mechanical response of noncancerous cells. Among the few studies on cancer cells, those reviewed by Suresh [28] suggest that individual cancer cells with high metastatic potential are more compliant and less viscous in their response than untransformed ones. On the other hand, because of the higher ECM content of tumors, the cells in them also develop larger focal adhesions and more prominent and contractile cytoskeletons [24]. Because of the difficulty in teasing out the complex emergent behavior from these dynamics, experimental studies have focused on the mechanical response of entire tumor spheroids. We review a few studies later.

13.2.2.1 Compressive Stress Suppresses Tumor Growth

A significant and early study in this regard was presented by Helmlinger et al. [17]. In experiments on tumor spheroids of a colon adenocarcinoma cell line (LS174T) growing in chemically inert agarose gels, the authors found that tumor growth was suppressed under compressive stresses of the order of 45–120 mmHg. Over the period of ~30 days, tumors growing stress-free attained average diameters of 800 μm. However, tumors growing within a confined gel developed a compressive stress because of their growth-related expansion. Under this compressive stress, the grown diameters over the 30-day period were 100–400 μm depending on the magnitude of the constraining stress. Notably, this was not merely an elastic constraint in operation; when released from the compressive stress, the growth-suppressed tumors did not instantly spring to the larger diameter of the stress-free tumors. Their diameters remained significantly smaller upon release, and their growth rates increased only gradually. Tumors released from the compressive stress at day 30 grew to diameters of ~700 μm by day 45 and had a positive rate of growth at that time, in comparison with the stress-free growing tumors, which had plateaued out at ~900 μm by the same time. The empirical evidence suggested that the compressive stress had interfered with the biochemistry of growth. Interestingly, closer examination using cell viability and cell death assays revealed that death rates declined under the compressive stress, while proliferation rates remained about the same; however, it was suggested that a greater number of cells were driven into quiescence under the stress. The use of agarose gels was important because the cells form no adhesions to agarose (effectively, the ECM), allowing the mechanical response to be studied separately from any chemical influences. The authors did not, however, report on the means of tumor seeding, leaving open some questions on the effect of intercellular environment, and of competition for transported nutrients and oxygen. They also did not report the tumor spheroids' shapes (whether truly spheroidal, or more ellipsoidal), a question with some bearing on a detailed understanding of the inferred stress-induced growth suppression.

At the single-cell level, the biochemical effect of stress has been investigated by Chang et al. [3]. The authors found that flow-driven shear stress induced increased expressions of cyclin B1 and p21^{CIP1} and decreased expressions of cyclins A, D1, and E, cyclin-dependent protein kinases (Cdk)-1, Cdk2, Cdk4, and Cdk6, and p27^{KIP1} as well as a decrease in Cdk1 activity. While the specific stress states were different, it may be surmised that related effects lie behind tumor growth suppression described in the previous paragraph.

13.2.2.2 Other Effects of Stress on Tumor Growth

Other studies followed; notable among them was one by Koike et al. [19], on aggressive Dunning R3327 rat prostate carcinoma AT3.1 cells as well as on the less aggressive AT2.1 cell line. In this study, the authors found that, in general, compressive stress prevented cell escape from the cancer cell lines with greater metastatic potential, while having less influence on those with lower metastatic potential. Weaver et al. [30] and Suresh [28] have reported on single-cell studies that may be implicated in these tumor-scale observations.

More recently, Cheng et al. [4], motivated by Helmlinger, described the shape attained by tumors growing in gels. They found ellipsoidal shapes to occur in some cases, a phenomenon that they reasoned was due to unequal stresses along different axes of the ellipsoid. The shorter ellipsoidal axes, the authors suggested, experienced higher compressive stress, leading to greater suppression of growth. The stresses were estimated by tracking beads embedded in the gel and were correlated with levels of caspase, a cell death-related enzyme, in the tumor. They did not, however, speculate on the origin of the stress that caused growth suppression. This is an important question, because if the stress was externally applied, its spatially varying tensorial form would need to be such as to lead to a higher compressive component along the orientation of the minor axis of the observed tumor, going by the results of Helmlinger et al. [17]. The authors did not report how the chosen tumor was oriented relative to any purported loading of the system. If, on the other hand, stress in the tumor–gel system was caused due to suppression of the kinematic expansion of growing tumors, the stress fields are more complex, but do not in general support the authors' conclusions. This is because it can be shown that the compressive stress components would always be aligned such as to cause a suppression of growth along precisely the directions that the authors found the tumors to be elongated.

In our own work [21], we have found that ellipsoidal and spheroidal tumors arise under specific, repeatable conditions—see Figures 13.1 and 13.2. Of these, Figure 13.1 illustrates a distribution of tumors with different aspect ratios. Figure 13.2 illustrates the evolution of diameter and minor/major axes in spheroidal and ellipsoidal tumors. This holds for several cancer cell lines, including the LS174T cell line in agarose gels. The larger tumors (major axes ~1000 μm) show necrotic cores; however, the necrotic cores seem to have no

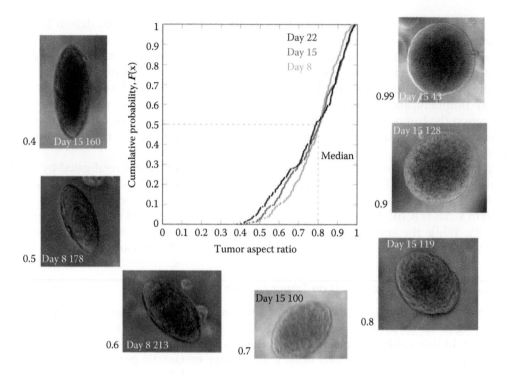

FIGURE 13.1
The distribution of aspect ratios of tumors grown in gels. The aspect ratios (≤1), age of the tumor, and tumor number are shown.

influence on the development of ellipsoidal shapes. The shape of growing tumors could have important implications for their metastatic potential and hence for prognosis. In other work [20], we have described methods for characterizing the mechanics of tumors, using techniques applicable to soft, rate-dependent materials.

13.2.2.3 Free Energy Flows during Tumor Growth

A relatively less-explored direction for tumor growth experiments is the characterization of the free energy rates associated with the constituent bio-chemo-mechanical processes. The underlying principle is that a tumor must make use of its ATP stores of energy to perform many functions: the division, growth, and migration of cells and the mechanical work of expansion against a constraining stress, to name just the few that are currently accessible by continuum models of tumor growth. There are many others, including the complex processes underlying cytoskeleton remodeling, adhesion, intracellular transport, and protein transcription, indeed, almost too many to enumerate here. As biophysical models improve in their sophistication and multiscale character, by incorporating many more hierarchical processes of the cell, it will gradually become possible to address more of them.

However, even with the preliminary state of the existing models, if the rates of the dominant processes are known, they can be compared to reveal whether there are certain processes that consume a significant proportion of the energy obtained from glycolysis. Such processes, which we define as energy intensive, in theory, may be targets for drug

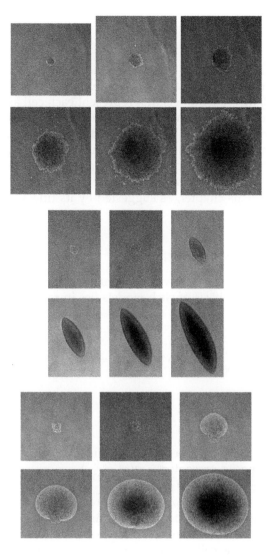

FIGURE 13.2
The growth of tumors as spheroids (top row, tumors growing in 0.3% agarose gel at day 0, 2, 4, 6, 8, 10) and ellipsoids (middle and bottom rows, tumors growing in 1% agarose gel at day 0, 2, 6, 8, 10, 12). The scale bar is 250 µm.

interventions, with the aim of starving the tumor of energy, *thereby preventing its uncontrolled growth*. While now in its infancy, this approach could hold promise. Such a program of study would require much experimental characterization: cell division, growth, and death rates, the mechanisms of cytoskeletal remodeling, intracellular transport, glucose and oxygen transport and reaction rates, cell migration velocities, and the mechanical properties of the tumor and gel. The remaining data either are standard quantities in the literature (e.g., the energy derived by consuming a single molecule of glucose via the glycolytic pathway) or can be computed (e.g., the actual glucose and oxygen consumption rates as dependent on developing cell concentrations, initial and boundary conditions, and supply). Our group has embarked on such a long-term program of study [23]. More details are forthcoming in Section 13.4.

13.3 Models

Tumor growth models can be broadly classified into two types: (1) continuum descriptions that employ partial differential equations of reaction–transport and mechanics over the tumor and (2) discrete models that represent each cell as either a node or compartment, and rules for cell division and motion [6,26]. Also included in the latter category are cellular automata models. These are models in which each cell is represented as an individual entity that interacts with others through, usually simple, rules that are typically empirical. While this level of cellular automata models is often lacking in physics, they are known to result in qualitatively correct collective behavior and are used in many fields of science. This chapter, as suggested by its title, will focus on continuum physics models.

13.3.1 Continuum Physics of Tumor Growth

A large class of models for tumor growth are actually directed at studying the effects of drugs on the tumor [18]. The basis of such models lies in reaction–diffusion equations of the following form:

$$\frac{\partial \rho^\alpha}{\partial t} = D^\alpha \nabla^2 \rho^\alpha + \pi^\alpha, \quad \alpha = 1, 2, \ldots, \text{ in } \Omega \tag{13.1}$$

where
   ρ^α is the concentration of a species
   D^α is its diffusivity
   π^α is the reaction term that acts as a source or a sink for species α, which can, however, depend on all the species present: $\pi^\alpha = \pi^\alpha(\rho^\alpha, \rho^\beta, \ldots)$
   Ω is the region of interest in $\mathbb{R}^3$ (3D space)

The cells themselves can be represented by extending this formulation to include spatio-temporally varying cell concentration as one of the species, for example, $\alpha = c$, denoting cells. In such a treatment, the cells can be either nonmigratory, $D^\alpha = 0$, or cell motion can be modeled as a random walk. The net cell production rate per unit volume is then π^c. It includes cell creation and death terms and depends on the other species, which could include nutrients, oxygen, and other soluble factors, including drugs and in some examples also includes ECM (collagen) production. The mathematical form of π^c is detailed in Equation 13.6. We note that while individual cells divide, grow, and die at timescales ranging from hours to days, the continuum description using a cell concentration averages this out to a single rate term, π^c. Also, while the biochemical reactions of glucose and oxygen consumption operate on timescales of less than a second, the concentration of these species varies on timescales of hours in culture.

13.3.1.1 Continuum Theory of Mixtures

The set of partial differential equations to solve for a typical problem of tumor growth including cells, ECM ($\alpha = m$), glucose ($\alpha = g$), oxygen ($\alpha = o$), and mechanics includes

$$\frac{\partial \rho^c}{\partial t} = -\nabla \cdot \boldsymbol{j}^c + \pi^c - \pi_d^c$$

$$\frac{\partial \rho^m}{\partial t} = \pi^m - \pi_d^m$$

$$\frac{\partial \rho^g}{\partial t} = -\nabla \cdot \boldsymbol{j}^g + \pi^g \qquad (13.2)$$

$$\frac{\partial \rho^o}{\partial t} = -\nabla \cdot \boldsymbol{j}^o + \pi^o$$

$$\nabla \cdot \sigma + \boldsymbol{f} = 0$$

where j^α is the flux of species α. The constitutive forms of these fluxes are detailed later. The last partial differential equation in (13.2) is for the quasistatic balance of linear momentum, since mechanics attains equilibrium at much shorter timescales than do the reaction–transport equations. The previous set of equations holds in the deformed configuration of the tumor, Ω, which implies that each ρ^α is the respective concentration defined as mass per unit deformed volume. Cell death is modeled by the sink term $-\pi_d^c$ and ECM degradation by $-\pi_d^m$. In Rudraraju et al. [27], we have modeled cell death by a sink term $-\pi_d^c$ that depends on glucose and oxygen concentration dropping below critical values as well as on the stress σ (stress-induced cell death).

The form of the previous equations can be systematically derived by starting with a theory of mixtures' representation of the tumor [13]: the equations of mass transport are written for each species without imposing conservation of mass, since reactions consume and produce chemical species, while cells and ECM are also produced, cells die and the ECM suffers degradation. Conservation of total mass, however, does hold, since ultimately, glucose and oxygen are converted into cells and ECM (neglecting other chemical and biological entities). This condition gives rise to a constraint on the source and sink terms when summed over all species. Likewise, the balance of linear momentum equation is written in its full form as the balance of linear momentum for each species that has nonnegligible momentum ($\alpha = c$, m, and then reduced to a constraint condition between the interspecies interaction forces, due to the conservation of total linear momentum in the absence of external forces. Since the momentum of the other species undergoing transport ($\alpha = g$, o) in this setting is negligible, their individual balances of linear momentum can be dropped from the formulation.

The same approach also is taken in writing the balance of energy of each species and the entropy inequality. This aspect of the thermodynamic treatment will not be outlined here, because it requires very detailed arguments and several stages of working through equations. The reader is directed to Garikipati et al. [13] for a full treatment.

13.3.1.2 Continuum Mechanics of Growth: The Elasto-Growth Decomposition

The partial differential equation-based treatment in (13.1) has been used by Jackson and Byrne [18] to study the regulation of tumor growth by drugs and matrix (collagen) encapsulation. However, the kinematics of growth was not considered. Its inclusion brings the treatment of tumor growth entirely within the domain of continuum mechanics, coupled with reaction–transport. This approach is based, typically, on finite strains, because tumor volume growth ratios are typically far in excess of 1; place the problem well outside the regime of the infinitesimal strain theory. The body, occupying a region $\Omega_0 \subset \mathbb{R}^3$, undergoes a deformation, $\varphi(X, t)$ (a point-to-point map that is a vector), where $X \in \Omega_0$ is a position vector, the reference position of a material point, and t is time. Importantly, the deformation could arise in response to external loads as well as material growth. The deformation gradient tensor, which is the fundamental kinematic quantity in finite strain mechanics, is defined in the usual manner: $F = \partial\varphi/\partial X$. The kinematics of growth is incorporated by the multiplicative, elasto-growth decomposition, $F = F^e F^g$. Here, F^g is the centerpiece of the continuum mechanical treatment of growth. It is the so-called growth tensor, which represents the *shape* of growth at a point in the tumor. It can vary from point to point, and of course, with time. It is important to understand that F^g is the shape that a small neighborhood would attain if able to grow stress-free, that is, without the constraint of the surrounding material. Its counterpart, F^e, also a tensor, is the elastic part of the deformation gradient. This component of the kinematics is responsible for the storage of elastic energy in a growing material and therefore for stress. Being elastic, it represents the constraint of the material surrounding the point that is growing according to F^g. In general, F^g and F^e are not themselves gradients of vector fields. Consequently, incompatibilities are possible in the *grown configuration* attained by the action of F^g. This means that if F^g alone were applied, adjacent neighborhoods would not *fit together* (see Figure 13.3).

We note that the brief history of this field has witnessed some controversy over the correctness of the elasto-growth decomposition—see Garikipati [12] and Ambrosi et al. [1], for reviews. For volumetric growth, that is, not at surfaces, the elasto-growth decomposition is now generally accepted as correct. At surfaces, and in limiting cases of dissolution/nucleation of the solid component, it also is understood that the *continuum, tensorial* representation implied by $F = F^e F^g$ is not intended to be applied.

The most direct, and perhaps simplest, definition can be introduced for F^g in the following form:

$$F^g = \left(\frac{[\rho^c + \rho^m]}{\tilde{\rho}} \right)^{1/3} \mathbf{1} \tag{13.3}$$

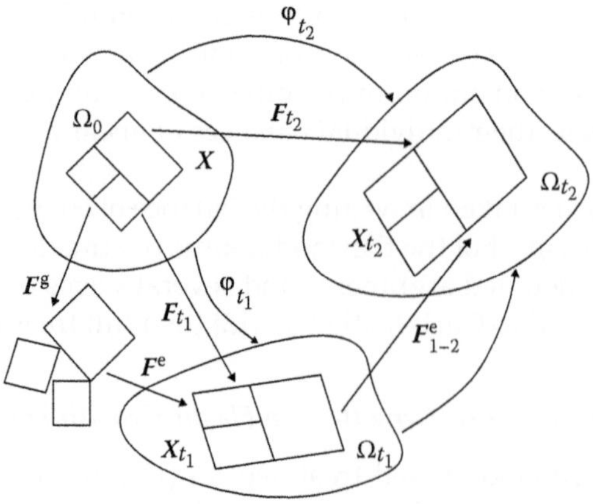

FIGURE 13.3
The kinematics of the elasto-growth decomposition. The configuration Ω_{t_1} at time t_1 has undergone growth in addition to elastic deformation: $F_{t_1} = F^e F^g$ The growth tensor, F^g, is incompatible. However, only elastic deformation has taken place between times t_1 and t_2. The deformation gradient from Ω_0 to Ω_{t_2} is $F_{t_2} = F^e_{1-2} F^e F^g$.

where **1** is the second-order isotropic tensor. This represents the isotropic swelling of neighborhoods that is the result of cell and ECM concentrations differing from a baseline of $\tilde{\rho}$, which also serves as normalization. This form can be generalized to anisotropic growth. It assumes that saturation always holds, since $\rho^c + \rho^m \neq \tilde{\rho}$ results in either swelling or shrinkage. Instead, a saturation function can be introduced, $F^g = s(\rho^c, \rho^m)\mathbf{1}$, where

$$s(\rho^c, \rho^m) = \begin{cases} 1 & \rho^c + \rho^m \leq \bar{\rho} \\ \left(\dfrac{[\rho^c + \rho^m - \bar{\rho}]}{\tilde{\rho}} \right)^{1/3} & \text{otherwise} \end{cases}$$

This elasto-growth decomposition is used, in the simplest case, with a hyperelastic material model, where the (possibly anisotropic) strain energy density function is $\psi = \psi(C^e)$, where $C^e = F^{eT}F^e$ is the elastic right Cauchy–Green tensor. Alternately, viscoelastic material models may be used. The stresses would be obtained by using standard results in finite strain mechanics: the first Piola–Kirchhoff stress tensor, $P = \partial\psi/\partial F^e$, and the Cauchy stress $\sigma = \det(F)PF^{eT}$.

13.3.1.3 Fluid–Solid Interaction during Growth

For tumors that have a significant fluid content, the aforementioned treatment needs to be extended to include the mechanics of fluid–solid interaction, and the fluid's role in modifying species transport [22]. The fluid concentration ρ^f is subject to a mass balance equation, while the fluid stress, σ^f, is governed by a volume-averaged form of the Navier–Stokes equations, in its most general form. Glucose and oxygen diffuse relative to the fluid velocity, v^f, which itself is defined relative to the solid component (cells and ECM). This is expressed as

$$v^\alpha = v^f + \tilde{v}^\alpha, \quad \alpha = c, m$$

$$\text{where} \quad \tilde{v}^\alpha = \frac{j^\alpha}{\rho^\alpha} \tag{13.4}$$

Complicated conditions arise concerning saturation of the tumor's solid phase by the fluid, and Equation 13.4 needs to be further extended to account for fluid-induced swelling. The fluid itself may be incompressible, which is the physically correct regime for tumor mechanics. The case of saturation of the tumor's solid phase by an incompressible fluid leads to the greatest simplification and a transport equation of the following form:

$$\frac{d\rho^\alpha}{dt} = D^\alpha \nabla^2 \rho^\alpha + \pi^\alpha - v^f \cdot \nabla\rho^\alpha + \rho^\alpha \frac{v^f \cdot \nabla\rho^f}{\rho^f} \tag{13.5}$$

where the last two terms arise due to advection of species α by the fluid flow.

For the treatment outlined in this section, where the mechanics of the fluid and solid components are separately modeled, very different effective mechanical responses are obtained depending on whether the fluid–solid interaction is treated by homogenization of strain or stress. Narayanan et al. [22] demonstrated that homogenization of stress leads to a more compliant effective response and one that enjoys greater numerical stability.

13.3.1.4 Cell Production and Nutrient Consumption for Tumor Growth

In Narayanan et al. [23] and Rudraraju et al. [27], we have specialized the source, sink, and transport terms in Equation 13.2 to model the cell production, cell death, and glucose and oxygen consumption that characterize the early, prevascular stage of tumor growth. These source and sink terms are

$$\pi^c(\rho^c,\rho^g,\rho^o) = \frac{\rho^c}{\tau^c(\rho^g,\rho^o)} e^{t/\tau^c(\rho^g,\rho^o)}$$

$$\text{where} \quad \tau^c(\rho^g,\rho^o) = \frac{t^c_{dbl}(\rho^g,\rho^o)}{\log_e 2}$$

$$(13.6)$$

with t^c_{dbl} being the cell doubling time. Equation 13.6 models exponential cell division, which is a natural model, pointwise. The effect of nutrient and oxygen supply lies in the form of $t^c_{dbl}(\rho^g,\rho^o)$, which we obtained from the work of Casciari et al. [2] and specialized to the pointwise, continuum treatment:

$$t^c_{dbl}(\rho^g,\rho^o) = \frac{t^c_{opt}}{0.014} \frac{\rho^g + 1.2\times10^{-2}}{\rho^g} \frac{\rho^o + 7.3\times10^3}{\rho^o} (\rho^{H^+})^{0.46} \qquad (13.7)$$

with t^c_{opt} being the optimal doubling time. The cell death term π^c_d includes the effects of nutrient and oxygen depletion as well as stress-induced death:

$$\pi^c_d(\rho^c,\rho^g,\rho^o,\sigma) = \rho^c \kappa e^{2-(\rho^o_0/\rho^o)^2-(\rho^g_0/\rho^g)^2} + \rho^c \kappa(1-e^{-(tr\sigma/p)^2}) \qquad (13.8)$$

The first term on the right-hand side in Equation 13.8 represents a smooth increase in cell death rate as the glucose and oxygen concentration is depleted locally. The second term represents a smooth increase in cell death rate with the trace of the Cauchy stress. This is a purely phenomenological model, which is in need of improvement by comparison with experiments. Note that it considers the influence of the stress that is complementary to the shear stress dependence studied by Chang et al. [3].

ECM production rates are surprisingly difficult to come by in the experimental literature. We have considered a simple linear dependence on the cell concentration, such that a cell produces 5% of its mass as ECM in a week, guided by observations in the laboratories of our collaborators.

$$\pi^m(\rho^c) = 8.27\times10^{-8}\rho^c \qquad (13.9)$$

The nutrient and oxygen consumption rates are

$$\pi^g(\rho^c,\rho^g,\rho^o) = \rho^c\left(1.14\times10^{-10} + \frac{3.65\times10^{-17}}{\rho^o}\right)\left(\frac{\rho^g}{\rho^g + 1.47\times10^{-4}}\right)(\rho^{H^+})^{-1.2}$$

$$\pi^o(\rho^c,\rho^g,\rho^o) = \rho^c\left(7.68\times10^{-7} + \frac{3.84\times10^{-15}}{\rho^g(\rho^{H^+})^{0.92}}\right)\left(\frac{\rho^o}{\rho^o + 7.21\times10^{-3}}\right)$$

$$(13.10)$$

Casciari and coworkers assumed Michaelis–Menten kinetics for the consumption of glucose and oxygen by the cells and used this assumption to fit their empirical forms for t_{dbl}^c as well as for π^g and π^o. This has resulted in the specific forms of the ρ^g, ρ^o-dependence of t_{dbl}^c in Equation 13.7, and the ρ^g-dependence of π^g, and the ρ^o-dependence of π^o in Equation 13.10.

13.3.1.5 Transport and Strain Energy Density Functions

Cell migration has a random walk component to it. Additionally, cells are known to migrate toward stiffer ECM, which corresponds well with higher concentrations of ρ^m. This is the phenomenon of haptotaxis. Chemotaxis, the migration of cells toward a chemoattractant, has not been considered by us, because our models of tumor growth have not yet been extended to considerations of cell escape to initiate metastasis. There are no chemoattractants among the chemicals we have considered in our models; cells are not known to follow glucose or oxygen gradients to escape from tumors. Accordingly, we have included only the random walk and haptotactic components in the transport laws for cell migration:

$$j^c = -D^c \nabla \rho^c + \beta \rho^c \nabla \rho^m \tag{13.11}$$

where
  D^c is an effective diffusivity for the random walk of a cell
  β is a haptotactic coefficient

Glucose and oxygen transport coefficients in tumors are simply the corresponding diffusivities in water:

$$j^g = -D^g \nabla \rho^g, \quad j^o = -D^o \nabla \rho^o \tag{13.12}$$

which are widely available in the literature.

The hyperelastic response of the soft, isotropic material of the tumor has been modeled by us using the Mooney–Rivlin strain energy density function [23,27]:

$$\psi_{MR}(C^e) = \frac{1}{2} \kappa (J^e - 1)^2 + \frac{1}{2} \mu (\bar{I}_1 - 3) \tag{13.13}$$

$$\text{where} \quad J^e = \sqrt{\det C^e}, \quad \bar{I}_1 = \mathrm{tr}(J^{e^{-2/3}} C^e) \tag{13.14}$$

where
  κ is the bulk modulus
  J^e is the elastic part of the volume change ratio
  tr is the trace operator applied to a tensor

Other alternatives are possible, of course, such as the neo-Hookean functions, Ogden models, and exponential functions in the principal stretches. Yet others based on micromechanical treatments of the long-chain molecular structure of collagenous soft tissue [14], or based on the wormlike chain model, are admissible also.

The viscoelastic response of the soft tissue can be incorporated as outlined in Narayanan et al. [23]. The viscous stress is defined as

$$\sigma_v = (J^e)^{-5/3} F^e Q F^{e^T} \tag{13.15}$$

where Q, a stress-like quantity, is governed by

$$\dot{Q} + \frac{Q}{\tau} = \frac{\gamma}{\tau} \text{dev}\left[\frac{\partial \bar{\psi}}{\partial \bar{C}^e}\right] \tag{13.16}$$

$$\text{where} \quad \bar{\psi}(\bar{C}^e) = \frac{1}{2}\mu(\bar{I}_1 - 3)$$

In this equations, dev is the deviatoric component of the corresponding tensor, γ is a non-dimensional modulus, and τ is a relaxation time. However, as argued in Narayanan et al. [23], the timescale of viscous relaxation of soft tissue is on the order of ~1000 s, which is negligible when compared with the order of 30 days (~3×10^6 s) for the growth of tumors in our experiments [20]. Viscous effects can therefore be neglected over these longer timescales, although they remain of relevance in any computation at the shorter timescales of soft tissue relaxation.

13.4 General Computational Results and Specific Studies

Our computations [23,27] are finite element based, in which the tumor is defined as a region with different properties than the surrounding gel. Accordingly, the elastic properties of the two media are different, with the tumor being either stiffer or more compliant than the gel. Glucose and oxygen diffuse through the gel and tumor, with the same diffusivities in both media, but their consumption rates vanish over the gel. Cells and ECM are present only in the tumor region at the initial state, but cells can migrate into the gel and deposit ECM there. The gel is thus defined by the subdomain where $\rho^c, \rho^m \rightarrow 0$. We first show results to provide examples of the field quantities that are typically studied using continuum models of tumor growth (see Figures 13.4 and 13.5).

The results presented here are 2D computations. See Rudraraju et al. [27] for results in three dimensions. The initial tumor seed is a uniform spherical concentration of cells ($\rho^c = 250$ fg μm^{-3} over a circle of radius 50 μm), with elastic properties (Young's modulus, $E = 6$ kPa, $\nu = 0.49$), and the surrounding matrix is modeled as a compliant elastic material ($E = 24$ kPa, and Poisson ratio $\nu = 0.49$). Based on the experimental observations of change in tumor volume, the optimal cell doubling times are assumed to be around 1 day ($t_{dbl}^c = 1.002 \times 10^5$ s). In the Introduction, we have referred to how cell populations within the same tumor can come into competition with each other, specifically in the context of diffusion limitation, resulting in cell death, necrosis, and an increase in acidity of the tumor environment. Such an effect also was revealed in our tumor growth models. Our models were extended to include a cancer *stem* cell population ($\alpha = s$), motivated by our observation of faster division of cells in some sections of the tumor:

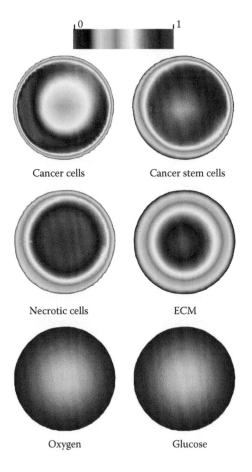

Cancer cells

Cancer stem cells

Necrotic cells

ECM

Oxygen

Glucose

FIGURE 13.4
Field solutions of the indicated quantities from a typical computation of tumor growth. The concentration of each field is normalized by its maximum value over the domain. Note the eccentric distribution of the normal and cancer cells, the cancer stem cells, and the necrotic (dead cells). Attention also is drawn to the lower concentrations of oxygen and glucose in the core of the tumor. See text for details.

$$\frac{\partial \rho^s}{\partial t} = -\nabla \cdot j^s + \pi^s - \pi_d^s \tag{13.17}$$

A simple model was adopted in which the stem cells divide with slightly less than the doubling time of normal cells: $t_{dbl}^s = 0.9 \times t_{dbl}^c$ (Figure 13.4). The initial distribution of cancer stem cells has a concentration $\rho^s = 250$ fg μm^{-3} located eccentrically with respect to the primary cancer cell population. The concentration field of cancer stem cells is also shown in Figures 13.4 and 13.5.

Figure 13.4 shows the distribution of various fields at the end of 10 days. We draw attention to the eccentric distribution of the normal cancer cells and the cancer stem cells. The eccentricity of the cancer stem cells is the result of their initial distribution, which was located closer to the upper right of the tumor. As these stem cells have a shorter doubling time than the normal cancer cells (see previous paragraph), they proliferate more rapidly, and outcompete the normal cancer cells of glucose and oxygen. As a result, the normal

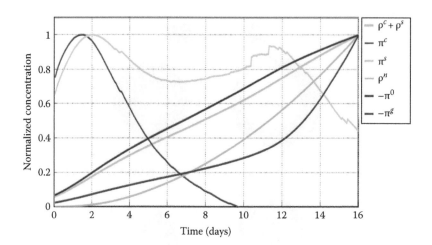

FIGURE 13.5
The evolution of field quantities with time, from a typical computation of tumor growth. The fields are shown at a point that is located at three quarters of the initial radius of the tumor. Each quantity is normalized by its maximum value over the 16-day period. Note the decline of normal cancer cells' proliferation rate to zero, while the cancer stem cells' proliferation rate initially decreases, recovers, and decreases again. See text for details.

cancer cells, whose initial distribution was radially symmetric, have seen their concentration dwindle in the core and restricted to the diametrically opposite side of the tumor (bottom left). Next, consider the glucose and oxygen concentrations. Their supply to the cells has become diffusion limited as the tumor has grown, resulting in the lower concentrations in the core of the tumor. This has largely driven cell death in this computation, resulting in the high concentration of necrotic (dead) cells in the center. The reason for the slight eccentricity in this population is that more stem cells have died and their eccentric location has shifted the distribution of dead cells also.

Figure 13.5 shows the evolution of the same fields as in Figure 13.4 at a point that is located at three quarters of the initial radius of the tumor. It is notable that π^c and π^s both reach their maxima around 2 days. While π^c then monotonically declines to zero due to diffusion limitation of glucose and oxygen, π^s also follows a similar decline until π^c is shut off. Then, having outcompeted the normal cells, the stem cells increase their proliferation rate, until the glucose and oxygen, although still increasing, are insufficient for the stem cells in the last few days of the computation, thereby causing π^s to go into decline again. Over this time, the increase in total cell number, $\rho^c + \rho^s$, is mainly driven by the cancer stem cell population, ρ^s. This is a computation in which, at the chosen point in the tumor, the proximity to the boundary ensures a monotonic increase in glucose and oxygen concentrations.

13.4.1 Computation of the Free Energy Flows of Tumor Growth

The aforementioned constitutive relations are largely based on the experimental literature—an approach that is different from the formal use of the thermodynamic dissipation inequality to specify restrictions on constitutive relations in continuum theories. Instead, the dissipation inequality can be used to pose the question of energy-intensive processes, as suggested in Section 13.2.2.3. This was the crux of the paper by Narayanan et al. [23]. The study computed the free energy consumption rates that were accessible to the authors' continuum model of the prevascular growth of a tumor from an initial radius of 50 μm. The initial and boundary value problem considered a single-cell population (see later for alternatives) and did not model cell death. However, random motion and haptotactic cell migration were included. Glucose and oxygen were considered, and the influence of mechanics was accounted for.

The study revealed that the negative free energy rates associated with creation of cells and ECM (indicating a decrease in free energy to create biomass) as well as the storage of free energy in the newly formed cells and ECM were of the same order of magnitude as the rate of free energy extraction from glucose by glycolysis. The magnitudes of these free energy rates were orders greater than the mechanical free energy rates associated with work done against stress by the growing tumor and the rate of loss of free energy as the cells moved down the free energy gradient. This pointed on one hand to the inefficiency of biochemical processes that underlie the synthesis of new cells and ECM as well as to the futility of trying to starve a tumor of its energy by intervening with the mechanical processes (growth against a stress and cell migration).

The authors also dwelt in detail upon the uncertainty induced by the available experimental data and the models used. Specifically, it was recognized that these estimates for free energy rates did not resolve the subcellular processes of cytoskeletal remodeling, adhesion and de-adhesion between the cell and ECM, and all of the cell signaling and protein transport that underlie cell functions. Further hierarchical studies were thus suggested by this preliminary, macroscopic, continuum investigation. It was noted however that such studies of free energy rates are perhaps the only unifying approach to study the complicated, coupled biochemical and mechanical processes on a common basis in understanding the totality of tumor growth. Furthermore, such studies are not possible using experimental techniques exclusively, and thus present a compelling argument to justify computational studies.

13.4.2 Computational Studies of Tumor Shape

Recently, we have considered a number of mechanisms that lead to symmetry breaking from spheroidal to ellipsoidal shapes in tumors growing in agarose gels, as observed in our experiments [21], also alluded in Section 13.2.2.2. These computational studies were reported in Rudraraju et al. [27].

13.4.2.1 Stem Cell Populations

The computations with the stem cell population included were aimed at testing whether, with such a population arising at an eccentric location in the tumor, their faster proliferation itself could cause a shape change. The computations indicated that the cancer stem cells prevailed in the competition for resources against the normal cancer cells, which were restricted to one edge of the tumor. While this led to a slightly ellipsoidal shape, it was not as pronounced as seen in our experiments (Figure 13.2).

13.4.2.2 Elastically Compliant Layers in the Gel

A second mechanism that we considered for the shape change was the existence of microscopic planes in the gel with greater elastic compliance, along which the growing tumor could bulge out against the gel, thereby developing an ellipsoidal shape. For this purpose, an initial and boundary value problem of tumor growth in a gel was solved, with a thin layer, $0.1 \times$ the thickness of the gel, where the gel's shear and bulk moduli μ_{gel}, κ_{gel} were lower. This did have the hypothesized effect of promoting an ellipsoidal tumor shape.

13.4.2.3 Stress-Driven Migration of Cells

A third mechanism considered for symmetry breaking was stress-driven migration of cells down a gradient of compressive stress. This effect was modeled by modifying j^c in (13.11):

$$j^c = -D^c \nabla \rho^c + M \rho^c \nabla[\operatorname{tr} \sigma] \qquad (13.18)$$

where M represents an effective mobility of the cells in the presence of the stress gradient driving their migration. The motivation was to examine, from computations, whether local stresses in the tumor–gel system could be driving cells away from regions of compression (negative tr[σ]). These computations also included the surface tension of the tumor–gel interface, modeled as

$$\sigma_{srf} = \gamma_{srf}(\nabla \rho^c)\left(1 - n \otimes n\right)$$

$$\lim_{|\nabla \rho^c| \to 0} = 0, \quad n = \frac{\nabla \rho^c}{|\nabla \rho^c|} \qquad (13.19)$$

where
  $\gamma_{srf}(\nabla \rho^c)$ is the surface tension function
  n is the unit normal to the tumor–gel interface

The computations revealed that this hypothesis was also about as reasonable for symmetry breaking as the elastically compliant layer.

The three mechanisms considered earlier were found to have differing degrees of effectiveness in causing symmetry breaking from spheroidal to ellipsoidal shapes. A detailed discussion is available in [27]. We note that this study of symmetry breaking, also, is not accessible via experimental methods alone, providing another justification for the use of modeling to study tumor growth. Work in progress in our group is examining a more compelling explanation for shape symmetry breaking in growing tumors [21].

13.4.2.4 Models Including the Competition between Surface Tension and Internal Turgor Pressure and the Bifurcation to Angiogenesis

The aforementioned continuum tumor growth models do not incorporate a mechanism of cell escape. This is, obviously, a mechanism of much importance to the continued development to a tumor toward metastasis of the cancer. It also does not treat the process of angiogenesis, a crucial stage by which the tumor escapes an otherwise nutrient-starved state. Such treatments also are very well advanced in the tumor growth literature. The associated models are a combination of continuum physics (including mechanics) and agent-based models [5,11]. Of particular importance is the treatment that such models make possible of cell escape as the outcome of competition between internal turgor pressure of the tumor and the surface tension in its wall. This is a bifurcation problem, whose instabilities are well connected with the breakdown of the tumor wall and resultant cell escape.

13.5 Epilogue

Continuum physics modeling of tumor growth is a rich topic with room for rather sophisticated models of reaction–transport and mechanics. It also has the attraction of being able to pose and examine solutions to certain questions on tumor growth that are difficult to access using experimental methods alone. However, the imperative of experimental biophysical investigations cannot be understated.

Glossary

Glycolysis: A cascade of biochemical reactions by which the cell metabolizes a molecule of glucose to produce two ATP molecules.

Metastasis: The process by which cancer cells that migrate away from the parent tumor are transported by the vascular or lymphatic systems and seed a new tumor in distant organs.

Oxidative phosphorylation: A more efficient metabolic pathway than glycolysis (see above for definition of glycolysis) in cells, by which 30–38 molecules of ATP are produced from a single molecule of glucose.

Tumor spheroid: An in vitro tumor model in which cancer cells agglomerate into a spheroidal mass.

References

1. D. Ambrosi, G.A. Ateshian, E.M. Arruda, S.C. Cowin, J. Dumais, A. Goriely, G.A. Holzapfel et al. Perspectives on biological growth and remodeling. *J. Mech. Phys. Solids*, 59:863–883, 2011.
2. J.J. Casciari, S.V. Sotirchos, and R.M. Sutherland. Variations in tumor cell growth rates and metabolism with oxygen concentration, glucose concentration, and extracellular pH. *J. Cell. Physiol.*, 151:386–394, 1992.
3. S.F. Chang, C.A. Chang, D.-Y. Lee, Y.-M. Yeh, C.-R. Yeh, C.-K. Cheng, S. Chien, and J.J. Chen. Tumor cell cycle arrest by shear stress: Roles of integrins and smad. *Proc. Natl. Acad. Sci. USA*, 105:3927–3932, 2008.
4. G. Cheng, J. Tse, R.K. Jain, and L.L. Munn. Micro-environmental mechanical stress controls tumor spheroid size and morphology by suppressing proliferation and inducing apoptosis in cancer cells. *PLoS One*, 4:e4632-1–e4632-11, 2009.
5. V. Cristini, X. Li, J.S. Lowengrub, and S.M. Wise. Nonlinear simulations of solid tumor growth using a mixture model. *J. Math. Biol.*, 58:723–763, 2009.
6. G.P. Figueredo, T.V. Joshi, J.M. Osborne, H.M. Byrne, and M.R Owen. On-lattice agent-based simulation of populations of cells within the open-source Chaste framework. *Interface Focus*, 3:20120081, 2013.
7. J.P. Freyer. Decreased mitochondrial function in quiescent cells isolated from multicellular tumor spheroids. *J. Cell. Physiol.*, 176:138–149, 1998.
8. J.P. Freyer, J.P. Schor, K.A. Jarrett, M. Neeman, and L.O. Sillerud. Cellular energetics measured by phosphorus nuclear magnetic resonance spectroscopy are not correlated with chronic nutrient deficiency in multicellular tumor spheroids. *Cancer Res.*, 51:3831–3837, 1991.

9. J.P. Freyer and R.M. Sutherland. A reduction in the rates of in situ oxygen and glucose consumption in emt6/ro spheroids during growth. *J. Cell. Physiol.*, 124:516–524, 1985.

10. J.P. Freyer and R.M. Sutherland. Regulation of growth saturation and development of necrosis in emt6/ro multicellular spheroids by glucose and oxygen supply. *Cancer Res.*, 46:3504–3512, 1986.

11. H.B. Frieboes, X. Zheng, C.-H. Sun, B. Tromberg, R. Gatenby, and V. Cristini. An integrated computational/experimental model of tumor invasion. *Cancer Res.*, 66:1597–1604, 2006.

12. K. Garikipati. The kinematics of biological growth. *Appl. Mech. Rev.*, 62:030801, 2009.

13. K. Garikipati, E.M. Arruda, K. Grosh, H. Narayanan, and S. Calve. A continuum treatment of growth in biological tissue: The coupling of mass transport and mechanics. *J. Mech. Phys. Solids*, 52:1595–1625, 2004.

14. K. Garikipati, S. Göktepe, and C. Miehe. Elastica-based strain energy functions for soft biological tissue. *J. Mech. Phys. Solids*, 56:1693–1713, 2008.

15. R.A. Gatenby and R.J. Gillies. Why do cancers have high aerobic glycolysis? *Nat. Rev. Cancer*, 4:891–899, 2004.

16. B. Geiger, A. Bershadsky, R. Pankov, and K. Yamada. Transmembrane extracellular matrix-cytoskeleton crosstalk. *Nat. Rev. Mol. Cell Biol.*, 2:793–805, 2001.

17. G. Helmlinger, P.A. Netti, H.C. Lichtenbald, R.J. Melder, and R.K. Jain. Solid stress inhibits the growth of multicellular tumor spheroids. *Nat. Biotechnol.*, 15:778–783, 1997.

18. T.L. Jackson and H.M. Byrne. A mechanical model of tumor encapsulation and transcapsular spread. *Math. Biosci.*, 180:307–328, 2002.

19. C. Koike, T.D. McKee, A. Pluen, S. Ramanujan, K. Burton, L.L. Munn, Y. Boucher, and R.K. Jain. Solid stress facilitates spheroid formation: Potential involvement of hyaluronan. *Br. J. Cancer*, 86:947–953, 2002.

20. K.L. Mills, K. Garikipati, and R. Kemkemer. Experimental characterization of tumor spheroids for studies of energetics of tumor growth. *Int. J. Mater. Res.*, 102:889–895, 2011.

21. K.L. Mills, R. Kekmemer, S.S. Rudraraju, and K. Garikipati. Elasticity and the shape of prevascular solid tumors. arxiv.org/abs/1405.3293.

22. H. Narayanan, E.M. Arruda, K. Grosh, and K. Garikipati. The micromechanics of fluid–solid interactions during growth in porous soft biological tissue. *Biomech. Model. Mechanobiol.*, 8:167–181, 2009.

23. H. Narayanan, S.N. Verner, K.L. Mills, R. Kemkemer, and K. Garikipati. *In silico* estimates of the free energy rates of growing tumor spheroids. *J. Phys. Cond. Matt.*, 22:194122-1–194122-16, 2010.

24. M.J. Paszek, Z. Nastaran, K.R. Johnson, J.N. Lakins, G.I. Rozenberg, A. Gefen, C.A. Reinhart-King et al. Tensional homeostasis and the malignant phenotype. *Cancer Cell*, 8:241–254, 2005.

25. P. Roca-Cusachs, T. Iskratsch, and M.P. Sheetz. Finding the weakest link—Exploring integrin-mediated mechanical molecular pathways. *J. Cell Sci.*, 125:3025–3038, 2012.

26. T. Roose, S.J. Chapman, and P.K. Maini. Mathematical models of avascular tumor growth. *SIAM Rev.*, 49:179–208, 2007.

27. S. Rudraraju, K.L. Mills, R. Kemkemer, and K. Garikipati. Multiphysics modeling of reactions, mass transport and mechanics of tumor growth. In *IUTAM Proceedings on Computer Models in Biomechanics—From Nano to Macro*, pp. 293–304. Springer, 2012. Palo Alto, California.

28. S. Suresh. Biomechanics and biophysics of cancer cells. *Acta Biomater.*, 3:413–438, 2007.

29. O. Warburg, F. Wind, and E. Negelein. The metabolism of tumors in the body. *J. Gen. Physiol.*, 8:519–530, 1927.

30. V.M. Weaver, O.W. Petersen, F. Wang, C.A. Larabell, P. Briand, C. Damsky, and M.J. Bissell. Reversion of the malignant phenotype of human breast cells in three-dimensional culture and in vivo by integrin blocking antibodies. *J. Cell. Biol.*, 137:231–245, 1997.

31. R.A. Weinberg. *The Biology of Cancer*. Garland Science, Taylor & Francis Group, New York, 2007.

32. H. Yu, J.K. Mouw, and V.M. Weaver. Forcing form and function: Biomechanical regulation of tumor evolution. *Trends Cell Biol.*, 21:47–56, 2011.

14

Cell Force–Mediated Collagen Remodeling in Cancer Metastasis

Paolo P. Provenzano

CONTENTS

14.1 Introduction

Metastasis, the development of additional tumor(s) in organs other than the primary tumor, is the leading cause of death from cancer. In general, the metastatic cascade associated with solid tumors is summarized in stages including focal invasion where cells invade through the stroma, the process of intravasation where cells enter into the vasculature (or may enter the lymphatic system), survival in circulation, and extravasation where cells exit the vasculature and move into a host site where they form another tumor (Gupta and Massague 2006; Sahai 2007). While our understanding of each of these processes is incomplete, it is clear that the early switch from noninvasive cell behavior toward an invasive cell phenotype that ultimately results in cell invasion through the stroma is a critical step in the metastasis cascade. In this context, the features of the cellular microenvironment and the features of the tumor stroma in general play a dominant role in dictating the fate of a genetically aberrant cancerous cell. Not every genetically abnormal cell throughout the tumor behaves the same, largely due to heterogeneities that exist throughout solid tumors. Where one cell may invade throughout the stroma and successfully negotiate the stages of metastasis to develop a new tumor, another tumor cell may never proliferate robustly, become robustly invasive, or leave the tumor. This is due in large part to spatial and temporal variations in the microenvironment surrounding these cells. The tumor microenvironment contains a wide range of cells including fibroblasts, immune cells, and the tumor supporting vasculature as well as a robust extracellular matrix (ECM), most notably collagen, which are associated with a robust fibroinflammatory response, or desmoplasia, found in many solid tumors. The spatial distribution of these elements is not random or uniform throughout the tumor, while the dynamic interactions among these components coordinate to promote tumor progression (Folkman 1971; Mintz and Illmensee 1975; Teicher et al. 1990; Elenbaas et al. 2001; Shekhar et al. 2003; Pollard 2004; Condeelis et al. 2005; Orimo et al. 2005; Condeelis and Pollard 2006; Gaggioli et al. 2007; Sahai 2007; Calvo et al. 2013). Here, in this chapter, the focus will be primarily directed toward the discussion of work that elucidates the physical processes associated with focal invasion at the single cell level in 3D microenvironments and the importance of cell contractile forces that reorganize the stromal ECM in the tumor microenvironment to provide mechanical and spatial cues to help facilitate cancer cell invasion.

14.2 Mechanical Microenvironment

Cells in their local ECM microenvironment encounter complex, simultaneously active, biochemical, architectural, and mechanical stimuli that profoundly influence cell behavior. In this context, the mechanical properties of the microenvironment and the forces present through the cell due to cell contractility play a role in regulating the development and normal function of essentially every tissue in the human body. Indeed, the mechanical interactions between a cell or group of cells and its microenvironment are emerging as a unifying principle that connects nano-, micro-, and macroscale behaviors to regulate tissue function (reviewed in Discher et al. 2005; Vogel and Sheetz 2006; Wang et al. 2009; Wozniak and Chen 2009). For example, cell and/or ECM mechanics are known to influence in cell differentiation (Engler et al. 2006, 2008; Kshitiz et al. 2012a,b), migration (Lo et al. 2000; Gardel et al. 2008; Hadjipanayi et al. 2009), morphogenesis (Sahai and Marshall 2002; Wozniak et al. 2003; Paszek et al. 2005; Alcaraz et al. 2008; Provenzano et al. 2008a; Gehler et al. 2009; Kshitiz et al. 2012a; McLeod et al. 2013), and proliferation (Wang et al. 2000; Williams et al. 2008; Provenzano et al. 2009b; Ulrich et al. 2009; Tilghman et al. 2010), all of which are critical cell behaviors that can profoundly influence organ function in normal and diseased states. In fact, within this context, the operant mechanics between the cell and its microenvironment that drive resulting shifts in cell architecture and signaling (via mechanotransduction signals where cells transduce mechanical stimuli into a biochemical response) are known to influence numerous human diseases (e.g., Grinnell 2003; Robling et al. 2006; Hahn and Schwartz 2009; Wong et al. 2012), not the least of which is tumor formation and progression (Wozniak et al. 2003; Paszek et al. 2005; Provenzano et al. 2009b; Ulrich et al. 2009; Goetz et al. 2011; Samuel et al. 2011).

Mechanical stress and cell strain can arise from an externally applied load or deformation. In the context of cell biology, this stimulus is aptly named *outside-in* mechanical signaling (Provenzano and Keely 2011) and is commonplace in load-bearing connective tissues such as ligament, tendon and bone, and the cardiovascular system (Robling et al. 2006; Woo et al. 2006; Butler et al. 2009; Humphrey and Holzapfel 2012). Alternatively, or in concert with outside-in stimuli, ECM-attached cells also physically probe their microenvironment, most often by *tugging* on the ECM, and integrate complex heterogeneous mechanical cues in the microenvironment in order to sense and respond to variations in ECM stiffness. Analogous to the *outside-in* nomenclature, this behavior is termed *inside-out* mechanical stimuli. This cell behavior is dependent on adhesion to the ECM where receptors such as integrins, and associated focal adhesion complexes, act as a linkage to transmit force between the intracellular cytoskeleton and the extracellular ECM, where myosin-based contractility acts as

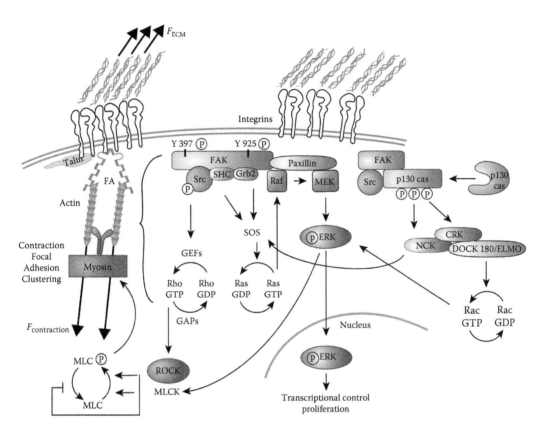

FIGURE 14.1
Inside–out contractile force as a regulator of cell signaling. Stiffness of the ECM influences the magnitude of the contractile force of the cell that is transmitted to the ECM through integrins. As such, within the range for which a given cell can effectively respond, elevated stiffness of the ECM results in elevated forces transmitted across the cell–matrix interface due to increased Rho-mediated intracellular contractile force, which promotes 3D matrix adhesion formation and maturation that leads to the activation of highly dynamic signaling networks that regulate fundamental cell processes such as proliferation and migration. (From Provenzano, P.P. and Keely, P.J., *J. Cell Sci.*, 124, 1195, 2011. With permission.)

a primary regulator of cellular traction (contractile) forces (Geiger et al. 2001; Provenzano and Keely 2011). Furthermore, the small GTPase Rho is a key regulator of the myosin-driven cell traction force axis, largely through its effector Rho-associated protein kinase (ROCK) (Figure 14.1). In fact, inhibition of Rho or ROCK abolishes cell contractility as well as the ability of the cell to respond to increases in matrix stiffness (e.g., Wozniak et al. 2003; Shiu et al. 2004; Paszek et al. 2005; Provenzano et al. 2008b, 2009b; Samuel et al. 2011). Thus, the Rho–ROCK pathway through myosin is critical for the generation and maintenance of cell contractile forces and resultant matrix reorganization.

Importantly, cells also possess the ability to modulate the magnitude of their traction force in response to 2D substrate stiffness or 3D microenvironments. For both fibroblasts and epithelial cells, increasing substrate stiffness results in elevated traction force (Lo et al. 2000; Paszek et al. 2005; Saez et al. 2005). Furthermore, it has been reported that cells such as fibroblasts can alter their internal cell stiffness to match the stiffness of their local substrate (Solon et al., 2007) and that actin stress fiber organization and kinetics are strongly influenced by intra and extracellular mechanical interplay (Deshpande et al., 2006;

Elson and Genin, 2013; Hsu et al., 2009; Kaunas et al., 2006; Lee et al., 2012; McGarry et al., 2009; Wei et al., 2008; Zemel et al., 2010a; Zemel et al., 2010b). Thus, a critical determinant of cell behavior is the development of a feedback loop by which cells sense the stiffness of their microenvironment and exert contractile force at a magnitude that scales with this stiffness. Under normal physiological conditions, this dynamic force balance relationship, often termed tensional homeostasis, helps maintain normal cell behavior (Yamada and Cukierman 2007; Klein et al. 2009; Mammoto and Ingber 2010). Alternatively, during pathologies such as cancer where the ECM stiffness is uncharacteristically elevated, an abnormally elevated tensional homeostasis develops that can result in atypical cellular behaviors including hyperactivation of focal adhesion signaling, Rho GTPase activity, and alignment of the collagenous stroma that drives invasion of breast carcinoma cells (Provenzano et al. 2006, 2008a,b, 2009b; Conklin et al. 2011). Hence, in addition to the biochemical intricacies of the cellular microenvironment, it is now clear that the stiffness and architecture of the ECM and the corresponding cellular response to these cues play a fundamental role in regulation of cancer progression.

14.3 Matrix Reorganization and Contact Guidance in Cancer

Cell migration is a process guided by soluble biochemical stimuli and physical cues whereby both physical and chemical cues can be provided by the ECM (Lauffenburger and Horwitz 1996; Petrie et al. 2009; Cukierman and Bassi 2010). Cell migration/invasion within 3D microenvironments is an important in vivo event during cancer progression and metastasis. In contrast to cell migration on 2D surfaces, 3D migration requires that cells progress through a dense ECM by varying cell morphology and actively remodeling the matrix (Sahai and Marshall 2003; Wolf et al. 2003, 2007; Even-Ram and Yamada 2005; Wyckoff et al. 2006; Zaman et al. 2006). In addition to the role of proteases in these processes (Sabeh et al. 2009; Friedl and Wolf 2010; Wolf and Friedl 2011), an active matrix remodeling program involves cell contractility to reorganize the tumor stroma (Provenzano et al. 2006, 2008a; Leventhal et al. 2009; Goetz et al. 2011). Interestingly, both carcinoma cells and carcinoma-associated fibroblasts have robust ability to contract and remodel collagen matrices through generation of cell traction forces (Wozniak et al. 2003; Paszek et al. 2005; Provenzano et al. 2008b, 2009b; Gehler et al. 2009; Goetz et al. 2011; Samuel et al. 2011; Calvo et al. 2013). As a result, specific cell-directed patterns emerge over the course of disease progression. In particular, specific collagen architectures in the tumor stroma such radial (perpendicular to the cell boundary) realignment of this matrix develop in vivo that facilitate local invasion (Provenzano et al. 2006), which is consistent with findings that metastatic epithelial cells directly migrate along stromal collagen fibers (Wang et al. 2002). Furthermore, cells in regions where the collagen matrix remains organized parallel to the tumor boundary are noninvasive (Provenzano et al. 2006). More specifically, using multiphoton excitation microscopy coupled with second harmonic generation (SHG) imaging (see Section 14.4) to image live mammary tumors, three *tumor-associated collagen signatures* (TACS) were defined (Provenzano et al. 2006, 2008a), namely, TACS-1, TACS-2, and TACS-3 (Figure 14.2). *TACS-1* is the presence of locally dense collagen within the globally increased collagen concentration surrounding early tumors; *TACS-2*, straightened (taut) collagen fibers stretched around the carcinoma cells in a tangential orientation (i.e., parallel to the cell boundary) constraining the cell cluster volume; and *TACS-3*, radially aligned

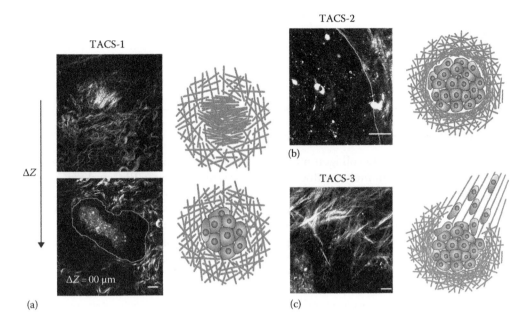

FIGURE 14.2
(a–c) Multiphoton excitation and SHG imaging of TACS as described in Section 14.3. Note the contact guidance cues provided by the reorganized collagen matrix in (c) that promote carcinoma cell invasion. Scale bars = 25 μm. (Adapted from Provenzano, P.P. et al., *BMC Med.*, 4, 38, 2006; Provenzano, P.P. et al., *Clin. Exp. Metastasis*, 26, 357, 2009a. With permission.)

(i.e., perpendicular to the cell boundary) collagen fibers that facilitate local invasion. This finding has profound implications for disease progression since human patients with higher TACS-3 scores have poorer prognosis with shorter survival (Conklin et al. 2011).

The phenomenon by which ECM architectures influence cell polarity, migration direction, and persistence is known as contact guidance. For instance, contact guidance from topographical features on 2D substrates (Teixeira et al. 2003; Doyle et al. 2009; Kim et al. 2012) or 3D microenvironments (Guido and Tranquillo 1993; Dickinson et al. 1994) provides directional cues that influence cell orientation and migration direction. In cancer, these ECM anisotropies in the local microenvironment specifically influence invasion and disease progression. Interestingly, mechanistic studies in tissue-engineered 3D microenvironments designed to recapitulate the extant features of carcinoma cells in vivo revealed a dependence on cell traction force for the matrix remodeling necessary to generate ECM architecture that provides contact guidance (Provenzano et al. 2008b). In the absence of sufficient force generating capacity, matrix remodeling does not occur and cell invasion is significantly inhibited (Provenzano et al. 2008b; Carey et al. 2013). Inhibition of either Rho, ROCK, or myosin renders the cells incapable of generating contact guidance (Provenzano et al. 2008b). Alternatively, in tumors with greater collagen density, a scenario that increases focal adhesion signaling, Rho GTPase activation, and dependent cell contractile force, matrix reorganization to provide contact guidance cues occurs earlier in disease and more robustly (Provenzano et al. 2008a). Likewise, it was later shown that matrix alignment is increased when collagen cross-linking increases the stiffness of the collagenous stroma (Levental et al. 2009). Thus, complex feedback loops appear to exist between ECM mechanical properties and architecture and signaling between focal adhesion-associated proteins and the Rho–ROCK–myosin axis to regulate cellular force magnitude that profoundly influences disease progression and patient outcome.

14.4 Imaging Stromal Architecture

To date, imaging of the 3D collagen matrix in vitro or in vivo has relied heavily on two microscopy modalities, namely, confocal reflectance microscopy (CRM) and SHG microscopy. CRM of collagen is a technique that utilizes reflected laser light, that is, backscattered light, to image collagen based on differences in the refractive index. It does not require any labeling of collagen and can be easily implemented with simultaneous imaging of fluorescent probes for cell imaging making it quite useful for imaging thin samples. However, CRM of collagen may not detect fibers at particular orientations in the sample (Jawerth et al. 2010), and its imaging depth is consistent with confocal microscopy and thus remains relatively low (~50–100 µm). Yet, CRM has been used with good effect to image in vitro collagen matrices and aid our understanding of the relationship between matrix architecture, gel mechanics, and cell behavior, including cell force–mediated matrix contraction (see, e.g., Wolf et al. 2003; Roeder et al. 2009; Yang and Kaufman 2009; Yang et al. 2009, 2010; Stevenson et al. 2010; Harjanto et al. 2011).

An alternate approach is to utilize multiphoton laser-scanning microscopy (MPLSM) (Denk et al. 1990), which has been used with great success to view live tumor cells within 3D microenvironments both in vitro and in vivo to substantially enhance our understanding of the processes of local invasion and metastasis (Brown et al. 2001, 2003; Jain et al. 2002; Wang et al. 2002; Sahai et al. 2005; Skala et al. 2005; Provenzano et al. 2006, 2012). For instance, MPLSM facilitated the detection of TACS (Provenzano et al. 2006), and Condeelis and colleagues employed intravital multiphoton imaging of mammary carcinoma xenografts expressing fluorescent reporters in conjunction with collagen imaging to study cancer cell invasion in vivo (Wang et al. 2002; Sahai et al. 2005). These authors demonstrate a novel interaction of mammary carcinoma cells with collagen, as invading cells migrate along collagen fibers toward the vasculature (Wang et al. 2002, 2007) as well as a fundamental set of signaling pathways associated with invasive subpopulations of cells (Wang et al. 2002, 2004, 2006, 2007). Without the ability to image the live tumor environment, it seems unlikely that such an analysis of a defined metastatic subpopulation of cells could be obtained and directly correlated with motility behavior.

Multiphoton excitation of fluorescence with MPLSM builds upon the advantages of confocal microscopy (Brakenhoff et al. 1985; White et al. 1987; Diaspro and Sheppard 2002a; Pawley 2006), particularly optical sectioning, but uses short pulses of longer wavelength light (near IR), such that two or more photons (at twice the normal excitation wavelength for two-photon excitation) are absorbed to excite a fluorophore. Since the probability of two or more photons simultaneously exciting a fluorophore outside the focal volume is extremely low, excitation naturally remains restricted to the plane of focus (i.e., optically sectioned) (Denk et al. 1990; Centonze and White 1998). In addition, MPLSM has an effective imaging depth much greater than conventional confocal microscopy, deposits less energy to the sample reducing photo-induced toxicity (Squirrell et al. 1999; Provenzano et al. 2009a), and has a quadratic dependence for fluorescent intensity (Denk et al. 1990; Diaspro and Sheppard 2002b; Helmchen and Denk 2002):

$$I_f(t) \propto \delta_2 P(t)^2 \left[\pi \frac{(NA)^2}{hc\lambda} \right]^2 \tag{14.1}$$

where
$I_f(t)$ is the fluorescence intensity
δ_2 is the molecular cross section
$P(t)$ is the laser power
NA is the numerical aperture
h is Planck's constant
c is the speed of light
λ is the wavelength

A particularly appealing aspect of MPLSM is that it can be used to produce SHG signals from biological materials, such as fibrillar collagen (Campagnola et al. 2002; Zoumi et al. 2002; Cox et al. 2003), and therefore is a valuable platform for understanding collagen architectures in the 3D cellular microenvironment. SHG is a coherent process that results from the laser field interacting with a noncentrosymmetric ordered structure such as collagen to produce a nonlinear, second-order, polarization (Freund and Deutsch 1986; Campagnola et al. 2002; Stoller et al. 2002; Cox et al. 2003; Mohler et al. 2003; Williams et al. 2005; Plotnikov et al. 2006) described as

$$\mathbf{P} = \varepsilon_0 \left(\chi^{(1)}\mathbf{E} + \chi^{(2)}\mathbf{E}^2 + \chi^{(3)}\mathbf{E}^3 + \cdots \right) \tag{14.2}$$

where polarization ($\mathbf{P}$) and electric field ($\mathbf{E}$) are vectors, and the nonlinear susceptibilities, $\chi^{(i)}$, are tensors with second term in Equation 14.2 accounting for SHG. Thus, the emission wavelength is exactly half of the incident wavelength and therefore can be separated from multiphoton excitation (MPE)-generated fluorescent signals originating from the same excitation wavelength. Furthermore, in addition to providing direct information of collagen fiber network architecture (see Figure 14.2), SHG can provide direct quantitative information about collagen fiber content and composition (Brown et al. 2003; Nadiarnykh et al. 2007, 2010; Lacomb et al. 2008). Consequently, MPLSM is very well suited for imaging cell-mediated matrix reorganization and provides a powerful tool for gaining spatial and quantitative information about the collagen and its interactions with cells in solid tumors.

14.5 Modeling Changes in the 3D Cell Microenvironment due to Cell Contractile Force

A number of interesting models have been developed that may provide insight into local matrix density and architecture in the 3D cellular microenvironment (Murray et al. 1983; Murray and Oster 1984a,b; Perelson et al. 1986; Ngwa and Maini 1995; Barocas and Tranquillo 1997; Dallon and Sherratt 1998; Dallon et al. 1999; Olsen et al. 1999; Aghvami et al. 2013; Reinhardt et al. 2013; Sander 2013). While not always generated for the purpose of understanding cancer cell contractility and generation of matrix architectures such as those that produce contact guidance, these models are very applicable to the questions discussed in this chapter. A brief description of these models and their utility will be provided here.

Due to cellular contractile forces, the density of the local matrix surrounding a cell or group of cells increases during active cell traction as the matrix is pulled toward the cell (Yamato et al. 1995; Provenzano et al. 2008b, 2009b; Evans and Barocas 2009; Stevenson et al. 2010). As discussed in more detail earlier, this can have a profound influence on carcinoma cells as they are exquisitely sensitive to the stiffness of the ECM in their microenvironment (Paszek et al. 2005; Provenzano et al. 2009b). In fact, it has been reported that higher levels of local collagen density surrounding carcinoma cells elicit elevated signaling through 3D matrix adhesion complexes to upregulate the Rho GTPase pathway resulting in increased cellular contractile forces and collagen reorganization to architectures that provide contact guidance (Provenzano et al. 2008a,b, 2009b; Carey et al. 2013). To address the question of pericellular collagen density surrounding cells and obtain a predictive platform to inter-pret and guide experiments, Gooch and coworkers developed a mathematical framework that describes collagen density adjacent to contracting cells (Stevenson et al. 2010). Under the assumption that the ECM is compacted (i.e., gel contraction) as a result of cell traction forces that remodel the pericellular ECM, with the magnitude of cellular force genera-tion governed by the properties of the local pericellular ECM, the model captured global (i.e., macroscopic) collagen gel contraction from human umbilical vein endothelial cells (HUVECs) and NIH 3T3 fibroblasts with great fidelity. In this model, cells are assumed to be discrete elements, in contrast to models that consider cells as a continuous phase that sense the average mechanical properties of the ECM gel (Barocas and Tranquillo 1997). The key assumption is that cells sense the properties in their local microenvironment and upon reaching a cell-specific set point stop further compacting the matrix and maintain these conditions. As such, this model has utility to understand changes in collagen density fol-lowing cell traction–mediated matrix reorganization, which is described as

$$\frac{d\rho_f}{d\rho_0} = \frac{(1 - \theta - m^* c_0/\rho_0)}{(1 - \theta)^2} \tag{14.3}$$

where ρ is the collagen density with subscripts f and 0 denoting the final and initial densi-ties, respectively. The parameter θ requires more explanation. Assuming the volume occu-pied by cells is negligible, following mass conservation, the authors (Stevenson et al. 2010) propose that the final volume of the contracted matrix is composed of ECM surrounding cells that undergoes compaction and unaffected collagen in regions distant from the cells (i.e., $V_f = V_c + V_u$), where V is the volume and subscripts f, c, and u denote the final, pericellu-lar region, and unaffected region volumes, respectively. From this framework, the authors generate a *fraction of the volume compaction* metric, θ, ranging from 0 to 1 where 0 indicates no contraction has taken place, while 1 conversely indicates compaction to zero volume:

$$\theta = 1 - \left(\frac{V_f}{V_0}\right) = m^* \cdot \left(\frac{c_0}{\rho_0}\right) - V^* \cdot c_0 \tag{14.4}$$

where
  V_0 is the initial gel volume
  m^* is the mass of collagen compacted in the pericellular volumes
  c_0 equals N/V_0, where N is the number of cells
  V^* is the volume of collagen compacted around each cell
  ρ_0 is the initial collagen density

Thus, this model provides an estimate of local collagen density surrounding cells following cell-mediated contraction of the collagen matrix and may provide a framework for use with additional models to parse out the relationship between matrix density, stiffness, and alignment in the cellular microenvironment.

In addition to characterizing changes in collagen density, several models have been recently proposed that more directly address cell traction and the generation of matrix anisotropy for contact guidance that will be discussed here (Aghvami et al. 2013; Reinhardt et al. 2013; Sander 2013). In addition, a number of earlier studies have been fundamental to our knowledge of cell traction forces and ECM remodeling. For example, Barocas and Tranquillo (1997) presented a model based on the application of anisotropic biphasic theory and strain-induced fiber alignment to model the stresses involved in inhomogeneous matrix contraction that leads to ECM alignment, which in turn influence cell orientation via contact guidance. This model and models of this type, some of which are described later in this section, account for stresses in the ECM network and cell traction stress that acts anisotropically according to cell orientation (represented as a tensor). A relationship between the cell orientation tensor and ECM orientation tensor is likewise defined leading to a description of strain-induced ECM alignment and its contact guidance. Readers who are interested in modeling matrix remodeling and genesis of contact guidance cues are strongly encouraged to examine this model by Barocas and Tranquillo as well as the fundamental work by Dallon, Sherratt, and Oster and others (e.g., Murray et al. 1983; Murray and Oster 1984a,b; Perelson et al. 1986; Ngwa and Maini 1995; Dallon and Sherratt 1998; Dallon et al. 1999; Olsen et al. 1999).

More recently, a number of models have been proposed that also examine cell traction and the generation of matrix anisotropy for contact guidance. One interesting approach is the development of a multiscale computation model to simulate the influence of cell traction on matrix reorganization (Aghvami et al. 2013; Figure 14.3). This model has particular applicability to tissue engineering since it accounts for external mechanical stimuli (i.e., tensile or compressive deformation), which are commonly applied to tissue-engineered constructs to produce favorable mechanical properties. However, this model

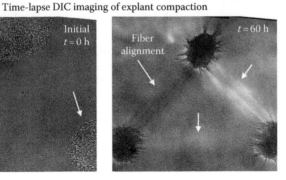
(a)

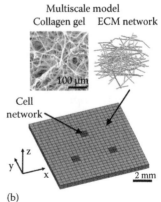
(b)

FIGURE 14.3
(a) Fiber alignment between cell explants after 60 h. (b) Multiscale modeling strategy where cell networks produce traction forces to reorganize the adjacent ECM. (Reproduced from Aghvami, M. et al., *J. Biomech. Eng.*, 135, 71004, 2013. With permission.)

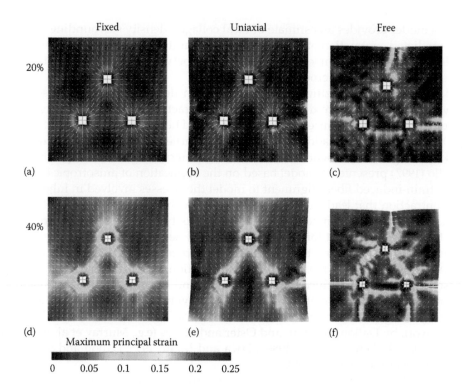

FIGURE 14.4
Principal strain patterns under different applied boundary conditions. Strain in fixed (a and d; top, bottom, left, and right fixed), uniaxial (b and e; top and bottom fixed), and free (c and f) during 20% and 40% compaction for three explants. As this model can capture cell traction–mediated matrix remodeling that aligns regions of the ECM when no external loading regime is applied, it is an appealing approach for modeling these behaviors during cancer invasion. White arrows show principal direction. (Reproduced from Aghvami, M. et al., *J. Biomech. Eng.*, 135, 71004, 2013. With permission.)

also demonstrates the ability to capture cell traction–mediated matrix remodeling that aligns regions of the ECM when no external loading regime is applied (Figure 14.4), making it a rather appealing approach for modeling these behaviors in cancer studies. The fundamental components of the models are a constitutive relationship for the ECM fibers, a volume-averaged stress across a network of linked fibers, and an account for the macroscopic stress balance. Represented mathematically, the force in an ECM fiber is described as

$$F = \frac{E_f A_f}{B}[\exp(B\varepsilon_f) - 1] \tag{14.5}$$

and the volume-averaged Cauchy stress $\langle \sigma_{ij} \rangle$ of the network and macroscopic force balance are, respectively,

$$\langle \sigma_{ij} \rangle = \frac{1}{V} \int_V \sigma_{ij} \, dV = \frac{1}{V} \sum_{BN} x_i F_j \tag{14.6}$$

and

$$\langle \sigma_{ij,i} \rangle = \frac{1}{V} \oint_V \left(\sigma_{ij} - \langle \sigma_{ij} \rangle \right) u_{k,i} n_k \, dA \qquad (14.7)$$

where
  F is the force
  E_f is Young's modulus
  A_f is the fiber cross-sectional area
  ε_f is Green's strain
  B is a fitting parameter to account for nonlinearity
  V is the network volume
  σ_{ij} is the local microscopic stress
  u is the displacement
  n is the unit normal vector
  BN signifies boundary nodes

In addition, see references Chandran and Barocas (2006), Stylianopoulos and Barocas (2007a,b), and Sander et al. (2009) for additional details of the equations and implementation of the model. Interestingly, and as seen in Figure 14.4, this model is able to capture the generation of contact guidance–inducing ECM reorganization by fibroblasts with great fidelity, and thus it seems likely that this approach would capture similar reported behavior in carcinoma cells and tumor explants that generate similar contact guidance structures that promote invasion (Provenzano et al. 2008b). In fact, such a model may aid our understanding of the relationship between cancer cell force generation capacity, matrix reorganization, and its potential for invasion and thus may additionally offer a predictive capacity to distinguish weakly from strongly invasive cells or cell populations.

An alternate approach to understand ECM alignment adjacent to cells employs theory from classical linear elasticity (Sander 2013). While this approach is certainly less complex than the model previously described, it captures, perhaps surprisingly, the matrix realignment behavior within a critical radius around the cell that provides contact guidance reasonably well and is specifically motivated by the case of invasion of cells from tumor spheroids in regions of matrix alignment (Figure 14.5). Furthermore, results from this model suggest regional variations in the degree to which the ECM architecture enhances invasiveness, a conclusion that certainly warrants further study in tumors, which by their very nature are extremely heterogeneous in both composition and organization. Another interesting approach to elucidate the relationship between cell traction forces and matrix deformation has been presented to better understand cancer invasion in the 3D environment (Koch et al. 2012). Using fluorescent microscopy to track fluorescent beads that are tightly embedded in a collagen matrix, the authors calculate the strain energy and matrix deformation using a finite element analysis of fluorescent bead displacement (i.e., nodes in the finite element mesh) assuming linear elastic material behavior. Anisotropy of the strain energy density distribution around a cell was determined by calculating the second moment of the strain energy density with an index of anisotropy defined as the ratio of the maximum and minimum eigenvalues of these moments. Of note, this analysis reveals that the invasive carcinoma cell lines from both lung and breast cancers demonstrate considerably higher 3D traction force than noninvasive cell lines from these same types of cancers. Furthermore, invasive cells possessed a more spindle-like morphology than their noninvasive counterparts that results in an anisotropic distribution of cell tractions around the cells with ECM displacements

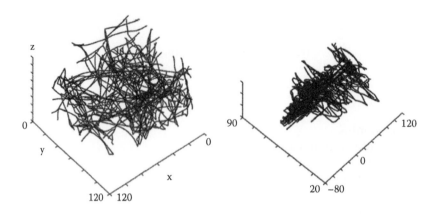

FIGURE 14.5
Three-dimensional plots of the collagen network between two simulated tumor spheroids using finite element elasticity calculations where the left panel is the zero displacement condition and the right panel shows model predicted alignment. Distances are in microns. (Reproduced from Sander, L.M., *J. Biomech. Eng.*, 135, 71006, 2013. With permission.)

aligned with the long axis of the cell (Figure 14.6). While not specifically examining individual ECM fibers and their alignment, certainly, these results suggest increased ECM alignment in these regions considering what we have already discussed about cell traction and generation of contact guidance ECM architectures. Consistent with this concept, the study does also show that changes in cell direction during invasion through matrix are preceded by increases in the strain energy density along these vectors of invasion. It seems likely that combining these analyses with models that provide information about ECM fiber orientation as a function of cell traction stress could provide additional insight into the relationship between these factors and the physical mechanisms that regulate cancer cell invasion.

A significant, but largely open, question regarding contact guidance as a driver of cancer cell invasion is the development and influence of durotactic cues in the matrix. On 2D substrates, cells have been shown to migrate in the direction of increasing stiffness (Lo et al. 2000; Hadjipanayi et al. 2009), a process termed durotaxis. Furthermore, elevated matrix stiffness, or elevated collagen density that increases matrix stiffness, increases cell force–dependent matrix remodeling to promote collagen alignment and cell invasion (Provenzano et al. 2008a, 2009b; Levental et al. 2009). Thus, a relationship appears to exist between matrix stiffness, cell contractile force, and the generation of contact guidance cues. However, this relationship is, to date, not well understood. In particular the relationship between heterogeneous ECM stiffness and contact guidance for the case where the matrix has already been aligned and cells are now invading through it is largely unknown. Using an agent-based approach, Gooch and coworkers recently presented a model to examine these behaviors (Reinhardt et al. 2013). During the simulations, ECM was pulled toward the cells to increase local matrix density and align ECM fibers with lines of ECM formed between cells while migrating faster in regions of higher collagen density (Figure 14.7), consistent with the reports discussed earlier. Interestingly, when the model was adapted to examine the influence of a stiffness gradient along the uniform matrix, the simulated cells on this matrix migrated preferentially toward increasing matrix stiffness, suggesting that if these findings translate to the fully 3D situation, then the spatial heterogeneities in the stiffness of aligned matrix may further regulate 3D cell migration. Many questions remain, however, regarding this complex relationship between contact guidance cues and durotactic cues including the relative influence of these

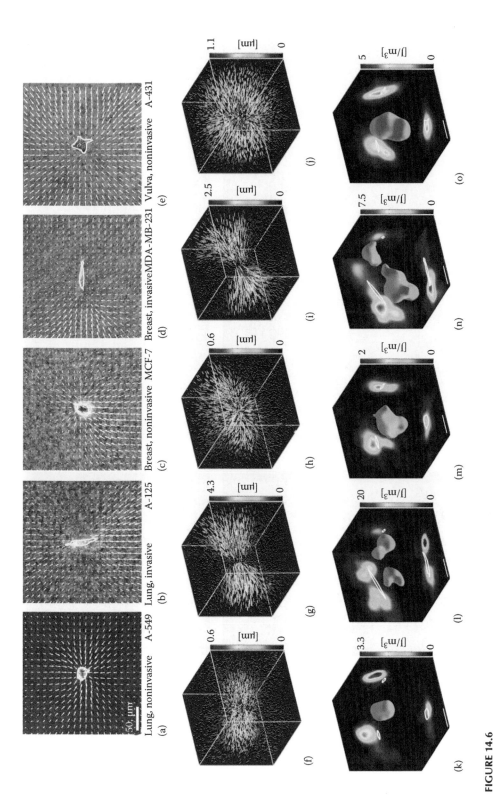

FIGURE 14.6

(a–j) Normalized displacement fields in the x–y plane (a–e) or in 3D (f–j) around invasive and noninvasive carcinoma cell lines. Note the increased anisotropy associated with invasive cells. (k–o) Strain energy density around the cells shown in (a–j). Again note the anisotropic distribution associated with invasive carcinoma cells. (Reproduced from Koch, T.M. et al., *PLoS One*, 7, e33476, 2012. With permission.)

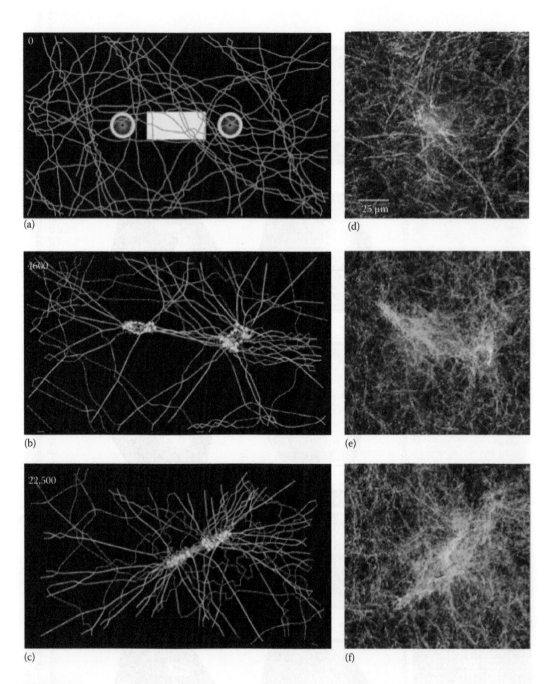

FIGURE 14.7
(a) Two-cell free-floating matrix model, (b) aligned fibers between cells show elevated strain, and (c) the case where reorganization is sufficient for macroscopic matrix compaction. (d–f) Collagen remodeling over time at (d) 0, (e) 4, and (f) 16 h. Note the changes in alignment and density. Scale bar = 25 μm. (Reproduced from Reinhardt, J.W. et al., *J. Biomech. Eng.*, 135, 71003, 2013. With permission. Experimental data in d–f is from McLeod, C. et al., *J. Biomech. Eng.*, 135, 71002, 2013.)

factors on cell polarity and the direction, persistence, and velocity of 3D cell migration. Thus, while this study certainly marks an important contribution to these efforts, the superposition of the factors in the full 3D case was not examined, and it is clear that additional experimental and theoretical work is warranted to parse out the influence of these key features on 3D cell motility and in particular the complex microenvironment surrounding cancerous cells.

14.6 Conclusion

Despite the fact that our understanding of the metastatic cascade remains incomplete, a more robust understanding of these complex processes is emerging as we elucidate the mechanical and architectural signals that are in play in conjunction with soluble factors. While these factors help drive disease and therefore present additional challenges to impeding disease progression, they may also provide opportunities to identify new therapeutic strategies. With recent advances in our understanding of the intracellular signal transduction networks regulating focal adhesion and GTPase behavior and their role in modulating cell traction forces, a picture is emerging of the physical and molecular mechanisms that initiate, promote, and maintain matrix remodeling associated with this cancer cell invasion. Indeed, with our current and evolving multidisciplinary imaging, tissue engineering, modeling and cell biology platforms the field is well positioned to rapidly advance our knowledge though a deeper quantitative understanding of the spatial and temporal behavior of cells within complex environments. By understanding how these signals conspire to drive invasion new therapeutic vulnerabilities are likely to be identified.

Acknowledgment

Dr. Provenzano is supported by start-up funds from the University of Minnesota College of Science and Engineering and the Masonic Cancer Center, the NIH (P50 CA101955 UMN/UAB Pancreatic Cancer SPORE Career Development Award), and the University of Minnesota Institute for Engineering in Medicine. Dr. Provenzano thanks the many authors who shared figures and other materials for this chapter and the Provenzano lab for helpful discussions regarding this work.

References

Aghvami M, Barocas VH, Sander EA. 2013. Multiscale mechanical simulations of cell compacted collagen gels. *Journal of Biomechanical Engineering* 135: 71004.

Alcaraz J, Xu R, Mori H, Nelson CM, Mroue R, Spencer VA, Brownfield D, Radisky DC, Bustamante C, Bissell MJ. 2008. Laminin and biomimetic extracellular elasticity enhance functional differentiation in mammary epithelia. *EMBO Journal* 27: 2829–2838.

Barocas VH, Tranquillo RT. 1997. An anisotropic biphasic theory of tissue-equivalent mechanics: The interplay among cell traction, fibrillar network deformation, fibril alignment, and cell contact guidance. *Journal of Biomechanical Engineering* 119: 137–145.

Brakenhoff GJ, van der Voort HT, van Spronsen EA, Linnemans WA, Nanninga N. 1985. Three-dimensional chromatin distribution in neuroblastoma nuclei shown by confocal scanning laser microscopy. *Nature* **317**: 748–749.

Brown E, McKee T, diTomaso E, Pluen A, Seed B, Boucher Y, Jain RK. 2003. Dynamic imaging of collagen and its modulation in tumors in vivo using second-harmonic generation. *Nature Medicine* **9**: 796–800.

Brown EB, Campbell RB, Tsuzuki Y, Xu L, Carmeliet P, Fukumura D, Jain RK. 2001. In vivo measurement of gene expression, angiogenesis and physiological function in tumors using multiphoton laser scanning microscopy. *Nature Medicine* **7**: 864–868.

Butler DL, Goldstein SA, Guldberg RE, Guo XE, Kamm R, Laurencin CT, McIntire LV, Mow VC, Nerem RM, Sah RL et al. 2009. The impact of biomechanics in tissue engineering and regenerative medicine. *Tissue Engineering Part B: Reviews* **15**: 477–484.

Calvo F, Ege N, Grande-Garcia A, Hooper S, Jenkins RP, Chaudhry SI, Harrington K, Williamson P, Moeendarbary E, Charras G et al. 2013. Mechanotransduction and YAP-dependent matrix remodelling is required for the generation and maintenance of cancer-associated fibroblasts. *Nature Cell Biology* **15**: 637–646.

Campagnola PJ, Millard AC, Terasaki M, Hoppe PE, Malone CJ, Mohler WA. 2002. Three-dimensional high-resolution second-harmonic generation imaging of endogenous structural proteins in biological tissues. *Biophysical Journal* **82**: 493–508.

Carey SP, Starchenko A, McGregor AL, Reinhart-King CA. 2013. Leading malignant cells initiate collective epithelial cell invasion in a three-dimensional heterotypic tumor spheroid model. *Clinical & Experimental Metastasis* **30**: 615–630.

Centonze VE, White JG. 1998. Multiphoton excitation provides optical sections from deeper within scattering specimens than confocal imaging. *Biophysical Journal* **75**: 2015–2024.

Chandran PL, Barocas VH. 2006. Affine versus non-affine fibril kinematics in collagen networks: Theoretical studies of network behavior. *Journal of Biomechanical Engineering* **128**: 259–270.

Condeelis J, Pollard JW. 2006. Macrophages: Obligate partners for tumor cell migration, invasion, and metastasis. *Cell* **124**: 263–266.

Condeelis J, Singer RH, Segall JE. 2005. The great escape: When cancer cells hijack the genes for chemotaxis and motility. *Annual Review of Cell and Developmental Biology* **21**: 695–718.

Conklin MW, Eickhoff JC, Riching KM, Pehlke CA, Eliceiri KW, Provenzano PP, Friedl A, Keely PJ. 2011. Aligned collagen is a prognostic signature for survival in human breast carcinoma. *American Journal of Pathology* **178**: 1221–1232.

Cox G, Kable E, Jones A, Fraser I, Manconi F, Gorrell MD. 2003. 3-Dimensional imaging of collagen using second harmonic generation. *Journal of Structural Biology* **141**: 53–62.

Cukierman E, Bassi DE. 2010. Physico-mechanical aspects of extracellular matrix influences on tumorigenic behaviors. *Seminars in Cancer Biology* **20**: 139–145.

Dallon JC, Sherratt JA. 1998. A mathematical model for fibroblast and collagen orientation. *Bulletin of Mathematical Biology* **60**: 101–129.

Dallon JC, Sherratt JA, Maini PK. 1999. Mathematical modelling of extracellular matrix dynamics using discrete cells: Fiber orientation and tissue regeneration. *Journal of Theoretical Biology* **199**: 449–471.

Denk W, Strickler JH, Webb WW. 1990. Two-photon laser scanning fluorescence microscopy. *Science* **248**: 73–76.

Deshpande VS, McMeeking RM, Evans AG. 2006. A bio-chemo-mechanical model for cell contractility. *Proceedings of the National Academy of Sciences of the United States of America* **103**: 14015–14020.

Diaspro A, Sheppard CJR. 2002a. Two-photon microscopy: Basic principles and architectures. In *Confocal and Two-Photon Microscopy: Foundations, Applications, and Advances* (Diaspro A, ed.). Wiley-Liss, Inc., New York.

Diaspro A, Sheppard CJR. 2002b. Two-photon excitation fluorescence microscopy. In *Confocal and Two-Photon Microscopy: Foundations, Applications, and Advances* (Diaspro A, ed.), pp. 39–73. Wiley-Liss, Inc., New York.

Dickinson RB, Guido S, Tranquillo RT. 1994. Biased cell migration of fibroblasts exhibiting contact guidance in oriented collagen gels. *Annals of Biomedical Engineering* **22**: 342–356.

Discher DE, Janmey P, Wang YL. 2005. Tissue cells feel and respond to the stiffness of their substrate. *Science (New York, NY)* **310**: 1139–1143.

Doyle AD, Wang FW, Matsumoto K, Yamada KM. 2009. One-dimensional topography underlies three-dimensional fibrillar cell migration. *The Journal of Cell Biology* **184**: 481–490.

Elenbaas B, Spirio L, Koerner F, Fleming MD, Zimonjic DB, Donaher JL, Popescu NC, Hahn WC, Weinberg RA. 2001. Human breast cancer cells generated by oncogenic transformation of primary mammary epithelial cells. *Genes & Development* **15**: 50–65.

Elson EL, Genin GM. 2013. The role of mechanics in actin stress fiber kinetics. *Experimental cell research* **319**: 2490–2500.

Engler AJ, Carag-Krieger C, Johnson CP, Raab M, Tang HY, Speicher DW, Sanger JW, Sanger JM, Discher DE. 2008. Embryonic cardiomyocytes beat best on a matrix with heart-like elasticity: Scar-like rigidity inhibits beating. *Journal of Cell Science* **121**: 3794–3802.

Engler AJ, Sen S, Sweeney HL, Discher DE. 2006. Matrix elasticity directs stem cell lineage specification. *Cell* **126**: 677–689.

Evans MC, Barocas VH. 2009. The modulus of fibroblast-populated collagen gels is not determined by final collagen and cell concentration: Experiments and an inclusion-based model. *Journal of Biomechanical Engineering* **131**: 101014.

Even-Ram S, Yamada KM. 2005. Cell migration in 3D matrix. *Current Opinion in Cell Biology* **17**: 524–532.

Folkman J. 1971. Tumor angiogenesis: Therapeutic implications. *New England Journal of Medicine* **285**: 1182–1186.

Freund I, Deutsch M. 1986. Second-harmonic microscopy of biological tissue. *Optics Letters* **11**: 94–96.

Friedl P, Wolf K. 2010. Plasticity of cell migration: A multiscale tuning model. *The Journal of Cell Biology* **188**: 11–19.

Gaggioli C, Hooper S, Hidalgo-Carcedo C, Grosse R, Marshall JF, Harrington K, Sahai E. 2007. Fibroblast-led collective invasion of carcinoma cells with differing roles for RhoGTPases in leading and following cells. *Nature Cell Biology* **9**: 1392–1400.

Gardel ML, Sabass B, Ji L, Danuser G, Schwarz US, Waterman CM. 2008. Traction stress in focal adhesions correlates biphasically with actin retrograde flow speed. *The Journal of Cell Biology* **183**: 999–1005.

Gehler S, Baldassarre M, Lad Y, Leight JL, Wozniak MA, Riching KM, Eliceiri KW, Weaver VM, Calderwood DA, Keely PJ. 2009. Filamin A-beta1 integrin complex tunes epithelial cell response to matrix tension. *Molecular Biology of the Cell* **20**: 3224–3238.

Geiger B, Bershadsky A, Pankov R, Yamada KM. 2001. Transmembrane crosstalk between the extracellular matrix—Cytoskeleton crosstalk. *Nature Reviews* **2**: 793–805.

Goetz JG, Minguet S, Navarro-Lerida I, Lazcano JJ, Samaniego R, Calvo E, Tello M, Osteso-Ibanez T, Pellinen T, Echarri A et al. 2011. Biomechanical remodeling of the microenvironment by stromal caveolin-1 favors tumor invasion and metastasis. *Cell* **146**: 148–163.

Grinnell F. 2003. Fibroblast biology in three-dimensional collagen matrices. *Trends in Cell Biology* **13**: 264–269.

Guido S, Tranquillo RT. 1993. A methodology for the systematic and quantitative study of cell contact guidance in oriented collagen gels. Correlation of fibroblast orientation and gel birefringence. *Journal of Cell Science* **105 (Pt 2)**: 317–331.

Gupta GP, Massague J. 2006. Cancer metastasis: Building a framework. *Cell* **127**: 679–695.

Hadjipanayi E, Mudera V, Brown RA. 2009. Guiding cell migration in 3D: A collagen matrix with graded directional stiffness. *Cell Motility and the Cytoskeleton* **66**: 121–128.

Hahn C, Schwartz MA. 2009. Mechanotransduction in vascular physiology and atherogenesis. *Nature Reviews* **10**: 53–62.

Harjanto D, Maffei JS, Zaman MH. 2011. Quantitative analysis of the effect of cancer invasiveness and collagen concentration on 3D matrix remodeling. *PLoS ONE* **6**: e24891.

Helmchen F, Denk W. 2002. New developments in multiphoton microscopy. *Current Opinion in Neurobiology* **12**: 593–601.

Hsu HJ, Lee CF, Kaunas R. 2009. A dynamic stochastic model of frequency-dependent stress fiber alignment induced by cyclic stretch. *PloS one* **4**: e4853.

Humphrey JD, Holzapfel GA. 2012. Mechanics, mechanobiology, and modeling of human abdominal aorta and aneurysms. *Journal of Biomechanics* **45**: 805–814.

Jain RK, Munn LL, Fukumura D. 2002. Dissecting tumour pathophysiology using intravital microscopy. *Nature Reviews Cancer* **2**: 266–276.

Jawerth LM, Munster S, Vader DA, Fabry B, Weitz DA. 2010. A blind spot in confocal reflection microscopy: The dependence of fiber brightness on fiber orientation in imaging biopolymer networks. *Biophysical Journal* **98**: L1–L3.

Kaunas R, Usami S, Chien S. 2006. Regulation of stretch-induced JNK activation by stress fiber orientation. *Cellular signalling* **18**: 1924–1931.

Kim DH, Provenzano PP, Smith CL, Levchenko A. 2012. Matrix nanotopography as a regulator of cell function. *The Journal of Cell Biology* **197**: 351–360.

Klein EA, Yin L, Kothapalli D, Castagnino P, Byfield FJ, Xu T, Levental I, Hawthorne E, Janmey PA, Assoian RK. 2009. Cell-cycle control by physiological matrix elasticity and in vivo tissue stiffening. *Current Biology* **19**: 1511–1518.

Koch TM, Munster S, Bonakdar N, Butler JP, Fabry B. 2012. 3D Traction forces in cancer cell invasion. *PLoS ONE* **7**: e33476.

Kshitiz, Hubbi ME, Ahn EH, Downey J, Afzal J, Kim DH, Rey S, Chang C, Kundu A, Semenza GL et al. 2012a. Matrix rigidity controls endothelial differentiation and morphogenesis of cardiac precursors. *Science Signaling* **5**: ra41.

Kshitiz, Park J, Kim P, Helen W, Engler AJ, Levchenko A, Kim DH. 2012b. Control of stem cell fate and function by engineering physical microenvironments. *Integrative Biology: Quantitative Biosciences from Nano to Macro* **4**: 1008–1018.

Lacomb R, Nadiarnykh O, Campagnola PJ. 2008. Quantitative second harmonic generation imaging of the diseased state osteogenesis imperfecta: Experiment and simulation. *Biophysical Journal* **94**: 4504–4514.

Lauffenburger DA, Horwitz AF. 1996. Cell migration: A physically integrated molecular process. *Cell* **84**: 359–369.

Lee SL, Nekouzadeh A, Butler B, Pryse KM, McConnaughey WB, Nathan AC, Legant WR, Schaefer PM, Pless RB, Elson EL, Genin GM. 2012. Physically-induced cytoskeleton remodeling of cells in three-dimensional culture. *PloS one* **7**: e45512.

Levental KR, Yu H, Kass L, Lakins JN, Egeblad M, Erler JT, Fong SF, Csiszar K, Giaccia A, Weninger W et al. 2009. Matrix crosslinking forces tumor progression by enhancing integrin signaling. *Cell* **139**: 891–906.

Lo CM, Wang HB, Dembo M, Wang YL. 2000. Cell movement is guided by the rigidity of the substrate. *Biophysical Journal* **79**: 144–152.

Mammoto T, Ingber DE. 2010. Mechanical control of tissue and organ development. *Development (Cambridge, England)* **137**: 1407–1420.

McGarry JP, Fu J, Yang MT, Chen CS, McMeeking RM, Evans AG, Deshpande VS. 2009. Simulation of the contractile response of cells on an array of micro-posts. *Philosophical transactions Series A, Mathematical, physical, and engineering sciences* **367**: 3477–3497.

McLeod C, Higgins J, Miroshnikova Y, Liu R, Garrett A, Sarang-Sieminski AL. 2013. Microscopic matrix remodeling precedes endothelial morphological changes during capillary morphogenesis. *Journal of Biomechanical Engineering* **135**: 71002.

Mintz B, Illmensee, K. 1975. Normal genetically mosaic mice produced from malignant teratocarcinoma cells. *Proceedings of the National Academy of Sciences of the United States of America* **72**: 3585–3589.

Mohler W, Millard AC, Campagnola PJ. 2003. Second harmonic generation imaging of endogenous structural proteins. *Methods* **29**: 97–109.

Murray JD, Oster GF. 1984a. Cell traction models for generating pattern and form in morphogenesis. *Journal of Mathematical Biology* **19**: 265–279.

Murray JD, Oster GF. 1984b. Generation of biological pattern and form. *IMA Journal of Mathematics Applied in Medicine and Biology* **1**: 51–75.

Murray JD, Oster GF, Harris AK. 1983. A mechanical model for mesenchymal morphogenesis. *Journal of Mathematical Biology* **17**: 125–129.

Nadiarnykh O, LaComb RB, Brewer MA, Campagnola PJ. 2010. Alterations of the extracellular matrix in ovarian cancer studied by Second Harmonic Generation imaging microscopy. *BMC Cancer* **10**: 94.

Nadiarnykh O, Plotnikov S, Mohler WA, Kalajzic I, Redford-Badwal D, Campagnola PJ. 2007. Second harmonic generation imaging microscopy studies of osteogenesis imperfecta. *Journal of Biomedical Optics* **12**: 051805.

Ngwa GA, Maini PK. 1995. Spatio-temporal patterns in a mechanical model for mesenchymal morphogenesis. *Journal of Mathematical Biology* **33**: 489–520.

Olsen L, Maini PK, Sherratt JA, Dallon J. 1999. Mathematical modelling of anisotropy in fibrous connective tissue. *Mathematical Biosciences* **158**: 145–170.

Orimo A, Gupta PB, Sgroi DC, Arenzana-Seisdedos F, Delaunay T, Naeem R, Carey VJ, Richardson AL, Weinberg RA. 2005. Stromal fibroblasts present in invasive human breast carcinomas promote tumor growth and angiogenesis through elevated SDF-1/CXCL12 secretion. *Cell* **121**: 335–348.

Paszek MJ, Zahir N, Johnson KR, Lakins JN, Rozenberg GI, Gefen A, Reinhart-King CA, Margulies SS, Dembo M, Boettiger D et al. 2005. Tensional homeostasis and the malignant phenotype. *Cancer Cell* **8**: 241–254.

Pawley JB. 2006. *Handbook of Biological Confocal Microscopy*. Springer, New York.

Perelson AS, Maini PK, Murray JD, Hyman JM, Oster GF. 1986. Nonlinear pattern selection in a mechanical model for morphogenesis. *Journal of Mathematical Biology* **24**: 525–541.

Petrie RJ, Doyle AD, Yamada KM. 2009. Random versus directionally persistent cell migration. *Nature Reviews* **10**: 538–549.

Plotnikov SV, Millard AC, Campagnola PJ, Mohler WA. 2006. Characterization of the myosin-based source for second-harmonic generation from muscle sarcomeres. *Biophysical Journal* **90**: 693–703.

Pollard JW. 2004. Tumour-educated macrophages promote tumour progression and metastasis. *Nature Reviews Cancer* **4**: 71–78.

Provenzano PP, Cuevas C, Chang AE, Goel VK, Von Hoff DD, Hingorani SR. 2012. Enzymatic targeting of the stroma ablates physical barriers to treatment of pancreatic ductal adenocarcinoma. *Cancer Cell* **21**: 418–429.

Provenzano PP, Eliceiri KW, Campbell JM, Inman DR, White JG, Keely PJ. 2006. Collagen reorganization at the tumor-stromal interface facilitates local invasion. *BMC Medicine* **4**: 38.

Provenzano PP, Eliceiri KW, Keely PJ. 2009a. Multiphoton microscopy and fluorescence lifetime imaging microscopy (FLIM) to monitor metastasis and the tumor microenvironment. *Clinical & Experimental Metastasis* **26**: 357–370.

Provenzano PP, Inman DR, Eliceiri KW, Keely PJ. 2009b. Matrix density-induced mechanoregulation of breast cell phenotype, signaling and gene expression through a FAK-ERK linkage. *Oncogene* **28**: 4326–4343.

Provenzano PP, Inman DR, Eliceiri KW, Knittel JG, Yan L, Rueden CT, White JG, Keely PJ. 2008a. Collagen density promotes mammary tumor initiation and progression. *BMC Medicine* **6**: 11.

Provenzano PP, Inman DR, Eliceiri KW, Trier SM, Keely PJ. 2008b. Contact guidance mediated three-dimensional cell migration is regulated by Rho/ROCK-dependent matrix reorganization. *Biophysical Journal* **95**: 5374–5384.

Provenzano PP, Keely PJ. 2011. Mechanical signaling through the cytoskeleton regulates cell proliferation by coordinated focal adhesion and Rho GTPase signaling. *Journal of Cell Science* **124**: 1195–1205.

Reinhardt JW, Krakauer DA, Gooch KJ. 2013. Complex matrix remodeling and durotaxis can emerge from simple rules for cell–matrix interaction in agent-based models. *Journal of Biomechanical Engineering* **135**: 71003.

Robling AG, Castillo AB, Turner CH. 2006. Biomechanical and molecular regulation of bone remodeling. *Annual Review of Biomedical Engineering* **8**: 455–498.

Roeder BA, Kokini K, Voytik-Harbin SL. 2009. Fibril microstructure affects strain transmission within collagen extracellular matrices. *Journal of Biomechanical Engineering* **131**: 031004.

Sabeh F, Shimizu-Hirota R, Weiss SJ. 2009. Protease-dependent versus-independent cancer cell invasion programs: Three-dimensional amoeboid movement revisited. *The Journal of Cell Biology* **185**: 11–19.

Saez A, Buguin A, Silberzan P, Ladoux B. 2005. Is the mechanical activity of epithelial cells controlled by deformations or forces? *Biophysical Journal* **89**: L52–L54.

Sahai E. 2007. Illuminating the metastatic process. *Nature Reviews Cancer* **7**: 737–749.

Sahai E, Marshall CJ. 2002. ROCK and Dia have opposing effects on adherens junctions downstream of Rho. *Nature Cell Biology* **4**: 408–415.

Sahai E, Marshall CJ. 2003. Differing modes of tumour cell invasion have distinct requirements for Rho/ROCK signalling and extracellular proteolysis. *Nature Cell Biology* **5**: 711–719.

Sahai E, Wyckoff J, Philippar U, Segall JE, Gertler F, Condeelis J. 2005. Simultaneous imaging of GFP, CFP and collagen in tumors in vivo using multiphoton microscopy. *BMC Biotechnology* **5**: 14.

Samuel MS, Lopez JI, McGhee EJ, Croft DR, Strachan D, Timpson P, Munro J, Schroder E, Zhou J, Brunton VG et al. 2011. Actomyosin-mediated cellular tension drives increased tissue stiffness and beta-catenin activation to induce epidermal hyperplasia and tumor growth. *Cancer Cell* **19**: 776–791.

Sander EA, Stylianopoulos T, Tranquillo RT, Barocas VH. 2009. Image-based biomechanics of collagen-based tissue equivalents. *IEEE Engineering in Medicine and Biology Magazine: The Quarterly Magazine of the Engineering in Medicine & Biology Society* **28**: 10–18.

Sander LM. 2013. Alignment localization in nonlinear biological media. *Journal of Biomechanical Engineering* **135**: 71006.

Shekhar MP, Pauley R, Heppner G, Werdell J, Santner SJ, Pauley RJ, Tait L. 2003. Host microenvironment in breast cancer development: Extracellular matrix-stromal cell contribution to neoplastic phenotype of epithelial cells in the breast. *Breast Cancer Research* **5**: 130–135. Epub 2003 February 2020.

Shiu YT, Li S, Marganski WA, Usami S, Schwartz MA, Wang YL, Dembo M, Chien S. 2004. Rho mediates the shear-enhancement of endothelial cell migration and traction force generation. *Biophysical Journal* **86**: 2558–2565.

Skala MC, Squirrell JM, Vrotsos KM, Eickhoff JC, Gendron-Fitzpatrick A, Eliceiri KW, Ramanujam N. 2005. Multiphoton microscopy of endogenous fluorescence differentiates normal, precancerous, and cancerous squamous epithelial tissues. *Cancer Research* **65**: 1180–1186.

Solon J, Levental I, Sengupta K, Georges PC, Janmey PA. 2007. Fibroblast adaptation and stiffness matching to soft elastic substrates. *Biophysical Journal* **93**: 4453–4461.

Squirrell JM, Wokosin DL, White JG, Bavister BD. 1999. Long-term two-photon fluorescence imaging of mammalian embryos without compromising viability. *Nature Biotechnology* **17**: 763–767.

Stevenson MD, Sieminski AL, McLeod CM, Byfield FJ, Barocas VH, Gooch KJ. 2010. Pericellular conditions regulate extent of cell-mediated compaction of collagen gels. *Biophysical Journal* **99**: 19–28.

Stoller P, Kim BM, Rubenchik AM, Reiser KM, Da Silva LB. 2002. Polarization-dependent optical second-harmonic imaging of a rat-tail tendon. *Journal of Biomedical Optics* **7**: 205–214.

Stylianopoulos T, Barocas VH. 2007a. Multiscale, structure-based modeling for the elastic mechanical behavior of arterial walls. *Journal of Biomechanical Engineering* **129**: 611–618.

Stylianopoulos T, Barocas VH. 2007b. Volume-averaging theory for the study of the mechanics of collagen networks. *Computer Methods in Applied Mechanics and Engineering* **196**: 2981–2990.

Teicher BA, Herman TS, Holden SA, Wang YY, Pfeffer MR, Crawford JW, Frei E, 3rd. 1990. Tumor resistance to alkylating agents conferred by mechanisms operative only in vivo. *Science* **247**: 1457–1461.

Teixeira AI, Abrams GA, Bertics PJ, Murphy CJ, Nealey PF. 2003. Epithelial contact guidance on well-defined micro- and nanostructured substrates. *Journal of Cell Science* **116**: 1881–1892.

Tilghman RW, Cowan CR, Mih JD, Koryakina Y, Gioeli D, Slack-Davis JK, Blackman BR, Tschumperlin DJ, Parsons JT. 2010. Matrix rigidity regulates cancer cell growth and cellular phenotype. *PLoS ONE* **5**: e12905.

Ulrich TA, de Juan Pardo EM, Kumar S. 2009. The mechanical rigidity of the extracellular matrix regulates the structure, motility, and proliferation of glioma cells. *Cancer Research* **69**: 4167–4174.

Vogel V, Sheetz M. 2006. Local force and geometry sensing regulate cell functions. *Nature Reviews* **7**: 265–275.

Wang HB, Dembo M, Wang YL. 2000. Substrate flexibility regulates growth and apoptosis of normal but not transformed cells. *American Journal of Physiology* **279**: C1345–C1350.

Wang N, Tytell JD, Ingber DE. 2009. Mechanotransduction at a distance: Mechanically coupling the extracellular matrix with the nucleus. *Nature Reviews* **10**: 75–82.

Wang W, Goswami S, Lapidus K, Wells AL, Wyckoff JB, Sahai E, Singer RH, Segall JE, Condeelis JS. 2004. Identification and testing of a gene expression signature of invasive carcinoma cells within primary mammary tumors. *Cancer Research* **64**: 8585–8594.

Wang W, Mouneimne G, Sidani M, Wyckoff J, Chen X, Makris A, Goswami S, Bresnick AR, Condeelis JS. 2006. The activity status of cofilin is directly related to invasion, intravasation, and metastasis of mammary tumors. *The Journal of Cell Biology* **173**: 395–404.

Wang W, Wyckoff JB, Frohlich VC, Oleynikov Y, Huttelmaier S, Zavadil J, Cermak L, Bottinger EP, Singer RH, White JG et al. 2002. Single cell behavior in metastatic primary mammary tumors correlated with gene expression patterns revealed by molecular profiling. *Cancer Research* **62**: 6278–6288.

Wang W, Wyckoff JB, Goswami S, Wang Y, Sidani M, Segall JE, Condeelis JS. 2007. Coordinated regulation of pathways for enhanced cell motility and chemotaxis is conserved in rat and mouse mammary tumors. *Cancer Research* **67**: 3505–3511.

Wei Z, Deshpande VS, McMeeking RM, Evans AG. 2008. Analysis and interpretation of stress fiber organization in cells subject to cyclic stretch. *Journal of biomechanical engineering* **130**: 031009.

White JG, Amos WB, Fordham M. 1987. An evaluation of confocal versus conventional imaging of biological structures by fluorescence light microscopy. *The Journal of Cell Biology* **105**: 41–48.

Williams CM, Engler AJ, Slone RD, Galante LL, Schwarzbauer JE. 2008. Fibronectin expression modulates mammary epithelial cell proliferation during acinar differentiation. *Cancer Research* **68**: 3185–3192.

Williams RM, Zipfel WR, Webb WW. 2005. Interpreting second-harmonic generation images of collagen I fibrils. *Biophysical Journal* **88**: 1377–1386.

Wolf K, Friedl P. 2011. Extracellular matrix determinants of proteolytic and non-proteolytic cell migration. *Trends in Cell Biology* **21**: 736–744.

Wolf K, Mazo I, Leung H, Engelke K, von Andrian UH, Deryugina EI, Strongin AY, Brocker EB, Friedl P. 2003. Compensation mechanism in tumor cell migration: Mesenchymal-amoeboid transition after blocking of pericellular proteolysis. *The Journal of Cell Biology* **160**: 267–277.

Wolf K, Wu YI, Liu Y, Gieger J, Tam E, Overall C, Stack MS, Friedl P. 2007. Multi-step pericellular proteolysis controls the transition from individual to collective cancer cell invasion. *Nature Cell Biology* **9**: 893–904.

Wong VW, Longaker MT, Gurtner GC. 2012. Soft tissue mechanotransduction in wound healing and fibrosis. *Seminars in Cell & Developmental Biology* **23**: 981–986.

Woo SL, Abramowitch SD, Kilger R, Liang R. 2006. Biomechanics of knee ligaments: Injury, healing, and repair. *Journal of Biomechanics* **39**: 1–20.

Wozniak MA, Chen CS. 2009. Mechanotransduction in development: A growing role for contractility. *Nature Reviews* **10**: 34–43.

Wozniak MA, Desai R, Solski PA, Der CJ, Keely PJ. 2003. ROCK-generated contractility regulates breast epithelial cell differentiation in response to the physical properties of a three-dimensional collagen matrix. *The Journal of Cell Biology* **163**: 583–595.

Wyckoff JB, Pinner SE, Gschmeissner S, Condeelis JS, Sahai E. 2006. ROCK- and myosin-dependent matrix deformation enables protease-independent tumor-cell invasion in vivo. *Current Biology* **16**: 1515–1523.

Yamada KM, Cukierman E. 2007. Modeling tissue morphogenesis and cancer in 3D. *Cell* **130**: 601–610.

Yamato M, Adachi E, Yamamoto K, Hayashi T. 1995. Condensation of collagen fibrils to the direct vicinity of fibroblasts as a cause of gel contraction. *Journal of Biochemistry* **117**: 940–946.

Yang YL, Kaufman LJ. 2009. Rheology and confocal reflectance microscopy as probes of mechanical properties and structure during collagen and collagen/hyaluronan self-assembly. *Biophysical Journal* **96**: 1566–1585.

Yang YL, Leone LM, Kaufman LJ. 2009. Elastic moduli of collagen gels can be predicted from two-dimensional confocal microscopy. *Biophysical Journal* **97**: 2051–2060.

Yang YL, Motte S, Kaufman LJ. 2010. Pore size variable type I collagen gels and their interaction with glioma cells. *Biomaterials* **31**: 5678–5688.

Zaman MH, Trapani LM, Sieminski AL, Mackellar D, Gong H, Kamm RD, Wells A, Lauffenburger DA, Matsudaira P. 2006. Migration of tumor cells in 3D matrices is governed by matrix stiffness along with cell-matrix adhesion and proteolysis. *Proceedings of the National Academy of Sciences of the United States of America* **103**: 10889–10894.

Zemel A, Rehfeldt F, Brown AE, Discher DE, Safran SA. 2010a. Cell shape, spreading symmetry and the polarization of stress-fibers in cells. *Journal of physics Condensed matter: an Institute of Physics journal* **22**: 194110.

Zemel A, Rehfeldt F, Brown AE, Discher DE, Safran SA. 2010b. Optimal matrix rigidity for stress fiber polarization in stem cells. *Nature physics* **6**: 468–473.

Zoumi A, Yeh A, Tromberg BJ. 2002. Imaging cells and extracellular matrix in vivo by using second-harmonic generation and two-photon excited fluorescence. *Proceedings of the National Academy of Sciences of the United States of America* **99**: 11014–11019.

Index

Printed and bound by CPI Group (UK) Ltd, Croydon, CR0 4YY

18/10/2024

01776254-0008